W0106153

Progress in Mathematics

Volume 178

Series Editors

H. Bass

J. Oesterlé

A. Weinstein

Lluís Puig

On the Local Structure of Morita and Rickard Equivalences between Brauer Blocks

Springer Basel AG

Author:

Lluís Puig Carreres
CNRS, Institut de Mathématiques de Jussieu
Université Denis Diderot (Paris VII)
F-75251 Paris Cedex 05

1991 Mathematics Subject Classification 20C11

A CIP catalogue record for this book is available from the Library of Congress, Washington D.C., USA

Deutsche Bibliothek Cataloging-in-Publication Data

Puig, Lluís:
On the local structure of Morita and Rickard equivalences between Brauer blocks / Lluís Puig. - Basel ; Boston ; Berlin : Birkhäuser, 1999
(Progress in mathematics ; Vol. 178)
ISBN 978-3-0348-9732-7 ISBN 978-3-0348-8693-2 (eBook)
DOI 10.1007/978-3-0348-8693-2

This work is subject to copyright. All rights are reserved, whether the whole or part of the material is concerned, specifically the rights of translation, reprinting, re-use of illustrations, broadcasting, reproduction on microfilms or in other ways, and storage in data banks. For any kind of use whatsoever, permission from the copyright owner must be obtained.

© 1999 Springer Basel AG
Originally published by Birkhäuser Verlag in 1999
Softcover reprint of the hardcover 1st edition 1999
Printed on acid-free paper produced of chlorine-free pulp. TCF ∞

ISBN 978-3-0348-9732-7

9 8 7 6 5 4 3 2 1

a tots els meus amics, per tota llur confiança

Contents

1 Introduction

1.1 This work began by answering the implicit question that Michel Broué raises in [6, 6.C, (pu 1,2,3)], namely by proving that two blocks have the same defect group and isomorphic source algebras if and only if they are Morita equivalent via a bimodule with trivial source (actually, at that time this answer was already known to Leonard Scott: see Remark 7.5 below). In the last part of that paper, Broué attempts to formulate, in terms of derived categories, the old idea of building the structure of a block from the collection of "local data": the origin of this idea goes back to fifteen years ago when Michel and I tried to extend to the so-called *Frobenius blocks with abelian kernel* - blocks with abelian defect groups on which the inertial quotients act freely - the methods employed successfully in the *nilpotent blocks* [9]. Our first approach had been to switch from virtual characters to "virtual modules" and [22] has to be understood as a contribution to this effort: there we show that, indeed, the "construction" of virtual characters described in [8] and [19] can be performed in full generality with "virtual modules".

1.2 But even in the well-known blocks with cyclic defect groups - the easiest case of Frobenius blocks - the description via "virtual modules" that Broué developed tentatively was powerless to provide a complete alternative argument, and, as a matter of fact, the suggestion of Leonard Scott to improve this approach - formulated in a lecture by Brouè - had been to enrich the "virtual modules" with suitable "differential maps", in other words, to replace them by suitable complexes of modules.

1.3 Consequently, when in his thesis Jeremy Rickard develops a "Morita theory" for the equivalences between the so-called *derived categories* of the categories of modules over two algebras A and B (cf. [30]) and exhibits such an equivalence in the case where A is the algebra associated with a block with cyclic defect groups and B the group algebra of the corresponding Frobenius group (cf. [29]), we are both ready to recognize in this equivalence the relationship that Broué was seeking: namely, we both become convinced that there should exist an equivalence between the derived categories of the categories of modules over the algebra of a Frobenius block with abelian kernel and the group algebra of the corresponding Frobenius group, lifting the equivalence between the corresponding *stable categories* that we obtain in [24]. More generally, Broué reveals in [4], [5], [6] and [7] some evidence for the existence of a particular kind of such equivalences (we are more explicit below) when A is the algebra of a block with abelian defect groups and B is a suitable twisted group algebra of the semidirect product of a defect group by an inertial quotient.

1.4 Yet, in order to apply our old idea of inductive construction from the collection of "local data", when A and B are the algebras associated with two blocks, it was necessary to analyse the relationship between the "local data" of these blocks. A first attempt in that direction is Rickard's result in [33] on the so-called *splendid equivalences*, which is effective essentially for principal blocks, and Linckelmann's generalization to the nonprincipal ones [16]. The main purpose of our work is to provide a systematic treatment of this relationship, which exhibits actually some "local structure" of the equivalences themselves, namely the so-called *local tracing triples* introduced in Section 16; in particular, in Section 19, we improve Rickard and Linckelmann's results by showing the redundancy of most of their hypotheses.

1.5 To start such a treatment, we face two kinds of difficulty. The first one - which appears already in the ordinary equivalences between the categories of modules over the algebras associated with two blocks - is that both Morita and Rickard's results on their equivalences are formulated in terms of $A \otimes_{\mathcal{O}} B^\circ$-modules, whereas all our machinery is supported by $\mathcal{O}$-algebras (cf. [19] and [22], for instance). The second one is that in Rickard's case we have to develop that machinery in a differential $\mathbf{Z}$-graded context. In order to comment on it, let us fix firstly some notation (we employ freely the terminology recalled in Section 2); as usual, p is a prime number and $\mathcal{O}$ a complete discrete valuation ring having an algebraically closed residue field k of characteristic p; let G and G' be finite groups, consider blocks b of G and b' of G', choose maximal local pointed groups P_γ on $\mathcal{O}Gb$ and $P'_{\gamma'}$ on $\mathcal{O}G'b'$, and denote respectively by b° and $(b')^\circ$ the blocks of G and G' such that $(\mathcal{O}Gb)^\circ \cong \mathcal{O}Gb^\circ$ and $(\mathcal{O}G'b')^\circ \cong \mathcal{O}G'(b')^\circ$.

1.6 Recall that, in a widely employed terminology, b and b' are *Morita equivalent* if there is an $\mathcal{O}$-finite $\mathcal{O}$-free $\mathcal{O}(G \times G')$-module M'' associated with $b \otimes (b')^\circ$ such that, denoting by $(M'')^*$ the dual $\mathcal{O}$-module $\mathrm{Hom}_{\mathcal{O}}(M'', \mathcal{O})$, which is an $\mathcal{O}$-finite $\mathcal{O}$-free $\mathcal{O}(G \times G')$-module associated with $b^\circ \otimes b'$, we have respectively $\mathcal{O}(G \times G)$- and $\mathcal{O}(G' \times G')$-module isomorphisms

1.6.1 $$M'' \underset{\mathcal{O}G'}{\otimes} (M'')^* \cong \mathcal{O}Gb \quad \text{and} \quad (M'')^* \underset{\mathcal{O}G}{\otimes} M'' \cong \mathcal{O}G'b' \quad .$$

A first step towards an $\mathcal{O}$-algebra formulation is to notice that, since this condition implies that the restrictions of M'' to $\mathcal{O}(G \times 1)$ and to $\mathcal{O}(1 \times G')$ are both projective, the $\mathcal{O}(G \times G)$-module $M'' \otimes_{\mathcal{O}G'} (M'')^*$ is actually isomorphic to $\mathrm{End}_{\mathcal{O}}(M'')^{1 \times G'}$ which has an obvious structure of $\mathcal{O}G$*-interior algebra* (or interior G-algebra: as we explain in 2.3 below, here we modify slighty our previous terminology), and that moreover the first isomorphism in 1.6.1 forces an $\mathcal{O}G$-interior algebra isomorphism (see Proposition 6.5)

1.6.2 $$\mathcal{O}Gb \cong \mathrm{End}_{\mathcal{O}}(M'')^{1 \times G'} \quad ;$$

but, although this isomorphism "connects" already $\mathcal{O}Gb$ with $\mathcal{O}G'b'$, it is not yet clear, for instance, how to relate the source algebras $(\mathcal{O}Gb)_\gamma$ and $(\mathcal{O}G'b')_{\gamma'}$.

1.7 To illustrate what can be expected, let us recall the cases we currently know of Morita equivalences between blocks: b and b' are Morita equivalent in the following three situations:

1.7.1 The blocks b and b' are *nilpotent* and the defect groups P and P' are isomorphic (cf. [21, Main Theorem]).

1.7.2 The groups G and G' are p-solvable and the $\mathscr{k}^*$-groups obtained by iterating *Fong's reduction* are isomorphic (cf. [25]).

1.7.3 The group G is a Chevalley group over a finite field of characteristic different from p and, for some parabolic subgroup H of G, some block e of H such that $eb = e$, some Levi complement L in H and some block f of L such that $fe = e$, formed by *cuspidal* irreducible characters, we have $G' = N_G(L, f)$, $b' = f$, $P' \cdot C_G(P') \subset L$ and $\mathrm{Br}_{P'}(e) \neq 0$ (cf. [27, Corollary 5.10]).

In all them, P and P' are isomorphic, so that we may assume that $P' = P$, and there is an indecomposable *endopermutation* $\mathcal{O}P$-module N of vertex P such that we have an $\mathcal{O}P$-interior algebra embedding

1.7.4
$$(\mathcal{O}Gb)_\gamma \longrightarrow \mathrm{End}_\mathcal{O}(N) \underset{\mathcal{O}}{\otimes} (\mathcal{O}G'b')_{\gamma'} \quad ;$$

we emphasize that, from this embedding, the relationship between all the invariants associated with b and b' is easily obtained (see 7.6 and 7.7 below). On the other hand notice that, although a $\mathscr{k}$-algebra isomorphism $\mathscr{k}P \cong \mathscr{k}P'$ between the group $\mathscr{k}$-algebras of two finite p-groups is obviously an example of Morita equivalence between blocks, in this situation - at our knowledge - the relationship between P and P' is far from be clear.

1.8 In 1.6 the $\mathcal{O}(G \times G')$-module M'' is clearly indecomposable and the link between isomorphisms 1.6.2 and embedding 1.7.4 is provided by a vertex P'' and an $\mathcal{O}P''$-source N'' of M''. Indeed, since M'' is a direct summand of the induced $\mathcal{O}(G \times G')$-module $\mathrm{Ind}_{P''}^{G \times G'}(N'')$, from isomorphism 1.6.2 we get an $\mathcal{O}G$-interior algebra embedding (cf. 2.6.5)

1.8.1
$$\mathcal{O}G\,b \longrightarrow \mathrm{Ind}_{P''}^{G \times G'}\big(\mathrm{End}_\mathcal{O}(N'')\big)^{1 \times G'} \quad .$$

At this point, the main step towards the announced formulation is the introduction of an induction procedure, via *any* finite group homomorphism $\varphi : H \to H'$, of any $\mathcal{O}H$-interior algebra B to obtain an $\mathcal{O}H'$-interior algebra $\mathrm{Ind}_\varphi(B)$ (see Section 3): this induction allow us "to compute" the right end of embedding 1.8.1 and then, denoting respectively by $\rho : P'' \to G$ and $\rho' : P'' \to G'$ the restrictions of the first and the second projection maps for $G \times G'$, we get a new $\mathcal{O}G$-interior algebra embedding

1.8.2
$$\mathcal{O}G\,b \longrightarrow \mathrm{Ind}_\rho\big(\mathrm{End}_\mathcal{O}(N'') \underset{\mathcal{O}}{\otimes} \mathrm{Res}_{\rho'}(\mathcal{O}G'b')\big)$$

which is already a good formulation to start the "local analysis" and leads to embedding 1.7.4 whenever ρ and ρ' are injective.

1.9 Let us discuss now on the second kind of difficulty. Following Cartan and Eilenberg, a *differential* $\mathcal{O}$-module is just a module over the *ring of dual numbers* $\mathcal{O} \oplus \mathcal{O}d$ where $d^2 = 0$ (cf. [10, Chap. IV, §2]); in the same spirit, a **Z**-*graduation* in an $\mathcal{O}$-finite $\mathcal{O}$-module M is just a pairwise orthogonal idempotent decomposition $\{i_z^M\}_{z\in\mathbf{Z}}$ of id_M in $\mathrm{End}_{\mathcal{O}}(M)$ (notice that the finiteness condition on M forces $i_z^M = 0$ for all but a finite number of $z \in \mathbf{Z}$) and therefore, denoting by $\mathcal{F}$ the ring of all the $\mathcal{O}$-valued functions over $\mathbf{Z}$, M becomes an $\mathcal{F}$-module; thus, if we confine attention to the $\mathcal{O}$-finite $\mathcal{O}$-modules, a differential $\mathbf{Z}$-graded $\mathcal{O}$-module M is just a $\mathcal{D}$-module where $\mathcal{D} = \mathcal{F} \oplus \mathcal{F}d$ endowed with a suitable product (see Section 9 for all the detail). Fortunately enough, according to Rickard's results (cf. [31, 5.5]), we *can* confine attention to the differential $\mathbf{Z}$-graded $\mathcal{O}$-modules which are indeed finitely generated as $\mathcal{O}$-modules. Consequently, our successful guess has been that it suffices to replace the $\mathcal{O}(G \times G')$-module M'' above by a suitable *$\mathcal{D}(G \times G')$-module* - which we denote still by M'' - to switch from Morita's equivalences to Rickard's equivalences; coherently, we then replace the $\mathcal{O}G$-interior algebras by the *$\mathcal{D}G$-interior algebras* which, mimicking the case of the algebras of all the $\mathcal{O}$-linear endomorphisms of $\mathcal{D}G$-modules, are $\mathcal{O}$-finite $\mathcal{O}$-algebras A endowed with a unitary $\mathcal{O}$-algebra homomorphism

1.9.1 $$\mathcal{D}G \longrightarrow A$$

1.10 A technical problem is that $\mathcal{F}$ has no coproduct inducing the tensor product of two $\mathbf{Z}$-graded $\mathcal{O}$-modules; *a posteriori*, Pierre Cartier explained to us Grothendieck's point of view relating the $\mathbf{Z}$-graduations with the action of the one-dimensional torus (see [12, Chap. II, §2, 2.5]), which exploits the fact that $\mathcal{F}$ is the $\mathcal{O}$-dual of a Hopf $\mathcal{O}$-algebra, namely of the group algebra of $\mathbf{Z}$ over $\mathcal{O}$, and Pareigis' contribution [18] showing that $\mathcal{D}$ is also the $\mathcal{O}$-dual of a Hopf $\mathcal{O}$-algebra $\mathcal{C}$ (see Remark 9.13 below). Here we notice that the subring $\mathcal{F}_\circ$ of all the periodic functions in $\mathcal{F}$ *has* the convenient Hopf $\mathcal{O}$-algebra structure and is *dense* in $\mathcal{F}$ in the sense that it covers any $\mathcal{O}$-finite quotient of $\mathcal{F}$ (see Proposition 9.7); then, the subring $\mathcal{D}_\circ = \mathcal{F}_\circ \oplus \mathcal{F}_\circ d$ of $\mathcal{D}$ has the analogous property, which allow us to define the tensor product of $\mathcal{O}$-finite $\mathcal{D}G$-interior algebras (see 11.11 below).

1.11 As Rickard notices in [31], any equivalence between the derived categories (as *triangulated* categories) of the categories of $\mathcal{O}$-finite $\mathcal{O}Gb$- and $\mathcal{O}G'b'$-modules comes from suitable equivalences between the so-called *homotopic categories* of $\mathcal{O}Gb$ and $\mathcal{O}G'b'$, which are just quotients of the categories of $\mathcal{O}$-finite $\mathcal{D}Gb$-and $\mathcal{D}G'b'$-modules, namely the quotients obtained from the *homotopic equivalence* between maps (see 10.7 below). Actually, here we do *not* consider the derived categories but only the homotopic categories of $\mathcal{O}Gb$ and $\mathcal{O}G'b'$, and the equivalences we analyse are those, between these homotopy categories, exhibited by Rickard which, following Broué [7], we call *Rickard equivalences* (see 18.2.2 for the precise definition). But, in the same way that there is a particular type of Morita equivalences between blocks, namely those corresponding in 1.8 above to the case where ρ and ρ' are injective or, equivalently, $\mathrm{End}_{\mathcal{O}}(N'')$ has a P''-stable $\mathcal{O}$-basis (see Corollary 7.4), Broué guesses the existence of a suitable type of Rickard equivalence which should appear underlying the so-called *Deligne-Lusztig induction*, for Chevalley groups over finite fields of characteristic different from p, in some well-chosen situations (cf. [4, §6], [6, §6.C] and [7, §4]).

1.12 Explicitly, according to Michel Broué, the *p-adic cohomology* of the *Deligne-Lusztig varieties* should come from suitable $\mathcal{D}(G \times G')$-modules which restricted to any p-subgroup of $G \times G'$ would have a stable $\mathcal{O}$-basis; moreover, the pairs of blocks foreseen by Broué have isomorphic defect groups, equivalent *Brauer categories* and, in the centralizers of the corresponding *Brauer pairs*, the pairs of blocks are again in Broué's foresight. In [32] Jeremy Rickard proves that the proposed p-adic cohomology comes indeed from the expected kind of $\mathcal{D}(G \times G')$-modules and in [33] completed with Linckelmann's contribution in [16] it is proved that, assuming the isomorphism of the defect groups, a condition on relative projectivity and the equivalence of the Brauer categories, this special type of Rickard equivalence - that Rickard denominates *splendid* - is inherited by the corresponding blocks in the centralizers of the corresponding p-subgroups; as known example, let us mention the quite simple $\mathcal{D}(G \times G')$-module constructed by Raphaël Rouquier in [35], when P is cyclic and G' is the corresponding Frobenius group. As a matter of fact, there is a weaker hypothesis on stable $\mathcal{O}$-bases which specialized to the Morita equivalence requires only a P''-stable $\mathcal{O}$-basis in $\mathrm{End}_{\mathcal{O}}(N'')$ and which *implies* the isomorphism of the defect groups and the equivalence of the Brauer categories, together with the local inheritance (see Section 19): we call the Rickard equivalences fulfilling our hypothesis *basic* since, as far as we know, the condition on relative projectivity could be not fulfilled and because of both its involved $\mathcal{O}$-bases and its fundamental expected role.

1.13 This memoir is divided in nineteen sections and two appendices; a partial account of its contents, without proofs, has been written down in [28]. Although Morita equivalences are now a particular case of Rickard equivalences, here we decided to start by Morita's situation avoiding to cumulate both kind of difficulties mentioned above; moreover, this choice allows us to discuss a particular kind of stable equivalence which, as in [24], is the expected one when recollecting "local Morita equivalences" and which, according to Broué [6, 5.2], is forced by the Rickard equivalences. However, we try to avoid redundancy by postponing most of the developments until we are able to consider them for the differential structures. After collecting in Section 2 all the usual notation and most of the auxiliary results we need in the sequel, in Section 3 we introduce the non-necessarily injective induction of $\mathcal{O}H$-interior algebras, which is extended to the $\mathcal{D}H$-interior algebras in Section 12. As we mention in 1.8 above, this induction allows us to "compute" certain $\mathcal{O}G$-interior algebras, which we call *Hecke $\mathcal{O}G$-interior algebras*, consisting of fixed points in ordinarily induced $\mathcal{O}(G \times G')$-interior algebras: we do it in Section 4 and extended it to $\mathcal{D}$ in Section 15. In Section 5 and 16, we study the "local structure" of Hecke $\mathcal{O}G$- and $\mathcal{D}G$-interior algebras, which is the central subject of this memoir: in Section 16 we introduce the *local tracing triples* mentioned above.

1.14 In Section 6 we consider the so-called *Morita stable equivalences* (called *stable equivalences of Morita type* in [6]) between b and b', which are the equivalences between the stable categories of $\mathcal{O}Gb$ and $\mathcal{O}G'b'$ of the type mentioned above, and we apply our previous results to describe the relationship between the source algebras of b and b', obtaining an equivalent local formulation of these equivalences. Our *basic* hypothesis in 1.12 makes sense for Morita stable equivalences too and in Section 7 we get equivalent formulations of it which are specific to Morita's situation, and analyse its consequences in the relationship between the local structure of b and b'; as a matter of fact, when $\mathcal{O}$ has characteristic zero all the known Morita stable equivalences between blocks are basic, but may be this signifies just a lack of knowledge. Section 8 is devoted to a by-product of our analysis of Morita stable equivalences, namely that *if* $\mathrm{char}(\mathcal{O}) = 0$ *then a block which is Morita stable equivalent to a nilpotent block is nilpotent* which answers in the affirmative - in characteristic zero - the question we raise in [21, 1.8]; by the way, we get also that *if* $\mathrm{char}(\mathcal{O}) = 0$ *then all the Morita stable equivalences between nilpotent blocks are basic* but Markus Linckelmann proves a stronger result in [15]. The main tool for the proof of these statements is a criterion on *permutation* $\mathcal{O}P$*-modules* proved by Alfred Weiss in [39] under the hypothesis that $\mathcal{O}$ is unramified and by Klaus Roggenkamp in [34] without any restriction on $\mathcal{O}$; since, at the time we prove it, we ignored the existence of Roggenkamp's paper but were convinced that it should not depend on the ramification of $\mathcal{O}$, we issued our own proof of Weiss' criterion, that it turns out to be shorter: for the reader's convenience we give it in Appendix 1.

1.15 In Section 9 the second part of this memoir begins with the introduction of the $\mathcal{O}$-algebra $\mathcal{D}$ mentioned above and of its $\mathcal{O}$-subalgebra $\mathcal{D}_\circ$ endowed with a suitable Hopf $\mathcal{O}$-algebra structure. Sections 10 and 11 are devoted to the introduction of the $\mathcal{D}$-structures we need and, together with Section 9, to fix our new notation and new terminology: as we explain in Section 11, we consider *$\mathcal{O}$-finite $\mathcal{D}$-interior algebras* rather than the *differential $\mathbb{Z}$-graded $\mathcal{O}$-algebras* considered by Bernhard Keller in [13]. As in [10, Chap. IV, 2.3], in this context it is still true that *contractiblity* is relative projectivity from $\mathcal{O}$ to $\mathcal{D}$ (cf. Proposition 10.8) and the *$\mathcal{D}$-interiority* allows the existence of an induction procedure from $\mathcal{F}$- to $\mathcal{D}$-interior algebras (cf. 11.13). Moreover, although we confine attention to $\mathcal{O}$-finite $\mathcal{D}$-modules and $\mathcal{O}$-finite $\mathcal{D}$-interior algebras, we explain in Remarks 10.5 and 11.12 how to extend our definitions when the $\mathcal{O}$-finiteness is removed. In the same spirit, to illustrate the usefulness of this point of view, we apply it to the construction of the *tensor induction* of $\mathcal{D}H$-modules in Appendix 2; more generally, in this appendix we construct the tensor induction of *$\mathcal{D}$-interior H-algebras* (see 11.4 for the definition) since, anyway, for any $\mathcal{D}H$-module N we have to consider the $\mathcal{D}H$-interior algebra $\mathrm{End}_{\mathcal{O}}(N)$ and the tensor induction need not preserve the interiorness of the group.

1.16 One of the main difficulties in dealing with a $\mathcal{D}G$-interior algebra A is that the *Brauer section* of A at a p-subgroup Q of G need not be immediately related to the *local points* of Q on A (see 14.5 for the definition), so that the description of the local structure of A does not necessarily give a good knowledge on its Brauer sections; on the other hand, since $(\mathcal{O}G)(Q) \cong kC_G(Q)$, in order to study the local inheritance of Rickard equivalences mentioned in 1.12, we are also interested in the relationship between Brauer sections throughout a Rickard equivalence; but, as a general rule, Brauer sections can be efficaciously considered only in the presence of suitable stable $\mathcal{O}$-bases. In Section 13, we prepare the study of this relationship by considering, under a suitable "basic" hypothesis, a significant property of the Brauer sections of the *induced* $\mathfrak{D}H$*-interior algebras* introduced in Section 12; actually, in order to argue by induction in the proof of Theorem 17.12, we have to consider a slightly more general situation, as described in 13.3.

1.17 This preparation is continued in Section 17 where we introduce the *basic local tracing triples* and study the local structure of a *basic Hecke* $\mathfrak{D}G$*-interior algebra*, namely a Hecke $\mathfrak{D}G$-interior algebra where all the local tracing triples are basic (but this terminology is scarcely employed); a significant fact is that, in the relationship between this local structure and the local structure of $\mathcal{O}G'b'$, we cannot involve the *local pointed groups* on $\mathcal{O}G'b'$ but only the (G', b')*-Brauer pairs*; actually, it was already clear *a priori* that the connexion between local pointed groups on $\mathcal{O}Gb$ and on $\mathcal{O}G'b'$ should be loose since a Rickard equivalence between b and b' need not give an easy connexion between the sets of simple kGb- and $kG'b'$-modules, which correspond bijectively with the local points of the trivial subgroup; but we have no criterion *a posteriori* to guarantee that our results exhibit the tightest connexion. We apply all these results in Section 19, where we give the precise definition of the *basic Rickard equivalences* between b and b', which generalizes the corresponding Morita equivalences considered in Section 7, and prove the facts announced in 1.12 above.

1.18 Section 14 has a wide content; the first half is devoted to extend from $\mathcal{O}G$- to $\mathfrak{D}G$-interior algebras the notion of *pointed groups* together with all the corresponding developments. One of these developments is the so-called *Higman embeddings* relating the *embedded algebras* associated with pointed groups "before" and "after" induction; usually, we prove the uniqueness of the corresponding *exoembeddings* using the restriction to suitable embedded algebras (cf. [19, Proposition 3.6]) but it can be also proved using an embedding in a bigger one (cf. [1, Theorem 6.4]); it happens that in the *noninjective induction*, introduced in Sections 3 and 12, the starting algebra need not produce any embedding in the induced one (cf. 3.4), and we succeed in proving the uniqueness of the new Higman exoembeddings - we emphasize that we have no *succedaneum* of Higman's critarion for them and do not employ this terminology - only by embedding them in a bigger algebra, which can be done in a systematic way with the so-called *Higman envelope* of a $\mathfrak{D}G$-interior algebra (cf. Proposition 14.7). In the second half of this section, we introduce the new Higman embeddings for local pointed groups in the so-called *local tracing pairs* on induced algebras, which seem to be a good means for studying their local structure; these pairs become the local tracing triples on Hecke $\mathfrak{D}G$-interior algebras in Section 16 (cf. Proposition 16.7) but we do not succeed in deriving all the properties of the triples from the corresponding ones of the pairs. Finally, in Section 18 we consider the Rickard equivalences between blocks and, as in Section 6 for the Morita stable equivalences, describe the relationship between the source algebras of the blocks, obtaining an equivalent local formulation.

Paris, December 1996.

2 General notation, terminology and quoted results

2.1 Throughout the paper p is a prime number, k a field of characteristic p, assumed algebraically closed except in Appendix 1, and $\mathcal{O}$ a complete discrete valuation ring with residue field k (we allow the possibility $\mathcal{O} = k$). All the $\mathcal{O}$-algebras we consider are associative and unitary except in Remark 11.12 below, but the homomorphisms between $\mathcal{O}$-algebras are not required to be unitary. The unity elements are simply denoted by 1 whenever they have no previous name as idempotents in bigger $\mathcal{O}$-algebras; abusively, 1 denotes also either the trivial element or the trivial subgroup in any group. All the $\mathcal{O}$-modules and $\mathcal{O}$-algebras we consider are assumed to be finitely generated as $\mathcal{O}$-modules (in short *$\mathcal{O}$-finite*) except when we mention explicitely the contrary; a significant exception is the $\mathcal{O}$-algebras $\mathcal{F}$ and $\mathcal{D}$ that we introduce in Section 9 and employ in all the subsequent sections. If A is an $\mathcal{O}$-algebra, we denote by $\mathrm{Aut}(A)$ the group of automorphisms of A, by A^* the group of invertible elements of A, by $J(A)$ the Jacobson radical of A, by $Z(A)$ the center of A and by A° the opposite $\mathcal{O}$-algebra; recall that

2.1.1
$$A = \sum_{a\in A^*} \mathcal{O}a \quad .$$

Usually, we consider A-modules on the left; if M is an A-module, we denote by M° the corresponding right A°-module, and *A-bimodule* stands for $A \otimes_{\mathcal{O}} A^\circ$-module.

2.2 Let G be a finite group; following Green, a G-algebra is an $\mathcal{O}$-algebra A endowed with a group homomorphism $G \to \mathrm{Aut}(A)$ and, for any $x \in G$ and any $a \in A$, we denote by a^x the image of a by the automorphism of A determined by x^{-1}; if H is a subgroup of G then A^H denotes the $\mathcal{O}$-subalgebra of elements of A fixed by H and, if K is a subgroup of H, we denote by A_K^H the two-sided ideal of A^H formed by the *relative traces* $\mathrm{Tr}_K^H(a) = \Sigma_x a^{x^{-1}}$ where $a \in A^K$ and x runs on a set of representatives for H/K in H; notice that A^H/A_K^H is a torsion $\mathcal{O}$-algebra. Following [8], if P is a p-subgroup of G, we set

2.2.1
$$A(P) = k \underset{\mathcal{O}}{\otimes} \left(A^P \Big/ \sum_Q A_Q^P\right)$$

where Q runs on the set of proper subgroups of P, and we denote by

2.2.2
$$\mathrm{Br}_P : A^P \longrightarrow A(P) \quad ,$$

or by Br_P^A to avoid confusion, the canonical map: we call this k-algebra *Brauer section* of A at P and this map *Brauer homomorphism*; notice that $A(P)$ is a $\bar{N}_G(P)$-algebra, where $\bar{N}_G(P) = N_G(P)/P$, and that we have (cf. [38, (11.9)])

2.2.3
$$\mathrm{Br}_P(A_P^G) = A(P)_1^{\bar{N}_G(P)} \quad .$$

Actually, all this notation makes sense for an $\mathcal{O}G$-module M and the natural action of $\mathrm{End}_{\mathcal{O}\mathcal{P}}(M)$ on M^P induces a unitary $\bar{N}_G(P)$-algebra homomorphism (see 2.4 below)

2.2.4 $$\big(\mathrm{End}_{\mathcal{O}}(M)\big)(P) \longrightarrow \mathrm{End}_{k}\big(M(P)\big) \quad ;$$

it is easily checked that (cf. [38, (27.6)]):

2.2.5 *If M has a P-stable $\mathcal{O}$-basis X then* $\mathrm{Br}_P(X \cap M^P)$ *is a k-basis of $M(P)$ and homomorphism 2.2.4 is bijective.*

We call an $\mathcal{O}G$-module which has a G-stable $\mathcal{O}$-basis (so that, in particular, is $\mathcal{O}$-free) *permutation $\mathcal{O}G$-module.* It is now quite clear that (cf. [38,(17.4)]):

2.2.6 *If M is projective then, for any nontrivial p-subgroup Q of G, we have $M(Q) = \{0\}$ and, in particular, $M^G = M_1^G$.*

Following Dade, we call an $\mathcal{O}P$-module N such that $\mathrm{End}_{\mathcal{O}}(N)$ has a P-stable $\mathcal{O}$-basis *endopermutation $\mathcal{O}P$-module* and, in that case, we say that $\mathrm{End}_{\mathcal{O}}(N)$ is a *Dade P-algebra* if $\big(\mathrm{End}_{\mathcal{O}}(N)\big)(P) \neq \{0\}$ (cf. [23, 1.3]); it is still true that (cf. [38, (28.6)]):

2.2.7 *If N is an endopermutation $\mathcal{O}P$-module then the k algebra $\big(\mathrm{End}_{\mathcal{O}}(N)\big)(P)$ is zero or simple.*

A homomorphism $f : A \longrightarrow A'$ between two G-algebras A and A' is an $\mathcal{O}$-algebra homomorphism such that $f(a^x) = f(a)^x$ for any $a \in A$ and any $x \in G$.

2.3 In most G-algebras we consider, the action of G on the G-algebra A comes from a group homomorphism $G \longrightarrow A^*$ and then it is worth to include this homomorphism as a *datum* in a suitable definition considering "interiority". But here we modify slightly our habitual terminology-namely "interior G-algebra" introduced in [19] - in order to consider the more general "mixed" situation where only the action of some normal subgroup H of G comes from a group homomorphism $H \longrightarrow A^*$; precisely, if H is a normal subgroup of G, an *$\mathcal{O}H$-interior G-algebra* is a G-algebra A endowed with a unitary G-algebra homomorphism

2.3.1 $$\mathcal{O}H \longrightarrow A$$

lifting the action of H on A; that is to say, for any $a \in A$ and any $y \in H$, denoting by $y{\cdot}a$ and $a{\cdot}y$ the products of a and the image of y in A^*, we have

2.3.2 $$a^y = y^{-1}{\cdot}a{\cdot}y \quad ;$$

in that case, denoting by $\Delta(G)$ the diagonal subgroup of $G \times G$ and considering the product $(H \times H){\cdot}\Delta(G)$ in $G \times G$, A has an evident structure of $\mathcal{O}\big((H \times H){\cdot}\Delta(G)\big)$-module. Notice that, if F is a subgroup of G then A can be considered as an $\mathcal{O}(H \cap F)$-interior F-algebra that we denote by $\mathrm{Res}_F^G(A)$, and A^F has an evident structure of $\mathcal{O}C_H(F)$-interior $N_G(F)$-algebra; similarly, if P is a p-subgroup of G then $A(P)$ is a $kC_H(P)$-interior $N_G(P)$-algebra; moreover, assuming that $F \subset H$ and, for any $\sigma \in \mathrm{Aut}(F)$, denoting by $N_A^\sigma(F)$ the set of $a \in A$ such that $a{\cdot}x = \sigma(x){\cdot}a$ for any $x \in F$, the sum

2.3.3 $$N_A(F) = \sum_{\sigma \in \mathrm{Aut}(F)} N_A^\sigma(F)$$

has a structure of $\mathcal{O}N_H(F)$-interior $N_G(F)$-algebra and, for any $\sigma \in \mathrm{Aut}(F)$, we set

2.3.4 $$N_A^\sigma(F)^* = N_A^\sigma(F) \cap N_A(F)^* \quad .$$

When $H = 1$ we recover the G-algebras, whereas when $H = G$ we say simply that A is an *$\mathcal{O}G$-interior algebra*; actually, we will consider usually this particular case, switching to the general point of view whenever some inductive argument demands it; the interested reader will complete easily this lack of generality.

2.4 A *homomorphism* $f : A \longrightarrow A'$ between two $\mathcal{O}H$-interior G-algebras A and A' is a G-algebra homomorphism such that $f(y{\cdot}a{\cdot}y') = y{\cdot}f(a){\cdot}y'$ for any $a \in A$ and any $y, y' \in H$; we say that f is an *embedding* if $\mathrm{Ker}(f) = \{0\}$ and $\mathrm{Im}(f) = f(1)A'f(1)$; notice that if i is an idempotent of A^G then iAi has a unique $\mathcal{O}H$-interior G-algebra structure such that the inclusion $iAi \subset A$ becomes an $\mathcal{O}H$-interior G-algebra embedding. Following [19], the *exterior homomorphism* or, in short, the *exomorphism* determined by f is the set $\tilde{f}$ of homomorphisms from A to A' obtained by composing f with all the conjugations by elements of $(A^G)^*$ and $(A'^G)^*$; actually, $\tilde{f}$ is already obtained from the elements of $(A'^G)^*$ (cf. [38, (8.1)]); consequently we have that:

2.4.1 *The elementwise composition of exomorphisms is still an exomorphism.*

The fact of employing exomorphisms means that we are considering the $\mathcal{O}H$-interior G-algebras as objects of the corresponding quotient category. Coherently, we say that $\tilde{f}$ is an *exoembedding*, an *exoisomorphism*, etc. whenever f is respectively an embedding, an isomorphism, etc. If F is a subgroup of G, we denote by $\tilde{f}^F : A^F \longrightarrow A'^F$ the exomorphism of $\mathcal{O}C_H(F)$-interior $N_G(F)$-algebras determined by the restriction of any representative of $\tilde{f}$, and by $\mathrm{Res}_F^G(\tilde{f})$ the $\mathcal{O}(H \cap F)$-interior F-algebra exomorphism from $\mathrm{Res}_F^G(A)$ to $\mathrm{Res}_F^G(A')$ containing $\tilde{f}$; it is not difficult to adapt the proof of Lemma 3.7 in [19] to get (see also [38, (12.11)]):

2.4.2 *If f and g are $\mathcal{O}H$-interior G-algebra homomorphisms from A to A' and we have $G = H{\cdot}F$ then $\tilde{f} = \tilde{g}$ is equivalent to* $\mathrm{Res}_F^G(\tilde{f}) = \mathrm{Res}_F^G(\tilde{g})$.

Similarly, if P is a p-subgroup of G, we denote by

2.4.3 $$\tilde{f}(P) : A(P) \longrightarrow A'(P)$$

the $\mathcal{O}C_H(P)$-interior $N_G(P)$-algebra exomorphism induced by any representative of $\tilde{f}$. We denote by $\tilde{\mathrm{A}}\mathrm{ut}(A)$ the group of *exoautomorphisms* of A and call a homomorphism from a group K to $\tilde{\mathrm{A}}\mathrm{ut}(A)$ *exterior action* of K on A. Analogously, an *exomorphism* $\tilde{\varphi} : G \longrightarrow G'$ from G to a second group G' is the set of homomorphisms from G to G' obtained by composing one of them with all the inner automorphisms of G and G', and we denote respectively by $\mathrm{Aut}(G)$, $\mathrm{Inn}(G)$ and $\tilde{\mathrm{A}}\mathrm{ut}(G)$ the groups of automorphisms, inner automorphisms and outer automorphisms of G.

2.5 If A and A' are $\mathcal{O}H$-interior G-algebras, the diagonal map $\Delta : G \longrightarrow G \times G$ determines $\mathcal{O}H$-interior G-algebra structures in the direct and the tensor products $A \times A'$ and $A \otimes_\mathcal{O} A'$; these products of $\mathcal{O}H$-interior G-algebras are clearly commutative and associative. In particular, if X is a finite set and $\{A_x\}_{x\in X}$ a family of

$\mathcal{O}H$-interior G-algebras, we denote by $\otimes_{x\in X} A_x$ the tensor product of this family, which can be defined independently of any total order relation on X by considering $\otimes_{x\in X} A_x$ as the quotient of the $\mathcal{O}$-free $\mathcal{O}$-module over the set of families $\{a_x\}_{x\in X}$, where $a_x \in A_x$ for any $x \in X$, by the $\mathcal{O}$-submodule generated by all the elements

2.5.1 $$\lambda'\cdot\{a'_x\}_{x\in X} + \lambda''\cdot\{a''_x\}_{x\in X} - \{a_x\}_{x\in X}$$

such that we have $\lambda' a'_x + \lambda'' a''_x = a_x$ for some $x \in X$, where $\lambda', \lambda'' \in \mathcal{O}$, and $a'_y = a''_y = a_y$ for any $y \in X - \{x\}$; when $A_x = A$ for any $x \in X$, we write $T_X(A)$ instead of $\otimes_{x\in X} A$; coherently, we denote by $\Pi_{x\in X} A_x$ the direct product of that family and write A^X when $A_x = A$ for any $x \in X$. Then, if M and M' are $\mathcal{O}G$-modules, $\mathrm{End}_{\mathcal{O}}(M)$ and $\mathrm{End}_{\mathcal{O}}(M')$ are clearly $\mathcal{O}G$-interior algebras and the canonical unitary $\mathcal{O}$-algebra homomorphism

2.5.2 $$\mathrm{End}_{\mathcal{O}}(M) \underset{\mathcal{O}}{\otimes} \mathrm{End}_{\mathcal{O}}(M') \longrightarrow \mathrm{End}_{\mathcal{O}}(M \underset{\mathcal{O}}{\otimes} M') \quad ,$$

which is bijective whenever M or M' is $\mathcal{O}$-free, determines an $\mathcal{O}G$-module structure on $M \otimes_{\mathcal{O}} M'$, so that homomorphism 2.5.2 becomes an $\mathcal{O}G$-interior algebra homomorphism. Moreover recall that, considering the right $\mathcal{O}H$-module M°, $M^\circ \otimes_{\mathcal{O}H} M'$ still has an $\mathcal{O}(G/H)$-module structure and we will omit the opposite mark "o"; in particular notice that, identifying G to $\Delta(G)$ and to $(G\times G)/(1\times G)$, and denoting by $\pi' : G\times G \longrightarrow G$ the second projection map, we have a canonical $\mathcal{O}G$-module isomorphism

2.5.3 $$\mathrm{End}_{\mathcal{O}}(M \underset{\mathcal{O}}{\otimes} M') \cong \mathrm{Ind}_{\Delta(G)}^{G\times G}(M) \underset{\mathcal{O}(1\times G)}{\otimes} \mathrm{Res}_{\pi'}(M')$$

mapping $m \otimes m'$ on $(1,1) \otimes m \otimes m'$ for any $m \in M$ and any $m' \in M'$.

2.6 Let F be a subgroup of G such that $H\cdot F = G$ and B an $\mathcal{O}K$-interior F-algebra where $K = H \cap F$; the *induced $\mathcal{O}H$-interior G-algebra* $\mathrm{Ind}_K^H(B)$ is the $\mathcal{O}F$-module

2.6.1 $$\mathrm{Ind}_K^H(B) = \mathcal{O}H \underset{\mathcal{O}K}{\otimes} B \underset{\mathcal{O}K}{\otimes} \mathcal{O}H$$

endowed with the F-stable distributive product defined by

2.6.2 $$(x\otimes b\otimes y)(x'\otimes b'\otimes y') = \begin{cases} x\otimes b\cdot yx'\cdot b'\otimes y' & \text{if } yx' \in K \\ 0 & \text{otherwise} \end{cases}$$

where $x, y, x', y' \in H$ and $b, b' \in B$, and with the F-algebra homomorphism

2.6.3 $$\mathcal{O}H \longrightarrow \mathrm{Ind}_K^H(B)$$

mapping $x \in H$ on $\Sigma_{y\in Y} xy \otimes 1 \otimes y^{-1}$, where Y is a set of representatives for H/K in H. Notice that $\{y \otimes 1 \otimes y^{-1}\}_{y\in Y}$ is a G-stable pairwise orthogonal idempotent decomposition of the unity element in $\mathrm{Ind}_K^H(B)$; conversely, it is easily checked that (see also [38, (16.6)]):

2.6.4 *If A is an $\mathcal{O}H$-interior G-algebra and I a G-stable pairwise orthogonal idempotent decomposition of the unity element in A where H acts transitively, denoting by F the stabilizer of an element i of I and setting $K = H \cap F$, there is a unique $\mathcal{O}H$-interior G-algebra isomorphism $A \cong \mathrm{Ind}_K^H(iAi)$ mapping $a \in iAi$ on $1 \otimes a \otimes 1$.*
In particular, if $F = K$ and $B = \mathrm{End}_{\mathcal{O}}(N)$ for some $\mathcal{O}K$-module N, we get easily from 2.6.4 a canonical $\mathcal{O}H$-interior algebra isomorphism (cf. [38, (16.4)])

2.6.5 $$\mathrm{Ind}_K^H\big(\mathrm{End}_{\mathcal{O}}(N)\big) \cong \mathrm{End}_{\mathcal{O}}\big(\mathrm{Ind}_K^H(N)\big) \quad .$$

We denote by

2.6.6 $$d_K^H(B) : B \longrightarrow \mathrm{Res}_F^G\big(\mathrm{Ind}_K^H(B)\big)$$

the $\mathcal{O}K$-interior F-algebra embedding mapping $b \in B$ on $1 \otimes b \otimes 1$. On the other hand, an $\mathcal{O}K$-interior F-algebra homomorphism $g : B \longrightarrow B'$ between two $\mathcal{O}K$-interior F-algebras B and B' determines a unique $\mathcal{O}H$-interior G-algebra homomorphism

2.6.7 $$\mathrm{Ind}_K^H(g) : \mathrm{Ind}_K^H(B) \longrightarrow \mathrm{Ind}_K^H(B')$$

such that $\mathrm{Ind}_K^H(g) \circ d_K^H(B) = d_K^H(B') \circ g$, which is bijective if and only if g is so; moreover, we denote by $\mathrm{Ind}_K^H(\tilde{g})$ the corresponding exomorphism (cf. [38, page 129]). Notice that, if A is an $\mathcal{O}H$-interior G-algebra, we have a unique $\mathcal{O}H$-interior G-algebra isomorphism (cf. [38, (16.5)])

2.6.8 $$\mathrm{Ind}_K^H(B) \underset{\mathcal{O}}{\otimes} A \cong \mathrm{Ind}_K^H\big(B \underset{\mathcal{O}}{\otimes} \mathrm{Res}_F^G(A)\big)$$

mapping $(1 \otimes b \otimes 1) \otimes a$ on $1 \otimes (b \otimes a) \otimes 1$ for any $a \in A$ and any $b \in B$.

2.7 Let A be an $\mathcal{O}H$-interior G-algebra; a *pointed group* K_β on A is a pair formed by a subgroup K of G and a conjugacy class β of primitive idempotents of A^K (cf. [19, Definition 1.1]); we say also that β is a *point* of K on A or a *point* of A^K and we denote by $\mathcal{P}_A(K)$ the set of them. For any $i \in \beta$, the $\mathcal{O}$-algebra iAi inherits an $\mathcal{O}(H \cap K)$-interior K-algebra structure given by the restricted action of K and the group homomorphism $H \cap K \longrightarrow (iAi)^*$ mapping $x \in H \cap K$ on $x{\cdot}i$; it is clear that the conjugation in A induces *unique* exoisomorphisms between these $\mathcal{O}(H \cap K)$-interior K-algebras corresponding to the different choices of i (cf. [20, 1.6]) and we denote by A_β one of them and by $\tilde{f}_\beta : A_\beta \longrightarrow \mathrm{Res}_K^G(A)$ the exoembedding determined by the inclusion: we call the pair $(A_\beta, \tilde{f}_\beta)$ *embedded algebra* associated with K_β (cf. [22, 2.13]). Notice that the kernels of the structural maps

2.7.1 $$K \longrightarrow \mathrm{Aut}(A_\beta) \quad \text{and} \quad H \cap K \longrightarrow (A_\beta)^*$$

do not depend on the choice of i and therefore the stabilizer $N_G(K_\beta)$ of β in $N_G(K)$ normalizes them. We say that a second pointed group L_ε on A *is contained in* K_β, and write $L_\varepsilon \subset K_\beta$, if $L \subset K$ and there is an $\mathcal{O}(H \cap L)$-interior L-algebra exoembedding

2.7.2 $$\tilde{f}_\varepsilon^\beta : A_\varepsilon \longrightarrow \mathrm{Res}_L^K(A_\beta)$$

such that $\tilde{f}_\varepsilon = \mathrm{Res}_L^K(\tilde{f}_\beta) \circ \tilde{f}_\varepsilon^\beta$ (cf. [20, 1.8]).

2.8 On the other hand, β determines a unique simple quotient $A(K_\beta)$ of A^K, called the *multiplicity algebra* of K_β, which has an evident structure of $\mathcal{k}C_H(K)$-interior $N_G(K_\beta)$-algebra; we denote by

2.8.1 $$s_\beta : A^K \longrightarrow A(K_\beta)$$

the canonical map, so that $s_\beta(\beta) \neq \{0\}$ and this condition determines $A(K_\beta)$ among the simple quotients of A^K; it is quite clear that (cf. [38, (13.3)]):

2.8.2 *If K_β and L_ε are pointed groups on A, we have $L_\varepsilon \subset K_\beta$ if and only if we have $L \subset K$ and $s_\varepsilon(\beta) \neq \{0\}$.*

Notice that, since $A(K_\beta)$ has only inner automorphisms, the action of $\bar{N}_G(K_\beta) = N_G(K_\beta)/K$ on $A(K_\beta)$ determines a central extension $\hat{\bar{N}}_G(K_\beta)$ of $\bar{N}_G(K_\beta)$ by $\mathcal{k}^*$, or $\mathcal{k}^*$-group with $\mathcal{k}^*$-quotient $\bar{N}_G(K_\beta)$ (cf. [22, 5.2]), and a $\mathcal{k}$-algebra homomorphism from the corresponding twisted group algebra $\mathcal{k}_*\hat{\bar{N}}_G(K_\beta)$ to $A(K_\beta)$ (cf. [22, 5.12 and 6.4]); moreover, the natural map $C_H(K) \longrightarrow \bar{N}_G(K_\beta)$ and the homomorphism s_β determine a canonical group homomorphism

2.8.3 $$C_H(K) \longrightarrow \hat{\bar{N}}_G(K_\beta) \quad .$$

In particular, up to an obvious extension of our terminology, $A(K_\beta)$ becomes a $\mathcal{k}_*\hat{\bar{N}}_G(K_\beta)$*-interior algebra* (cf. [22, 5.10]) and any simple $A(K_\beta)$-module becomes a $\mathcal{k}_*\hat{\bar{N}}_G(K_\beta)$-module noted $V_A(K_\beta)$ and called *multiplicity module* of K_β (cf. [22, 6.4] or [38, page 104]); we set $m_\beta(A) = \dim_{\mathcal{k}}\big(V_A(K_\beta)\big)$ and call it *multiplicity* of K_β on A; notice that (cf. [38, (4.16)]):

2.8.4 *Two idempotents i and j of A^K are conjugate in A^K if and only if, for any point β of K on A, we have*

$$\dim_{\mathcal{k}}\big(s_\beta(i)\cdot V_A(K_\beta)\big) = \dim_{\mathcal{k}}\big(s_\beta(j)\cdot V_A(K_\beta)\big) \quad .$$

2.9 We say that a pointed group P_γ on A is *local* if $\gamma \not\subset A^P_Q$ for any proper subgroup Q of P, which forces P to be a p-subgroup of G; this condition is equivalent to $\mathrm{Br}_P(\gamma) \neq \{0\}$ (cf. [38, (14.4)]) and therefore, denoting by $\mathcal{LP}_A(P)$ the set of local points of P on A, we have that (cf. [38, (14.5)]):

2.9.1 *The Brauer homomorphism* Br_P *determines a bijection between $\mathcal{LP}_A(P)$ and the set of points of $A(P)$.*

For any local pointed group P_γ on A, we have (cf. [19, Proposition 1.3])

2.9.2 $$s_\gamma(A^G_P) = A(P_\gamma)_1^{\bar{N}_G(P_\gamma)}$$

and, moreover, denoting by $A^P\cdot\gamma\cdot A^P$ the two-sided ideal of A^P generated by γ, a point α of G on A such that $P_\gamma \subset G_\alpha$ is contained in $\mathrm{Tr}^G_P(A^P\cdot\gamma\cdot A^P)$ if and only if $s_\gamma(\alpha)$ is a point of $A(P_\gamma)^{\bar{N}_G(P_\gamma)}$ contained in $A(P_\gamma)_1^{\bar{N}_G(P_\gamma)}$ (cf. [38, (18.3) and (19.1)]) or, equivalently by Higman's criterion, if and only if $V_{A_\alpha}(P_\gamma)$ is an indecomposable projective $k_*\hat{N}_G(P_\gamma)$-module (cf. [38, (17.8)]); it is now clear that (cf. [26, (1.4.1)]):

2.9.3 *The map s_γ induces a bijection between the set of points of G on A contained in $\mathrm{Tr}_P^G(A^P{\cdot}\gamma{\cdot}A^P) - \mathrm{Ker}(s_\gamma)$ and the set of isomorphism classes of indecomposable projective direct summands of the multiplicity module $V_A(P_\gamma)$ over $k_*\hat{N}_G(P_\gamma)$.*

For any pointed group K_β on A, we call a local pointed group P_γ on A which is maximal contained in K_β *defect pointed group* of K_β, and P is called a *defect group* of K_β; then, we call the $\mathcal{O}(H \cap P)$-interior P-algebra A_γ *source algebra* of A_β. Notice that (cf. [38, (18.3)]):

2.9.4 *A pointed group P_γ on A is a defect pointed group of K_β if and only if it is minimal fulfilling $P \subset K$ and $\beta \subset \mathrm{Tr}_P^K(A^P{\cdot}\gamma{\cdot}A^P)$.*

2.9.5 *K acts transitively on the set of defect pointed groups of K_β.*

Analogously to the $\mathcal{O}G$-module case, when A is $\mathcal{O}$-free we say that K_β or β is *projective* if K_β has a trivial defect group or, equivalently, if $\beta \subset A_1^K$.

2.10 When $H = G$ and $A = \mathrm{End}_\mathcal{O}(M)$, where M is an $\mathcal{O}G$-module, the points of a subgroup K of G on A can be identified to the isomorphism classes of the indecomposable direct summands of $\mathrm{Res}_K^G(M)$ and, if N is such a direct summand, we will denote occasionally by K_N the corresponding pointed group on A; in that case, the embedded algebra associated with K_N is just $\mathrm{End}_\mathcal{O}(N)$ and we denote simply by $V_M(K_N)$ the multiplicity $k_*\hat{N}_G(K_N)$-module of K_N. Customarily, a defect group P of K_N is called a *vertex* of N and the indecomposable $\mathcal{O}P$-module associated with a source algebra of $\mathrm{End}_\mathcal{O}(N)$ is called an $\mathcal{O}P$-source of N. Notice that if $e : B \longrightarrow A$ is an $\mathcal{O}G$-interior algebra embedding then we have $B = \mathrm{End}_\mathcal{O}(L)$ for a suitable direct summand L of M. Moreover, since M is a projective A-module, it is easily checked that (cf. [38, (10.7)]):

2.10.1 *For any $\mathcal{O}G$-module M', we have an $\mathcal{O}G$-interior algebra isomorphism $A \cong \mathrm{End}_\mathcal{O}(M')$ if and only if we have an $\mathcal{O}G$-module isomorphism $M \cong M'$.*

2.11 Let A and A' be $\mathcal{O}H$-interior G-algebras and $f : A \longrightarrow A'$ an $\mathcal{O}H$-interior G-algebra homomorphism; for any subgroup K of G, any point β of K on A and any point β' of K on A', we denote by $m(f)_\beta^{\beta'}$ the *multiplicity* of f at (β, β') (cf. [19, Definition 2.2]) defined by

2.11.1 $$m(f)_\beta^{\beta'} = \dim_k\Big(s_{\beta'}\big(f(i)\big){\cdot}V_{A'}(K_{\beta'})\Big)$$

where $i \in \beta$. Notice that if f is an embedding then, for any subgroup K of G, the restriction $f^K : A^K \longrightarrow A'^K$ of f is still an embedding and it is quite clear that (cf. [38, (8.5)]):

2.11.2 *If f is an embedding then all the multiplicities of f are equal to* 0 *or* 1 *and there is an injective map, preserving localness, from the set of points β of K on A into the set of points β' of K on A' mapping β on β' if and only if $m(f)_\beta^{\beta'} = 1$.*

That is to say, in that case, if $K_{\beta'}$ is a pointed group on A' and $f^{-1}(\beta') \neq \phi$ then $f^{-1}(\beta')$ is a point of K on A and usually we denote both β' and $f^{-1}(\beta')$ by the *same* letter: as a matter of fact, this abuse never makes confusion and simplifies dramatically the notation. For instance, if F is a subgroup of G such that $H \cdot F = G$ and, setting $K = H \cap F$, B is an $\mathcal{O}K$-interior F-algebra, any pointed group on B is also considered as a pointed group on the induced $\mathcal{O}H$-interior G-algebra $\mathrm{Ind}_K^H(B)$ via the canonical embedding $d_K^H(B)$ and it is easily checked that (cf. [19, Proposition 3.9]):

2.11.3 *For any local pointed group P_γ on $\mathrm{Ind}_K^H(B)$ there is $x \in H$ such that $P^x \subset F$ and γ^x comes from a local point of P^x on B.*

Moreover, exoembeddings are monomorphisms in the corresponding category and, whenever G has the same number of points on A and A', they are also epimorphisms; explicitely, we have the following broadly employed statement (when $H = G$, it follows from [38, (8.6) and (8.7)] and 2.4.2 above):

2.11.4 *If f is an embedding, for any $\mathcal{O}H$-interior G-algebra B and any pair of $\mathcal{O}H$-interior G-algebra homomorphisms $g, h : B \longrightarrow A$ such that $\tilde{f} \circ \tilde{g} = \tilde{f} \circ \tilde{h}$, we have $\tilde{g} = \tilde{h}$. If moreover G has the same number of points on A and A', for any pair of $\mathcal{O}H$-interior G-algebra homomorphisms $g', h' : A' \longrightarrow B$ such that $\tilde{g}' \circ \tilde{f} = \tilde{h}' \circ \tilde{f}$, we have $\tilde{g}' = \tilde{h}'$.*

Actually, this statement remains true with an extra differential structure and we will prove it in this general situation in Proposition 11.10 below. In Section 18, we need the following easy remark:

2.11.5 *If f admits an $\mathcal{O}G$-module section then, for any subgroup K of G, we have $f(A^K) = A'^K$ and, in particular, f induces a bijection between the set of pointed groups K_β on A such that $\beta \not\subset \mathrm{Ker}(f)$ and the set of pointed groups on A', which preserves inclusion and localness.*

2.12 Let A be an $\mathcal{O}G$-interior algebra; recall that Higman's criterion on relative projectivity for indecomposable modules can be extended to pointed groups as follows (cf. [19, Proposition 3.6]):

2.12.1 *If H_β and L_ε are pointed groups on A such that $L_\varepsilon \subset H_\beta$ then we have $\beta \subset \mathrm{Tr}_L^H(A^L{\cdot}\varepsilon{\cdot}A^L)$ if and only if there is an $\mathcal{O}H$-interior algebra exoembedding $\tilde{h} : A_\beta \longrightarrow \mathrm{Ind}_L^H(A_\varepsilon)$ such that the following diagram is commutative*

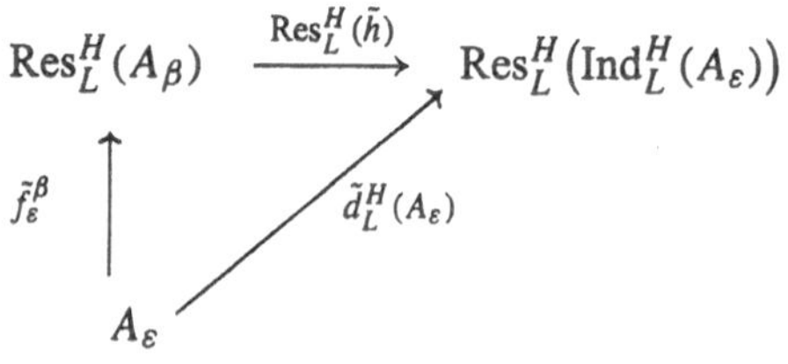

Moreover, in that case, $\tilde{h}$ is unique.

We denote by $\tilde{h}_\beta^\varepsilon$ this unique exoembedding and we call a representative of $\tilde{h}_\beta^\varepsilon$ *Higman embedding* associated with H_β and L_ε; it is quite clear that it induces a

Morita equivalence between A_β and A_ε (cf. [19, Proposition 2.7]); furthermore, it follows from Green's indecomposability theorem generalized to $\mathcal{O}G$-interior algebras (cf. [38, (23.5)]) that:

2.12.2 *Any Higman embedding associated with two pointed p-groups is an isomorphism.*

Notice that if $A = \mathrm{End}_\mathcal{O}(M)$ for a suitable $\mathcal{O}G$-module M then the diagram above is automatically commutative since, in that case, $\tilde{d}_L^H(A_\varepsilon)$ is the unique $\mathcal{O}L$-interior algebra exoembedding from A_ε to $\mathrm{Res}_L^H\big(\mathrm{Ind}_L^H(A_\varepsilon)\big)$ (cf. 2.4.2 above and [19, Proposition 2.3]). In particular, we have Higman's embeddings associated with any pointed group on A and its defect pointed groups (cf. 2.9.4) and consequently, according to 2.9.2 and 2.9.3, the following isomorphism (cf. [22, Lemma 9.12]) is quite significant (cf. [26, Theorem 3.4]):

2.12.3 *For any local pointed group P_γ on A there is a unique $k_*\hat{\bar{N}}_G(P_\gamma)$-interior algebra isomorphism*

$$\big(\mathrm{Ind}_P^G(A_\gamma)\big)(P_\gamma) \cong \mathrm{Ind}_{k^*}^{\hat{\bar{N}}_G(P_\gamma)}(k) = \mathrm{End}_k\big(k_*\hat{\bar{N}}_G(P_\gamma)\big)$$

mapping $s_\gamma(1 \otimes 1 \otimes 1)$ *on* $1 \otimes 1 \otimes 1$.

Actually, in Section 14, Higman embeddings will be included in the more general framework of the *Higman envelope* of $\mathcal{O}G$-interior algebras that, to avoid redundancy, we develop directly in the differential situation.

2.13 Let H_β and L_ε be pointed groups on A; following [20], we denote by $E_G(L_\varepsilon, H_\beta)$ the set of all the group exomorphisms from L to H which are induced by conjugation by an element x of G such that $L_\varepsilon \subset (H_\beta)^x$. More generally, we consider the set $F_A(L_\varepsilon, H_\beta)$ of all the injective group exomorphisms $\tilde{\varphi} : L \to H$ - called *A-fusions* from L_ε to H_β - such that, for a representative φ of $\tilde{\varphi}$, there is an $\mathcal{O}L$-interior algebra exoembedding $\tilde{f}_\varphi : A_\varepsilon \to \mathrm{Res}_\varphi(A_\beta)$ fulfilling (cf. [20, Definition 2.5])

2.13.1 $$\mathrm{Res}_1^H(\tilde{f}_\beta) \circ \mathrm{Res}_1^L(\tilde{f}_\varphi) = \mathrm{Res}_1^L(\tilde{f}_\varepsilon) \qquad ;$$

then, denoting by $E_G(L, H)$ the set of all the group exomorphisms from L to H induced by conjugation by elements of G, we have (cf. [20, 2.10] where there is a misprint: replace $\subset$ by $=$)

2.13.2 $$E_G(L_\varepsilon, H_\beta) = F_A(L_\varepsilon, H_\beta) \cap E_G(L, H) \qquad .$$

In particular, if P_γ is a local pointed group on A, notice that $N_G(P_\gamma)$ is the converse image of $F_A(P_\gamma, P_\gamma)$ - in short, noted $F_A(P_\gamma)$ - in $N_G(P)$ and that the image of $C_G(P)$ by homomorphism 2.8.3 is a normal subgroup of $\hat{\bar{N}}_G(P_\gamma)$ which has a trivial intersection with the image of k^* (cf. [22, Proposition 6.5]); hence, the corresponding quotient is still a k^*-group with k^*-quotient $N_G(P_\gamma)/P{\cdot}C_G(P)$ which is canonically isomorphic to $E_G(P_\gamma, P_\gamma) = E_G(P_\gamma)$ and we denote it by $\hat{E}_G(P_\gamma)$ (cf. [22, 6.6]). Moreover, we have that (cf. [20, Proposition 2.12]):

2.13.3 *For any $\varphi \in \mathrm{Aut}(P)$, $\tilde{\varphi}$ is an A-fusion from P_γ to P_γ if and only if we have $N^{\varphi}_{A_\gamma}(P)^* \neq \emptyset$.*

That is to say, we have a canonical surjective group homomorphism

2.13.4 $$F_A(P_\gamma) \longrightarrow \Big(\bigcup_{\varphi\in\mathrm{Aut}(P)} N^{\varphi}_{A_\gamma}(P)^*\Big)\Big/P\cdot(A^P_\gamma)^*$$

and we denote by $\hat{F}_A(P_\gamma)$ the k^*-group of k^*-quotient $F_A(P_\gamma)$ lifting this homomorphism to a surjective k^*-group homomorphism (cf. [22, 6.7])

2.13.5 $$\hat{F}_A(P_\gamma) \longrightarrow \Big(\bigcup_{\varphi\in\mathrm{Aut}(P)} N^{\varphi}_{A_\gamma}(P)^*\Big)\Big/P\cdot\big(1+J(A^P_\gamma)\big) \quad ;$$

then, the existence of a canonical k^*-group homomorphism $\hat{E}_G(P_\gamma)^\circ \to \hat{F}_A(P_\gamma)$ (cf. [22, Proposition 6.12]) gives the following description of $\hat{E}_G(P_\gamma)^\circ$, employed in Section 5:

2.13.6 *$\hat{E}_G(P_\gamma)^\circ$ is isomorphic to the k^*-group of all the pairs $(\tilde{\varphi}, \bar{a})$ where $\tilde{\varphi}$ runs on $E_G(P_\gamma)$ and, for such a $\tilde{\varphi}$, $\bar{a}$ runs on the set*

$$\Big(\bigcup_{\varphi\in\tilde{\varphi}} N^{\varphi}_{A_\gamma}(P)^*\Big)\Big/P\cdot\big(1+J(A^P_\gamma)\big) \quad .$$

Finally notice that if A' is a second $\mathcal{O}G$-interior algebra and $f : A \to A'$ an $\mathcal{O}G$-interior algebra homomorphism admitting an $\mathcal{O}(G\times G)$-module section, it follows from 2.11.5 that $\gamma' = f(\gamma)$ is a local point of P on A' whenever $\gamma \not\subset \mathrm{Ker}(f)$, and then it is easily checked from 2.13.3 that f induces a k^*-group isomorphism

2.13.7 $$\hat{F}_A(P_\gamma) \cong \hat{F}_{A'}(P_{\gamma'}) \quad .$$

2.14 We call a primitive idempotent b of $Z(\mathcal{O}G)$ *block* of G; notice that, since $Z(\mathcal{O}G)$ is isomorphic to the $\mathcal{O}$-algebra of endomorphisms of the $\mathcal{O}G$-bimodule $\mathcal{O}G$, $\mathcal{O}Gb$ is an indecomposable $\mathcal{O}G$-bimodule. As we explain in 1.17 above, when $A = \mathcal{O}Gb$ we are not only interested in the local pointed groups on A, but also in the pairs (P, e) - called *(G, b)-Brauer pairs* - formed by a p-subgroup P of G such that $\mathrm{Br}_P(b) \neq 0$ and by a block e of $C_G(P)$ such that $\mathrm{Br}_P(be) = \mathrm{Br}_P(e)$ (cf. [8, Definition 1.6]). Notice that, since Br_P induces an $\mathcal{O}C_G(P)$-interior $N_G(P)$-algebra isomorphism (cf. 2.2.5)

2.14.1 $$(\mathcal{O}G)(P) \cong kC_G(P) \quad ,$$

any local point γ of P on $\mathcal{O}G$ determines an isomorphism class of simple $kC_G(P)$-modules (cf. 2.9.1) and therefore it determines also a block of $C_G(P)$ - that we denote by $b(\gamma)$ - fulfilling $\mathrm{Br}_P\big(b(\gamma)\gamma\big) = \mathrm{Br}_P(\gamma)$; in particular, if $s_\gamma(b) \neq 0$ then $\big(P, b(\gamma)\big)$ is a (G, b)-Brauer pair and it is clear that any (G, b)-Brauer pair has this form; more generally, the key point when dealing with (G, b)-Brauer pairs is that (cf. [8, Theorem 1.8]):

2.14.2 *If (P, e) is a (G, b)-Brauer pair and Q is a subgroup of P then there is a unique block f of $C_G(Q)$ fulfilling* $\mathrm{Br}_Q(f\gamma) = \mathrm{Br}_Q(\gamma)$ *for any local point γ of P on A such that $b(\gamma) = e$.*

In that case (Q, f) is also a (G, b)-Brauer pair and we write

2.14.3 $$(Q, f) \subset (P, e)$$

saying that (Q, f) *is contained in* (P, e) (cf. [8. Definition 1.7]); clearly, this is an order relation which is completely determined by the inclusion between local pointed groups (cf. [19, page 267]) and coincides with the ordinary subgroup inclusion inside a fixed (G, b)-Brauer pair (cf. [38, (40.9)]).

2.15 As in 2.13, if (P, e) and (Q, f) are (G, b)-Brauer pairs, we denote by $E_G\big((Q, f), (P, e)\big)$ the set of all the group exomorphisms from Q to P which are induced by conjugation by an element x of G such that $(Q, f) \subset (P, e)^x$; actually, for any local point γ of P on A such that $b(\gamma) = e$, we have

2.15.1 $$E_G\big((Q, f), (P, e)\big) = \bigcup_{\delta} E_G(Q_\delta, P_\gamma)$$

where δ runs on the set of local points of Q on A such that $b(\delta) = f$(cf. [19, page 267]); following [3], we call the category where the objects are the (G, b)-Brauer pairs and the set of morphisms from (Q, f) to (P, e) is $E_G\big((Q, f), (P, e)\big)$, endowed with the composition of group exomorphisms, *Brauer category* of the block b of G whereas, following [20], in the *local category* of b the objects are the local pointed groups on $\mathcal{O}Gb$ and $E_G(Q_\delta, P_\gamma)$ is the set of morphisms from Q_δ to P_γ, the composition being the same; moreover, we call the full subcategory of the local category of b formed by all the nontrivial local pointed groups on $\mathcal{O}Gb$ *stable local category* of b (we consider it in Section 7). On the other hand, since $\mathrm{Br}_Q(f)$ is primitive in $A(Q)^{C_G(Q)} = \mathrm{Br}_Q(A^{Q\cdot C_G(Q)})$ (cf. 2.14.1), it is quite clear that (cf. 2.8.2 above and [38, (37.7)]):

2.15.2 *There is a unique point β of $Q{\cdot}C_G(Q)$ on A such that* $\mathrm{Br}_Q(\beta) = \{\mathrm{Br}_Q(f)\}$ *and then, for any local point δ of Q on A such that $b(\delta) = f$, β is the unique point of $Q{\cdot}C_G(Q)$ on A such that $Q_\delta \subset Q{\cdot}C_G(Q)_\beta$.*

This statement has the following useful consequence:

2.15.3 *If $(Q, f) \subset (P, e)$ and P contains a defect group R of f then we have $Q_\delta \subset P_\gamma$ for any local points γ of P and δ of Q on A such that $b(\gamma) = e$ and $b(\delta) = f$.*

Indeed, there is a local point ε of $Q{\cdot}R$ on A such that $Q{\cdot}R_\varepsilon \subset P_\gamma$ (cf. [8, Proposition 1.5]) and then, once again, there is a local point δ' of Q on A such that $Q_{\delta'} \subset Q{\cdot}R_\varepsilon$, so that $Q_{\delta'} \subset P_\gamma$; now, since $b(\gamma) = e$, the uniqueness of f implies that $b(\delta') = f$ and then the uniqueness of β forces $Q{\cdot}R_\varepsilon \subset Q{\cdot}C_G(Q)_\beta$ (cf. 2.15.2), so that $Q{\cdot}R_\varepsilon$ is a defect pointed group of $Q{\cdot}C_G(Q)_\beta$ since $f \in \big(\mathcal{O}C_G(Q)\big)_R^{C_G(Q)}$ (cf. 2.9.4); consequently, since $b(\delta) = f$ implies that $Q_\delta \subset Q{\cdot}C_G(Q)_\beta$ (cf. 2.15.2), it implies also that $Q_\delta \subset Q{\cdot}R_\varepsilon$ (cf. 2.9.5), so that $Q_\delta \subset P_\gamma$ as announced. Furthermore notice that, if $(Q, f) \subset (P, e)$ and Q is normal in P, (P, e) is still a (H, f)-Brauer pair for any subgroup H of $N_G(Q, f)$ containing $P{\cdot}C_G(Q)$ since $C_H(Q) = C_G(Q)$ and f is still a block of H. Similarly, since the Brauer homomorphism $\mathrm{Br}_Q : (kG)^Q \longrightarrow kC_G(Q)$ (cf. 2.14.1) has an evident $kC_G(Q)$-module section, it follows easily from 2.11.5 that:

2.15.4 *If $P = Q{\cdot}C_P(Q)$, for any point ν of P on A such that* $\mathrm{Br}_Q(\nu) \neq \{0\}$, $\mathrm{Br}_Q(\nu)$ *is a point of $C_P(Q)$ on $A(Q)$.*

Finally, an easy consequence of 2.14.2 is that:

2.15.5 *If $(Q, f) \subset (P, e)$, Q is normal in P and γ is a local point of P on A such that $b(\gamma) = e$, then there is $i \in \gamma$ such that $fi = i = if$.*

Indeed, in that situation P fixes f and we have $\mathrm{Br}_P(f\gamma) = Br_P(\gamma)$ (cf. [8, Theorem 1.8]), so that it suffices to choose a pairwise orthogonal primitive idempotent decomposition I of the unity element in A^P such that $f \in \Sigma_{i\in I}\mathcal{O}i$, and to pick i in $\gamma \cap I$.

2.16 Let Q_δ be a local pointed group on $A = \mathcal{O}Gb$, set $f = b(\delta)$ and denote by β the point of $Q{\cdot}C_G(Q)$ on A such that $\mathrm{Br}_Q(\beta) = \{\mathrm{Br}_Q(f)\}$; we say that Q_δ and (Q, f) are *selfcentralizing* if Q_δ is a defect pointed group of $Q{\cdot}C_G(Q)_\beta$ or, equivalently, if $Z(Q)$ is a defect group of f (cf. [38, (41.1)]); in that case, δ is the unique local point of Q on A such that $b(\delta) = f$ and consequently, we get that (cf. [38, (41.2)]):

2.16.1 *The natural map from the set of local pointed groups on A onto the set of (G, b)-Brauer pairs determines a bijection between the sets of all the selfcentralizing ones, which preserves inclusion.*

Actually, we have that:

2.16.2 *Any defect pointed group P_γ of $Q{\cdot}C_G(Q)_\beta$ is selfcentralizing and, for any (G, b)-Brauer pair (R, g) containing (Q, f), $|C_R(Q)| \le |C_P(Q)|$.*

Indeed, since $Q_\delta \subset P_\gamma$ (cf. 2.9.5) and β is the unique point of $Q{\cdot}C_G(Q)$ on A such that $Q_\delta \subset Q{\cdot}C_G(Q)_\beta$ (cf. 2.15.2), we have

2.16.3 $$Q_\delta \subset P_\gamma \subset P{\cdot}C_G(P)_\alpha \subset Q{\cdot}C_G(Q)_\beta \quad ,$$

where α is the point of $P{\cdot}C_G(P)$ on A such that $\mathrm{Br}_P(\alpha) = \Big\{\mathrm{Br}_P\big(b(\gamma)\big)\Big\}$, so that P_γ is a local pointed group on A which is obviously maximal fulfilling $P_\gamma \subset P{\cdot}C_G(P)_\alpha$; similarly, by 2.14.2 there is a block h of $C_G\big(Q{\cdot}C_R(Q)\big)$ such that $(Q, f) \subset \big(Q{\cdot}C_R(Q), h\big) \subset (R, g)$ and then $\big(Q{\cdot}C_R(Q), h\big)$ is also a $\big(Q{\cdot}C_G(Q), f\big)$-Brauer pair, so that for a suitable $z \in C_G(Q)$ we have

$\big(Q{\cdot}C_R(Q), h\big)^z \subset \big(P, b(\gamma)\big)$ (cf. [38, (40.13)]) which proves the inequality. Finally, the following alternative definition justifies the terminology (cf. [38, (41.3)]):

2.16.4 *Q_δ is selfcentralizing if and only if $C_P(Q) \subset Q$ for any local pointed group P_γ on A such that $Q_\delta \subset P_\gamma$ or, equivalently, for any (G, b)-Brauer pair (P, e) such that $(Q, f) \subset (P, e)$.*

In particular, any local pointed group on A containing a selfcentralizing one is also selfcentralizing (so that in 2.15.3 above the point γ is actually unique).

3 Noninjective induction of $\mathcal{O}G$-interior algebras

3.1 Let G and G' be finite groups, $\varphi : G \to G'$ a group homomorphism and A an $\mathcal{O}G$-interior algebra. When φ is injective it is well-known how to construct the induced $\mathcal{O}G'$-interior algebra $\mathrm{Ind}_\varphi(A)$ (cf. 2.6) by mimicking the evident construction of $\mathrm{End}_\mathcal{O}(\mathrm{Ind}_\varphi(M))$ from $\mathrm{End}_\mathcal{O}(M)$ whenever M is an $\mathcal{O}$-free $\mathcal{O}G$-module. Although perhaps it is less evident, it turns out that such a mimicry is always possible, even when φ is not injective; we will apply this general construction in the following section to give an alternative description of the *Hecke $\mathcal{O}G$-interior algebras*, which will appear naturally when dealing with Morita equivalences between Brauer blocks in Section 6 below.

3.2 Set $K = \mathrm{Ker}(\varphi)$ and consider A as an $\mathcal{O}K$-module by left multiplication; the tensor product $\mathcal{O} \otimes_{\mathcal{O}K} A$ has an evident right A-module structure

3.2.1 $$(\mathcal{O} \underset{\mathcal{O}K}{\otimes} A) \times A \longrightarrow \mathcal{O} \underset{\mathcal{O}K}{\otimes} A$$

which is clearly compatible with the actions by conjugation of K on A and on $\mathcal{O} \otimes_{\mathcal{O}K} A$ (where it coincides with the right multiplication); on the other hand, if $a, b \in A$ and K fixes $1 \otimes a$ in $\mathcal{O} \otimes_{\mathcal{O}K} A$, we get

3.2.2 $$(1 \otimes a)\cdot b = (1 \otimes a)\cdot(x\cdot b)$$

for any $x \in K$; consequently, the map 3.2.1 restricted to $(\mathcal{O} \otimes_{\mathcal{O}K} A)^K$ factorizes throughout $\mathcal{O} \otimes_{\mathcal{O}K} A$ giving a new K-stable map

3.2.3 $$(\mathcal{O} \underset{\mathcal{O}K}{\otimes} A)^K \times (\mathcal{O} \underset{\mathcal{O}K}{\otimes} A) \longrightarrow \mathcal{O} \underset{\mathcal{O}K}{\otimes} A$$

which induces by restriction a product in $(\mathcal{O} \otimes_{\mathcal{O}K} A)^K$

3.2.4 $$(\mathcal{O} \underset{\mathcal{O}K}{\otimes} A)^K \times (\mathcal{O} \underset{\mathcal{O}K}{\otimes} A)^K \longrightarrow (\mathcal{O} \underset{\mathcal{O}K}{\otimes} A)^K$$

mapping $(1 \otimes a, 1 \otimes b)$ on $1 \otimes ab$ for any $a, b \in A$ such that K fixes $1 \otimes a$ and $1 \otimes b$ in $\mathcal{O} \otimes_{\mathcal{O}K} A$. This product is clearly associative and distributive so that $(\mathcal{O} \otimes_{\mathcal{O}K} A)^K$ becomes an $\mathcal{O}$-algebra. Moreover, the structural $\mathcal{O}$-algebra homomorphism $\mathcal{O}G \longrightarrow A$ composed with the canonical map $A \longrightarrow \mathcal{O} \otimes_{\mathcal{O}K} A$ factorizes throughout $\mathcal{O}(G/K)$ and is compatible with the actions of K by conjugation; hence, we get a map

3.2.5 $$\mathcal{O}(G/K) \longrightarrow (\mathcal{O} \underset{\mathcal{O}K}{\otimes} A)^K$$

which is actually an $\mathcal{O}$-algebra homomorphism. In conclusion, we have that:

3.2.6 *The $\mathcal{O}G$-interior algebra structure of A induces an $\mathcal{O}(G/K)$-interior algebra structure in $(\mathcal{O} \otimes_{\mathcal{O}K} A)^K$.*

3.3 We are ready to define the *induced $\mathcal{O}G'$-interior algebra* $\mathrm{Ind}_\varphi(A)$ from the ordinary induction (cf. 2.6). Considering $(\mathcal{O}\otimes_{\mathcal{O}K}A)^K$ as an $\mathcal{O}(\varphi(G))$-interior algebra throughout the canonical group isomorphism $G/K\cong\varphi(G)$, we set

3.3.1
$$\mathrm{Ind}_\varphi(A)=\mathrm{Ind}_{\varphi(G)}^{G'}\big((\mathcal{O}\underset{\mathcal{O}K}{\otimes}A)^K\big)$$

and, in particular, if φ is surjective then we identify $\mathrm{Ind}_\varphi(A)$ with $(\mathcal{O}\otimes_{\mathcal{O}K}A)^K$. That is to say, consider $\mathcal{O}G'$ and $(\mathcal{O}\otimes_{\mathcal{O}K}A)^K$ as $\mathcal{O}(G/K)$-bimodules throughout φ and homomorphism 3.2.5; then, $\mathrm{Ind}_\varphi(A)$ is the $\mathcal{O}G'$-bimodule

3.3.2
$$\mathcal{O}G'\underset{\mathcal{O}(G/K)}{\otimes}(\mathcal{O}\underset{\mathcal{O}K}{\otimes}A)^K\underset{\mathcal{O}(G/K)}{\otimes}\mathcal{O}G'$$

endowed with the distributive product that, for any $x',s',y',t'\in G'$ and any $a,b\in A$ such that K fixes $1\otimes a$ and $1\otimes b$ in $\mathcal{O}\otimes_{\mathcal{O}K}A$, maps $\big(x'\otimes(1\otimes a)\otimes s',\,y'\otimes(1\otimes b)\otimes t'\big)$ either on zero or on $x'\otimes(1\otimes a{\cdot}z{\cdot}b)\otimes t'$ whenever $s'y'=\varphi(z)$ for a suitable $z\in G$; in particular, the $\mathcal{O}G'$-bimodule structure determines an $\mathcal{O}$-algebra homomorphism $\mathcal{O}G'\to\mathrm{Ind}_\varphi(A)$ mapping $x'\in G'$ on

3.3.3
$$x'{\cdot}\mathrm{Tr}_{\varphi(G)}^{G'}\big(1\otimes(1\otimes 1)\otimes 1\big)=\mathrm{Tr}_{\varphi(G)}^{G'}\big(1\otimes(1\otimes 1)\otimes 1\big){\cdot}x' \quad .$$

3.4 Notice that, even if A has no $\mathcal{O}$-torsion, $\mathrm{Ind}_\varphi(A)$ is not necessarily $\mathcal{O}$-free since $\mathcal{O}\otimes_{\mathcal{O}K}A$ can occasionally have $\mathcal{O}$-torsion; however, it is clear that:

3.4.1 *If A is a projective $\mathcal{O}K$-module by left multiplication then $\mathrm{Ind}_\varphi(A)$ is $\mathcal{O}$-free.*

On the other hand, whereas in the injective induction A is naturally embedded in $\mathrm{Res}_\varphi\big(\mathrm{Ind}_\varphi(A)\big)$ (cf. 2.6.6), in general we have only an $\mathcal{O}G$-interior algebra homomorphism (cf. 2.3.3)

3.4.2
$$d_\varphi(A):N_A(K)\longrightarrow\mathrm{Res}_\varphi\big(\mathrm{Ind}_\varphi(A)\big)$$

mapping $a\in N_A(K)$ on $1\otimes(1\otimes a)\otimes 1$; indeed, it is quite clear that, for any $a\in N_A(K)$, K fixes $1\otimes a$ in $\mathcal{O}\otimes_{\mathcal{O}K}A$. Moreover, it is clear that

3.4.3
$$d_\varphi(A_1^K)=1\otimes(\mathcal{O}\underset{\mathcal{O}K}{\otimes}A)_1^K\otimes 1 \quad ;$$

in particular, it follows that:

3.4.4 *If $\mathcal{O}\otimes_{\mathcal{O}K}A$ is a projective $\mathcal{O}K$-module then $d_\varphi(A)$ induces an $\mathcal{O}G$-interior algebra embedding from $N_A(K)/\mathrm{Ker}\big(d_\varphi(A)\big)$ to $\mathrm{Res}_\varphi\big(\mathrm{Ind}_\varphi(A)\big)$.*

3.5 Let B be a second $\mathcal{O}G$-interior algebra and $f:A\to B$ an $\mathcal{O}G$-interior algebra homomorphism; it is clear that $f\big(N_A(K)\big)\subset N_B(K)$ and that the canonical map

3.5.1
$$\mathcal{O}\underset{\mathcal{O}K}{\otimes}f:\mathcal{O}\underset{\mathcal{O}K}{\otimes}A\longrightarrow\mathcal{O}\underset{\mathcal{O}K}{\otimes}B$$

induces an $\mathcal{O}(G/K)$-interior algebra homomorphism

3.5.2 $$(\mathcal{O} \underset{\mathcal{O}K}{\otimes} A)^K \longrightarrow (\mathcal{O} \underset{\mathcal{O}K}{\otimes} B)^K \quad ;$$

consequently, we get an $\mathcal{O}G'$-interior algebra homomorphism

3.5.3 $$\mathrm{Ind}_\varphi(f) : \mathrm{Ind}_\varphi(A) \longrightarrow \mathrm{Ind}_\varphi(B)$$

and we have clearly the following commutative diagram of $\mathcal{O}G$-interior algebra homomorphisms

3.5.4 $$\begin{array}{ccc} \mathrm{Ind}_\varphi(A) & \xrightarrow{\mathrm{Ind}_\varphi(f)} & \mathrm{Ind}_\varphi(B) \\ {\scriptstyle d_\varphi(A)}\uparrow & & \uparrow{\scriptstyle d_\varphi(B)} \\ N_A(K) & \xrightarrow{\quad f \quad} & N_B(K) \end{array} \quad .$$

For Remark 3.9 below it is useful to notice that:

3.5.5 *If $b \in B^G$ centralizes $\mathrm{Im}(f)$ then $\mathrm{Tr}_{\varphi(G)}^{G'}\big(d_\varphi(B)(b)\big)$ centralizes $\mathrm{Im}\big(\mathrm{Ind}_\varphi(f)\big)$.*

Moreover, if C is a third $\mathcal{O}G$-interior algebra and $g : B \to C$ a second $\mathcal{O}G$-interior algebra homomorphism, it is obvious that

3.5.6 $$\mathrm{Ind}_\varphi(g \circ f) = \mathrm{Ind}_\varphi(g) \circ \mathrm{Ind}_\varphi(f) \quad .$$

3.6 If M is an $\mathcal{O}G$-module, we can look at $\mathrm{End}_\mathcal{O}(M)$ as an $\mathcal{O}G$-interior algebra, so that it makes sense to consider now $\mathrm{Ind}_\varphi\big(\mathrm{End}_\mathcal{O}(M)\big)$, and we will relate this induced $\mathcal{O}G'$-interior algebra to the $\mathcal{O}G'$-interior algebra $\mathrm{End}_\mathcal{O}\big(\mathrm{Ind}_\varphi(M)\big)$, where $\mathrm{Ind}_\varphi(M)$ denotes the usual coefficient extension $\mathcal{O}G' \otimes_{\mathcal{O}G} M$; it is clear that the coefficient extension functor throughout φ determines an $\mathcal{O}$-algebra homomorphism from $\mathrm{End}_\mathcal{O}(M)^G$ to $\mathrm{End}_\mathcal{O}\big(\mathrm{Ind}_\varphi(M)\big)^{G'}$ and we denote by $\mathrm{Ind}_\varphi(f)$ the image of $f \in \mathrm{End}_\mathcal{O}(M)^G$.

Proposition 3.7 *For any $\mathcal{O}G$-module M there is a unique $\mathcal{O}G'$-interior algebra homomorphism*

3.7.1 $$\mathrm{ind}_{\varphi,M} : \mathrm{Ind}_\varphi\big(\mathrm{End}_\mathcal{O}(M)\big) \longrightarrow \mathrm{End}_\mathcal{O}\big(\mathrm{Ind}_\varphi(M)\big)$$

which, for any $x', s' \in G'$ and any $f \in \mathrm{End}_\mathcal{O}(M)$ such that K fixes $1 \otimes f$ in $\mathcal{O} \otimes_{\mathcal{O}K} \mathrm{End}_\mathcal{O}(M)$, maps $x' \otimes (1 \otimes f) \otimes s'$ on the $\mathcal{O}$-linear endomorphism of $\mathrm{Ind}_\varphi(M)$ mapping $y' \otimes m \in \mathrm{Ind}_\varphi(M)$, where $y' \in G'$ and $m \in M$, either on zero or on $x' \otimes f(z{\cdot}m)$ whenever $s'y' = \varphi(z)$ for a suitable $z \in G$. Moreover, if M is $\mathcal{O}$-free then $\mathrm{ind}_{\varphi,M}$ is an isomorphism.

Remark 3.8 If $f \in \mathrm{End}_\mathcal{O}(M)^G$ then it is easily checked that

3.8.1 $$\mathrm{Ind}_\varphi(f) = \mathrm{Tr}_{\varphi(G)}^{G'}\Big(\mathrm{ind}_{\varphi,M}\Big(d_\varphi\big(\mathrm{End}_\mathcal{O}(M)\big)(f)\Big)\Big) \quad .$$

Proof: It follows from 2.6.5 and from definition 3.3.1 that it suffices to prove the proposition when φ is surjective. In that case, denoting by $I(\mathcal{O}K)$ the augmentation ideal of $\mathcal{O}K$ and identifying $\mathrm{Hom}_{\mathcal{O}}\big(M, I(\mathcal{O}K)\cdot M\big)$ to its canonical image in $\mathrm{End}_{\mathcal{O}}(M)$, it is clear that

3.7.2 $$I(\mathcal{O}K)\cdot\mathrm{End}_{\mathcal{O}}(M) \subset \mathrm{Hom}_{\mathcal{O}}\big(M, I(\mathcal{O}K)\cdot M\big)$$

and that we have the equality whenever M is $\mathcal{O}$-free; consequently, we have canonical $\mathcal{O}G$-bimodule homomorphisms

3.7.3 $$\begin{aligned}\mathrm{End}_{\mathcal{O}}(M)/I(\mathcal{O}K)\cdot\mathrm{End}_{\mathcal{O}}(M) &\longrightarrow \mathrm{End}_{\mathcal{O}}(M)/\mathrm{Hom}_{\mathcal{O}}\big(M, I(\mathcal{O}K)\cdot M\big)\\ &\longrightarrow \mathrm{Hom}_{\mathcal{O}}\big(M, \mathrm{Ind}_{\varphi}(M)\big)\end{aligned}$$

since $\mathrm{Ind}_{\varphi}(M) \cong M/I(\mathcal{O}K)\cdot M$, and they are clearly bijective whenever M is $\mathcal{O}$-free. On the other hand, it follows from definition 3.3.1, up to suitable identifications, that

3.7.4 $$\mathrm{Ind}_{\varphi}\big(\mathrm{End}_{\mathcal{O}}(M)\big) = \big(\mathrm{End}_{\mathcal{O}}(M)/I(\mathcal{O}K)\cdot\mathrm{End}_{\mathcal{O}}(M)\big)^K$$

and, identifying $\mathrm{End}_{\mathcal{O}}\big(\mathrm{Ind}_{\varphi}(M)\big)$ to its image in $\mathrm{Hom}_{\mathcal{O}}\big(M, \mathrm{Ind}_{\varphi}(M)\big)$, it is easily checked that

3.7.5 $$\mathrm{End}_{\mathcal{O}}\big(\mathrm{Ind}_{\varphi}(M)\big) = \mathrm{Hom}_{\mathcal{O}}\big(M, \mathrm{Ind}_{\varphi}(M)\big)^K \quad .$$

In conclusion, the composed $\mathcal{O}G$-bimodule homomorphism in 3.7.3 determines from 3.7.4 and 3.7.5 an $\mathcal{O}G'$-bimodule homomorphism

3.7.6 $$\mathrm{Ind}_{\varphi}\big(\mathrm{End}_{\mathcal{O}}(M)\big) \longrightarrow \mathrm{End}_{\mathcal{O}}\big(\mathrm{Ind}_{\varphi}(M)\big)$$

which is bijective whenever M is $\mathcal{O}$-free; then, it is easily checked that it coincides with the announced homomorphism and, therefore, that it is an $\mathcal{O}$-algebra homomorphism too.

Remark 3.9 Since any A-module M has, in particular, an $\mathcal{O}G$-module structure given by the structural map $A \to \mathrm{End}_{\mathcal{O}}(M)$, which becomes an $\mathcal{O}G$-interior algebra homomorphism, we can induce it throughout φ and compose the induced $\mathcal{O}G'$-interior algebra homomorphism with $\mathrm{ind}_{\varphi,M}$

3.9.1 $$\mathrm{Ind}_{\varphi}(A) \longrightarrow \mathrm{Ind}_{\varphi}\big(\mathrm{End}_{\mathcal{O}}(M)\big) \longrightarrow \mathrm{End}_{\mathcal{O}}\big(\mathrm{Ind}_{\varphi}(M)\big)$$

endowing $\mathrm{Ind}_{\varphi}(M)$ with an $\mathrm{Ind}_{\varphi}(A)$-module structure. Moreover, if N is a second A-module and $f : M \to N$ an A-module homomorphism then the map

3.9.2 $$\mathrm{Ind}_{\varphi}(f) : \mathrm{Ind}_{\varphi}(M) \longrightarrow \mathrm{Ind}_{\varphi}(N)$$

is an $\mathrm{Ind}_{\varphi}(A)$-module homomorphism too, as it follows from Remark 3.8 and from 3.5.5 applied to the structural map $A \to \mathrm{End}_{\mathcal{O}}(M \oplus N)$.

3.10 Let G'' be a third finite group and $\varphi' : G' \to G''$ a second group homomorphism; we will discuss now the relationship between the $\mathcal{O}G'$-interior algebras $\mathrm{Ind}_{\varphi'}\big(\mathrm{Ind}_\varphi(A)\big)$ and $\mathrm{Ind}_{\varphi'\circ\varphi}(A)$. Set $K' = \mathrm{Ker}(\varphi')$ and $L = \mathrm{Ker}(\varphi'\circ\varphi) = \varphi^{-1}(K')$; since $(\mathcal{O}\otimes_{\mathcal{O}K} A)^K$ is an $\mathcal{O}(G/K)$-interior algebra and $\varphi(L) \subset K'$, the canonical embedding $(\mathcal{O}\otimes_{\mathcal{O}K} A)^K \to \mathrm{Ind}_\varphi(A)$, which, for any $a \in A$ such that K fixes $1\otimes a$ in $\mathcal{O}\otimes_{\mathcal{O}K} A$, maps $1\otimes a$ on $1\otimes(1\otimes a)\otimes 1$, induces an $\mathcal{O}$-linear map

3.10.1 $$\mathcal{O}\underset{\mathcal{O}(L/K)}{\otimes}(\mathcal{O}\underset{\mathcal{O}K}{\otimes} A)^K \longrightarrow \mathcal{O}\underset{\mathcal{O}(\varphi(L))}{\otimes}\mathrm{Ind}_\varphi(A) \longrightarrow \mathcal{O}\underset{\mathcal{O}K'}{\otimes}\mathrm{Ind}_\varphi(A)$$

mapping $1\otimes(1\otimes a)$ on $1\otimes\big(1\otimes(1\otimes a)\otimes 1\big)$ in $\mathcal{O}\otimes_{\mathcal{O}K'}\mathrm{Ind}_\varphi(A)$; this map is clearly compatible with the actions of L/K by conjugation and, in particular, it determines an $\mathcal{O}$-linear map

3.10.2 $$r_{\varphi,K}(A) : \big(\mathcal{O}\underset{\mathcal{O}(L/K)}{\otimes}(\mathcal{O}\underset{\mathcal{O}K}{\otimes} A)^K\big)^{L/K} \longrightarrow \big(\mathcal{O}\underset{\mathcal{O}K'}{\otimes}\mathrm{Ind}_\varphi(A)\big)^{\varphi(L)} \quad ;$$

moreover, K' acts by conjugation on $\mathcal{O}\otimes_{\mathcal{O}K'}\mathrm{Ind}_\varphi(A)$ and since $\varphi(L)\subset K'$ we have the relative trace map

3.10.3 $$\mathrm{Tr}_{\varphi(L)}^{K'} : \big(\mathcal{O}\underset{\mathcal{O}K'}{\otimes}\mathrm{Ind}_\varphi(A)\big)^{\varphi(L)} \longrightarrow \big(\mathcal{O}\underset{\mathcal{O}K'}{\otimes}\mathrm{Ind}_\varphi(A)\big)^{K'} \quad ;$$

finally, we consider the composed map $\mathrm{Tr}_{\varphi(L)}^{K'}\circ r_{\varphi,K}(A)$.

Proposition 3.11 *With the notation above, the $\mathcal{O}$-linear map*

3.11.1 $$\big(\mathcal{O}\underset{\mathcal{O}(L/K)}{\otimes}(\mathcal{O}\underset{\mathcal{O}K}{\otimes} A)^K\big)^{L/K} \longrightarrow \big(\mathcal{O}\underset{\mathcal{O}K'}{\otimes}\mathrm{Ind}_\varphi(A)\big)^{K'}$$

mapping $1\otimes(1\otimes a)$ on $\mathrm{Tr}_{\varphi(L)}^{K'}\Big(1\otimes\big(1\otimes(1\otimes a)\otimes 1\big)\Big)$, for any $a\in A$ such that K fixes $1\otimes a$ in $\mathcal{O}\otimes_{\mathcal{O}K} A$ and L/K fixes $1\otimes(1\otimes a)$ in $\mathcal{O}\otimes_{\mathcal{O}(L/K)}(\mathcal{O}\otimes_{\mathcal{O}K} A)^K$, induces an $\mathcal{O}G''$-interior algebra isomorphism

3.11.2 $$\mathrm{Ind}_{\varphi'(\varphi(G))}^{G''}\Big(\big(\mathcal{O}\underset{\mathcal{O}(L/K)}{\otimes}(\mathcal{O}\underset{\mathcal{O}K}{\otimes} A)^K\big)^{L/K}\Big) \cong \mathrm{Ind}_{\varphi'}\big(\mathrm{Ind}_\varphi(A)\big) \quad .$$

Proof: Assume first that φ is injective, so that $K = 1$, and identify G to its image in G', so that φ is the inclusion map; in that case, the map 3.11.1 becomes

3.11.3 $$(\mathcal{O}\underset{\mathcal{O}L}{\otimes} A)^L \longrightarrow \big(\mathcal{O}\underset{\mathcal{O}K'}{\otimes}\mathrm{Ind}_G^{G'}(A)\big)^{K'}$$

and maps $1 \otimes a$ on $\mathrm{Tr}_L^{K'}\big(1 \otimes (1 \otimes a \otimes 1)\big)$ for any $a \in A$ such that L fixes $1 \otimes a$ in $\mathcal{O} \otimes_{\mathcal{O}L} A$; more explicitly, denoting by T' a set of representatives for K'/L in K', it maps $1 \otimes a$ on $\Sigma_{t' \in T'} 1 \otimes (1 \otimes a \otimes t'^{-1})$ which does not depend on the choice of T' and, since $L = K' \cap G$, it is easily checked that the map 3.11.3 is an $\mathcal{O}$-algebra homomorphism and, for any $x, y \in G$, it maps $1 \otimes x{\cdot}a{\cdot}y^{-1}$ on the element

3.11.4
$$\begin{aligned}\sum_{t' \in T'} 1 \otimes (1 \otimes x{\cdot}a{\cdot}y^{-1} \otimes t'^{-1}) &= \sum_{t' \in T'} 1 \otimes (x \otimes a \otimes y^{-1}t'^{-1}) \\ &= \sum_{t' \in T'} 1 \otimes x{\cdot}\big(1 \otimes a \otimes (t'^{y})^{-1}\big){\cdot}y^{-1} \\ &= x{\cdot}\Big(\sum_{t' \in T'} 1 \otimes (1 \otimes a \otimes t'^{-1})\Big){\cdot}y \quad .\end{aligned}$$

In particular, it maps the unity element $1 \otimes 1$ on the idempotent

3.11.5
$$i = \sum_{t' \in T'} 1 \otimes (1 \otimes 1 \otimes t'^{-1}) = 1 \otimes \mathrm{Tr}_G^{K'.G}(1 \otimes 1 \otimes 1)$$

since $L = K' \cap G$, and therefore the image of this homomorphism is contained in

3.11.6
$$i.\big(\mathcal{O} \underset{\mathcal{O}K'}{\otimes} \mathrm{Ind}_G^{G'}(A)\big).i \cong \mathcal{O} \underset{\mathcal{O}K'}{\otimes} \mathrm{Ind}_G^{K'.G}(A) \qquad ;$$

moreover, we have the evident $\mathcal{O}$-linear isomorphisms (cf. 2.6)

3.11.7
$$\begin{aligned}\mathcal{O} \underset{\mathcal{O}K'}{\otimes} \mathrm{Ind}_G^{K'.G}(A) &\cong \mathcal{O} \underset{\mathcal{O}K'}{\otimes} \big(\mathcal{O}(K'.G) \underset{\mathcal{O}G}{\otimes} A \underset{\mathcal{O}G}{\otimes} \mathcal{O}(K'.G)\big) \\ &\cong \mathcal{O}(G/L) \underset{\mathcal{O}G}{\otimes} A \underset{\mathcal{O}G}{\otimes} \mathcal{O}(K'.G) \\ &\cong \mathcal{O} \underset{\mathcal{O}L}{\otimes} A \underset{\mathcal{O}L}{\otimes} \mathcal{O}K' \cong \bigoplus_{t' \in T'} \mathcal{O} \underset{\mathcal{O}L}{\otimes} (A \otimes t'^{-1})\end{aligned}$$

and any direct summand is isomorphic to $\mathcal{O} \otimes_{\mathcal{O}L} A$. Consequently, the map 3.11.1 is an $\mathcal{O}(G/L)$-interior algebra embedding.

On the other hand, according to definition 3.3.1, $\big(\mathcal{O} \otimes_{\mathcal{O}K'} \mathrm{Ind}_G^{G'}(A)\big)^{K'}$ is an $\mathcal{O}(G'/K')$-interior algebra and, denoting by $\bar{X}'$ a set or representatives for $G'/K'{\cdot}G$ in G'/K', it is clear that $\{i^{\bar{x}'}\}_{\bar{x}' \in \bar{X}'}$ is a pairwise orthogonal idempotent decomposition of the unity element in it. Hence, it follows from 2.6.4 that embedding 3.11.1 determines an $\mathcal{O}(G'/K')$-interior algebra isomorphism

3.11.8
$$\mathrm{Ind}_{\varphi'(G)}^{\varphi'(G')}\big((\mathcal{O} \underset{\mathcal{O}L}{\otimes} A)^L\big) \cong \big(\mathcal{O} \underset{\mathcal{O}K'}{\otimes} \mathrm{Ind}_G^{G'}(A)\big)^{K'}$$

and inducing it to G'' we obtain isomorphism 3.11.2 when φ is injective. Finally, it is easily checked that the general cases reduces to that one applied to the $\mathcal{O}(G/K)$-interior algebra $(\mathcal{O} \otimes_{\mathcal{O}K} A)^K$ (cf. 3.3.1).

3.12 Notice that the inclusion of $(\mathcal{O} \otimes_{\mathcal{O}K} A)^K$ in $\mathcal{O} \otimes_{\mathcal{O}K} A$ determines an $\mathcal{O}$-linear map

3.12.1 $$\mathcal{O} \underset{\mathcal{O}(L/K)}{\otimes} (\mathcal{O} \underset{\mathcal{O}K}{\otimes} A)^K \longrightarrow \mathcal{O} \underset{\mathcal{O}(L/K)}{\otimes} (\mathcal{O} \underset{\mathcal{O}K}{\otimes} A) \cong \mathcal{O} \underset{\mathcal{O}L}{\otimes} A$$

which is compatible with the actions of L by conjugation in both ends and therefore it restricts to an $\mathcal{O}$-linear map

3.12.2 $$s_{K,L}(A) : \big(\mathcal{O} \underset{\mathcal{O}(L/K)}{\otimes} (\mathcal{O} \underset{\mathcal{O}K}{\otimes} A)^K\big)^{L/K} \longrightarrow (\mathcal{O} \underset{\mathcal{O}L}{\otimes} A)^L$$

which is actually an $\mathcal{O}(G/L)$-interior algebra homomorphism.

Corollary 3.13 *With the notation above, there is a unique $\mathcal{O}G''$-interior algebra homomorphism*

3.13.1 $$t_{\varphi',\varphi}(A) : \mathrm{Ind}_{\varphi'}\big(\mathrm{Ind}_{\varphi}(A)\big) \longrightarrow \mathrm{Ind}_{\varphi'\circ\varphi}(A)$$

mapping $1 \otimes \mathrm{Tr}_{\varphi(L)}^{K'}\Big(1 \otimes \big(1 \otimes (1 \otimes a) \otimes 1\big)\Big) \otimes 1$ *on* $1 \otimes (1 \otimes a) \otimes 1$ *for any* $a \in A$ *such that* K *fixes* $1 \otimes a$ *in* $\mathcal{O} \otimes_{\mathcal{O}K} A$ *and* L *fixes* $1 \otimes (1 \otimes a)$ *in* $\mathcal{O} \otimes_{\mathcal{O}(L/K)} (\mathcal{O} \otimes_{\mathcal{O}K} A)^K$. *Moreover,* $t_{\varphi',\varphi}(A)$ *is an isomorphism if and only if* $s_{K,L}(A)$ *is an isomorphism. In particular, if either* φ *or* φ' *is injective, or* A *is a* $1 \times L$*-projective* $\mathcal{O}(L \times L)$*-module by left and right multiplication then* $t_{\varphi',\varphi}(A)$ *is an isomorphism.*

Remark 3.14 If M is an $\mathcal{O}G$-module, by Proposition 3.7 we have now the following diagram of $\mathcal{O}G''$-interior algebra homomorphisms

3.14.1 $$\begin{array}{ccc}
\mathrm{End}_{\mathcal{O}}\Big(\mathrm{Ind}_{\varphi'}\big(\mathrm{Ind}_{\varphi}(M)\big)\Big) & \cong & \mathrm{End}_{\mathcal{O}}\big(\mathrm{Ind}_{\varphi'\circ\varphi}(M)\big) \\
\uparrow {\scriptstyle \mathrm{ind}_{\varphi',\mathrm{Ind}_{\varphi}(M)}} & & \\
\mathrm{Ind}_{\varphi'}\Big(\mathrm{End}_{\mathcal{O}}\big(\mathrm{Ind}_{\varphi}(M)\big)\Big) & & \uparrow {\scriptstyle \mathrm{ind}_{\varphi'\circ\varphi,M}} \\
\uparrow {\scriptstyle \mathrm{Ind}_{\varphi'}(\mathrm{ind}_{\varphi,M})} & & \\
\mathrm{Ind}_{\varphi'}\Big(\mathrm{Ind}_{\varphi}\big(\mathrm{End}_{\mathcal{O}}(M)\big)\Big) & \xrightarrow{t_{\varphi',\varphi}(\mathrm{End}_{\mathcal{O}}(M))} & \mathrm{Ind}_{\varphi'\circ\varphi}\big(\mathrm{End}_{\mathcal{O}}(M)\big)
\end{array}$$

which is easily checked to be commutative; in particular, if M and $\mathrm{Ind}_{\varphi}(M)$ are $\mathcal{O}$-free then $t_{\varphi',\varphi}\big(\mathrm{End}_{\mathcal{O}}(M)\big)$ is an isomorphism. Similarly, for $\mathcal{O}G$-interior algebras in general, we have the following criterion: assume that there are two families $\{X_i\}_{i\in I}$ and $\{Y_i\}_{i\in I}$ of $\mathcal{O}L$-modules such that A is a direct summand of $\oplus_{i\in I} X_i \otimes_{\mathcal{O}} Y_i^{\circ}$ as

$\mathcal{O}L$-bimodules and that, for any $i \in I$, $\mathcal{O} \otimes_{\mathcal{O}K} X_i$ and $\mathcal{O} \otimes_{\mathcal{O}L} X_i$ are both $\mathcal{O}$-free; in that case, for any $i \in I$, we get the following right $\mathcal{O}(L/K)$-module isomorphisms

$$\begin{aligned} \mathcal{O} \underset{\mathcal{O}(L/K)}{\otimes} \big(\mathcal{O} \underset{\mathcal{O}K}{\otimes} (X_i \underset{\mathcal{O}}{\otimes} Y_i^\circ)\big)^K &\cong \mathcal{O} \underset{\mathcal{O}(L/K)}{\otimes} \big((\mathcal{O} \underset{\mathcal{O}K}{\otimes} X_i) \underset{\mathcal{O}}{\otimes} (Y_i^\circ)^K\big) \\ &\cong (\mathcal{O} \underset{\mathcal{O}L}{\otimes} X_i) \underset{\mathcal{O}}{\otimes} (Y_i^\circ)^K \\ &\cong \big(\mathcal{O} \underset{\mathcal{O}L}{\otimes} (X_i \underset{\mathcal{O}}{\otimes} Y_i^\circ)\big)^K \end{aligned} \tag{3.14.2}$$

and therefore $s_{K,L}(A)$ is an isomorphism.

Proof: Identifying G/L to $\varphi'\big(\varphi(G)\big)$, it is easily checked that the $\mathcal{O}G''$-interior algebra homomorphism

$$\mathrm{Ind}_{G/L}^{G''}\big(s_{K,L}(A)\big) : \mathrm{Ind}_{G/L}^{G''}\Big(\big(\mathcal{O} \underset{\mathcal{O}(L/K)}{\otimes} (\mathcal{O} \underset{\mathcal{O}K}{\otimes} A)^K\big)^{L/K}\Big) \to \mathrm{Ind}_{\varphi'\circ\varphi}(A) \tag{3.13.2}$$

composed with the inverse of isomorphism 3.11.2 is the announced homomorphism $t_{\varphi',\varphi}(A)$ and it is an isomorphism if and only if $s_{K,L}(A)$ is so (cf. 2.6). Moreover if either $K = 1$ or $K = L$ then $s_{K,L}(A)$ is clearly an isomorphism. Finally, if A is a $1 \times L$-projective $\mathcal{O}(L \times L)$-module then A is a direct summand of $\mathrm{Ind}_{1\times L}^{L\times L}(A) \cong \mathcal{O}L \otimes_{\mathcal{O}} A$ as $\mathcal{O}L$-bimodules and therefore it fulfills the criterion in Remark 3.14, so that $s_{K,L}(A)$ is an isomorphism too.

3.15 Let B be a second $\mathcal{O}G$-interior algebra and $f : A \to B$ an $\mathcal{O}G$-interior algebra homomorphism; it is clear from the description of $t_{\varphi',\varphi}(A)$ and $t_{\varphi',\varphi}(B)$ in Corollary 3.13 that we have the following commutative diagram of $\mathcal{O}G''$-interior algebra homomorphisms.

$$\begin{array}{ccc} \mathrm{Ind}_{\varphi'}\big(\mathrm{Ind}_{\varphi}(A)\big) & \xrightarrow{\ \mathrm{Ind}_{\varphi'}(\mathrm{Ind}_{\varphi}(f))\ } & \mathrm{Ind}_{\varphi'}\big(\mathrm{Ind}_{\varphi}(B)\big) \\ {\scriptstyle t_{\varphi',\varphi}(A)}\big\downarrow & & \big\downarrow{\scriptstyle t_{\varphi',\varphi}(B)} \\ \mathrm{Ind}_{\varphi'\circ\varphi}(A) & \xrightarrow{\ \mathrm{Ind}_{\varphi'\circ\varphi}(f)\ } & \mathrm{Ind}_{\varphi'\circ\varphi}(B) \end{array} \tag{3.15.1}$$

On the other hand, if G''' is a fourth finite group and $\varphi'' : G'' \to G'''$ a third group homomorphism, the following diagram of $\mathcal{O}G'''$-interior algebra homomorphisms is also commutative

$$\begin{array}{ccc} \mathrm{Ind}_{\varphi''}\Big(\mathrm{Ind}_{\varphi'}\big(\mathrm{Ind}_{\varphi}(A)\big)\Big) & \xrightarrow{\ \mathrm{Ind}_{\varphi''}(t_{\varphi',\varphi}(A))\ } & \mathrm{Ind}_{\varphi''}\big(\mathrm{Ind}_{\varphi'\circ\varphi}(A)\big) \\ {\scriptstyle t_{\varphi'',\varphi'}(\mathrm{Ind}_{\varphi}(A))}\big\downarrow & & \big\downarrow{\scriptstyle t_{\varphi'',\varphi'\circ\varphi}} \\ \mathrm{Ind}_{\varphi''\circ\varphi'}\big(\mathrm{Ind}_{\varphi}(A)\big) & \xrightarrow{\ t_{\varphi''\circ\varphi',\varphi}(A)\ } & \mathrm{Ind}_{\varphi''\circ\varphi'\circ\varphi}(A) \end{array} \tag{3.15.2}$$

Indeed, setting $K'' = \mathrm{Ker}(\varphi'')$, $L' = \mathrm{Ker}(\varphi'' \circ \varphi')$ and $M = \mathrm{Ker}(\varphi'' \circ \varphi' \circ \varphi)$, we get from 3.3.1, Proposition 3.11 and Corollary 3.13 the following evident $\mathcal{O}G'''$-interior algebra isomorphisms

$$\mathrm{Ind}_{\varphi''\circ\varphi'\circ\varphi}(A) \cong \mathrm{Ind}^{G'''}_{\varphi''(\varphi'(\varphi(G)))}\big((\mathcal{O} \underset{\mathcal{O}M}{\otimes} A)^M\big)$$

$$\mathrm{Ind}_{\varphi''}\big(\mathrm{Ind}_{\varphi'\circ\varphi}(A)\big) \cong \mathrm{Ind}^{G'''}_{\varphi''(\varphi'(\varphi(G)))}\Big(\big(\mathcal{O} \underset{\mathcal{O}(M/L)}{\otimes} (\mathcal{O} \underset{\mathcal{O}L}{\otimes} A)^L\big)^{M/L}\Big)$$

3.15.3 $$\mathrm{Ind}_{\varphi''\circ\varphi'}\big(\mathrm{Ind}_{\varphi}(A)\big) \cong \mathrm{Ind}^{G'''}_{\varphi''(\varphi'(\varphi(G)))}\Big(\big(\mathcal{O} \underset{\mathcal{O}(M/K)}{\otimes} (\mathcal{O} \underset{\mathcal{O}K}{\otimes} A)^K\big)^{M/K}\Big)$$

$$\mathrm{Ind}_{\varphi''}\Big(\mathrm{Ind}_{\varphi'}\big(\mathrm{Ind}_{\varphi}(A)\big)\Big) \cong \mathrm{Ind}_{\varphi''}\left(\mathrm{Ind}^{G''}_{\varphi'(\varphi(G))}\Big(\big(\mathcal{O} \underset{\mathcal{O}(L/K)}{\otimes} (\mathcal{O} \underset{\mathcal{O}K}{\otimes} A)^K\big)^{L/K}\Big)\right)$$

$$\cong \mathrm{Ind}^{G'''}_{\varphi''(\varphi'(\varphi(G)))}\left(\Big(\mathcal{O} \underset{\mathcal{O}(M/L)}{\otimes} \big(\mathcal{O} \underset{\mathcal{O}(L/K)}{\otimes} (\mathcal{O} \underset{\mathcal{O}K}{\otimes} A)^K\big)^{L/K}\Big)^{M/L}\right)$$

and there is not difficult (but painful!) to check that diagram 3.15.2 becomes the image by $\mathrm{Ind}^{G'''}_{\varphi''(\varphi'(\varphi(G)))}$ of the commutative diagram involved with the equality

3.15.4 $$s_{L,M}(A) \circ \big(\mathcal{O} \underset{\mathcal{O}(M/L)}{\otimes} s_{K,L}(A)\big) = s_{K,M}(A) \circ s_{L,M}\big((\mathcal{O} \underset{\mathcal{O}K}{\otimes} A)^K\big) \quad .$$

Proposition 3.16 *For any $\mathcal{O}G'$-interior algebra A' there is a unique $\mathcal{O}G'$-interior algebra homomorphism*

3.16.1 $$F_\varphi(A, A') : \mathrm{Ind}_\varphi(A) \underset{\mathcal{O}}{\otimes} A' \longrightarrow \mathrm{Ind}_\varphi\big(A \underset{\mathcal{O}}{\otimes} \mathrm{Res}_\varphi(A')\big)$$

mapping $\big(1 \otimes (1 \otimes a) \otimes 1\big) \otimes a'$ on $1 \otimes \big(1 \otimes (a \otimes a')\big) \otimes 1$ for any $a \in A$ such that K fixes $1 \otimes a$ in $\mathcal{O} \otimes_{\mathcal{O}K} A$, and any $a' \in A'$. Moreover, if either A' is $\mathcal{O}$-free or A is a 1×1-projective $\mathcal{O}(K \times K)$-module by left and right multiplication then $F_\varphi(A, A')$ is an isomorphism.

Remark 3.17 Notice that $N_A(K) \otimes \mathrm{Res}_\varphi(A') \subset N_{A\otimes_\mathcal{O}\mathrm{Res}_\varphi(A')}(K)$ and for any $a \in N_A(K)$ and any $a' \in A'$, we have

3.17.1 $$d_\varphi\big(A \underset{\mathcal{O}}{\otimes} \mathrm{Res}_\varphi(A')\big)(a \otimes a') = F_\varphi(A, A')\big(d_\varphi(A)(a) \otimes a'\big) \quad .$$

Remark 3.18 If M is an $\mathcal{O}G$-module such that $\mathrm{Res}^G_K(M)$ is projective and M' is an $\mathcal{O}G'$-module then the isomorphism $F_\varphi\big(\mathrm{End}_\mathcal{O}(M), \mathrm{End}_\mathcal{O}(M')\big)$ above together with Proposition 3.7 determine an $\mathcal{O}G'$-module isomorphism (cf. 2.10.1)

3.18.1 $$\mathrm{Ind}_\varphi(M) \underset{\mathcal{O}}{\otimes} M' \cong \mathrm{Ind}_\varphi\big(M \underset{\mathcal{O}}{\otimes} \mathrm{Res}_\varphi(M')\big)$$

which, in particular, induces the so-called *Frobenius reciprocity isomorphism*; indeed, we have the following evident $\mathcal{O}$-linear isomorphisms

3.18.2
$$\begin{aligned}\mathrm{Hom}_{\mathcal{O}G}\big(\mathrm{Ind}_\varphi(M^*), M'\big) &\cong \big(\mathrm{Ind}_\varphi(M) \underset{\mathcal{O}}{\otimes} M'\big)^{G'} \\ &\cong \Big(\mathrm{Ind}_\varphi\big(M \underset{\mathcal{O}}{\otimes} \mathrm{Res}_\varphi(M')\big)\Big)^{G'} \quad ;\end{aligned}$$

moreover, for any $\mathcal{O}(G/K)$-module $\bar{M}$ it is easily checked that the relative trace map $\mathrm{Tr}_{\varphi(G)}^{G'}$ on $\mathrm{Ind}_{\varphi(G)}^{G'}(\bar{M})$ induces an $\mathcal{O}$-linear isomorphism

3.18.3
$$\bar{M}^{G/K} \cong \mathrm{Ind}_{\varphi(G)}^{G'}(\bar{M})^{G'}$$

and on the other hand, since $\mathrm{Res}_K^G(M)$ is projective, the relative trace map Tr_1^K on $M \otimes_{\mathcal{O}} \mathrm{Res}_\varphi(M')$ induces an $\mathcal{O}(G/K)$-module isomorphism

3.18.4
$$\mathcal{O} \underset{\mathcal{O}K}{\otimes} \big(M \underset{\mathcal{O}}{\otimes} \mathrm{Res}_\varphi(M')\big) \cong \big(M \underset{\mathcal{O}}{\otimes} \mathrm{Res}_\varphi(M')\big)^K \quad ;$$

consequently, we have also the following $\mathcal{O}$-linear isomorphisms

3.18.5
$$\begin{aligned}\Big(\mathrm{Ind}_\varphi\big(M \underset{\mathcal{O}}{\otimes} \mathrm{Res}_\varphi(M')\big)\Big)^{G'} &\cong \big(M \underset{\mathcal{O}}{\otimes} \mathrm{Res}_\varphi(M')\big)^G \\ &\cong \mathrm{Hom}_{\mathcal{O}G}\big(M^*, \mathrm{Res}_\varphi(M')\big) \quad .\end{aligned}$$

Proof: The associativity of the tensor product determines an $\mathcal{O}$-module isomorphism

3.16.2
$$\big(\mathcal{O} \underset{\mathcal{O}K}{\otimes} A\big) \underset{\mathcal{O}}{\otimes} \mathrm{Res}_\varphi(A') \cong \mathcal{O} \underset{\mathcal{O}K}{\otimes} \big(A \underset{\mathcal{O}}{\otimes} \mathrm{Res}_\varphi(A')\big)$$

mapping $(1 \otimes a) \otimes a'$ on $1 \otimes (a \otimes a')$ for any $a \in A$ and any $a' \in A'$ which is evidently compatible with the actions of G; in particular, it induces an $\mathcal{O}$-linear map

3.16.3
$$\big(\mathcal{O} \underset{\mathcal{O}K}{\otimes} A\big)^K \underset{\mathcal{O}}{\otimes} \mathrm{Res}_{\varphi(G)}^{G'}(A') \longrightarrow \Big(\mathcal{O} \underset{\mathcal{O}K}{\otimes} \big(A \underset{\mathcal{O}}{\otimes} \mathrm{Res}_\varphi(A')\big)\Big)^K$$

which, up to identify G/K to $\varphi(G)$, is actually an $\mathcal{O}(G/K)$-interior algebra homomorphism as it is easily checked, and is clearly bijective whenever either A' is $\mathcal{O}$-free or A is a 1×1-projective $\mathcal{O}(K \times K)$-module (since we have $\big(\mathcal{O} \otimes_{\mathcal{O}K} \mathrm{Ind}_{1\times 1}^{K\times K}(X)\big)^K \cong X$ for any $\mathcal{O}$-module X.) On the other hand, by 2.6.8 we have the canonical $\mathcal{O}G'$-interior algebra isomorphism

3.16.4
$$\mathrm{Ind}_{\varphi(G)}^{G'}\big((\mathcal{O} \underset{\mathcal{O}K}{\otimes} A)^K\big) \underset{\mathcal{O}}{\otimes} A' \cong \mathrm{Ind}_{\varphi(G)}^{G'}\big((\mathcal{O} \underset{\mathcal{O}K}{\otimes} A)^K \underset{\mathcal{O}}{\otimes} \mathrm{Res}_{\varphi(G)}^{G'}(A')\big) \quad .$$

Hence, inducing homomorphism 3.16.3 from $\varphi(G)$ to G' and composing the induced homomorphism with isomorphism 3.16.4, we get from 3.3.1 the announced homomorphism 3.16.1, which is indeed bijective whenever either A' is $\mathcal{O}$-free or A is a 1×1-projective $\mathcal{O}(K \times K)$-module.

3.19 With the notation of Proposition 3.16, let B and B' be respectively $\mathcal{O}G$- and $\mathcal{O}G'$-interior algebras and, correspondingly, $f : A \to B$ and $f' : A' \to B'$ $\mathcal{O}G$- and $\mathcal{O}G'$-interior algebra homomorphisms; we have clearly the following commutative diagram of $\mathcal{O}G'$-interior algebra homomorphisms

3.19.1
$$\begin{array}{ccc} \mathrm{Ind}_\varphi(A) \otimes_{\mathcal{O}} A' & \xrightarrow{F_\varphi(A,A')} & \mathrm{Ind}_\varphi\big(A \otimes_{\mathcal{O}} \mathrm{Res}_\varphi(A')\big) \\ {\scriptstyle \mathrm{Ind}_\varphi(f)\otimes f'}\downarrow & & \downarrow{\scriptstyle \mathrm{Ind}_\varphi(f\otimes f')} \\ \mathrm{Ind}_\varphi(B) \otimes_{\mathcal{O}} B' & \xrightarrow{F_\varphi(B,B')} & \mathrm{Ind}_\varphi\big(B \otimes_{\mathcal{O}} \mathrm{Res}_\varphi(B')\big) \end{array} .$$

On the other hand, if G'' is a third finite group, A'' an $\mathcal{O}G''$-interior algebra and $\varphi' : G' \to G''$ a group homomorphism, it is easily checked that the following diagram of $\mathcal{O}G''$-interior algebra homomorphisms is commutative

3.19.2
$$\begin{array}{ccc} \mathrm{Ind}_{\varphi'}\big(\mathrm{Ind}_\varphi(A)\big) \otimes_{\mathcal{O}} A'' & \xrightarrow{t_{\varphi,\varphi'}(A)\otimes \mathrm{id}} & \mathrm{Ind}_{\varphi'\circ\varphi}(A) \otimes_{\mathcal{O}} A'' \\ {\scriptstyle F_{\varphi'}(\mathrm{Ind}_\varphi(A),A'')}\downarrow & & \\ \mathrm{Ind}_{\varphi'}\big(\mathrm{Ind}_\varphi(A) \otimes_{\mathcal{O}} \mathrm{Res}_{\varphi'}(A'')\big) & & \downarrow{\scriptstyle F_{\varphi'\circ\varphi}(A,A'')} \\ {\scriptstyle \mathrm{Ind}_{\varphi'}(F_\varphi(A,\mathrm{Res}_{\varphi'}(A'')))}\downarrow & & \\ \mathrm{Ind}_{\varphi'}\Big(\mathrm{Ind}_\varphi\big(A \otimes_{\mathcal{O}} \mathrm{Res}_{\varphi'\circ\varphi}(A'')\big)\Big) & \xrightarrow{t_{\varphi,\varphi'}(A\otimes \mathrm{Res}_{\varphi'\circ\varphi}(A''))} & \mathrm{Ind}_{\varphi'\circ\varphi}\big(A \otimes_{\mathcal{O}} \mathrm{Res}_{\varphi'\circ\varphi}(A'')\big) \end{array} .$$

3.20 Let H' be a finite group and $\eta' : H' \to G'$ a group homomorphism, and consider the following pull-back

3.20.1
$$\begin{array}{ccc} G & \xrightarrow{\varphi} & G' \\ {\scriptstyle \eta}\uparrow & & \uparrow{\scriptstyle \eta'} \\ H & \xrightarrow{\psi} & H' \end{array} ;$$

that is to say, H is the subgroup of $G \times H'$ formed by the pairs (x, y') such that $\varphi(x) = \eta'(y')$ and we denote respectively by η and ψ the restrictions to H of the first and the second projection maps; in particular, η maps bijectively $\mathrm{Ker}(\psi)$ onto $K = \mathrm{Ker}(\varphi)$. More generally, for any $x' \in G'$, we can consider the group homomorphisms $\eta'_{x'} : H' \to G'$ mapping $y' \in H'$ on $\eta'(y')^{x'}$ and the corresponding pull-back

3.20.2
$$\begin{array}{ccc} G & \xrightarrow{\varphi} & G' \\ {\scriptstyle \eta_{x'}}\uparrow & & \uparrow{\scriptstyle \eta'_{x'}} \\ H_{x'} & \xrightarrow{\psi_{x'}} & H' \end{array} ;$$

recall that $i_{x'} = \mathrm{Tr}^{H'}_{\psi_{x'}(H_{x'})}\big(1 \otimes (1 \otimes 1) \otimes 1\big)$ is the unity element in the $\mathcal{O}H'$-interior algebra $\mathrm{Ind}_{\psi_{x'}}\big(\mathrm{Res}_{\eta_{x'}}(A)\big)$.

Proposition 3.21 *With the notation above, let X' be a set of representatives for $\eta'(H')\backslash G'/\varphi(G)$ in G'. There is a unique $\mathcal{O}H'$-interior algebra embedding*

3.21.1 $$d_{\eta',\varphi}(A) : \mathrm{Ind}_{\psi}\big(\mathrm{Res}_{\eta}(A)\big) \longrightarrow \mathrm{Res}_{\eta'}\big(\mathrm{Ind}_{\varphi}(A)\big)$$

mapping $1 \otimes (1 \otimes a) \otimes 1 \in \mathrm{Ind}_{\psi}\big(\mathrm{Res}_{\eta}(A)\big)$ *on* $1 \otimes (1 \otimes a) \otimes 1 \in \mathrm{Ind}_{\varphi}(A)$ *for any $a \in A$ such that K fixes $1 \otimes a$ in $\mathcal{O} \otimes_{\mathcal{O}K} A$. Moreover, the set*

3.21.2 $$\big\{d_{\eta'_{x'},\varphi}(A)(i_{x'})^{x'^{-1}}\big\}_{x' \in X'}$$

is a pairwise orthogonal idempotent decomposition of the unity element in $\mathrm{Ind}_{\varphi}(A)$ *which does not depend on the choice of X'.*

Proof: Denoting by $\eta'' : H' \to \eta'(H')$ the group homomorphism determined by η', we have from 3.3.1

3.21.3 $$\mathrm{Res}_{\eta'}\big(\mathrm{Ind}_{\varphi}(A)\big) = \mathrm{Res}_{\eta''}\Big(\mathrm{Res}^{G'}_{\eta'(H')}\Big(\mathrm{Ind}^{G'}_{\varphi(G)}\big((\mathcal{O} \underset{\mathcal{O}K}{\otimes} A)^K\big)\Big)\Big)$$

and therefore we get from 2.6.4 an $\mathcal{O}H'$-interior algebra embedding from the $\mathcal{O}H'$-interior algebra

3.21.4 $$\mathrm{Res}_{\eta''}\Big(\mathrm{Ind}^{\eta'(H')}_{\eta'(H')\cap\varphi(G)}\Big(\mathrm{Res}^{\varphi(G)}_{\eta'(H')\cap\varphi(G)}\big((\mathcal{O} \underset{\mathcal{O}K}{\otimes} A)^K\big)\Big)\Big)$$

into the right member of 3.21.3; but it is clear that

3.21.5 $$\eta'(H') \cap \varphi(G) = \varphi\big(\eta(H)\big) = \eta'\big(\psi(H)\big) \qquad ;$$

hence, denoting by $\nu' : \psi(H) \to \varphi(G)$ the group homomorphism determined by η', the $\mathcal{O}H'$-interior algebra 3.21.4 is canonically isomorphic to the $\mathcal{O}H'$-interior algebra

3.21.6 $$\mathrm{Ind}^{H'}_{\psi(H)}\Big(\mathrm{Res}_{\nu'}\big((\mathcal{O} \underset{\mathcal{O}K}{\otimes} A)^K\big)\Big)$$

and, since η maps bijectively $L = \mathrm{Ker}(\psi)$ onto K, there is an $\mathcal{O}(H/L)$-interior algebra isomorphism

3.21.7 $$\mathrm{Res}_{\nu'}\big((\mathcal{O} \underset{\mathcal{O}K}{\otimes} A)^K\big) \cong \big(\mathcal{O} \underset{\mathcal{O}L}{\otimes} \mathrm{Res}_{\eta}(A)\big)^L$$

mapping $1 \otimes a \in (\mathcal{O} \otimes_{\mathcal{O}K} A)^K$ on $1 \otimes a \in \big(\mathcal{O} \otimes_{\mathcal{O}L} \mathrm{Res}_{\eta}(A)\big)^L$ for any $a \in A$ such that K fixes $1 \otimes a$ in $\mathcal{O} \otimes_{\mathcal{O}K} A$; consequently, the $\mathcal{O}H'$-interior algebra 3.21.6 becomes

3.21.8 $$\mathrm{Ind}^{H'}_{\psi(H)}\Big(\big(\mathcal{O} \underset{\mathcal{O}L}{\otimes} \mathrm{Res}_{\eta}(A)\big)^L\Big) = \mathrm{Ind}_{\psi}\big(\mathrm{Res}_{\eta}(A)\big) \qquad .$$

Finally, it is easily checked that the set 3.21.2 is just the set of sums of the element of any orbit of $\eta'(H')$ in the set of the idempotents $x' \otimes (1 \otimes 1) \otimes x'^{-1}$ of $\mathrm{Ind}_\varphi(A)$ where x' runs on a set of representatives for $G'/\varphi(G)$ in G'.

3.22 Let B be a second $\mathcal{O}G$-interior algebra and $f : A \to B$ an $\mathcal{O}G$-interior algebra homomorphism; we have the following commutative diagram of $\mathcal{O}H'$-interior algebra homomorphisms

3.22.1
$$\begin{array}{ccc} \mathrm{Ind}_\psi\big(\mathrm{Res}_\eta(A)\big) & \xrightarrow{d_{\eta',\varphi}(A)} & \mathrm{Res}_{\eta'}\big(\mathrm{Ind}_\varphi(A)\big) \\ {\scriptstyle \mathrm{Ind}_\psi(f)}\big\uparrow & & \big\uparrow{\scriptstyle \mathrm{Ind}_\varphi(f)} \\ \mathrm{Ind}_\psi\big(\mathrm{Res}_\eta(B)\big) & \xrightarrow{d_{\eta',\varphi}(B)} & \mathrm{Res}_{\eta'}\big(\mathrm{Ind}_\varphi(B)\big) \end{array} .$$

On the other hand, let G'' and H'' be finite groups, $\varphi' : G' \to G''$ and $\eta'' : H'' \to G''$ group homomorphisms, and

3.22.2
$$\begin{array}{ccccc} G & \xrightarrow{\varphi} & G' & \xrightarrow{\varphi'} & G'' \\ {\scriptstyle \eta}\big\uparrow & & {\scriptstyle \eta'}\big\uparrow & & {\scriptstyle \eta''}\big\uparrow \\ H & \xrightarrow{\psi} & H' & \xrightarrow{\psi'} & H'' \end{array}$$

two pull-backs of groups; then, it is clear that the diagram

3.22.3
$$\begin{array}{ccc} G & \xrightarrow{\varphi' \circ \varphi} & G'' \\ {\scriptstyle \eta}\big\uparrow & & \big\uparrow{\scriptstyle \eta''} \\ H & \xrightarrow{\psi' \circ \psi} & H'' \end{array}$$

is still a pull-back of groups and we have the following diagram of $\mathcal{O}H''$-interior algebra homomorphisms

3.22.4
$$\begin{array}{ccc} \mathrm{Ind}_{\psi'}\Big(\mathrm{Ind}_\psi\big(\mathrm{Res}_\eta(A)\big)\Big) & \xrightarrow{t_{\psi,\psi'}(\mathrm{Res}_\eta(A))} & \mathrm{Ind}_{\psi'\circ\psi}\big(\mathrm{Res}_\eta(A)\big) \\ {\scriptstyle \mathrm{Ind}_{\psi'}(d_{\eta',\psi}(A))}\big\downarrow & & \\ \mathrm{Ind}_{\psi'}\Big(\mathrm{Res}_{\eta'}\big(\mathrm{Ind}_\varphi(A)\big)\Big) & & \big\downarrow{\scriptstyle d_{\eta'',\varphi'\circ\varphi}(A)} \\ {\scriptstyle d_{\eta'',\varphi'}(\mathrm{Ind}_\varphi(A))}\big\downarrow & & \\ \mathrm{Res}_{\eta''}\Big(\mathrm{Ind}_{\varphi'}\big(\mathrm{Ind}_\varphi(A)\big)\Big) & \xrightarrow{\mathrm{Res}_{\eta''}(t_{\varphi',\varphi}(A))} & \mathrm{Res}_{\eta''}\big(\mathrm{Ind}_{\varphi'\circ\varphi}(A)\big) \end{array}$$

which is easily checked to be commutative.

4 Hecke $\mathcal{O}G$-interior algebras and noninjective induction

4.1 Let G and G' be finite groups, $H^{\cdot\cdot}$ a subgroup of $G \times G'$ and $B^{\cdot\cdot}$ an $\mathcal{O}H^{\cdot\cdot}$-interior algebra. For a suitable choice of $H^{\cdot\cdot}$ and $B^{\cdot\cdot}$, the $\mathcal{O}G$-interior algebra $\mathrm{Ind}_{H^{\cdot\cdot}}^{G\times G'}(B^{\cdot\cdot})^{1\times G'}$ plays a crucial role in the analysis of Morita equivalences between blocks, as we show in Section 6 below. For the sake of commentary, it is handy to give a name to this $\mathcal{O}G$-interior algebra: we call it *Hecke $\mathcal{O}G$-interior algebra* associated with G', $H^{\cdot\cdot}$ and $B^{\cdot\cdot}$, since when G is trivial and $B^{\cdot\cdot} = \mathrm{End}_{\mathcal{O}}(N^{\cdot\cdot})$ for some $\mathcal{O}H^{\cdot\cdot}$-module $N^{\cdot\cdot}$, it becomes an ordinary Hecke algebra. The main purpose of this section is to provide an alternative description of the Hecke $\mathcal{O}G$-interior algebra in terms of the non-necessarily injective induction introduced in Section 3. First of all we discuss the particular case where $H^{\cdot\cdot} = G \times G'$; denote respectively by π and π' the first and the second projection maps for $G \times G'$ and identify the $\mathcal{O}G'$-interior algebra $\mathrm{End}_{\mathcal{O}}(\mathcal{O}G')$ to $\mathrm{Ind}_1^{G'}(\mathcal{O})$, so that $1 \otimes 1 \otimes 1 : \mathcal{O}G' \longrightarrow \mathcal{O}G'$ maps the trivial element of G' on itself and all the others elements on zero; recall that, since π is surjective, we identify $\mathrm{Ind}_\pi(A^{\cdot\cdot})$ to $(\mathcal{O} \otimes_{\mathcal{O}(1\times G')} A^{\cdot\cdot})^{1\times G'}$ for any $\mathcal{O}(G \times G')$-interior algebra $A^{\cdot\cdot}$ (cf. 3.3), and moreover we consider $\mathrm{Res}_{G\times 1}^{G\times G'}(A^{\cdot\cdot})$ as an $\mathcal{O}G$-interior algebra.

Proposition 4.2 *Let $A^{\cdot\cdot}$ be an $\mathcal{O}(G \times G')$-interior algebra. There is a unique $\mathcal{O}G$-interior algebra isomorphism*

4.2.1 $$\mathrm{Res}_{G\times 1}^{G\times G'}(A^{\cdot\cdot}) \cong \mathrm{Ind}_\pi\Big(A^{\cdot\cdot} \underset{\mathcal{O}}{\otimes} \mathrm{Res}_{\pi'}\big(\mathrm{End}_{\mathcal{O}}(\mathcal{O}G')\big)\Big)$$

mapping $a^{\cdot\cdot} \in A^{\cdot\cdot}$ on $1 \otimes \mathrm{Tr}_{1\times 1}^{1\times G'}\big(a^{\cdot\cdot} \otimes (1 \otimes 1 \otimes 1)\big)$. In particular, this isomorphism restricts to an $\mathcal{O}G$-interior algebra isomorphism

4.2.2 $$H_{G,G'}(A^{\cdot\cdot}) : (A^{\cdot\cdot})^{1\times G'} \cong \mathrm{Ind}_\pi\big(A^{\cdot\cdot} \underset{\mathcal{O}}{\otimes} \mathrm{Res}_{\pi'}(\mathcal{O}G')\big)$$

which maps $a^{\cdot\cdot} \in (A^{\cdot\cdot})^{1\times G'}$ on $1 \otimes (a^{\cdot\cdot} \otimes 1)$.

Proof: Since $\mathrm{End}_{\mathcal{O}}(\mathcal{O}G') \cong \mathrm{Ind}_1^{G'}(\mathcal{O})$, it is clear that we have a canonical $\mathcal{O}(G \times G')$-interior algebra isomorphism

4.2.3 $$\mathrm{Res}_{\pi'}\big(\mathrm{End}_{\mathcal{O}}(\mathcal{O}G')\big) \cong \mathrm{Ind}_{G\times 1}^{G\times G'}(\mathcal{O}) \qquad ;$$

hence, by Proposition 3.16, we have an $\mathcal{O}(G \times G')$-interior algebra isomorphism

4.2.4 $$A^{\cdot\cdot} \underset{\mathcal{O}}{\otimes} \mathrm{Res}_{\pi'}\big(\mathrm{End}_{\mathcal{O}}(\mathcal{O}G')\big) \cong \mathrm{Ind}_{G\times 1}^{G\times G'}\big(\mathrm{Res}_{G\times 1}^{G\times G'}(A^{\cdot\cdot})\big)$$

mapping $a^{\cdot\cdot} \otimes (1 \otimes 1 \otimes 1)$ on $1 \otimes a^{\cdot\cdot} \otimes 1$ for any $a^{\cdot\cdot} \in A^{\cdot\cdot}$; moreover, it follows from Corollary 3.13 that there is an $\mathcal{O}G$-interior algebra isomorphism

4.2.5 $$\mathrm{Ind}_\pi\Big(\mathrm{Ind}_{G\times 1}^{G\times G'}\big(\mathrm{Res}_{G\times 1}^{G\times G'}(A^{\cdot\cdot})\big)\Big) \cong \mathrm{Res}_{G\times 1}^{G\times G'}(A^{\cdot\cdot})$$

mapping $1 \otimes \mathrm{Tr}_{1\times 1}^{1\times G'}(1 \otimes a^{\cdot\cdot} \otimes 1)$ on $a^{\cdot\cdot}$ for any $a^{\cdot\cdot} \in A^{\cdot\cdot}$; now, we get the announced isomorphism 4.2.1 by composing the inverses of isomorphism 4.2.5 and the corresponding induced isomorphism of isomorphism 4.2.4. In particular, isomorphism 4.2.1 maps any $a^{\cdot\cdot} \in (A^{\cdot\cdot})^{1\times G'}$ on $1 \otimes \big(a^{\cdot\cdot} \otimes \mathrm{Tr}_{1\times 1}^{1\times G'}(1 \otimes 1 \otimes 1)\big)$ and, since $\mathrm{Tr}_{1\times 1}^{1\times G'}(1 \otimes 1 \otimes 1) = \mathrm{id}_{\mathcal{O}G'}$, identifying $\mathcal{O}G'$ to $\mathcal{O}G' \cdot \mathrm{id}_{\mathcal{O}G'}$ in $\mathrm{End}_{\mathcal{O}}(\mathcal{O}G')$ we get that it maps $(A^{\cdot\cdot})^{1\times G'}$ into $\mathrm{Ind}_\pi\big(A^{\cdot\cdot} \otimes_{\mathcal{O}} \mathrm{Res}_{\pi'}(\mathcal{O}G')\big)$; conversely, any element of $\mathcal{O} \otimes_{\mathcal{O}(1\times G')} \big(A^{\cdot\cdot} \otimes_{\mathcal{O}} \mathrm{Res}_{\pi'}(\mathcal{O}G')\big)$ can be written in the form $1 \otimes (a^{\cdot\cdot} \otimes 1)$ for a unique $a^{\cdot\cdot} \in A^{\cdot\cdot}$, so that it belongs to $\mathrm{Ind}_\pi\big(A^{\cdot\cdot} \otimes_{\mathcal{O}} \mathrm{Res}_{\pi'}(\mathcal{O}G')\big)$ if and only if $1 \times G'$ fixes $a^{\cdot\cdot}$.

4.3 In the general situation, denote respectively by ρ and ρ' the restrictions of π and π' to $H^{\cdot\cdot}$; it is clear that

4.3.1 $\mathrm{Ker}(\rho) = H^{\cdot\cdot} \cap (1 \times G') = 1 \times K'$ and $\mathrm{Ker}(\rho') = H^{\cdot\cdot} \cap (G \times 1) = K \times 1$

for suitable subgroups K of G and K' of G'; moreover, consider $B^{\cdot\cdot} \otimes_{\mathcal{O}} \mathrm{Res}_{\rho'}(\mathcal{O}G')$ as $\mathcal{O}(1 \times K')$-bimodule by left and right multiplication and denote by

4.3.2 $$q_{G,G'}(B^{\cdot\cdot}) : \mathcal{O} \underset{\mathcal{O}(1\times K')}{\otimes} \big(B^{\cdot\cdot} \underset{\mathcal{O}}{\otimes} \mathrm{Res}_{\rho'}(\mathcal{O}G')\big) \longrightarrow \mathrm{Ind}_{H^{\cdot\cdot}}^{G\times G'}(B^{\cdot\cdot})$$

the $\mathcal{O}$-linear map which sends $1 \otimes (b^{\cdot\cdot} \otimes x')$ to $(1, x')^{-1} \otimes b^{\cdot\cdot} \otimes (1, 1)$ for any $b^{\cdot\cdot} \in B^{\cdot\cdot}$ and any $x' \in G'$; notice that this map is compatible with the actions by conjugation of $1 \times K'$ in both ends.

Theorem 4.4 *With the notation above, there is a unique $\mathcal{O}G$-interior algebra isomorphism*

4.4.1 $$H_{G,G'}(B^{\cdot\cdot}) : \mathrm{Ind}_{H^{\cdot\cdot}}^{G\times G'}(B^{\cdot\cdot})^{1\times G'} \cong \mathrm{Ind}_\rho\big(B^{\cdot\cdot} \underset{\mathcal{O}}{\otimes} \mathrm{Res}_{\rho'}(\mathcal{O}G')\big)$$

mapping $\mathrm{Tr}_{1\times K'}^{1\times G'}\big(q_{G,G'}(B^{\cdot\cdot})(1 \otimes a)\big)$ *on* $1 \otimes (1 \otimes a) \otimes 1$ *for any a in* $B^{\cdot\cdot} \otimes_{\mathcal{O}} \mathrm{Res}_{\rho'}(\mathcal{O}G')$ *such that* $1 \times K'$ *fixes* $1 \otimes a$ *in* $\mathcal{O} \otimes_{\mathcal{O}(1\times K')} \big(B^{\cdot\cdot} \otimes_{\mathcal{O}} \mathrm{Res}_{\rho'}(\mathcal{O}G')\big)$.

Remark 4.5 For any subgroup H of G, the isomorphism $H_{1,G}(\mathcal{O}H)$ induces the following $\mathcal{O}$-algebra isomorphism, already noticed by Broué

4.5.1 $$\begin{aligned} \mathrm{Ind}_H^G(\mathcal{O}H) &\cong \big(\mathcal{O} \underset{\mathcal{O}H}{\otimes} (\mathcal{O}H \underset{\mathcal{O}}{\otimes} \mathcal{O}G)\big)^H \\ &\cong \big((\mathcal{O} \underset{\mathcal{O}H}{\otimes} \mathcal{O}H) \underset{\mathcal{O}}{\otimes} \mathcal{O}G\big)^H \cong (\mathcal{O}G)^H \end{aligned}$$

which, for any $a \in (\mathcal{O}G)^H$, maps $\mathrm{Tr}_H^G(a \otimes 1 \otimes 1)$ on a.

Proof: First of all, we apply Proposition 4.2 to the $\mathcal{O}(G \times G')$-interior algebra $\mathrm{Ind}_{H^{\cdot\cdot}}^{G\times G'}(B^{\cdot\cdot})$, getting an $\mathcal{O}G$-interior algebra isomorphism

4.4.2 $$\mathrm{Ind}_{H^{\cdot\cdot}}^{G\times G'}(B^{\cdot\cdot})^{1\times G'} \cong \mathrm{Ind}_\pi\big(\mathrm{Ind}_{H^{\cdot\cdot}}^{G\times G'}(B^{\cdot\cdot}) \underset{\mathcal{O}}{\otimes} \mathrm{Res}_{\pi'}(\mathcal{O}G')\big)$$

which, for any $a \in B^{\cdot\cdot} \otimes_{\mathcal{O}} \mathrm{Res}_{\rho'}(\mathcal{O}G')$ such that $1 \times K'$ fixes $1 \otimes a$ in $\mathcal{O} \otimes_{\mathcal{O}(1\times K')} \big(B^{\cdot\cdot} \otimes_{\mathcal{O}} \mathrm{Res}_{\rho'}(\mathcal{O}G')\big)$ and, therefore, fixes $b = q_{G,G'}(B^{\cdot\cdot})(1 \otimes a)$ in $\mathrm{Ind}_{H^{\cdot\cdot}}^{G\times G'}(B^{\cdot\cdot})$, maps $\mathrm{Tr}_{1\times K'}^{1\times G'}(b)$ on the element

$$4.4.3 \qquad 1 \otimes \big(\mathrm{Tr}_{1\times K'}^{1\times G'}(b) \otimes 1\big) = 1 \otimes \big(\mathrm{Tr}_{1\times K'}^{1\times G'}(b \otimes 1)\big) \quad .$$

Secondly, it follows from Proposition 3.16 that there is an $\mathcal{O}(G \times G')$-interior algebra isomorphism

$$4.4.4 \qquad \mathrm{Ind}_{H^{\cdot\cdot}}^{G\times G'}(B^{\cdot\cdot}) \underset{\mathcal{O}}{\otimes} \mathrm{Res}_{\pi'}(\mathcal{O}G') \cong \mathrm{Ind}_{H^{\cdot\cdot}}^{G\times G'}\big(B^{\cdot\cdot} \underset{\mathcal{O}}{\otimes} \mathrm{Res}_{\rho'}(\mathcal{O}G')\big)$$

which, setting $a = \Sigma_{x'\in G'} a^{\cdot\cdot}_{x'} \otimes x'$ where $a^{\cdot\cdot}_{x'} \in B^{\cdot\cdot}$ for any $x' \in G'$, maps the element

$$\begin{aligned} b \otimes 1 &= q_{G,G'}(B^{\cdot\cdot})\Big(\sum_{x'\in G'} 1 \otimes (a^{\cdot\cdot}_{x'} \otimes x')\Big) \otimes 1 \\ 4.4.5 \qquad &= \sum_{x'\in G'} \big((1,x')^{-1} \otimes a^{\cdot\cdot}_{x'} \otimes (1,1)\big) \otimes 1 \\ &= \sum_{x'\in G'} (1,x')^{-1}\cdot\big(((1,1) \otimes a^{\cdot\cdot}_{x'} \otimes (1,1)) \otimes x'\big) \end{aligned}$$

on $\Sigma_{x'\in G'}(1,x')^{-1} \otimes (a^{\cdot\cdot}_{x'} \otimes x') \otimes (1,1)$ and therefore, denoting by Y' a set of representatives for G'/K' in G', the induced isomorphism

$$4.4.6 \quad \mathrm{Ind}_{\pi}\big(\mathrm{Ind}_{H^{\cdot\cdot}}^{G\times G'}(B^{\cdot\cdot}) \underset{\mathcal{O}}{\otimes} \mathrm{Res}_{\pi'}(\mathcal{O}G')\big) \cong \mathrm{Ind}_{\pi}\Big(\mathrm{Ind}_{H^{\cdot\cdot}}^{G\times G'}\big(B^{\cdot\cdot} \underset{\mathcal{O}}{\otimes} \mathrm{Res}_{\rho'}(\mathcal{O}G')\big)\Big)$$

maps $1 \otimes \mathrm{Tr}_{1\times K'}^{1\times G'}(b \otimes 1)$ on the element

$$\begin{aligned} &1 \otimes \sum_{(y',x')\in Y'\times G'} (1, y'x'^{-1}) \otimes (a^{\cdot\cdot}_{x'} \otimes x') \otimes (1, y'^{-1}) \\ 4.4.7 \qquad &= 1 \otimes \sum_{(y',x')\in Y'\times G'} (1,1) \otimes (a^{\cdot\cdot}_{x'} \otimes x') \otimes (1, y'^{-1}) \\ &= 1 \otimes \mathrm{Tr}_{1\times K'}^{1\times G'}\big((1,1) \otimes a \otimes (1,1)\big) \end{aligned}$$

since $\mathrm{Ker}\,(\pi) = 1 \times G'$. Finally, it follows from Corollary 3.13 that we have an $\mathcal{O}G$-interior algebra isomorphism

$$4.4.8 \qquad \mathrm{Ind}_{\pi}\Big(\mathrm{Ind}_{H^{\cdot\cdot}}^{G\times G'}\big(B^{\cdot\cdot} \underset{\mathcal{O}}{\otimes} \mathrm{Res}_{\rho'}(\mathcal{O}G')\big)\Big) \cong \mathrm{Ind}_{\rho}\big(B^{\cdot\cdot} \underset{\mathcal{O}}{\otimes} \mathrm{Res}_{\rho'}(\mathcal{O}G')\big)$$

mapping $1 \otimes \mathrm{Tr}_{1\times K'}^{1\times G'}\big((1,1) \otimes a \otimes (1,1)\big)$ on $1 \otimes (1 \otimes a) \otimes 1$. Consequently, we get the announced $\mathcal{O}G$-interior algebra isomorphism 4.4.1 by composing isomorphisms 4.4.2, 4.4.6 and 4.4.8.

4.6 Let $C^{\cdot\cdot}$ be a second $\mathcal{O}H^{\cdot\cdot}$-interior algebra and $f^{\cdot\cdot} : B^{\cdot\cdot} \longrightarrow C^{\cdot\cdot}$ an $\mathcal{O}H^{\cdot\cdot}$-interior algebra homomorphism; it is clear that we have the following commutative diagram of $\mathcal{O}G$-interior algebra homomorphisms

4.6.1
$$\begin{array}{ccc} \mathrm{Ind}_{H^{\cdot\cdot}}^{G\times G'}(B^{\cdot\cdot})^{1\times G'} & \cong & \mathrm{Ind}_\rho\big(B^{\cdot\cdot} \underset{\mathcal{O}}{\otimes} \mathrm{Res}_{\rho'}(\mathcal{O}G')\big) \\ {\scriptstyle \mathrm{Ind}_{H^{\cdot\cdot}}^{G\times G'}(f^{\cdot\cdot})}\big\downarrow & & \big\downarrow{\scriptstyle \mathrm{Ind}_\rho(f^{\cdot\cdot}\otimes \mathrm{id}_{\mathcal{O}G'})} \\ \mathrm{Ind}_{H^{\cdot\cdot}}^{G\times G'}(C^{\cdot\cdot})^{1\times G'} & \cong & \mathrm{Ind}_\rho\big(C^{\cdot\cdot} \underset{\mathcal{O}}{\otimes} \mathrm{Res}_{\rho'}(\mathcal{O}G')\big) \end{array} .$$

On the other hand, if G'' is a third finite group, $L^{\cdot\cdot}$ a subgroup of $H^{\cdot\cdot} \times G''$ and $D^{\cdot\cdot}$ an $\mathcal{O}L^{\cdot\cdot}$-interior algebra, it is quite clear that we have a canonical OG-interior algebra isomorphism

4.6.2 $$\mathrm{Ind}_{H^{\cdot\cdot}}^{G\times G'}\big(\mathrm{Ind}_{L^{\cdot\cdot}}^{H^{\cdot\cdot}\times G''}(D^{\cdot\cdot})^{1\times 1\times G''}\big)^{1\times G'} \cong \mathrm{Ind}_{L^{\cdot\cdot}}^{G\times G'\times G''}(D^{\cdot\cdot})^{1\times G'\times G''} \quad ;$$

in other terms, we obtain that:

4.6.3 *The Hecke $\mathcal{O}G$-interior algebra associated with G', $H^{\cdot\cdot}$ and the Hecke $\mathcal{O}H^{\cdot\cdot}$-interior algebra associated with G'', $L^{\cdot\cdot}$ and $D^{\cdot\cdot}$ is the Hecke $\mathcal{O}G$-interior algebra associated with $G' \times G''$, $L^{\cdot\cdot}$ and $D^{\cdot\cdot}$.*

4.7 The particular form of the second member of isomorphism 4.4.1 is the starting point to relate some Hecke $\mathcal{O}G$-interior algebras to Morita equivalences between blocks; in this direction, we describe here functors naturally associated with the Hecke $\mathcal{O}G$-interior algebra $\mathrm{Ind}_{H^{\cdot\cdot}}^{G\times G'}(B^{\cdot\cdot})^{1\times G'}$ or, more precisely, with the *data* G, G', $H^{\cdot\cdot}$ and $B^{\cdot\cdot}$, that we use in Section 6. First of all, fix an idempotent $\hat{\imath}$ of $\mathrm{Ind}_{H^{\cdot\cdot}}^{G\times G'}(B^{\cdot\cdot})^{G\times G'}$; notice that, for any $\mathcal{O}G'$-interior algebra A', we have the $\mathcal{O}G$-interior algebra $\mathrm{Ind}_\rho\big(B^{\cdot\cdot} \otimes_{\mathcal{O}} \mathrm{Res}_{\rho'}(A')\big)$ and that the structural map $\mathrm{st}_{A'} : \mathcal{O}G' \longrightarrow A'$ induces an $\mathcal{O}G$-interior algebra homomorphism

4.7.1 $$\mathrm{Ind}_\rho(\mathrm{id} \underset{\mathcal{O}}{\otimes} \mathrm{st}_{A'}) : \mathrm{Ind}_\rho\big(B^{\cdot\cdot} \underset{\mathcal{O}}{\otimes} \mathrm{Res}_{\rho'}(\mathcal{O}G')\big) \longrightarrow \mathrm{Ind}_\rho\big(B^{\cdot\cdot} \underset{\mathcal{O}}{\otimes} \mathrm{Res}_{\rho'}(A')\big) \quad ,$$

so that $\hat{\imath}$ determines an idempotent $\hat{\imath}_{A'}$ in $\mathrm{Ind}_\rho\big(B^{\cdot\cdot} \otimes_{\mathcal{O}} \mathrm{Res}_{\rho'}(A')\big)^G$; moreover, if B' is a second $\mathcal{O}G'$-interior algebra and $f' : A' \longrightarrow B'$ an $\mathcal{O}G'$-interior algebra homomorphism then we have the $\mathcal{O}G$-interior algebra homomorphism

4.7.2 $$\mathrm{Ind}_\rho(\mathrm{id} \underset{\mathcal{O}}{\otimes} f') : \mathrm{Ind}_\rho\big(B^{\cdot\cdot} \underset{\mathcal{O}}{\otimes} \mathrm{Res}_{\rho'}(A')\big) \longrightarrow \mathrm{Ind}_\rho\big(B^{\cdot\cdot} \underset{\mathcal{O}}{\otimes} \mathrm{Res}_{\rho'}(B')\big)$$

and it is quite clear that

4.7.3 $$\mathrm{Ind}_\rho(\mathrm{id} \underset{\mathcal{O}}{\otimes} f')(1)\,\hat{\imath}_{B'} = \mathrm{Ind}_\rho(\mathrm{id} \underset{\mathcal{O}}{\otimes} f')(\hat{\imath}_{A'}) = \hat{\imath}_{B'}\mathrm{Ind}_\rho(\mathrm{id} \underset{\mathcal{O}}{\otimes} f')(1) \quad ,$$

so that we get

4.7.4

$$\mathrm{Ind}_\rho(\mathrm{id} \underset{\mathcal{O}}{\otimes} f')\Big(\hat{\imath}_{A'}\mathrm{Ind}_\rho\big(B^{\cdot\cdot} \underset{\mathcal{O}}{\otimes} \mathrm{Res}_{\rho'}(A')\big)\hat{\imath}_{A'}\Big) \subset \hat{\imath}_{B'}\mathrm{Ind}_\rho\big(B^{\cdot\cdot} \underset{\mathcal{O}}{\otimes} \mathrm{Res}_{\rho'}(B')\big)\hat{\imath}_{B'} \; .$$

Consequently, denoting respectively by $\mathrm{Alg}_{\mathcal{O}G}$ and $\mathrm{Alg}_{\mathcal{O}G'}$ the categories of $\mathcal{O}G$- and $\mathcal{O}G'$-interior algebras, with the corresponding homomorphisms, we have a functor

4.7.5 $$\mathcal{F}_{G,G'}(B^{\cdot\cdot})_{\hat{\imath}} : \mathrm{Alg}_{\mathcal{O}G'} \longrightarrow \mathrm{Alg}_{\mathcal{O}G}$$

mapping A' on $\hat{\imath}_{A'}\mathrm{Ind}_\rho\big(B^{\cdot\cdot} \otimes_{\mathcal{O}} \mathrm{Res}_{\rho'}(A')\big)\hat{\imath}_{A'}$ and f' on the corresponding restriction of $\mathrm{Ind}_\rho(\mathrm{id} \otimes_{\mathcal{O}} f')$.

4.8 We are specially interested in the case where $B^{\cdot\cdot} = \mathrm{End}_{\mathcal{O}}(M^{\cdot\cdot})$ for some $\mathcal{O}H^{\cdot\cdot}$-module $M^{\cdot\cdot}$; then, for any $\mathcal{O}G'$-module M', we have the $\mathcal{O}G$-module $\mathrm{Ind}_\rho\big(M^{\cdot\cdot} \otimes_{\mathcal{O}} \mathrm{Res}_{\rho'}(M')\big)$ which is in fact the restriction to $\mathcal{O}G$ via the structural map of the $\mathrm{Ind}_\rho\big(B^{\cdot\cdot} \otimes_{\mathcal{O}} \mathrm{Res}_{\rho'}(\mathcal{O}G')\big)$-module given by the canonical $\mathcal{O}G$-interior algebra homomorphism (cf. Remark 3.9)

4.8.1 $$\mathrm{Ind}_\rho\big(B^{\cdot\cdot} \underset{\mathcal{O}}{\otimes} \mathrm{Res}_{\rho'}(\mathcal{O}G')\big) \longrightarrow \mathrm{End}_{\mathcal{O}}\Big(\mathrm{Ind}_\rho\big(M^{\cdot\cdot} \underset{\mathcal{O}}{\otimes} \mathrm{Res}_{\rho'}(M')\big)\Big)$$

and, once again, $\hat{\imath}$ determines an idempotent $\hat{\imath}_{M'}$ in $\mathrm{End}_{\mathcal{O}}\Big(\mathrm{Ind}_\rho\big(M^{\cdot\cdot} \otimes_{\mathcal{O}} \mathrm{Res}_{\rho'}(M')\big)\Big)$; moreover, if N' is a second $\mathcal{O}G'$-module and $g' : M' \longrightarrow N'$ an $\mathcal{O}G'$-module homomorphism then we have an $\mathcal{O}G$-module homomorphism

4.8.2 $$\mathrm{Ind}_\rho(\mathrm{id} \underset{\mathcal{O}}{\otimes} g') : \mathrm{Ind}_\rho\big(M^{\cdot\cdot} \underset{\mathcal{O}}{\otimes} \mathrm{Res}_{\rho'}(M')\big) \longrightarrow \mathrm{Ind}_\rho\big(M^{\cdot\cdot} \underset{\mathcal{O}}{\otimes} \mathrm{Res}_{\rho'}(N')\big)$$

and it follows from Remark 3.9 that this map is in fact a module homomorphism over $\mathrm{Ind}_\rho\big(B^{\cdot\cdot} \otimes_{\mathcal{O}} \mathrm{Res}_{\rho'}(\mathcal{O}G')\big)$, so that we have

4.8.3
$$\mathrm{Ind}_\rho(\mathrm{id} \underset{\mathcal{O}}{\otimes} g')\Big(\hat{\imath}_{M'}\Big(\mathrm{Ind}_\rho\big(M^{\cdot\cdot} \underset{\mathcal{O}}{\otimes} \mathrm{Res}_{\rho'}(M')\big)\Big)\Big) \subset \hat{\imath}_{N'}\Big(\mathrm{Ind}_\rho\big(M^{\cdot\cdot} \underset{\mathcal{O}}{\otimes} \mathrm{Res}_{\rho'}(N')\big)\Big).$$

Consequently, denoting respectively by $\mathrm{Mod}_{\mathcal{O}G}$ and $\mathrm{Mod}_{\mathcal{O}G'}$ the categories of $\mathcal{O}G$- and $\mathcal{O}G'$-modules, we have a functor

4.8.4 $$\mathcal{F}_{G,G'}(M^{\cdot\cdot})_{\hat{\imath}} : \mathrm{Mod}_{\mathcal{O}G'} \longrightarrow \mathrm{Mod}_{\mathcal{O}G}$$

mapping M' on $\hat{\imath}_{M'}\Big(\mathrm{Ind}_\rho\big(M^{\cdot\cdot} \otimes_{\mathcal{O}} \mathrm{Res}_{\rho'}(M')\big)\Big)$ and g' on the corresponding restriction of $\mathrm{Ind}_\rho(\mathrm{id} \otimes_{\mathcal{O}} g')$. Notice that, if $M^{\cdot\cdot}$ and M' are $\mathcal{O}$-free then by Proposition 3.7 we have an $\mathcal{O}G$-interior algebra isomorphism

4.8.5 $$\mathrm{End}_{\mathcal{O}}\big(\mathcal{F}_{G,G'}(M^{\cdot\cdot})_{\hat{\imath}}(M')\big) \cong \mathcal{F}_{G,G'}\big(\mathrm{End}_{\mathcal{O}}(M^{\cdot\cdot})\big)_{\hat{\imath}}\big(\mathrm{End}_{\mathcal{O}}(M')\big) \quad .$$

4.9 Actually, this functor is, up to isomorphism, just the well-known functor $\hat{\imath}\big(\mathrm{Ind}_{H^{\cdot\cdot}}^{G\times G'}(M^{\cdot\cdot})\big)\otimes_{\mathcal{O}G''}$ determined by the direct summand $\hat{\imath}\big(\mathrm{Ind}_{H^{\cdot\cdot}}^{G\times G'}(M^{\cdot\cdot})\big)$ of $\mathrm{Ind}_{H^{\cdot\cdot}}^{G\times G'}(M^{\cdot\cdot})$, where we identify $\hat{\imath}$ to its image in $\mathrm{End}_{\mathcal{O}(G\times G')}\big(\mathrm{Ind}_{H^{\cdot\cdot}}^{G\times G'}(M^{\cdot\cdot})\big)$ (cf. 2.6.5). Indeed, from the canonical $\mathcal{O}H^{\cdot\cdot}$-module isomorphism (cf. 2.5.3)

4.9.1 $$M^{\cdot\cdot} \underset{\mathcal{O}}{\otimes} \mathrm{Res}_{\rho'}(M') \cong \mathrm{Ind}_\Delta(M^{\cdot\cdot}) \underset{\mathcal{O}H^{\cdot\cdot}}{\otimes} \mathrm{Res}_{\rho'}(M')$$

where $\Delta : H^{\cdot\cdot} \longrightarrow H^{\cdot\cdot} \times H^{\cdot\cdot}$ is the diagonal map, we get the evident $\mathcal{O}G$-module isomorphisms

$$\begin{aligned}\mathrm{Ind}_\rho\big(M^{\cdot\cdot} \underset{\mathcal{O}}{\otimes} \mathrm{Res}_{\rho'}(M')\big) &\cong \mathcal{O}G \underset{\mathcal{O}H^{\cdot\cdot}}{\otimes} \big(\mathrm{Ind}_\Delta(M^{\cdot\cdot}) \underset{\mathcal{O}H^{\cdot\cdot}}{\otimes} \mathrm{Res}_{\rho'}(M')\big)\\ &\cong \big(\mathcal{O}G \underset{\mathcal{O}H^{\cdot\cdot}}{\otimes} \mathrm{Ind}_\Delta(M^{\cdot\cdot}) \underset{\mathcal{O}H^{\cdot\cdot}}{\otimes} \mathcal{O}G'\big) \underset{\mathcal{O}G'}{\otimes} M' \\ &\cong \mathrm{Ind}_{H^{\cdot\cdot}}^{G\times G'}(M^{\cdot\cdot}) \underset{\mathcal{O}G'}{\otimes} M'\end{aligned} \tag{4.9.2}$$

and it is easily checked that the composed $\mathcal{O}G$-module isomorphism is natural on M' and is compatible, via isomorphism 4.4.1, with the canonical actions of $\mathrm{Ind}_\rho\big(B^{\cdot\cdot} \otimes_{\mathcal{O}} \mathrm{Res}_{\rho'}(\mathcal{O}G')\big)$ on the left member and of $\mathrm{Ind}_{H^{\cdot\cdot}}^{G\times G'}(B^{\cdot\cdot})^{1\times G'}$ on the right member; hence, we obtain a natural $\mathcal{O}G$-module isomorphism

$$\begin{aligned}\mathcal{F}_{G,G'}(M^{\cdot\cdot})_{\hat{\imath}}(M') &= \hat{\imath}_{M'}\Big(\mathrm{Ind}_\rho\big(M^{\cdot\cdot} \underset{\mathcal{O}}{\otimes} \mathrm{Res}_{\rho'}(M')\big)\Big)\\ &\cong \hat{\imath}\big(\mathrm{Ind}_{H^{\cdot\cdot}}^{G\times G'}(M^{\cdot\cdot})\big) \underset{\mathcal{O}G'}{\otimes} M' \quad .\end{aligned} \tag{4.9.3}$$

4.10 Since Morita equivalences between blocks are obviously closed by composition and we claim that they coincide with the corresponding restrictions of some functors described above (see 6.7 below), some of those functors, up to restriction, should certainly be closed by composition too. Finally, we give a sufficient condition for closure by composition, up to restriction, of those functors; in order to state it, let G'' be a third finite group, $L^{\cdot\cdot}$ a subgroup of $G' \times G''$ and $C^{\cdot\cdot}$ an $\mathcal{O}L^{\cdot\cdot}$-interior algebra, denote respectively by σ' and σ'' the restrictions to $L^{\cdot\cdot}$ of the first and the second projection maps for $G' \times G''$ and set

$$\mathrm{Ker}(\sigma') = 1 \times J'' \quad \text{and} \quad \mathrm{Ker}(\sigma'') = J' \times 1 \tag{4.10.1}$$

for suitable subgroups J' of G' and J'' of G''; moreover, consider the pull-back of groups

$$\begin{array}{ccc} H^{\cdot\cdot} & \xrightarrow{\ \rho'\ } & G' \\ {\scriptstyle\sigma}\uparrow & & \uparrow{\scriptstyle\sigma'} \\ H^{\cdot\cdot} \underset{G'}{\times} L^{\cdot\cdot} & \xrightarrow{\ \rho''\ } & L^{\cdot\cdot} \end{array} \tag{4.10.2}$$

so that $H^{\cdot\cdot} \times_{G'} L^{\cdot\cdot}$ is, up to identification, a subgroup of $H^{\cdot\cdot} \times L^{\cdot\cdot}$; set $F^{\cdot\cdot} = (\rho \times \sigma'')\,(H^{\cdot\cdot} \times_{G'} L^{\cdot\cdot})$, so that $F^{\cdot\cdot}$ is a subgroup of $G \times G''$, and denote by $\rho \times_{G'} \sigma'' : H^{\cdot\cdot} \times_{G'} L^{\cdot\cdot} \longrightarrow F^{\cdot\cdot}$ the group homomorphism determined by $\rho \times \sigma''$, and by τ and τ'' the respective restrictions to $F^{\cdot\cdot}$ of the first and the second projection maps for $G \times G''$; notice that

$$\mathrm{Ker}(\tau) = 1 \times \sigma''\big(\sigma'^{-1}(K')\big) \quad \text{and} \quad \mathrm{Ker}(\tau'') = \rho\big(\rho'^{-1}(J')\big) \times 1 \quad . \tag{4.10.3}$$

Proposition 4.11 *With the notation above, assume that $B^{\cdot\cdot}$ and $C^{\cdot\cdot}$ considered respectively as $\mathcal{O}(\mathrm{Ker}(\rho))$- and $\mathcal{O}(\mathrm{Ker}(\sigma'))$-bimodules are both projective. Then the $\mathcal{O}F^{\cdot\cdot}$-interior algebra $\mathrm{Ind}_{\rho\times_{G'}\sigma''}(\mathrm{Res}_\sigma(B^{\cdot\cdot})\otimes_{\mathcal{O}}\mathrm{Res}_{\rho''}(C^{\cdot\cdot}))$ considered as an $\mathcal{O}(\mathrm{Ker}\,(\tau))$-bimodule is projective and we have a natural embedding*

4.11.1
$$\mathcal{F}_{G,G'}\Big(\mathrm{Ind}_{\rho\underset{G'}{\times}\sigma''}\big(\mathrm{Res}_\sigma(B^{\cdot\cdot})\underset{\mathcal{O}}{\otimes}\mathrm{Res}_{\rho''}(C^{\cdot\cdot})\big)\Big)_1\to\mathcal{F}_{G,G'}(B^{\cdot\cdot})_1\circ\mathcal{F}_{G,G'}(C^{\cdot\cdot})_1\quad .$$

Moreover, if $B^{\cdot\cdot}=\mathrm{End}_{\mathcal{O}}(M^{\cdot\cdot})$ for some $\mathcal{O}H^{\cdot\cdot}$-module $M^{\cdot\cdot}$ and $C^{\cdot\cdot}=\mathrm{End}_{\mathcal{O}}(N^{\cdot\cdot})$ for some $\mathcal{O}L^{\cdot\cdot}$-module $N^{\cdot\cdot}$ then we have a natural direct injection

4.11.2
$$\mathcal{F}_{G,G'}\Big(\mathrm{Ind}_{\rho\underset{G'}{\times}\sigma''}\big(\mathrm{Res}_\sigma(M^{\cdot\cdot})\underset{\mathcal{O}}{\otimes}\mathrm{Res}_{\rho''}(N^{\cdot\cdot})\big)\Big)_1\to\mathcal{F}_{G,G'}(M^{\cdot\cdot})_1\circ\mathcal{F}_{G,G'}(N^{\cdot\cdot})_1\quad .$$

Remark 4.12 According to 4.9, the last statement affirms just that the $\mathcal{O}(G\times G'')$-module $\mathrm{Ind}_{F^{\cdot\cdot}}^{G\times G''}\big(\mathrm{Ind}_{\rho\times_{G'}\sigma''}(\mathrm{Res}_\sigma(M^{\cdot\cdot})\otimes_{\mathcal{O}}\mathrm{Res}_{\rho''}(N^{\cdot\cdot}))\big)$ is a direct summand of $\mathrm{Ind}_{H^{\cdot\cdot}}^{G\times G'}(M^{\cdot\cdot})\otimes_{\mathcal{O}}\mathrm{Ind}_{L^{\cdot\cdot}}^{G\times G''}(N^{\cdot\cdot})$.

Proof: For any $\mathcal{O}G''$-interior algebra A'' we have to compute

4.11.3
$$\begin{aligned}&\mathcal{F}_{G,G'}(B^{\cdot\cdot})_1\big(\mathcal{F}_{G,G'}(C^{\cdot\cdot})_1(A'')\big)\\&\qquad=\mathrm{Ind}_\rho\Big(B^{\cdot\cdot}\underset{\mathcal{O}}{\otimes}\mathrm{Res}_{\rho'}\big(\mathrm{Ind}_{\sigma'}(C^{\cdot\cdot}\underset{\mathcal{O}}{\otimes}\mathrm{Res}_{\sigma''}(A''))\big)\Big)\quad ;\end{aligned}$$

but we get from Proposition 3.21 a natural $\mathcal{O}H^{\cdot\cdot}$-interior algebra embedding

4.11.4
$$\mathrm{Ind}_\sigma\big(\mathrm{Res}_{\rho''}(C^{\cdot\cdot}\underset{\mathcal{O}}{\otimes}\mathrm{Res}_{\sigma''}(A''))\big)\longrightarrow\mathrm{Res}_{\rho'}\big(\mathrm{Ind}_{\sigma'}(C^{\cdot\cdot}\underset{\mathcal{O}}{\otimes}\mathrm{Res}_{\sigma''}(A''))\big)$$

and from Proposition 3.16 a natural $\mathcal{O}H^{\cdot\cdot}$-interior algebra isomorphism

4.11.5
$$\begin{aligned}&B^{\cdot\cdot}\underset{\mathcal{O}}{\otimes}\mathrm{Ind}_\sigma\big(\mathrm{Res}_{\rho''}(C^{\cdot\cdot}\underset{\mathcal{O}}{\otimes}\mathrm{Res}_{\sigma''}(A''))\big)\\&\qquad\cong\mathrm{Ind}_\sigma\big(\mathrm{Res}_\sigma(B^{\cdot\cdot})\underset{\mathcal{O}}{\otimes}\mathrm{Res}_{\rho''}(C^{\cdot\cdot})\underset{\mathcal{O}}{\otimes}\mathrm{Res}_{\sigma''\circ\rho''}(A'')\big)\quad .\end{aligned}$$

On the other hand, we claim that $\mathrm{Res}_\sigma(B^{\cdot\cdot})\otimes_{\mathcal{O}}\mathrm{Res}_{\rho''}(C^{\cdot\cdot})$ considered as an $\mathcal{O}(\mathrm{Ker}\,(\rho\circ\sigma))$-bimodule is projective; indeed, since $B^{\cdot\cdot}$ considered as an $\mathcal{O}(\mathrm{Ker}(\rho))$-bimodule is projective, $\mathrm{Res}_\sigma(B^{\cdot\cdot})$ is a direct summand of

4.11.6
$$\mathrm{Ind}_{\mathrm{Ker}(\sigma)\times\mathrm{Ker}(\sigma)}^{\sigma^{-1}(\mathrm{Ker}(\rho))\times\sigma^{-1}(\mathrm{Ker}(\rho))}(\mathcal{O})^n$$

for a suitable n and therefore, since ρ'' induces an isomorphism between $\mathrm{Ker}(\sigma)$ and $\mathrm{Ker}(\sigma')$ (cf. 4.10.2), the $\mathcal{O}\big(\sigma^{-1}(\mathrm{Ker}(\rho))\big)$-bimodule $\mathrm{Res}_\sigma(B^{\cdot\cdot})\otimes_{\mathcal{O}}\mathrm{Res}_{\rho''}(C^{\cdot\cdot})$ becomes a direct summand of

4.11.7 $$\mathrm{Ind}_{\mathrm{Ker}(\sigma)\times\mathrm{Ker}(\sigma)}^{\sigma^{-1}(\mathrm{Ker}(\rho))\times\sigma^{-1}(\mathrm{Ker}(\rho))}\Big(\mathrm{Res}_{\rho''}\big(\mathrm{Res}_{\mathrm{Ker}(\sigma')\times\mathrm{Ker}(\sigma')}^{L^{\cdot\cdot}\times L^{\cdot\cdot}}(C^{\cdot\cdot})\big)\Big)^n$$

which proves our claim since $C^{\cdot\cdot}$ considered as an $\mathcal{O}\big(\mathrm{Ker}(\sigma')\big)$-bimodule is also projective. Consequently, it follows from Corollary 3.13 and Proposition 3.16 that we have natural $\mathcal{O}G$-interior algebra isomorphisms

$$\mathrm{Ind}_\rho\Big(\mathrm{Ind}_\sigma\big(\mathrm{Res}_\sigma(B^{\cdot\cdot})\underset{\mathcal{O}}{\otimes}\mathrm{Res}_{\rho''}(C^{\cdot\cdot})\underset{\mathcal{O}}{\otimes}\mathrm{Res}_{\sigma''\circ\rho''}(A'')\big)\Big)$$

$$\cong\mathrm{Ind}_{\rho\circ\sigma}\big(\mathrm{Res}_\sigma(B^{\cdot\cdot})\underset{\mathcal{O}}{\otimes}\mathrm{Res}_{\rho''}(C^{\cdot\cdot})\underset{\mathcal{O}}{\otimes}\mathrm{Res}_{\sigma''\circ\rho''}(A'')\big)$$

4.11.8 $$\cong\mathrm{Ind}_\tau\Big(\mathrm{Ind}_{\rho\underset{G'}{\times}\sigma''}\Big(\mathrm{Res}_\sigma(B^{\cdot\cdot})\underset{\mathcal{O}}{\otimes}\mathrm{Res}_{\rho''}(C^{\cdot\cdot})\underset{\mathcal{O}}{\otimes}\mathrm{Res}_{\rho\underset{G'}{\times}\sigma''}\big(\mathrm{Res}_{\tau''}(A'')\big)\Big)\Big)$$

$$\cong\mathrm{Ind}_\tau\Big(\mathrm{Ind}_{\rho\underset{G'}{\times}\sigma''}\big(\mathrm{Res}_\sigma(B^{\cdot\cdot})\underset{\mathcal{O}}{\otimes}\mathrm{Res}_{\rho''}(C^{\cdot\cdot})\big)\underset{\mathcal{O}}{\otimes}\mathrm{Res}_{\tau''}(A'')\Big)$$

$$=\mathcal{F}_{G,G'}\Big(\mathrm{Ind}_{\rho\underset{G'}{\times}\sigma''}\big(\mathrm{Res}_\sigma(B^{\cdot\cdot})\underset{\mathcal{O}}{\otimes}\mathrm{Res}_{\rho''}(C^{\cdot\cdot})\big)\Big)_1(A'')$$

since $\mathrm{Ind}_{\rho\times_{G'}\sigma''}\big(\mathrm{Res}_\sigma(B^{\cdot\cdot})\otimes_{\mathcal{O}}\mathrm{Res}_{\rho''}(C^{\cdot\cdot})\big)$ considered as an $\mathcal{O}\big(\mathrm{Ker}(\tau)\big)$-bimodule is still projective. In conclusion, if we tensor by $B^{\cdot\cdot}$ embedding 4.11.4 and induce via ρ the resulting embedding, this induced embedding composed with the composition of isomorphisms 4.11.8 gives the announced embedding 4.11.1 (the naturality is easily checked).

Moreover, if $B^{\cdot\cdot}=\mathrm{End}_{\mathcal{O}}(M^{\cdot\cdot})$ for some $\mathcal{O}H^{\cdot\cdot}$-module $M^{\cdot\cdot}$ and $C^{\cdot\cdot}=\mathrm{End}_{\mathcal{O}}(N^{\cdot\cdot})$ for some $\mathcal{O}L^{\cdot\cdot}$-module $N^{\cdot\cdot}$ (recall that, according to our hypothesis, $\mathrm{Res}_{\mathrm{Ker}(\rho)}^{H^{\cdot\cdot}}(M^{\cdot\cdot})$ and $\mathrm{Res}_{\mathrm{Ker}(\sigma')}^{L^{\cdot\cdot}}(N^{\cdot\cdot})$ are both projective) then it follows from isomorphisms 4.8.5 and 4.9.3 that we have an $\mathcal{O}G$-interior algebra isomorphism

4.11.9 $$\mathcal{F}_{G,G'}(B^{\cdot\cdot})_1\Big(\mathcal{F}_{G,G'}(C^{\cdot\cdot})_1\big(\mathrm{End}_{\mathcal{O}}(\mathcal{O}G'')\big)\Big)$$
$$\cong\mathrm{End}_{\mathcal{O}}\big(\mathrm{Ind}_{H^{\cdot\cdot}}^{G\times G'}(M^{\cdot\cdot})\underset{\mathcal{O}G'}{\otimes}\mathrm{Ind}_{L^{\cdot\cdot}}^{G'\times G''}(N^{\cdot\cdot})\big)\quad ;$$

similarly, since we have an $\mathcal{O}F^{\cdot\cdot}$-interior algebra isomorphism (cf. Proposition 3.7)

4.11.10 $$\mathrm{Ind}_{\rho\underset{G'}{\times}\sigma''}\big(\mathrm{Res}_\sigma(B^{\cdot\cdot})\underset{\mathcal{O}}{\otimes}\mathrm{Res}_{\rho''}(C^{\cdot\cdot})\big)$$
$$\cong\mathrm{End}_{\mathcal{O}}\Big(\mathrm{Ind}_{\rho\underset{G'}{\times}\sigma''}\big(\mathrm{Res}_\sigma(M^{\cdot\cdot})\underset{\mathcal{O}}{\otimes}\mathrm{Res}_{\rho''}(N^{\cdot\cdot})\big)\Big)\quad ,$$

we get again from isomorphisms 4.8.5 and 4.9.3

4.11.11
$$\begin{aligned}&\mathcal{F}_{G,G''}\Big(\mathrm{Ind}_{\rho \underset{G'}{\times} \sigma''}\big(\mathrm{Res}_\sigma(B^{\cdot\cdot}) \underset{\mathcal{O}}{\otimes} \mathrm{Res}_{\rho''}(C^{\cdot\cdot})\big)\Big)_1\big(\mathrm{End}_\mathcal{O}(\mathcal{O}G'')\big)\\ &\qquad \cong \mathrm{End}_\mathcal{O}\bigg(\mathrm{Ind}_{F^{\cdot\cdot}}^{G\times G''}\Big(\mathrm{Ind}_{\rho \underset{G'}{\times} \sigma''}\big(\mathrm{Res}_\sigma(M^{\cdot\cdot}) \underset{\mathcal{O}}{\otimes} \mathrm{Res}_{\rho''}(N^{\cdot\cdot})\big)\Big)\bigg) \quad ;\end{aligned}$$

consequently, embedding 4.11.1 applied to $\mathrm{End}_\mathcal{O}(\mathcal{O}G'')$ supplies an $\mathcal{O}(G \times G'')$-module direct injection

4.11.12
$$\mathrm{Ind}_{F^{\cdot\cdot}}^{G\times G''}\Big(\mathrm{Ind}_{\rho \underset{G'}{\times} \sigma''}\big(\mathrm{Res}_\sigma(M^{\cdot\cdot}) \underset{\mathcal{O}}{\otimes} \mathrm{Res}_{\rho''}(N^{\cdot\cdot})\big)\Big) \to \mathrm{Ind}_{H^{\cdot\cdot}}^{G\times G''}(M^{\cdot\cdot}) \underset{\mathcal{O}G'}{\otimes} \mathrm{Ind}_{L^{\cdot\cdot}}^{G\times G''}(N^{\cdot\cdot})$$

and therefore the last statement follows from 4.9.3.

5 On the local structure of Hecke $\mathcal{O}G$-interior algebras

5.1 Keep all the notation of Section 4, set

5.1.1 $$A^{\cdot\cdot} = \mathrm{Ind}_{H^{\cdot\cdot}}^{G\times G'}(B^{\cdot\cdot}) \text{ and } \hat{A} = \mathrm{Ind}_{H^{\cdot\cdot}}^{G\times G'}(B^{\cdot\cdot})^{1\times G'}$$

and consider a point $\hat{\alpha}$ of G on $\hat{A}$; in this section we determine a defect pointed group of $G_{\hat{\alpha}}$ and the corresponding source algebra and multiplicity module (cf. 2.8 and 2.9), from analogous local invariants of $B^{\cdot\cdot}$. This knowledge of the local structure of the Hecke $\mathcal{O}G$-interior algebra $\hat{A}$ will be enough to study Morita equivalences between blocks, and we postpone the systematic analysis of the local pointed groups on $\hat{A}$ until Section 16, in order to do it in the more general framework of the $\mathfrak{D}G$-interior algebras, except for the contents of Proposition 5.5 and Corollary 5.7 below which will apply to $\mathfrak{D}G$-interior algebras as it stands.

5.2 First of all notice that, since $\hat{A}^G = (A^{\cdot\cdot})^{G\times G'}$, $\hat{\alpha}$ can be also considered as a point of $G \times G'$ on $A^{\cdot\cdot}$; thus, let $P^{\cdot\cdot}_{\gamma^{\cdot\cdot}}$ be a defect pointed group of $(G \times G')_{\hat{\alpha}}$ on $A^{\cdot\cdot}$, set $P = \rho(P^{\cdot\cdot})$ and $P' = \rho'(P^{\cdot\cdot})$, and denote respectively by $\sigma : P^{\cdot\cdot} \longrightarrow P$ and $\sigma' : P^{\cdot\cdot} \longrightarrow P'$ the group homomorphisms determined by ρ and ρ'; actually, we may assume that $P^{\cdot\cdot} \subset H^{\cdot\cdot}$ and that $\gamma^{\cdot\cdot}$ comes from a local point of $P^{\cdot\cdot}$ on $B^{\cdot\cdot}$ (cf. 2.11.3) and recall that we have the Higman $\mathcal{O}(G\times G')$-interior algebra embedding (cf. 2.12.1)

5.2.1 $$(A^{\cdot\cdot})_{\hat{\alpha}} \longrightarrow \mathrm{Ind}_{P^{\cdot\cdot}}^{G\times G'}(B^{\cdot\cdot}_{\gamma^{\cdot\cdot}})$$

which determines, together with the Hecke isomorphism $H_{G,G'}(B^{\cdot\cdot}_{\gamma^{\cdot\cdot}})$ (cf. 4.4.1), an $\mathcal{O}G$-interior algebra embedding

5.2.2 $$\hat{A}_{\hat{\alpha}} \longrightarrow \mathrm{Ind}_P^G\Big(\mathrm{Ind}_\sigma\Big(B^{\cdot\cdot}_{\gamma^{\cdot\cdot}} \underset{\mathcal{O}}{\otimes} \mathrm{Res}_{\sigma'}(\mathrm{Res}_{P'}^{G'}(\mathcal{O}G'))\Big)\Big) \quad .$$

Once again notice that, since $\hat{A}^P = (A^{\cdot\cdot})^{P\times G'}$, any point $\hat{\gamma}$ of P on $\hat{A}$ can be also considered as a point of $P \times G'$ on $A^{\cdot\cdot}$ and the corresponding multiplicity algebras $\hat{A}(P_{\hat{\gamma}})$ and $(A^{\cdot\cdot})((P \times G')_{\hat{\gamma}})$ coincide (cf. 2.8); in particular, since $N_{G\times G'}(P \times G') = N_G(P) \times G'$, we may identify $\bar{N}_G(P)$ to $\bar{N}_{G\times G'}(P \times G')$ and then we get the equalities (the second one up to a choice!) (cf. 2.8)

5.2.3 $$\hat{\bar{N}}_G(P_{\hat{\gamma}}) = \hat{\bar{N}}_{G\times G'}((P \times G')_{\hat{\gamma}}) \text{ and } V_{\hat{A}}(P_{\hat{\gamma}}) = V_{A^{\cdot\cdot}}((P \times G')_{\hat{\gamma}}) \quad .$$

Proposition 5.3 *With the notation above, there are local points $\hat{\gamma}$ of P on $\hat{A}$ and γ' of P' on $\mathcal{O}G'$ such that $P_{\hat{\gamma}}$ is a defect pointed group of $G_{\hat{\alpha}}$, embedding 5.2.2 induces an $\mathcal{O}P$-interior algebra embedding*

5.3.1 $$\hat{A}_{\hat{\gamma}} \longrightarrow \mathrm{Ind}_\sigma\Big(B^{\cdot\cdot}_{\gamma^{\cdot\cdot}} \underset{\mathcal{O}}{\otimes} \mathrm{Res}_{\sigma'}((\mathcal{O}G')_{\gamma'})\Big)$$

and we have

5.3.2 $$\hat{\bar{N}}_G(P_{\hat{\gamma}}) = \hat{\bar{N}}_{G\times G'}\big((P\times G')_{\hat{\gamma}}\big) \text{ and } V_{\hat{A}_{\hat{\alpha}}}(P_{\hat{\gamma}}) = V_{A^{\cdot\cdot}_{\hat{\alpha}}}\big((P\times G')_{\hat{\gamma}}\big) \quad .$$

Remark 5.4 In the terminology of [26, 3.3], this proposition determines a *G-triple* parametrizing $\hat{A}_{\hat{\alpha}}$ from a $G\times G'$-triple parametrizing $A^{\cdot\cdot}_{\hat{\alpha}}$. Indeed, first of all notice that the fact that embedding 5.2.2 induces embedding 5.3.1, together with the commutativity of diagram 4.6.1 imply that the following diagram of $\mathcal{O}P$-interior algebra exoembeddings is commutative

5.4.1 $$\begin{array}{ccc} \mathrm{Res}^{G\times G'}_{P\times G'}(A^{\cdot\cdot}_{\hat{\alpha}})^{1\times G'} & \xrightarrow{\tilde{h}_{\hat{\alpha}}} & \mathrm{Res}^{G\times G'}_{P\times G'}\big(\mathrm{Ind}^{G\times G'}_{P^{\cdot\cdot}}(B^{\cdot\cdot}_{\gamma^{\cdot\cdot}})\big)^{1\times G'} \\ {\scriptstyle \tilde{f}^{\hat{\alpha}}_{\hat{\gamma}}}\big\uparrow & & \big\uparrow{\scriptstyle \tilde{d}} \\ \hat{A}_{\hat{\gamma}} & \xrightarrow{\tilde{h}_{\hat{\gamma}}} & \mathrm{Ind}^{P\times G'}_{P^{\cdot\cdot}}(B^{\cdot\cdot}_{\gamma^{\cdot\cdot}})^{1\times G'} \end{array}$$

where $d = d^{G\times G'}_{P\times G'}\big(\mathrm{Ind}^{P\times G'}_{P^{\cdot\cdot}}(B^{\cdot\cdot}_{\gamma^{\cdot\cdot}})\big)^{1\times G'}$, $\tilde{h}_{\hat{\alpha}}$ is the restriction of exoembedding 5.2.1 and $\tilde{h}_{\hat{\gamma}}$ the composition of exoembedding 5.3.1 with $H_{P,G'}(B^{\cdot\cdot}_{\gamma^{\cdot\cdot}})\circ\mathrm{Ind}_{\sigma}(\tilde{\mathrm{id}}_{B^{\cdot\cdot}_{\gamma^{\cdot\cdot}}}\otimes_{\mathcal{O}}\tilde{f}_{\gamma'})$; in particular, the image j of the unity element of $\hat{A}_{\hat{\gamma}}$ in $\mathrm{Ind}^{G\times G'}_{P^{\cdot\cdot}}(B^{\cdot\cdot}_{\gamma^{\cdot\cdot}})^{P\times G'}$ by suitable representatives of $\tilde{f}^{\hat{\alpha}}_{\hat{\gamma}}$ and $\tilde{h}_{\hat{\alpha}}$ fulfills

5.4.2 $$j\mathrm{Tr}^{P\times G'}_{P^{\cdot\cdot}}(1\otimes 1\otimes 1) = j = \mathrm{Tr}^{P\times G'}_{P^{\cdot\cdot}}(1\otimes 1\otimes 1)j$$

and therefore in $A^{\cdot\cdot}$ we get

5.4.3 $$\hat{\gamma}\subset \mathrm{Tr}^{P\times G'}_{P^{\cdot\cdot}}\big((A^{\cdot\cdot})^{P^{\cdot\cdot}}\cdot\gamma^{\cdot\cdot}\cdot(A^{\cdot\cdot})^{P^{\cdot\cdot}}\big) \quad .$$

Consequently, since (cf. 2.9.4)

5.4.4 $$\hat{\alpha}\subset \mathrm{Tr}^G_P(\hat{A}^P\cdot\hat{\gamma}\cdot\hat{A}^P) = \mathrm{Tr}^{G\times G'}_{P\times G'}\big((A^{\cdot\cdot})^{P\times G'}\cdot\hat{\gamma}\cdot(A^{\cdot\cdot})^{P\times G'}\big) \quad ,$$

$P^{\cdot\cdot}_{\gamma^{\cdot\cdot}}$ is a minimal pointed group on $A^{\cdot\cdot}$ fulfilling 5.4.3 (cf. 2.9.4) and therefore we have (cf. 2.9.4)

5.4.5 $$P^{\cdot\cdot}_{\gamma^{\cdot\cdot}}\subset (P\times G')_{\hat{\gamma}}\subset (G\times G')_{\hat{\alpha}} \quad ;$$

in particular, $s_{\gamma^{\cdot\cdot}}(\hat{\gamma})$ and $s_{\gamma^{\cdot\cdot}}(\hat{\alpha})$ are respectively projective points of $\bar{N}_{P\times G'}(P^{\cdot\cdot}_{\gamma^{\cdot\cdot}})$ and $\bar{N}_{G\times G'}(P^{\cdot\cdot}_{\gamma^{\cdot\cdot}})$ on $(A^{\cdot\cdot})(P^{\cdot\cdot}_{\gamma^{\cdot\cdot}})$ (cf. 2.9.2), we have

5.4.6 $$\bar{N}_{P\times G'}(P^{\cdot\cdot}_{\gamma^{\cdot\cdot}})_{s_{\gamma^{\cdot\cdot}}(\hat{\gamma})}\subset \bar{N}_{G\times G'}(P^{\cdot\cdot}_{\gamma^{\cdot\cdot}})_{s_{\gamma^{\cdot\cdot}}(\hat{\alpha})}$$

and $s_{\gamma^{\cdot\cdot}}$ induces a k^*-group isomorphism

5.4.7 $$\hat{\omega}: \hat{\bar{N}}_{G\times G'}\big((P\times G')_{\hat{\gamma}}\big) \cong \hat{\bar{N}}_{\bar{N}_{G\times G'}(P^{\cdot\cdot}_{\gamma^{\cdot\cdot}})}\big(\bar{N}_{P\times G'}(P^{\cdot\cdot}_{\gamma^{\cdot\cdot}})_{s_{\gamma^{\cdot\cdot}}(\hat{\gamma})}\big)$$

together with a $k_*\hat{\bar{N}}_{G\times G'}\big((P\times G')_{\hat\gamma}\big)$-module isomorphism

$$5.4.8 \qquad V_{A^{\cdot\cdot}}\big((P\times G')_{\hat\gamma}\big) \cong \mathrm{Res}_{\hat\omega}\Big(V_{(A^{\cdot\cdot})(P^{\cdot\cdot}_{\gamma^{\cdot\cdot}})}\big(\bar N_{P\times G'}(P^{\cdot\cdot}_{\gamma^{\cdot\cdot}})_{s_{\gamma^{\cdot\cdot}}(\hat\gamma)}\big)\Big) \quad .$$

But $V_{A^{\cdot\cdot}_{\hat\alpha}}(P^{\cdot\cdot}_{\gamma^{\cdot\cdot}})$ is an indecomposable projective $k_*\hat{\bar N}_{G\times G'}(P^{\cdot\cdot}_{\gamma^{\cdot\cdot}})$-module (cf. 2.9.3) and $\bar N_{P\times G'}(P^{\cdot\cdot}_{\gamma^{\cdot\cdot}})$ is a normal subgroup of $\bar N_{G\times G'}(P^{\cdot\cdot}_{\gamma^{\cdot\cdot}})$. Hence $\bar N_{G\times G'}(P^{\cdot\cdot}_{\gamma^{\cdot\cdot}})$ acts transitively on the set of isomorphism classes of indecomposable direct summands of the restriction of $V_{A^{\cdot\cdot}_{\hat\alpha}}(P^{\cdot\cdot}_{\gamma^{\cdot\cdot}})$ to $k_*\hat{\bar N}_{P\times G'}(P^{\cdot\cdot}_{\gamma^{\cdot\cdot}})$ and, according to Theorem 5.8 below, one of these classes determines $\hat\gamma$ as a local point of P on

$$\mathrm{Ind}_\sigma\Big(B^{\cdot\cdot}_{\gamma^{\cdot\cdot}}\otimes_{\mathcal{O}}\mathrm{Res}_{\sigma'}\big((\mathcal{O}G')_{\gamma'}\big)\Big) \quad ,$$

whereas from 5.3.2 and 5.4.8 we get a $k_*\hat{\bar N}_G(P_{\hat\gamma})$-module isomorphism between $V_{\hat A_{\hat\alpha}}(P_{\hat\gamma})$ and the restriction via $\hat\omega$ of a multiplicity module of the pointed group on $\mathrm{End}_k\big(V_{A^{\cdot\cdot}_{\hat\alpha}}(P^{\cdot\cdot}_{\gamma^{\cdot\cdot}})\big)$ determined by that class. Actually, Proposition 5.3 is a particular case of Corollary 16.10 below, and there we are able to formulate a uniqueness part.

Proof: The existence of embedding 5.2.2 shows that P contains a defect group Q of $G_{\hat\alpha}$ (cf. 2.11.3); then, since

$$5.3.3 \qquad (A^{\cdot\cdot}_{\hat\alpha})^{G\times G'} = (\hat A_{\hat\alpha})^G = (\hat A_{\hat\alpha})^G_Q = (A^{\cdot\cdot}_{\hat\alpha})^{G\times G'}_{Q\times G'} \quad ,$$

$Q\times G'$ contains $(P^{\cdot\cdot})^{(x,x')}$ for a suitable $(x,x')\in G\times G'$ (cf. 2.9.4 and 2.9.5), so that Q contains P^x, and therefore we get $Q=P$; in particular, if $\hat\gamma$ is a local point of P on $\hat A_{\hat\alpha}$ then $P_{\hat\gamma}$ is a defect pointed group of $G_{\hat\alpha}$ (cf. 2.9.5).

Let J' be a pairwise orthogonal primitive idempotent decomposition of the unity element in $(\mathcal{O}G')^{P'}$ and, for any $j'\in J'$, set

5.3.4

$$e_{j'} = 1\otimes\big(1\otimes(1\otimes j')\big)\otimes 1 \in \mathrm{Ind}_P^G\Big(\mathrm{Ind}_\sigma\big(B^{\cdot\cdot}_{\gamma^{\cdot\cdot}}\underset{\mathcal{O}}{\otimes}\mathrm{Res}_{\sigma'}(\mathrm{Res}^{G'}_{P'}(\mathcal{O}G'))\big)\Big)^P ;$$

it is quite clear that $\{e_{j'}\}_{j'\in J'}$ is still a set of pairwise orthogonal idempotents in $\Big(1\otimes\mathrm{Ind}_\sigma\big(B^{\cdot\cdot}_{\gamma^{\cdot\cdot}}\otimes_{\mathcal{O}}\mathrm{Res}_{\sigma'}(\mathrm{Res}^{G'}_{P'}(\mathcal{O}G'))\big)\otimes 1\Big)^P$ and therefore that $\{\mathrm{Tr}_P^G(e_{j'})\}_{j'\in J'}$ is also a set of pairwise orthogonal idempotents, this time fixed by G and fulfilling $\Sigma_{j'\in J'}\mathrm{Tr}_P^G(e_{j'})=1$; consequently, since the unity element in $(\hat A_{\hat\alpha})^G$ is primitive, there are a representative h of exoembedding 5.2.2 and an element j' of J' such that

$$5.3.5 \qquad h(1)\mathrm{Tr}_P^G(e_{j'}) = h(1) = \mathrm{Tr}_P^G(e_{j'})h(1)$$

and then it is easily checked that

$$5.3.6 \qquad \begin{aligned}&\mathrm{Tr}_P^G(e_{j'})\mathrm{Ind}_P^G\Big(\mathrm{Ind}_\sigma\big(B^{\cdot\cdot}_{\gamma^{\cdot\cdot}}\underset{\mathcal{O}}{\otimes}\mathrm{Res}_{\sigma'}(\mathrm{Res}^{G'}_{P'}(\mathcal{O}G'))\big)\Big)\mathrm{Tr}_P^G(e_{j'})\\ &\qquad = \mathrm{Ind}_P^G\Big(\mathrm{Ind}_\sigma\big(B^{\cdot\cdot}_{\gamma^{\cdot\cdot}}\underset{\mathcal{O}}{\otimes}\mathrm{Res}_{\sigma'}(j'\mathcal{O}G'j')\big)\Big) \quad ,\end{aligned}$$

so that h determines an $\mathcal{O}G$-interior algebra embedding

5.3.7 $$\hat{A}_{\hat{\alpha}} \longrightarrow \mathrm{Ind}_P^G\Big(\mathrm{Ind}_\sigma\big(B^{\cdot\cdot}_{\gamma^{\cdot\cdot}} \underset{\mathcal{O}}{\otimes} \mathrm{Res}_{\sigma'}(j'\mathcal{O}G'j')\big)\Big) \quad ;$$

hence, denoting by γ' the point of P' on $\mathcal{O}G'$ which contains j' (cf. 2.7) and modifying if necessary our choice of $\hat{\gamma}$, embedding 5.3.7 induces an $\mathcal{O}P$-interior algebra embedding (cf. 2.11.3)

5.3.8 $$\hat{A}_{\hat{\gamma}} \longrightarrow \mathrm{Ind}_\sigma\Big(B^{\cdot\cdot}_{\gamma^{\cdot\cdot}} \underset{\mathcal{O}}{\otimes} \mathrm{Res}_{\sigma'}\big((\mathcal{O}G')_{\gamma'}\big)\Big) \quad .$$

Now, we claim that γ' is a local point of p' on $\mathcal{O}G'$; indeed, the inverse of the Hecke $\mathcal{O}P$-interior algebra isomorphism $H_{P,G'}(B^{\cdot\cdot}_{\gamma^{\cdot\cdot}})$ (cf. 4.4.1) determines an $\mathcal{O}G$-interior algebra isomorphism

5.3.9 $$\mathrm{Ind}_P^G\bigg(\mathrm{Ind}_\sigma\Big(B^{\cdot\cdot}_{\gamma^{\cdot\cdot}} \underset{\mathcal{O}}{\otimes} \mathrm{Res}_{\sigma'}\big(\mathrm{Res}_{P'}^{G'}(\mathcal{O}G')\big)\Big)\bigg) \cong \mathrm{Ind}_{P^{\cdot\cdot}}^{G\times G'}(B^{\cdot\cdot}_{\gamma^{\cdot\cdot}})^{1\times G'}$$

which maps the idempotent $\mathrm{Tr}_P^G(e_{j'})$ on

5.3.10 $$\mathrm{Tr}_P^G\Big(\mathrm{Tr}_{\mathrm{Ker}(\sigma)}^{1\times G'}\big((1\otimes j'^{\circ})\otimes 1\otimes(1\otimes 1)\big)\Big) = \mathrm{Tr}_{P^{\cdot\cdot}}^{G\times G'}\big((1\otimes j'^{\circ})\otimes 1\otimes(1\otimes 1)\big)$$

where we identify $\mathcal{O}(G\times G')$ to $\mathcal{O}G\otimes_{\mathcal{O}}\mathcal{O}G'$ and j'° denotes the image of j' by the canonical $\mathcal{O}$-algebra isomorphism $\mathcal{O}G'\cong(\mathcal{O}G')^{\circ}$; hence, if $j'=\mathrm{Tr}_{Q'}^{P'}(a')$ for a suitable subgroup Q' of P' and a suitable element a' of $(\mathcal{O}G')^{Q'}$, we have

5.3.11 $$\mathrm{Tr}_{P^{\cdot\cdot}}^{G\times G'}\big((1\otimes j'^{\circ})\otimes 1\otimes(1\otimes 1)\big) = \mathrm{Tr}_{Q^{\cdot\cdot}}^{G\times G'}\big((1\otimes a'^{\circ})\otimes 1\otimes(1\otimes 1)\big)$$

where $Q^{\cdot\cdot}=\sigma'^{-1}(Q')$ and a'° is the image of a' by that isomorphism, and it follows from equalities 5.3.5 that the image of the unity element of $A^{\cdot\cdot}_{\hat{\alpha}}$ in $\mathrm{Ind}_{P^{\cdot\cdot}}^{G\times G'}(B^{\cdot\cdot}_{\gamma^{\cdot\cdot}})$ by embedding 5.2.1 belongs to $\mathrm{Ind}_{P^{\cdot\cdot}}^{G\times G'}(B^{\cdot\cdot}_{\gamma^{\cdot\cdot}})^{G\times G'}_{Q^{\cdot\cdot}}$ which forces $Q^{\cdot\cdot}=P^{\cdot\cdot}$, so $Q'=P'$ (actually, the localness of γ' is also a consequence of Remark 5.9 below).

Finally, equalities 5.3.2 are an immediate consequence of equalities 5.2.3.

5.5 According to Proposition 5.3, to find the local pointed groups on $\hat{A}_{\hat{\alpha}}$ which are maximal contained in $G_{\hat{\alpha}}$ for any point $\hat{\alpha}$ of G on $\hat{A}$, it suffices to consider the local pointed groups $P^{\cdot\cdot}_{\gamma^{\cdot\cdot}}$ on $B^{\cdot\cdot}$ which are maximal contained in $(G\times G')_{\alpha^{\cdot\cdot}}$ for any point $\alpha^{\cdot\cdot}$ of $G\times G'$ on $A^{\cdot\cdot}$, and then to determine the local points of $P=\rho(P^{\cdot\cdot})$ on $\mathrm{Ind}_\sigma\Big(B^{\cdot\cdot}_{\gamma^{\cdot\cdot}}\otimes_{\mathcal{O}}\mathrm{Res}_{\sigma'}\big(\mathrm{Res}_{P'}^{G'}(\mathcal{O}G')\big)\Big)$ as above. In fact, we prove in Section 16 below that the analogous claim for all the local pointed groups on $\hat{A}$ remains true provided we consider all the local pointed groups on $B^{\cdot\cdot}$. Consequently, consider a local pointed group $Q^{\cdot\cdot}_{\delta^{\cdot\cdot}}$ on $B^{\cdot\cdot}$, set $Q=\rho(Q^{\cdot\cdot})$, $Q'=\rho'(Q^{\cdot\cdot})$ and $1\times U'=Q^{\cdot\cdot}\cap(1\times K')$ for a suitable subgroup U' of K', and denote respectively by

$\tau : Q^{\cdot\cdot} \longrightarrow Q$, $\tau' : Q^{\cdot\cdot} \longrightarrow Q'$ and $\zeta' : Q^{\cdot\cdot} \longrightarrow G'$ the group homomorphisms determined by ρ, ρ' and ρ'; in Theorem 5.8 below, we give a classification of the local points of Q on (cf. 3.3)

5.5.1 $$\mathrm{Ind}_\tau\big(B^{\cdot\cdot}_{\delta^{\cdot\cdot}} \underset{\mathcal{O}}{\otimes} \mathrm{Res}_{\zeta'}(\mathcal{O}G')\big) = \Big(\mathcal{O} \underset{\mathcal{O}(1\times U')}{\otimes} \big(B^{\cdot\cdot}_{\delta^{\cdot\cdot}} \underset{\mathcal{O}}{\otimes} \mathrm{Res}_{\zeta'}(\mathcal{O}G')\big)\Big)^{1\times U'} .$$

We start by a lemma concerning a slightly more general situation than we need here, which will be useful again in Section 17.

Lemma 5.6 *With the notation above, let $C^{\cdot\cdot}$ be an $\mathcal{O}Q^{\cdot\cdot}$-interior algebra, X' a set of representatives for $U'\backslash G'$ in G' and $\{c^{\cdot\cdot}_{x'}\}_{x'\in X'}$ a family of elements of $C^{\cdot\cdot}$. Consider a subgroup R of Q and set $R^{\cdot\cdot} = \tau^{-1}(R)$ and $N' = \pi'\big(N_{1\times G'}(R^{\cdot\cdot})\big)$. If $R^{\cdot\cdot}$ fixes $a = \Sigma_{x'\in X'} 1\otimes (c^{\cdot\cdot}_{x'}\otimes x')$ in $\mathcal{O}\otimes_{\mathcal{O}(1\times U')}\big(C^{\cdot\cdot}\otimes_{\mathcal{O}}\mathrm{Res}_{\zeta'}(\mathcal{O}G')\big)$ then, for any $x'\in X'\cap N'$, $R^{\cdot\cdot}$ fixes $1\otimes(c^{\cdot\cdot}_{x'}\otimes x')$ and in $\Big(\mathrm{Ind}_\tau\big(C^{\cdot\cdot}\otimes_{\mathcal{O}}\mathrm{Res}_{\zeta'}(\mathcal{O}G')\big)\Big)(R)$ we have*

5.6.1 $$\mathrm{Br}_R(a) = \sum_{x'\in X'\cap N'} \mathrm{Br}_R\big(1\otimes(c^{\cdot\cdot}_{x'}\otimes x')\big) .$$

Proof: Denoting by γ' a set of representatives for $U'\backslash G'/U'$ in X' and, for any $y'\in Y'$, setting $\mathcal{O}(U'y'U') = \Sigma_{u',v'\in U'}\mathcal{O}u'y'v'$ in $\mathcal{O}G'$, we have clearly the following R-stable direct sum decomposition

5.6.2 $$\mathrm{Ind}_\tau\big(C^{\cdot\cdot}\underset{\mathcal{O}}{\otimes}\mathrm{Res}_{\zeta'}(\mathcal{O}G')\big) = \bigoplus_{y'\in Y'}\Big(\mathcal{O}\otimes\big(C^{\cdot\cdot}\otimes\mathcal{O}(U'y'U')\big)\Big)^{1\times U'}$$

and R stabilizes the direct summand determined by $y'\in Y'$ if and only if we have $(U'y'U')^{u'} = U'y'U'$ for any $u'\in R' = \pi'(R^{\cdot\cdot})$; since $U'\subset R'$, it is easily checked that this condition is equivalent to $[y', R']\subset U'$ and then we have $[(1,y'), R^{\cdot\cdot}]\subset 1\times U'\subset R^{\cdot\cdot}$, so that y' belongs to N'. Hence, if $R^{\cdot\cdot}$ fixes a then it fixes also $b = \Sigma_{x'\in X'\cap N'} 1\otimes(c^{\cdot\cdot}_{x'}\otimes x')$ and we have already $\mathrm{Br}_R(a) = \mathrm{Br}_R(b)$; more explicitely, for any $(u,u')\in R^{\cdot\cdot}$, we get

5.6.3 $$\begin{aligned} b &= \sum_{x'\in X'\cap N'} 1\otimes(c^{\cdot\cdot}_{x'}\otimes x')^{(u,u')} \\ &= \sum_{x'\in X'\cap N'} 1\otimes\big((c^{\cdot\cdot}_{x'})^{(u,u')}\otimes[u', x'^{-1}]x'\big) \end{aligned}$$

and, since $[N', R']\subset U'$, we obtain

5.6.4 $$b = \sum_{x'\in X'\cap N'} 1\otimes\big((u^{-1}, u'^{-1})^{(1,x'^{-1})}\cdot c^{\cdot\cdot}_{x'}\cdot(u,u')\otimes x'\big)$$

which is equivalent to $(u, x'u'x'^{-1})\cdot c^{\cdot\cdot}_{x'} = c^{\cdot\cdot}_{x'}\cdot(u,u')$ for any $x'\in X'\cap N'$; that is to say, for any $x'\in X'\cap N'$, $R^{\cdot\cdot}$ fixes $1\otimes(c^{\cdot\cdot}_{x'}\otimes x')$.

5.7 Recall that we identify $Q^{\cdot\cdot}_{\delta^{\cdot\cdot}}$ with the corresponding local pointed groups on $A^{\cdot\cdot}$ and on $\mathrm{Ind}_{Q^{\cdot\cdot}}^{Q\times G'}(B^{\cdot\cdot}_{\delta^{\cdot\cdot}})$ (cf. 2.11) and that in both cases we get the same normalizer $N_{Q\times G'}(Q^{\cdot\cdot}_{\delta^{\cdot\cdot}})$ (cf. 2.13.2); then, consider the canonical $\mathcal{O}C_{Q\times G'}(Q^{\cdot\cdot})$-interior $N_{Q\times G'}(Q^{\cdot\cdot}_{\delta^{\cdot\cdot}})$-algebra homomorphisms (cf. 2.8.1 and 2.12.3)

5.7.1 $$\mathrm{Ind}_{Q^{\cdot\cdot}}^{Q\times G'}(B^{\cdot\cdot}_{\delta^{\cdot\cdot}})^{Q^{\cdot\cdot}} \xrightarrow{s_{\delta^{\cdot\cdot}}} \left(\mathrm{Ind}_{Q^{\cdot\cdot}}^{Q\times G'}(B^{\cdot\cdot}_{\delta^{\cdot\cdot}})\right)(Q^{\cdot\cdot}_{\delta^{\cdot\cdot}}) \cong \mathrm{End}_{k}\left(k_*\hat{\bar{N}}_{Q\times G'}(Q^{\cdot\cdot}_{\delta^{\cdot\cdot}})\right)$$

and recall that their composition maps the $\mathcal{O}$-algebra

5.7.2 $$\mathrm{Ind}_{Q^{\cdot\cdot}}^{Q\times G'}(B^{\cdot\cdot}_{\delta^{\cdot\cdot}})^{Q\times G'} = \mathrm{Ind}_{Q^{\cdot\cdot}}^{Q\times G'}(B^{\cdot\cdot}_{\delta^{\cdot\cdot}})_{Q^{\cdot\cdot}}^{Q\times G'}$$

surjectively onto the k-algebra (cf. 2.9.2)

5.7.3 $$\mathrm{End}_{k}\left(k_*\hat{\bar{N}}_{Q\times G'}(Q^{\cdot\cdot}_{\delta^{\cdot\cdot}})\right)^{\bar{N}_{Q\times G'}(Q^{\cdot\cdot}_{\delta^{\cdot\cdot}})} \cong k_*\hat{\bar{N}}_{Q\times G'}(Q^{\cdot\cdot}_{\delta^{\cdot\cdot}})^{\circ} \quad .$$

Moreover, recall that the k^*-group $\hat{\bar{N}}_{Q\times G'}(Q^{\cdot\cdot}_{\delta^{\cdot\cdot}})^{\circ}$ can be described as the k^*-group of pairs $\left(\overline{(u,x')}, \overline{b^{\cdot\cdot}}\right)$, where $\overline{(u,x')}$ is the image of an element (u,x') of $N_{Q\times G'}(Q^{\cdot\cdot}_{\delta^{\cdot\cdot}})$ in $\bar{N}_{Q\times G'}(Q^{\cdot\cdot}_{\delta^{\cdot\cdot}})$ and, denoting by $\varphi_{(u,x')}$ the automorphism of $Q^{\cdot\cdot}$ mapping $(v,v')\in Q^{\cdot\cdot}$ on $(uvu^{-1}, x'v'x'^{-1})$, $\overline{b^{\cdot\cdot}}$ is the image of $b^{\cdot\cdot}\in N_{B^{\cdot\cdot}_{\delta^{\cdot\cdot}}}^{\varphi_{(u,x')}}(Q^{\cdot\cdot})^*$ in

$$N_{B^{\cdot\cdot}_{\delta^{\cdot\cdot}}}(Q^{\cdot\cdot})^*/Q^{\cdot\cdot}\cdot\left(1+J\left((B^{\cdot\cdot}_{\delta^{\cdot\cdot}})^{Q^{\cdot\cdot}}\right)\right)$$

(cf. 2.13.6). On the other hand, notice that the Hecke $\mathcal{O}Q$-interior algebra isomorphism $H_{Q,G'}(B^{\cdot\cdot}_{\delta^{\cdot\cdot}})$ determines an $\mathcal{O}$-algebra isomorphism (cf. 4.4.1)

5.7.4 $$H_{\delta^{\cdot\cdot}} : \mathrm{Ind}_{Q^{\cdot\cdot}}^{Q\times G'}(B^{\cdot\cdot}_{\delta^{\cdot\cdot}})^{Q\times G'} \cong \mathrm{Ind}_{\tau}\left(B^{\cdot\cdot}_{\delta^{\cdot\cdot}} \underset{\mathcal{O}}{\otimes} \mathrm{Res}_{\zeta'}(\mathcal{O}G')\right)^{Q}$$

and therefore induces a bijection between the sets of points of $Q\times G'$ on $\mathrm{Ind}_{Q^{\cdot\cdot}}^{Q\times G'}(B^{\cdot\cdot}_{\delta^{\cdot\cdot}})$ and of Q on $\mathrm{Ind}_{\tau}\left(B^{\cdot\cdot}_{\delta^{\cdot\cdot}} \otimes_{\mathcal{O}} \mathrm{Res}_{\zeta'}(\mathcal{O}G')\right)$.

Theorem 5.8 *With the notation above, the composed map $s_{\delta^{\cdot\cdot}}\circ(H_{\delta^{\cdot\cdot}})^{-1}$ induces a surjective k-algebra homomorphism*

5.8.1 $$\left(\mathrm{Ind}_{\tau}\left(B^{\cdot\cdot}_{\delta^{\cdot\cdot}} \underset{\mathcal{O}}{\otimes} \mathrm{Res}_{\zeta'}(\mathcal{O}G')\right)\right)(Q) \longrightarrow k_*\hat{\bar{N}}_{Q\times G'}(Q^{\cdot\cdot}_{\delta^{\cdot\cdot}})^{\circ}$$

which has a nilpotent kernel. In particular, this homomorphism determines a bijection between the set of local points of Q on $\mathrm{Ind}_{\tau}\left(B^{\cdot\cdot}_{\delta^{\cdot\cdot}} \otimes_{\mathcal{O}} \mathrm{Res}_{\zeta'}(\mathcal{O}G')\right)$ and the set of points $\varepsilon^{\cdot\cdot}$ of $Q\times G'$ on $\mathrm{Ind}_{Q^{\cdot\cdot}}^{Q\times G'}(B^{\cdot\cdot}_{\delta^{\cdot\cdot}})$ such that $Q^{\cdot\cdot}_{\delta^{\cdot\cdot}} \subset (Q\times G')_{\varepsilon^{\cdot\cdot}}$. Moreover, setting $N' = \pi'\left(N_{1\times G'}(Q^{\cdot\cdot})\right)$, we have

5.8.2 $$\left(\mathrm{Ind}_{\tau}\left(B^{\cdot\cdot}_{\delta^{\cdot\cdot}} \underset{\mathcal{O}}{\otimes} \mathrm{Res}_{\zeta'}(\mathcal{O}G')\right)\right)(Q) = \sum_{x'\in N'} \mathrm{Br}_{Q}\left(\mathcal{O}\otimes\left(N_{B^{\cdot\cdot}_{\delta^{\cdot\cdot}}}^{\varphi_{(1,x')}}(Q^{\cdot\cdot})\otimes x'\right)\right)$$

and, for any $x' \in N'$ and any $b^{\cdot\cdot} \in N_{B_{\delta^{\cdot\cdot}}}^{\varphi(1,x')}(Q^{\cdot\cdot})$, homomorphism 5.8.1 maps $Br_Q\big(1 \otimes (b^{\cdot\cdot} \otimes x')\big)$ either on $\big(\overline{(1,x')}, \overline{b^{\cdot\cdot}}\big)$ whenever $b^{\cdot\cdot}$ is invertible in $B^{\cdot\cdot}_{\delta^{\cdot\cdot}}$ and $(1,x')$ fixes $\delta^{\cdot\cdot}$, or on zero otherwise.

Remark 5.9 Notice that, denoting by $\overline{C_{G'}(Q')}$ the image of $1 \times C_{G'}(Q')$ in $\hat{\hat{N}}_{Q\times G'}(Q^{\cdot\cdot}_{\delta^{\cdot\cdot}})$ and by $\overline{\mathrm{Br}}_{Q'} : (\mathcal{O}G')^{Q'} \longrightarrow k\overline{C_{G'}(Q')}$ the composition of $\mathrm{Br}_{Q'}$ and the canonical map $(\mathcal{O}G')(Q') \cong kC_{G'}(Q') \longrightarrow k\overline{C_{G'}(Q')}$ (cf. 2.14.1), for any $a' \in (\mathcal{O}G')^{Q'}$ homomorphism 5.8.1 maps $\mathrm{Br}_Q\big(1 \otimes (1 \otimes a')\big)$ on $\overline{\mathrm{Br}}_{Q'}(a')$; indeed, by Lemma 5.6, we may assume that $a' = \mathrm{Tr}^{Q'}_{C_{Q'}(x')}(x')$ for some $x' \in N'$ and then it is easily checked that $1 \otimes (1 \otimes a') = 1 \otimes (b^{\cdot\cdot} \otimes x')$, where $b^{\cdot\cdot}$ is the image in $B^{\cdot\cdot}_{\delta^{\cdot\cdot}}$ of the sum $\Sigma_{u' \in [x'^{-1}, Q']}(1,u')$ in $\mathcal{O}(1 \times U')$, which is invertible only if $[x'^{-1}, Q'] = \{1\}$. In particular, if δ' is a nonlocal point of Q' on $\mathcal{O}G'$ then we have $\mathrm{Br}_Q\big(1\otimes(1\otimes\delta')\big) = \{0\}$.

Remark 5.10 Since $N_{G\times G'}(Q^{\cdot\cdot}_{\delta^{\cdot\cdot}})$ normalizes $N_{Q\times G'}(Q^{\cdot\cdot}_{\delta^{\cdot\cdot}})$ and acts on $(A^{\cdot\cdot})(Q^{\cdot\cdot}_{\delta^{\cdot\cdot}})$, it acts also on $\hat{\hat{N}}_{Q\times G'}(Q^{\cdot\cdot}_{\delta^{\cdot\cdot}})$ (cf. 2.8) and therefore $E_{G\times G'}(Q^{\cdot\cdot}_{\delta^{\cdot\cdot}})$ has a canonical *exterior* action on the k-algebra $k_*\hat{\hat{N}}_{Q\times G'}(Q^{\cdot\cdot}_{\delta^{\cdot\cdot}})^\circ$ (cf. 2.4). On the other hand, the set $\hat{N}^{\cdot\cdot}$ of pairs $\big((x,x'), b^{\cdot\cdot}\big)$, where (x,x') runs on $N_{G\times G'}(Q^{\cdot\cdot}_{\delta^{\cdot\cdot}})$ and, for such an element, $b^{\cdot\cdot}$ runs on $N_{B_{\delta^{\cdot\cdot}}}^{\varphi(x,x')}(Q^{\cdot\cdot})^*$ (up to an obvious extension of our notation), has an evident k^*-group structure and we have a k^*-group and a group homomorphisms

5.10.1 $$\hat{N}^{\cdot\cdot} \longrightarrow N_{\mathrm{Ind}_\tau(B^{\cdot\cdot}_{\delta^{\cdot\cdot}} \otimes_{\mathcal{O}} \mathrm{Res}_{\zeta'}(\mathcal{O}G'))}(Q)^* \quad \text{and} \quad \hat{N}^{\cdot\cdot} \longrightarrow E_{G\times G'}(Q^{\cdot\cdot}_{\delta^{\cdot\cdot}})$$

respectively mapping $\big((x,x'), b^{\cdot\cdot}\big) \in \hat{N}^{\cdot\cdot}$ on $1 \otimes (b^{\cdot\cdot} \otimes x')$ and on $\tilde{\varphi}_{(x,x')}$, the second being surjective since we have (cf. 2.13.2 and 2.13.3)

5.10.2 $$E_{G\times G'}(Q^{\cdot\cdot}_{\delta^{\cdot\cdot}}) = F_{A^{\cdot\cdot}}(Q^{\cdot\cdot}_{\delta^{\cdot\cdot}}) \cap E_{G\times G'}(Q^{\cdot\cdot}) \quad ;$$

it follows easily that $E_{G\times G'}(Q^{\cdot\cdot}_{\delta^{\cdot\cdot}})$ has also a canonical *exterior* action on the k-algebra $\Big(\mathrm{Ind}_\tau\big(B^{\cdot\cdot}_{\delta^{\cdot\cdot}} \otimes_{\mathcal{O}} \mathrm{Res}_{\zeta'}(\mathcal{O}G')\big)\Big)(Q)$. Notice that these exterior actions are compatible with the exomorphism determined by homomorphism 5.8.1.

Proof: Let X' be a set of representatives for $U'\backslash G'$ in G'; for any subgroup R of Q, the inverse of the Hecke $\mathcal{O}Q$-interior algebra isomorphism $H_{Q,G'}(B^{\cdot\cdot}_{\delta^{\cdot\cdot}})$ determines an $\mathcal{O}$-algebra isomorphism (cf. 4.4.1)

5.8.3 $$\mathrm{Ind}_\tau\big(B^{\cdot\cdot}_{\delta^{\cdot\cdot}} \underset{\mathcal{O}}{\otimes} \mathrm{Res}_{\zeta'}(\mathcal{O}G')\big)^R \cong \mathrm{Ind}_{Q^{\cdot\cdot}}^{Q\times G'}(B^{\cdot\cdot}_{\delta^{\cdot\cdot}})^{R\times G'}$$

which, for any family $\{b^{\cdot\cdot}_{x'}\}_{x'\in X'}$ of elements of $B^{\cdot\cdot}_{\delta^{\cdot\cdot}}$ such that $R^{\cdot\cdot} = \tau^{-1}(R)$ fixes $a = \Sigma_{x'\in X'} 1 \otimes (b^{\cdot\cdot}_{x'} \otimes x')$ in $\mathcal{O} \otimes_{\mathcal{O}(1\times U')} \big(B^{\cdot\cdot}_{\delta^{\cdot\cdot}} \otimes_{\mathcal{O}} \mathrm{Res}_{\zeta'}(\mathcal{O}G')\big)$, maps a on $a^{\cdot\cdot} = \mathrm{Tr}^{1\times G'}_{1\times U'}\big(\Sigma_{x'\in X'}(1,x')^{-1} \otimes b^{\cdot\cdot}_{x'} \otimes (1,1)\big)$; in particular, it is easily checked that it maps $\mathrm{Tr}_R^Q(a)$ on $\mathrm{Tr}_{R\times G'}^{Q\times G'}(a^{\cdot\cdot}) = \mathrm{Tr}_{R^{\cdot\cdot}}^{Q^{\cdot\cdot}}(a^{\cdot\cdot})$; hence, it follows from 5.7 that $s_{\delta^{\cdot\cdot}}$ and isomorphism 5.8.3 determine a surjective k-algebra homomorphism

5.8.4 $$\Big(\mathrm{Ind}_\tau\big(B_{\delta^{\cdot\cdot}}^{\cdot\cdot} \otimes_{\mathcal{O}} \mathrm{Res}_{\zeta'}(\mathcal{O}G')\big)\Big)(Q) \longrightarrow k_*\hat{\bar{N}}_{Q\times G'}(Q_{\delta^{\cdot\cdot}}^{\cdot\cdot})^\circ \quad .$$

On the other hand, when $R = Q$ it follows from Lemma 5.6 that, for any $x' \in X'$, $b_{x'}^{\cdot\cdot}$ belongs to $N_{B_{\delta^{\cdot\cdot}}^{\cdot\cdot}}^{\varphi(1,x')}(Q^{\cdot\cdot})$ and that we have

5.8.5 $$\mathrm{Br}_Q(a) = \sum_{x'\in X'\cap N'} \mathrm{Br}_Q\big(1\otimes(b_{x'}^{\cdot\cdot}\otimes x')\big)$$

which proves already equality 5.8.2; moreover, for any $x' \in X'\cap N'$, isomorphism 5.8.3 maps $1\otimes(b_{x'}^{\cdot\cdot}\otimes x')$ on

5.8.6 $$a_{x',b_{x'}^{\cdot\cdot}}^{\cdot\cdot} = \mathrm{Tr}_{Q^{\cdot\cdot}}^{Q\times G'}\big((1,x')^{-1}\otimes b_{x'}^{\cdot\cdot}\otimes(1,1)\big) \quad ;$$

hence, since $(1,x')^{-1}\otimes b_{x'}^{\cdot\cdot}\otimes(1,1)$ belongs to $\big((1,x')^{-1}\otimes B_{\delta^{\cdot\cdot}}^{\cdot\cdot}\otimes(1,1)\big)^{Q^{\cdot\cdot}}$, $s_{\delta^{\cdot\cdot}}$ maps $a_{x',b_{x'}^{\cdot\cdot}}^{\cdot\cdot}$ on zero whenever $(1,x')$ does not fix $\delta^{\cdot\cdot}$ whereas, if $(1,x')$ normalizes $Q_{\delta^{\cdot\cdot}}^{\cdot\cdot}$, there is $b^{\cdot\cdot} \in N_{B_{\delta^{\cdot\cdot}}^{\cdot\cdot}}^{\varphi(1,x')}(Q^{\cdot\cdot})^*$ (cf. 2.13.2, 2.13.3 and [20, Proposition 2.14]) and therefore $(b^{\cdot\cdot})^{-1}b_{x'}^{\cdot\cdot}$ belongs to $(B_{\delta^{\cdot\cdot}}^{\cdot\cdot})^{Q^{\cdot\cdot}}$, so that we have

5.8.7 $$\begin{aligned} 0 &\neq s_{\delta^{\cdot\cdot}}\big((1,x')^{-1}\otimes b_{x'}^{\cdot\cdot}\otimes(1,1)\big)\\ &= s_{\delta^{\cdot\cdot}}\big((1,x')^{-1}\otimes b^{\cdot\cdot}\otimes(1,1)\big)s_{\delta^{\cdot\cdot}}\big((1,1)\otimes(b^{\cdot\cdot})^{-1}b_{x'}^{\cdot\cdot}\otimes(1,1)\big)\end{aligned}$$

if and only if $b_{x'}^{\cdot\cdot}$ is invertible in $B_{\delta^{\cdot\cdot}}^{\cdot\cdot}$. Consequently, the set of elements

5.8.8 $$s_{\delta^{\cdot\cdot}}(a_{x',b^{\cdot\cdot}}^{\cdot\cdot}) = \mathrm{Tr}_{Q^{\cdot\cdot}}^{N_{Q\times G'}(Q_{\delta^{\cdot\cdot}}^{\cdot\cdot})}\Big(s_{\delta^{\cdot\cdot}}\big((1,x')^{-1}\otimes b^{\cdot\cdot}\otimes(1,1)\big)\Big) \quad ,$$

where x' runs over $\pi'\big(N_{1\times G'}(Q_{\delta^{\cdot\cdot}}^{\cdot\cdot})\big)$ and, for such an x', $b^{\cdot\cdot}$ runs over $N_{B_{\delta^{\cdot\cdot}}^{\cdot\cdot}}^{\varphi(1,x')}(Q^{\cdot\cdot})^*$, generates the k-algebra (cf. 5.7)

5.8.9 $$\big(\mathrm{Ind}_{Q^{\cdot\cdot}}^{Q\times G'}(B_{\delta^{\cdot\cdot}}^{\cdot\cdot})\big)(Q_{\delta^{\cdot\cdot}}^{\cdot\cdot})^{\bar{N}_{Q\times G'}(Q_{\delta^{\cdot\cdot}}^{\cdot\cdot})} \cong k_*\hat{\bar{N}}_{Q\times G'}(Q_{\delta^{\cdot\cdot}}^{\cdot\cdot})^\circ \quad ;$$

but it is easily checked that this set is a k^*-subgroup of $\big(k_*\hat{\bar{N}}_{Q\times G'}(Q_{\delta^{\cdot\cdot}}^{\cdot\cdot})^\circ\big)^*$ with k^*-quotient of order at most $|\bar{N}_{Q\times G'}(Q_{\delta^{\cdot\cdot}}^{\cdot\cdot})|$ and that the map from it to $\hat{\bar{N}}_{Q\times G'}(Q_{\delta^{\cdot\cdot}}^{\cdot\cdot})^\circ$ sending $s_{\delta^{\cdot\cdot}}(a_{x',b^{\cdot\cdot}}^{\cdot\cdot})$ to $\big(\overline{(1,x')},\overline{b^{\cdot\cdot}}\big)$ is a surjective k^*-group homomorphism; in conclusion, this map is a k^*-group isomorphism and we get the announced k-algebra homomorphism 5.8.1 which is clearly surjective.

Finally, setting $T^{\cdot\cdot} = \zeta'^{-1}(N')$ and $T = \tau(T^{\cdot\cdot})$, and denoting respectively by $\xi' : T^{\cdot\cdot} \longrightarrow N'$ and $\eta : T^{\cdot\cdot} \longrightarrow T$ the group homomorphisms determined by ζ' and

τ, since $1 \times U' \subset T^{\cdot\cdot}$, Proposition 3.21 and the inclusion $N' \subset G'$ determine an injective $\mathcal{O}T$-interior algebra homomorphism

5.8.10
$$\mathrm{Ind}_\eta\big(\mathrm{Res}^{Q^{\cdot\cdot}}_{T^{\cdot\cdot}}(B^{\cdot\cdot}_{\delta^{\cdot\cdot}}) \underset{\mathcal{O}}{\otimes} \mathrm{Res}_{\xi'}(\mathcal{O}N')\big) \longrightarrow \mathrm{Res}^{Q}_{T}\Big(\mathrm{Ind}_\tau\big(B^{\cdot\cdot}_{\delta^{\cdot\cdot}} \underset{\mathcal{O}}{\otimes} \mathrm{Res}_{\zeta'}(\mathcal{O}G')\big)\Big) \quad ;$$

but, since Q' normalizes N', the left end of homomorphism 5.8.10 is actually an $\mathcal{O}T$-interior Q-algebra and the map is an $\mathcal{O}T$-interior Q-algebra homomorphism which, according to Lemma 5.6, induces a k-algebra isomorphism between the corresponding Brauer sections at Q (cf. 2.2). Similarly, since $Q^{\cdot\cdot}{\cdot}(1 \times N') = Q^{\cdot\cdot}{\cdot}(T \times N')$, by 2.6 the canonical isomorphism (cf. Proposition 3.12)

5.8.11
$$\mathrm{Ind}^{T\times N'}_{T^{\cdot\cdot}}\big(\mathrm{Res}^{Q^{\cdot\cdot}}_{T^{\cdot\cdot}}(B^{\cdot\cdot}_{\delta^{\cdot\cdot}})\big) \cong \mathrm{Res}^{Q}_{T}\big(\mathrm{Ind}^{Q^{\cdot\cdot}\cdot(1\times N')}_{Q^{\cdot\cdot}}(B^{\cdot\cdot}_{\delta^{\cdot\cdot}})\big)$$

can be considered as an $\mathcal{O}(T \times N')$-interior $Q^{\cdot\cdot}{\cdot}(1 \times N')$-algebra isomorphism and it is easily checked that the Hecke $\mathcal{O}T$-interior algebra isomorphism (cf. 4.4.1)

5.8.12
$$\mathrm{Ind}^{T\times N'}_{T^{\cdot\cdot}}\big(\mathrm{Res}^{Q^{\cdot\cdot}}_{T^{\cdot\cdot}}(B^{\cdot\cdot}_{\delta^{\cdot\cdot}})\big)^{1\times N'} \cong \mathrm{Ind}_\eta\big(\mathrm{Res}^{Q^{\cdot\cdot}}_{T^{\cdot\cdot}}(B^{\cdot\cdot}_{\delta^{\cdot\cdot}}) \underset{\mathcal{O}}{\otimes} \mathrm{Res}_{\xi'}(\mathcal{O}N')\big)$$

is also a Q-algebra one. Consequently, in order to prove that the kernel of homomorphism 5.8.1 is nilpotent, setting $N^{\cdot\cdot} = Q^{\cdot\cdot}{\cdot}(1 \times N')$ it suffices to prove that so is the kernel of the homomorphism determined by $s_{\delta^{\cdot\cdot}}$ (cf. 2.8.1)

5.8.13
$$k \underset{\mathcal{O}}{\otimes} \mathrm{Ind}^{N^{\cdot\cdot}}_{Q^{\cdot\cdot}}(B^{\cdot\cdot}_{\delta^{\cdot\cdot}})^{N^{\cdot\cdot}} \longrightarrow \big(\mathrm{Ind}^{N^{\cdot\cdot}}_{Q^{\cdot\cdot}}(B^{\cdot\cdot}_{\delta^{\cdot\cdot}})\big)(Q^{\cdot\cdot}_{\delta^{\cdot\cdot}})^{\bar N_{N^{\cdot\cdot}}(Q^{\cdot\cdot}_{\delta^{\cdot\cdot}})}$$

which results from the following lemma.

Lemma 5.11 *Let H be a finite group, L a normal subgroup of H and D an $\mathcal{O}L$-interior algebra where L has a unique point ε and ε has multiplicity one. There is a unique $k_*\hat{\bar N}_H(L_\varepsilon)$-interior algebra isomorphism*

5.11.1
$$\big(\mathrm{Ind}^H_L(D)\big)(L_\varepsilon) \cong \mathrm{Ind}^{\hat{\bar N}_H(L_\varepsilon)}_{k^*}(k)$$

mapping $s_\varepsilon(1 \otimes 1 \otimes 1)$ on $1 \otimes 1 \otimes 1$, and the $\mathcal{O}$-algebra homomorphism

5.11.2
$$k \underset{\mathcal{O}}{\otimes} \mathrm{Ind}^H_L(D)^H \longrightarrow \big(\mathrm{Ind}^H_L(D)\big)(L_\varepsilon)^{\bar N_H(L_\varepsilon)}$$

determined by s_ε is surjective and has a nilpotent kernel.

Remark 5.12 Notice that, by 2.6.5, we have a canonical $k_*\hat{\bar N}_H(L_\varepsilon)$-interior algebra isomorphism

5.12.1
$$\mathrm{Ind}^{\hat{\bar N}_H(L_\varepsilon)}_{k^*}(k) \cong \mathrm{End}_k\big(k_*\hat{\bar N}_H(L_\varepsilon)\big)$$

and therefore a k-algebra isomorphism (cf. 2.12.3)

5.12.2
$$\big(\mathrm{Ind}^H_L(D)\big)(L_\varepsilon)^{\bar N_H(L_\varepsilon)} \cong k_*\hat{\bar N}_H(L_\varepsilon)^\circ \quad .$$

Proof: Set $C = \mathrm{Ind}_L^H(D)$; since $1 = \mathrm{Tr}_L^H(1\otimes 1\otimes 1)$ and, for any $x \in H$, $x^{-1}\otimes 1\otimes x$ belongs to ε^x which is still a point of L on C, the set $\{\varepsilon^x\}_{x\in H}$ contains all the points of L on C, and in the k-algebra $C(L_\varepsilon)$ we get $s_\varepsilon(1) = \mathrm{Tr}_1^{\bar{N}_H(L_\varepsilon)}\big(s_\varepsilon(1\otimes 1\otimes 1)\big)$. Now, since $\bar{N}_H(L_\varepsilon)$ stabilizes the set $\big\{s_\varepsilon(x^{-1}\otimes 1\otimes x)\big\}_{x\in N_H(L_\varepsilon)}$ and $C(L_\varepsilon)$ is a $k_*\hat{\bar{N}}_H(L_\varepsilon)$-interior algebra (cf. 2.8), up to replace $\hat{\bar{N}}_H(L_\varepsilon)$ by a finite subgroup $\bar{N}$ covering $\bar{N}_H(L_\varepsilon)$ (cf. [22, Lemma 5.5]), statement 2.6.4 applies and therefore there is a unique $k_*\hat{\bar{N}}_H(L_\varepsilon)$-interior algebra isomorphism

5.11.3 $$C(L_\varepsilon) \cong \mathrm{Ind}_{k^*}^{\hat{\bar{N}}_H(L_\varepsilon)}(k)$$

mapping $s_\varepsilon(1\otimes 1\otimes 1)$ on $1\otimes 1\otimes 1$. Moreover, for any $a \in C^L{\cdot}\varepsilon{\cdot}C^L$, it is easily checked that

5.11.4 $$s_\varepsilon\big(\mathrm{Tr}_L^H(a)\big) = \mathrm{Tr}_1^{\bar{N}_H(L_\varepsilon)}\big(s_\varepsilon(a)\big)$$

and therefore, since C_L^H and $C(L_\varepsilon)_1^{\bar{N}_H(L_\varepsilon)}$ contain the respective unity elements and we have $s_\varepsilon(C^L{\cdot}\varepsilon{\cdot}C^L) = C(L_\varepsilon)$, we get

5.11.5 $$s_\varepsilon(C^H) = C(L_\varepsilon)^{\bar{N}_H(L_\varepsilon)} \quad .$$

Finally, for any $x \in H$, we have

5.11.6 $$C^H \cap \mathrm{Ker}(s_\varepsilon) = \big(C^H \cap \mathrm{Ker}(s_\varepsilon)\big)^x = C^H \cap \mathrm{Ker}(s_{\varepsilon^x})$$

and, since $\{\varepsilon^y\}_{y\in H}$ contains all the points of L on C, we obtain

5.11.7 $$C^H \cap \mathrm{Ker}(s_\varepsilon) \subset \bigcap_{x\in H} \mathrm{Ker}(s_{\varepsilon^x}) = J(C^L) \quad .$$

5.13 As embedding 5.3.1 justifies, we are actually interested in the local points of Q on $D = \mathrm{Ind}_\tau\Big(B_{\ddot{\delta}}\otimes_{\mathcal{O}} \mathrm{Res}_{\tau'}\big((\mathcal{O}G')_{\delta'}\big)\Big)$, where δ' is a local point of Q' on $\mathcal{O}G'$; but, choosing $j' \in \delta'$ and $(\mathcal{O}G')_{\delta'} = j'\mathcal{O}G'j'$, it is clear that

5.13.1 $$D = \big(1\otimes(1\otimes j')\big)\mathrm{Ind}_\tau\big(B_{\ddot{\delta}} \underset{\mathcal{O}}{\otimes} \mathrm{Res}_{\zeta'}(\mathcal{O}G')\big)\big(1\otimes(1\otimes j')\big)$$

and, by Remark 5.9, homomorphism 5.8.1 maps $\mathrm{Br}_Q\big(1\otimes(1\otimes j')\big)$ on $\overline{\mathrm{Br}}_{Q'}(j')$; hence, by Theorem 5.8, homomorphism 5.8.1 induces a bijection between that set of local points and the set of points of the k-algebra

5.13.2 $$\overline{\mathrm{Br}}_{Q'}(j')\big(k_*\hat{\bar{N}}_{Q\times G'}(Q_{\ddot{\delta}})^\circ\big)\overline{\mathrm{Br}}_{Q'}(j') \quad ;$$

in our last result we obtain a refinement of this fact in terms of $\hat{E}_{Q\times G'}(Q_{\ddot{\delta}})$ and $\hat{E}_{G'}(Q'_{\delta'})$. As in 5.7 above, the k^*-group $\hat{E}_{G'}(Q'_{\delta'})^\circ$ can be described as the k^*-group of pairs $(\tilde{\varphi}', \bar{a}')$, where $\tilde{\varphi}' \in E_{G'}(Q'_{\delta'})$ and $\bar{a}'$ is the image in

$$N_{(\mathcal{O}G')_{\delta'}}(Q')^*/Q'{\cdot}\Big(1 + J\big((\mathcal{O}G')_{\delta'}^{Q'}\big)\Big)$$

of $a' \in N^{\varphi'}_{(\mathcal{O}G')_{\delta'}}(Q')^*$ for some φ' in $\tilde{\varphi}'$ (cf. 2.13.6) and, for any $x' \in N_{G'}(Q')$, we denote by $\varphi_{x'}$ the automorphism of Q' mapping $u' \in Q'$ on $x'u'x'^{-1}$; moreover, we denote by $\hat{E}_{Q\times G'}(Q^{\cdot\cdot}_{\delta^{\cdot\cdot}}) * \hat{E}_{G'}(Q'_{\delta'})$ the converse image in the central product $\hat{E}_{Q\times G'}(Q^{\cdot\cdot}_{\delta^{\cdot\cdot}}) \otimes \hat{E}_{G'}(Q'_{\delta'})$ (cf. [22, 5.8]) of the subgroup E of the direct product $E_{Q\times G'}(Q^{\cdot\cdot}_{\delta^{\cdot\cdot}}) \times E_{G'}(Q'_{\delta'})$ formed by the pairs $(\tilde{\varphi}^{\cdot\cdot}, \tilde{\varphi}')$ such that $\tilde{\varphi}' \circ \tilde{\tau}' = \tilde{\tau}' \circ \tilde{\varphi}^{\cdot\cdot}$ (actually E is isomorphic to its images by the projection maps).

Corollary 5.14 *With the notation above, let δ' be a local point of Q' on $\mathcal{O}G'$ and set $N' = \pi'\big(N_{1\times G'}(Q^{\cdot\cdot})\big)$. We have*

5.14.1
$$\Big(\mathrm{Ind}_\tau\Big(B^{\cdot\cdot}_{\delta^{\cdot\cdot}} \underset{\mathcal{O}}{\otimes} \mathrm{Res}_{\tau'}\big((\mathcal{O}G')_{\delta'}\big)\Big)\Big)(Q) = \sum_{x'\in N'} \mathrm{Br}_Q\Big(\mathcal{O} \otimes \big(N^{\varphi_{(1,x')}}_{B^{\cdot\cdot}_{\delta^{\cdot\cdot}}}(Q^{\cdot\cdot}) \otimes N^{\varphi_{x'}}_{(\mathcal{O}G')_{\delta'}}(Q')\big)\Big)$$

and there is a unique $\&$-algebra homomorphism

5.14.2
$$\Big(\mathrm{Ind}_\tau\Big(B^{\cdot\cdot}_{\delta^{\cdot\cdot}} \underset{\mathcal{O}}{\otimes} \mathrm{Res}_{\tau'}\big((\mathcal{O}G')_{\delta'}\big)\Big)\Big)(Q) \longrightarrow \&_*\big(\hat{E}_{Q\times G'}(Q^{\cdot\cdot}_{\delta^{\cdot\cdot}}) * \hat{E}_{G'}(Q'_{\delta'})\big)^\circ$$

which, for any $x' \in N'$, any $b^{\cdot\cdot} \in N^{\varphi_{(1,x')}}_{B^{\cdot\cdot}_{\delta^{\cdot\cdot}}}(Q^{\cdot\cdot})$ and any $a' \in N^{\varphi_{x'}}_{(\mathcal{O}G')_{\delta'}}(Q')$, maps $\mathrm{Br}_Q\big(1 \otimes (b^{\cdot\cdot} \otimes a')\big)$ either on $(\tilde{\varphi}_{(1,x')}, \overline{b^{\cdot\cdot}}) \otimes (\tilde{\varphi}_{x'}, \bar{a}')$ whenever $b^{\cdot\cdot}$ and a' are respectively invertible in $B^{\cdot\cdot}_{\delta^{\cdot\cdot}}$ and $(\mathcal{O}G')_{\delta'}$, and moreover $(1, x')$ and x' fix respectively $\delta^{\cdot\cdot}$ and δ', or on zero otherwise. In particular, this homomorphism is surjective and has a nilpotent kernel.

Proof: With the notation in 5.13.1, by 5.8.2 we get

5.14.3
$$D = \sum_{x'\in N'} \mathrm{Br}_Q\Big(\mathcal{O} \otimes \big(N^{\varphi_{(1,x')}}_{B^{\cdot\cdot}_{\delta^{\cdot\cdot}}}(Q^{\cdot\cdot}) \otimes j'x'j'\big)\Big)$$

and it is clear that $j'x'j'$ belongs to $N^{\varphi_{x'}}_{(\mathcal{O}G')_{\delta'}}(Q')$ which proves already equality 5.14.1; moreover, homomorphism 5.8.1 induces a surjective $\&$-algebra homomorphism with nilpotent kernel

5.14.4
$$D(Q) \longrightarrow \overline{\mathrm{Br}}_{Q'}(j')\big(\&_*\hat{\bar{N}}_{Q\times G'}(Q^{\cdot\cdot}_{\delta^{\cdot\cdot}})^\circ\big)\overline{\mathrm{Br}}_{Q'}(j') \quad .$$

But, choosing a finite subgroup H of $\hat{\bar{N}}_{Q\times G'}(Q^{\cdot\cdot}_{\delta^{\cdot\cdot}})$ which covers $\bar{N}_{Q\times G'}(Q^{\cdot\cdot}_{\delta^{\cdot\cdot}})$ (cf. [22, Lemma 5.5]), denoting by L the converse image in H of the image of $1 \times C_{G'}(Q')$ in $\bar{N}_{Q\times G'}(Q^{\cdot\cdot}_{\delta^{\cdot\cdot}})$ and by e the central idempotent of $\&H$, contained in $\&L$, such that (cf. [22, Proposition 5.15])

5.14.5
$$\&_*\hat{\bar{N}}_{Q\times G'}(Q^{\cdot\cdot}_{\delta^{\cdot\cdot}})^\circ = \&He \quad ,$$

and setting $f = \overline{\mathrm{Br}}_{Q'}(j')$ in $\mathscr{k}Le$, we have the $\mathscr{k}$-algebra isomorphisms (cf. 2.6.5)

5.14.6
$$\begin{aligned}\big(f(\mathscr{k}H)f\big)^{\circ} &\cong \mathrm{End}_{\mathscr{k}}(\mathscr{k}Hf)^{H} \cong \mathrm{End}_{\mathscr{k}}\big(\mathrm{Ind}_L^H(\mathscr{k}Lf)\big)^{H}\\ &\cong \mathrm{Ind}_L^H\big(\mathrm{End}_{\mathscr{k}}(\mathscr{k}Lf)\big)^{H}\end{aligned}$$

and therefore, since $\mathscr{k}Lf$ is an indecomposable $\mathcal{O}L$-module, Lemma 5.11 applies, so that by this lemma and Remark 5.12, denoting by ε the unique point of L on $\mathrm{End}_{\mathscr{k}}(\mathscr{k}Lf)^{\circ}$, the canonical homomorphism s_{ε} induces a surjective $\mathscr{k}$-algebra homomorphism with nilpotent kernel

5.14.7 $$\overline{\mathrm{Br}}_{Q'}(j')\big(\mathscr{k}_*\hat{\bar{N}}_{Q\times G'}(Q_{\ddot{\delta}})^{\circ}\big)\overline{\mathrm{Br}}_{Q'}(j') = f(\mathscr{k}H)f \longrightarrow \mathscr{k}_*\hat{\bar{N}}_H(L_{\varepsilon})^{\circ} \quad .$$

Moreover, denoting by $\mu_L : (\mathscr{k}Le)^{\circ} \cong \mathrm{End}_{\mathscr{k}}(\mathscr{k}Le)^{L}$ the $\mathscr{k}$-algebra isomorphism determined by right multiplication and considering the canonical embedding $\mathrm{End}_{\mathscr{k}}(\mathscr{k}Lf) \longrightarrow \mathrm{End}_{\mathscr{k}}(\mathscr{k}Le)$, ε can be identified with the point of L on $\mathrm{End}_{\mathscr{k}}(\mathscr{k}Le)^{\circ}$ containing $\mu_L(f)$ and we have clearly a $\mathscr{k}$-algebra isomorphism

5.14.8 $$\big(\mathrm{End}_{\mathscr{k}}(\mathscr{k}Le)^{\circ}\big)(L_{\varepsilon}) \cong (\mathcal{O}G')(Q'_{\delta'})$$

which, for any $a' \in (\mathcal{O}G')^{Q'}$, maps $s_{\varepsilon}\Big(\mu_L\big(\overline{\mathrm{Br}}_{Q'}(a')\big)\Big)$ on $s_{\delta'}(a')$, so that the stabilizers in H of ε and δ' coincide and we have the group isomorphisms

5.14.9 $$\bar{N}_H(L_{\varepsilon}) \cong \Big(\pi'\big(N_{Q\times G'}(Q_{\ddot{\delta}})\big) \cap N_{G'}(Q'_{\delta'})\Big)/Q'{\cdot}C_{G'}(Q') \cong E \quad .$$

On the other hand, if $x' \in N'$, $b^{\cdot\cdot} \in N_{B_{\ddot{\delta}}}^{\varphi_{(1,x')}}(Q^{\cdot\cdot})$ and $a' \in N_{(\mathcal{O}G')_{\delta'}}^{\varphi_{x'}}(Q')$, in $\Big(\mathrm{Ind}_{\tau}\big(B_{\ddot{\delta}} \otimes_{\mathcal{O}} \mathrm{Res}_{\zeta'}(\mathcal{O}G')\big)\Big)(Q)$ we have

5.14.10 $$\mathrm{Br}_Q\big(1\otimes(b^{\cdot\cdot}\otimes a')\big) = \mathrm{Br}_Q\big(1\otimes(b^{\cdot\cdot}\otimes x')\big)\mathrm{Br}_Q\big(1\otimes(1\otimes x'^{-1}a')\big) \quad ;$$

hence, the image of this element by homomorphism 5.8.1 is not zero only if $b^{\cdot\cdot}$ is invertible in $B_{\ddot{\delta}}$ and $(1,x')$ fixes $\delta^{\cdot\cdot}$, and then, according to Remark 5.9, it is equal to $\big(\overline{(1,x')},\overline{b^{\cdot\cdot}}\big)\overline{\mathrm{Br}}_{Q'}(x'^{-1}a')$ in $\mathscr{k}_*\hat{\bar{N}}_{Q\times G'}(Q_{\ddot{\delta}})^{\circ} = \mathscr{k}He$. Moreover, considering the induction from L to H of the canonical embedding $\mathrm{End}_{\mathscr{k}}(\mathscr{k}Lf) \longrightarrow \mathrm{End}_{\mathscr{k}}(\mathscr{k}Le)$ and the obvious extension to $(\mathscr{k}He)^{\circ}$ of isomorphisms 5.14.6, we get the following commutative diagram

5.14.11
$$\begin{array}{ccc}(\mathscr{k}He)^{\circ} & \cong & \mathrm{Ind}_L^H\big(\mathrm{End}_{\mathscr{k}}(\mathscr{k}Le)\big)^{H}\\ \uparrow & & \uparrow\\ \big(f(\mathscr{k}H)f\big)^{\circ} & \cong & \mathrm{Ind}_L^H\big(\mathrm{End}_{\mathscr{k}}(\mathscr{k}Lf)\big)^{H}\end{array}$$

and the images of f and $\overline{\mathrm{Br}}_{Q'}(x'^{-1}a')$ in $\mathrm{Ind}_L^H\big(\mathrm{End}_{\mathscr{k}}(\mathscr{k}Le)^{\circ}\big)^{H}$ are respectively equal to $\mathrm{Tr}_L^H\big(1\otimes\mu_L(f)\otimes 1\big)$ and to $\mathrm{Tr}_L^H\Big(1\otimes\mu_L\big(\overline{\mathrm{Br}}_{Q'}(x'^{-1}a')\big)\otimes 1\Big)$; now, identifying ε to the corresponding point of L on $\mathrm{Ind}_L^H\big(\mathrm{End}_{k}(kLe)^\circ\big)$ (cf. 2. 11), since $\overline{\mathrm{Br}}_{Q'}(x'^{-1}a')$ belongs to $f^{(\overline{(1,x')},\overline{b^{\cdot\cdot}})}(kH)f$ and we have $f^{(\overline{(1,x')},\overline{b^{\cdot\cdot}})} = \overline{\mathrm{Br}}_{Q'}(j'^{x'})$, the element $s_\varepsilon\Big(\mathrm{Tr}_L^H\Big(1\otimes\mu_L\big(\overline{\mathrm{Br}}_{Q'}(x'^{-1}a')\big)\otimes 1\Big)\Big)$ is not zero only if $\big(\overline{(1,x')},\overline{b^{\cdot\cdot}}\big)$ fixes ε or, equivalently, if x' fixes δ' and in that case, setting $m = \mu_L\big(\overline{\mathrm{Br}}_{Q'}(x'^{-1}a')\big)$, we have

5.14.12 $$s_\varepsilon\big(\mathrm{Tr}_L^H(1\otimes m\otimes 1)\big) = \mathrm{Tr}_L^{N_H(L_\varepsilon)}\big(1\otimes s_\varepsilon(m)\otimes 1\big)$$

which is not zero only if $s_\varepsilon(m)\neq 0$ or, equivalently according to isomorphism 5.14.8, if $s_{\delta'}(x'^{-1}a')\neq 0$; once again, since $x'^{-1}a'$ belongs to $\big(j'^{x'}(\mathcal{O}G')j'\big)^{Q'}$, the element $s_{\delta'}(x'^{-1}a')$ is not zero only if x' fixes δ' and moreover, choosing $c'\in\big((\mathcal{O}G')^{Q'}\big)^*$ such that $j'^{x'} = j'^{c'}$, if we have

5.14.13 $$s_{\delta'}(c'x'^{-1}a') = \lambda s_{\delta'}(j')$$

for a suitable $\lambda\in k^*$, so that a' is invertible in $(\mathcal{O}G')_{\delta'}$.

Consequently, denoting by $t_{(b^{\cdot\cdot},a')}$ the image of $\mathrm{Br}_Q\big(1\otimes(b^{\cdot\cdot}\otimes a')\big)$ in $\mathrm{Ind}_L^H\big(\mathrm{End}_{k}\,(kLf)^\circ\big)^H$ by the composition of homomorphism 5.14.4 and isomorphisms 5.14.6, and, for any $(\tilde\varphi^{\cdot\cdot},\tilde\varphi')\in E$, setting

5.14.14 $$\hat T_{(\tilde\varphi^{\cdot\cdot},\tilde\varphi')} = \big\{s_\varepsilon(t_{(b^{\cdot\cdot},a')})\,|\,b^{\cdot\cdot}\in N^{\varphi^{\cdot\cdot}}_{B^{\cdot\cdot}_{\delta^{\cdot\cdot}}}(Q^{\cdot\cdot})^*,\ a'\in N^{\varphi'}_{(OG')_{\delta'}}(Q')^*\big\}\quad ,$$

where $\varphi^{\cdot\cdot}$ and φ' are respectively representatives of $\tilde\varphi^{\cdot\cdot}$ and $\tilde\varphi'$ (notice that this set does not depend on the choices of $\varphi^{\cdot\cdot}$ and φ'), the union

5.14.15 $$\hat T = \bigcup_{(\tilde\varphi^{\cdot\cdot},\tilde\varphi')\in E}\hat T_{(\tilde\varphi^{\cdot\cdot},\tilde\varphi')}$$

generates the k-algebra (cf. 5. 12. 2)

5.14.16 $$\Big(\mathrm{Ind}_L^H\big(\mathrm{End}_{k}(kLf)^\circ\big)\Big)(L_\varepsilon)^{\bar N_H(L_\varepsilon)} \cong k_*\hat{\bar N}_H(L_\varepsilon)^\circ\quad .$$

But it is easily checked that $\hat T$ is, up to suitable identifications, a k^*-subgroup of $\big(k_*\hat{\bar N}_H(L_\varepsilon)^\circ\big)^*$ with k^*-quotient of order at most $|E|$, so of order equal to $|E|$ since $\hat T$ generates $k_*\hat{\bar N}_H(L_\varepsilon)$ and $\bar N_H(L_\varepsilon)\cong E$ (cf. 5.14.9), and that we have a map

5.14.17 $$\hat T \longrightarrow \hat E_{Q\times G'}(Q^{\cdot\cdot}_{\delta^{\cdot\cdot}})^\circ * \hat E_{G'}(Q'_{\delta'})^\circ$$

sending $s_\varepsilon(t_{(b^{\cdot\cdot},a')})$ to $(\tilde\varphi^{\cdot\cdot},\bar b)\otimes(\tilde\varphi',\bar{a}')$ for any $(\tilde\varphi^{\cdot\cdot},\tilde\varphi')\in E$, any $b^{\cdot\cdot}\in N^{\varphi^{\cdot\cdot}}_{B^{\cdot\cdot}_{\delta^{\cdot\cdot}}}(Q^{\cdot\cdot})^*$ and any $a'\in N^{\varphi'}_{(OG')_{\delta'}}(Q')^*$, which is a surjective k^*-group homomorphism; in conclusion, this map is a k^*-group isomorphism and we get the announced k-algebra homomorphism 5.14.2.

6 Morita stable equivalences between Brauer blocks

6.1 Let G and G' be finite groups, b and b' blocks of G and G' respectively and $M^{\cdot\cdot}$ an indecomposable $\mathcal{O}(G \times G')$-module associated with $b \otimes (b')^\circ$ (where $(\mathcal{O}G'b')^\circ \cong \mathcal{O}G'(b')^\circ$ as $\mathcal{O}G'$-interior algebras), which restricted to both $\mathcal{O}(G \times 1)$ and $\mathcal{O}(1 \times G')$ is projective. Set $A = \mathcal{O}Gb$ and $A' = \mathcal{O}G'b'$, so that $M^{\cdot\cdot}$ is also an $A \otimes_\mathcal{O} (A')^\circ$-module, and denote respectively by Mod_A and $\mathrm{Mod}_{A'}$ the categories of $\mathcal{O}$-free A- and A'-modules; it is clear that $M^{\cdot\cdot}$ determines a functor from $\mathrm{Mod}_{A'}$ to Mod_A, namely the functor

6.1.1 $$F_{M^{\cdot\cdot}} : \mathrm{Mod}_{A'} \longrightarrow \mathrm{Mod}_A$$

mapping any $\mathcal{O}$-free A'-module M' on $M^{\cdot\cdot} \otimes_{A'} M' = M^{\cdot\cdot} \otimes_{\mathcal{O}G'} M'$, which is $\mathcal{O}$-free since $M^{\cdot\cdot}$ restricted to $\mathcal{O}(1 \times G')$ is projective, and any A'-module homomorphism $f' : M' \to N'$, where N' is a second $\mathcal{O}$-free A'-module, on the A-module homomorphism

6.1.2 $$\mathrm{id}_{M^{\cdot\cdot}} \underset{A'}{\otimes} f' : M^{\cdot\cdot} \underset{A'}{\otimes} M' \longrightarrow M^{\cdot\cdot} \underset{A'}{\otimes} N' \quad .$$

We say that $M^{\cdot\cdot}$ *defines a Morita equivalence between b and b'* whenever this functor is an equivalence of categories: a well-known theorem of Kiiti Morita [17] affirms that any equivalence of categories between $\mathrm{Mod}_{A'}$ and Mod_A has this form.

6.2 It is clear that the $\mathcal{O}$-dual $(M^{\cdot\cdot})^*$ is an indecomposable $A' \otimes_\mathcal{O} A^\circ$-module and, since $\mathcal{O}G$ and $\mathcal{O}G'$ are symmetric $\mathcal{O}$-algebras, its restrictions to both $A' \otimes 1$ and $1 \otimes A^\circ$ are projective too; hence, it defines also a functor $\mathcal{F}_{(M^{\cdot\cdot})^*} : \mathrm{Mod}_A \longrightarrow \mathrm{Mod}_{A'}$. Then, it is well-known (and easily checked) that the projectivity of the restrictions of both $M^{\cdot\cdot}$ and $(M^{\cdot\cdot})^*$ to both $\mathcal{O}(G \times 1)$ and $\mathcal{O}(1 \times G')$ implies that the functors $\mathcal{F}_{M^{\cdot\cdot}}$ and $\mathcal{F}_{(M^{\cdot\cdot})^*}$ are both left and right adjoint to one another; consequently, $M^{\cdot\cdot}$ defines a Morita equivalence between b and b' if and only if $\mathcal{F}_{M^{\cdot\cdot}}$ and $\mathcal{F}_{(M^{\cdot\cdot})^*}$ are inverses of each other or, equivalently, if and only if we have respectively an A- and an A'-bimodule isomorphisms

6.2.1 $$M^{\cdot\cdot} \underset{A'}{\otimes} (M^{\cdot\cdot})^* \cong A \qquad \text{and} \qquad (M^{\cdot\cdot})^* \underset{A}{\otimes} M^{\cdot\cdot} \cong A' \quad .$$

Notice that if an $A \otimes_\mathcal{O} (A')^\circ$-module fulfills condition 6.2.1 then it is indecomposable and its restrictions to $A \otimes 1$ and to $1 \otimes (A')^\circ$ are both projective.

6.3 As a matter of fact, our analysis below covers, without extra-effort, a more general situation, namely the case of the *Morita stable equivalences*: we say that $M^{\cdot\cdot}$ *defines a Morita stable equivalence between b and b'* if we have respectively an A- and an A'-bimodule isomorphisms

6.3.1 $$M^{\cdot\cdot} \underset{A'}{\otimes} (M^{\cdot\cdot})^* \cong A \oplus U \quad \text{and} \quad (M^{\cdot\cdot})^* \underset{A}{\otimes} M^{\cdot\cdot} \cong A' \oplus U'$$

for suitable projective A- and A'-bimodules U and U'. In that case, $\mathcal{F}_{M^{\cdot\cdot}}$ induces an equivalence of categories between the so-called *stable quotients* of $\mathrm{Mod}_{A'}$ and Mod_A - that is to say, between the categories $\overline{\mathrm{Mod}}_{A'}$ and $\overline{\mathrm{Mod}}_A$ having respectively the same objects than $\mathrm{Mod}_{A'}$ and Mod_A, and the morphisms being respectively the equivalent classes of homomorphisms in $\mathrm{Mod}_{A'}$ and in Mod_A for the condition that the difference between related homomorphisms factorizes throughout a projective module - but the converse is not known to be true. In [31, Corollary 5.5], Rickard exhibits such an $A \otimes_{\mathcal{O}} (A')^{\circ}$-module when the so-called *derived categories* of $\mathrm{Mod}_{A'}$ and Mod_A are equivalent (as *triangulated* categories).

6.4 Our first proposition translates condition 6.3.1 in terms of $\mathcal{O}G$-interior algebras and will provide the link with Hecke $\mathcal{O}G$-interior algebras. In order to short statements, we introduce the following *ad hoc* terminology: we say that an $\mathcal{O}G$-interior algebra homomorphism $f : B \longrightarrow \hat{B}$ between two $\mathcal{O}$-free $\mathcal{O}G$-interior algebras B and $\hat{B}$ is a *stable embedding* if $\mathrm{Ker}(f)$ and $f(1)\hat{B}f(1)/f(B)$ are projective $\mathcal{O}G$-bimodules or, equivalently, if the $\mathcal{O}G$-bimodule homomorphism $f : B \longrightarrow f(1)\hat{B}f(1)$ determines an isomorphism in the stable quotient of $\mathrm{Mod}_{\mathcal{O}(G\times G)}$; if moreover f is unitary we say that f is a *stable isomorphism*. In this last case, the following exact sequence of $\mathcal{O}G$-bimodules splits

6.4.1
$$0 \longrightarrow \mathrm{Ker}(f) \longrightarrow B \xrightarrow{f} \hat{B} \longrightarrow \hat{B}/f(B) \longrightarrow 0$$

and therefore, for any subgroup H of G, f induces an $\mathcal{O}C_G(H)$-interior $N_G(H)$-algebra isomorphism $B^H/B_1^H \cong \hat{B}^H/\hat{B}_1^H$ (cf. 2.2.6), so that we get:

6.4.2 *For any nonprojective pointed group H_β on B there is a unique nonprojective point $\hat{\beta}$ of H on $\hat{B}$ such that $f(\beta) \subset \hat{\beta}+\hat{B}_1^H$ and then f induces an $\mathcal{O}C_G(H)$-interior $N_G(H_\beta)$-algebra isomorphism*

$$f(H_\beta) : B(H_\beta) \cong \hat{B}(H_{\hat{\beta}}) \quad ;$$

in particular, we have $N_G(H_\beta) = N_G(H_{\hat{\beta}})$ and $f(H_\beta)$ determines a k^*-group isomorphism

$$\hat{f}(H_\beta) : \hat{\bar{N}}_G(H_\beta) \cong \hat{\bar{N}}_G(H_{\hat{\beta}}) \quad .$$

Moreover, that correspondence determines a bijection between the sets of nonprojective pointed groups on B and $\hat{B}$ which preserves inclusion and localness.

If H is a subgroup of G and $g : C \to \hat{C}$ is an $\mathcal{O}H$-interior algebra stable embedding between two $\mathcal{O}$-free $\mathcal{O}H$-interior algebras C and $\hat{C}$, it is quite clear that the induced homomorphism

6.4.3
$$\mathrm{Ind}_H^G(g) : \mathrm{Ind}_H^G(C) \longrightarrow \mathrm{Ind}_H^G(\hat{C})$$

is an $\mathcal{O}G$-interior algebra stable embedding. Notice that the composition of two $\mathcal{O}G$-interior algebra stable embeddings is an $\mathcal{O}G$-interior algebra stable embedding too.

Proposition 6.5 *The $\mathcal{O}(G \times G')$-module $M^{\cdot\cdot}$ defines a Morita stable equivalence between b and b' if and only if the structural map $A \to \mathrm{End}_{\mathcal{O}}(M^{\cdot\cdot})^{1\times G'}$ is an $\mathcal{O}G$-interior algebra stable isomorphism and b' has defect zero whenever b has defect zero. Moreover, $M^{\cdot\cdot}$ defines a Morita equivalence between b and b' if and only if we have an $\mathcal{O}G$-interior algebra isomorphism*

6.5.1 $$A \cong \mathrm{End}_{\mathcal{O}}(M^{\cdot\cdot})^{1\times G'} \quad .$$

Proof: Since $M^{\cdot\cdot}$ restricted to $\mathcal{O}(1 \times G')$ is projective, it is $\mathcal{O}$-free and the trace map $\mathrm{Tr}_{1\times 1}^{1\times G'}$ on $M^{\cdot\cdot} \otimes_{\mathcal{O}} (M^{\cdot\cdot})^* \cong \mathrm{End}_{\mathcal{O}}(M^{\cdot\cdot})$ induces an $\mathcal{O}G$-bimodule isomorphism (cf. 2.2.6)

6.5.2 $$M^{\cdot\cdot} \underset{\mathcal{O}G'}{\otimes} (M^{\cdot\cdot})^* \cong \big(M^{\cdot\cdot} \underset{\mathcal{O}}{\otimes} (M^{\cdot\cdot})^*\big)^{1\times G'} \quad .$$

Then, if $M^{\cdot\cdot}$ defines a Morita stable equivalence between b and b', we have an A-bimodule isomorphism

6.5.3 $$A \oplus U \cong \mathrm{End}_{\mathcal{O}}(M^{\cdot\cdot})^{1\times G'}$$

for a suitable projective A-bimodule U, and we denote by

6.5.4 $$f : A \longrightarrow \mathrm{End}_{\mathcal{O}}(M^{\cdot\cdot})^{1\times G'} \text{ and } g : \mathrm{End}_{\mathcal{O}}(M^{\cdot\cdot})^{1\times G'} \longrightarrow A$$

the corresponding split injection and split surjection; since for any $x \in G$ we have

6.5.5 $$x{\cdot}f(b) = f(xb) = f(bx) = f(b){\cdot}x \quad ,$$

the multiplication by $f(b)$ on the right determines an $\mathcal{O}G$-bimodule endomorphism $m_{f(b)}$ of $\mathrm{End}_{\mathcal{O}}(M^{\cdot\cdot})^{1\times G'}$ and it is easily checked that f is the composition of the structural map $\mathrm{st}_A : A \longrightarrow \mathrm{End}_{\mathcal{O}}(M^{\cdot\cdot})^{1\times G'}$ and $m_{f(b)}$; consequently, we get

6.5.6 $$\mathrm{id}_A = g \circ f = (g \circ m_{f(b)}) \circ \mathrm{st}_A \quad ,$$

so that st_A is injective, $\mathrm{st}_A(A)$ is a direct summand of $\mathrm{End}_{\mathcal{O}}(M^{\cdot\cdot})^{1\times G'}$ as $\mathcal{O}G$-bimodules and any complement is isomorphic to U, which is zero whenever $M^{\cdot\cdot}$ defines a Morita equivalence between b and b'; moreover, if b has defect zero then any $\mathcal{O}$-free A-module is projective and therefore any $\mathcal{O}$-free A'-module is projective too, so that b' has also defect zero (notice that this argument works even when $\mathcal{O} = k$).

Conversely, assume that $\mathrm{st}_A : A \longrightarrow \mathrm{End}_{\mathcal{O}}(M^{\cdot\cdot})^{1\times G'}$ is an $\mathcal{O}G$-interior algebra stable isomorphism; since A is an indecomposable $\mathcal{O}G$-bimodule and st_A is unitary, st_A is injective and therefore, by 6.5.2, we have an $\mathcal{O}G$-bimodule isomorphism

6.5.7 $$M^{\cdot\cdot} \underset{A'}{\otimes} (M^{\cdot\cdot})^* \cong A \oplus U \quad ,$$

where $U = \mathrm{End}_{\mathcal{O}}(M^{\cdot\cdot})^{1\times G'}/\mathrm{st}_A(A)$ is projective. Moreover, assume that either $U = \{0\}$ (which occurs whenever b and b' have defect zero) or b has a nonzero defect; in that case, in order to prove the second isomorphism in 6.3.1, we adapt an argument of Rickard (cf. [33, Theorem 2.1]); precisely, we will prove that the structural homomorphism $\mathrm{st}_{A'} : A' \to \mathrm{End}_{\mathcal{O}}(M^{\cdot\cdot})^{G\times 1}$ is also injective and that its cokernel is a projective $\mathcal{O}G'$-bimodule. By symmetry, the trace map $\mathrm{Tr}_{1\times 1}^{G\times 1}$ on $\mathrm{End}_{\mathcal{O}}(M^{\cdot\cdot})$ induces also an $\mathcal{O}G'$-bimodule isomorphism

6.5.8 $$(M^{\cdot\cdot})^* \underset{\mathcal{O}G}{\otimes} M^{\cdot\cdot} \cong \big((M^{\cdot\cdot})^* \underset{\mathcal{O}}{\otimes} M^{\cdot\cdot}\big)^{G\times 1} \quad ;$$

in particular, since each member is the dual of the other one, we get a $G' \times G'$-stable nondegenerate symmetric scalar product in $\mathrm{End}_{\mathcal{O}}(M^{\cdot\cdot})^{G\times 1}$; since A' admits also a $G' \times G'$-stable nondegenerate symmetric scalar product, it makes sense to consider the adjoint map $(\mathrm{st}_{A'})^{\mathrm{ad}}$ and the composition $(\mathrm{st}_{A'})^{\mathrm{ad}} \circ \mathrm{st}_{A'}$ is an $\mathcal{O}G'$-bimodule endomorphism of A', so that it is the multiplication by a suitable $z' \in Z(A')$.

We claim that z' is invertible in $Z(A')$; indeed, $\mathrm{st}_{A'}$ and $(\mathrm{st}_{A'})^{\mathrm{ad}}$ correspond respectively to the unit and counit of the adjunctions $(\mathcal{F}_{M^{\cdot\cdot}}, \mathcal{F}_{(M^{\cdot\cdot})^*})$ and $(\mathcal{F}_{(M^{\cdot\cdot})^*}, \mathcal{F}_{M^{\cdot\cdot}})$, and therefore, up to suitable identifications, the $\mathcal{O}(G \times G')$-module homomorphisms

6.5.9 $$M^{\cdot\cdot} \xrightarrow{\mathrm{id}_{M^{\cdot\cdot}} \otimes \mathrm{st}_{A'}} M^{\cdot\cdot} \underset{A'}{\otimes} (M^{\cdot\cdot})^* \underset{A}{\otimes} M^{\cdot\cdot} \xrightarrow{\mathrm{id}_{M^{\cdot\cdot}} \otimes (\mathrm{st}_{A'})^{\mathrm{ad}}} M^{\cdot\cdot}$$

are respectively a split injection and a split surjection by the general properties of adjoint functors; moreover, by 6.5.7, we have an $\mathcal{O}(G \times G')$-module isomorphism

6.5.10 $$M^{\cdot\cdot} \underset{A'}{\otimes} (M^{\cdot\cdot})^* \underset{A}{\otimes} M^{\cdot\cdot} \cong M^{\cdot\cdot} \oplus (U \underset{A}{\otimes} M^{\cdot\cdot}) \quad ;$$

consequently, since $U \otimes_A M^{\cdot\cdot}$ is a projective $\mathcal{O}(G \times G')$-module, either the $\mathcal{O}G$-bimodule A is not projective, so that the $\mathcal{O}(G \times G')$-module $M^{\cdot\cdot}$ is not projective too (cf. 6.5.7), and both $\mathrm{id}_{M^{\cdot\cdot}} \otimes \mathrm{st}_{A'}$ and $\mathrm{id}_{M^{\cdot\cdot}} \otimes (\mathrm{st}_{A'})^{\mathrm{ad}}$ determine nonzero isomorphisms in the stable quotient of $\mathrm{Mod}_{\mathcal{O}(G\times G')}$, or $P = 1$ (cf. 2.12.1 and [38, (44.3)]), $U = \{0\}$ and both homomorphisms are bijective; in any case $z'\mathrm{id}_{M^{\cdot\cdot}}$ is invertible in $\mathrm{End}_{\mathcal{O}}(M^{\cdot\cdot})^{G\times G'}$ and therefore z' does not belong to $J\big(Z(A')\big)$.

In particular, $\mathrm{st}_{A'}$ is a split injection, so that we have an $\mathcal{O}G'$-bimodule isomorphism

6.5.11 $$(M^{\cdot\cdot})^* \underset{A}{\otimes} M^{\cdot\cdot} \cong A' \oplus U'$$

for a suitable $\mathcal{O}G'$-bimodule U'; hence, by 6.5.7 and 6.5.11, we get

6.5.12 $$\begin{aligned} A' \oplus U' \oplus U' \oplus (U' \underset{A'}{\otimes} U') &\cong (M^{\cdot\cdot})^* \underset{A}{\otimes} M^{\cdot\cdot} \underset{A'}{\otimes} (M^{\cdot\cdot})^* \underset{A}{\otimes} M^{\cdot\cdot} \\ &\cong \big((M^{\cdot\cdot})^* \underset{A}{\otimes} M^{\cdot\cdot}\big) \oplus \big((M^{\cdot\cdot})^* \underset{A}{\otimes} U \underset{A}{\otimes} M^{\cdot\cdot}\big) \\ &\cong A' \oplus U' \oplus \big((M^{\cdot\cdot})^* \underset{A}{\otimes} U \underset{A}{\otimes} M^{\cdot\cdot}\big) \end{aligned}$$

and therefore the $\mathcal{O}G'$-bimodule U' is a direct summand of $(M^{\cdot\cdot})^* \otimes_A U \otimes_A M^{\cdot\cdot}$ which is projective and vanishes with U.

6.6 Let $P^{\cdot\cdot}$ be a vertex of $M^{\cdot\cdot}$, $N^{\cdot\cdot}$ an $\mathcal{O}P^{\cdot\cdot}$-source of $M^{\cdot\cdot}$ and $V^{\cdot\cdot}$ a multiplicity $k_* \hat{\bar{N}}_{G\times G'}(P^{\cdot\cdot}_{N^{\cdot\cdot}})$-module of the local pointed group on $\mathrm{End}_{\mathcal{O}}(M^{\cdot\cdot})$ determined by $P^{\cdot\cdot}$ and $N^{\cdot\cdot}$, noted $P^{\cdot\cdot}_{N^{\cdot\cdot}}$ (cf. 2.10); that is to say, $P^{\cdot\cdot}$ is a subgroup of $G \times G'$, $N^{\cdot\cdot}$ an indecomposable direct summand of $\mathrm{Res}^{G\times G'}_{P^{\cdot\cdot}}(M^{\cdot\cdot})$ and $V^{\cdot\cdot}$ an indecomposable projective $k_*\hat{\bar{N}}_{G\times G'}(P^{\cdot\cdot}_{N^{\cdot\cdot}})$-module such that $M^{\cdot\cdot}$ is the direct summand of $\mathrm{Ind}^{G\times G'}_{P^{\cdot\cdot}}(N^{\cdot\cdot})$ corresponding to the direct summand $V^{\cdot\cdot}$ of the $k_*\hat{\bar{N}}_{G\times G'}(P^{\cdot\cdot}_{N^{\cdot\cdot}})$-module (cf. 2.9.3)

6.6.1 $$V_{\mathrm{Ind}^{G\times G'}_{P^{\cdot\cdot}}(N^{\cdot\cdot})}(P^{\cdot\cdot}_{N^{\cdot\cdot}}) \cong \mathrm{End}_{k}\big(k_*\hat{\bar{N}}_{G\times G'}(P^{\cdot\cdot}_{N^{\cdot\cdot}})\big) \quad .$$

In particular, setting

6.6.2 $$S^{\cdot\cdot} = \mathrm{End}_{\mathcal{O}}(N^{\cdot\cdot}),\ A^{\cdot\cdot} = \mathrm{Ind}^{G\times G'}_{P^{\cdot\cdot}}(S^{\cdot\cdot}) \text{ and } \hat{A} = (A^{\cdot\cdot})^{1\times G'}$$

we have the Higman $\mathcal{O}(G \times G')$-interior algebra exoembedding (cf. 2.12.1)

6.6.3 $$\tilde{h} : \mathrm{End}_{\mathcal{O}}(M^{\cdot\cdot}) \longrightarrow A^{\cdot\cdot}$$

which induces a unique $\mathcal{O}G$-interior algebra exoembedding

6.6.4 $$\mathrm{End}_{\mathcal{O}}(M^{\cdot\cdot})^{1\times G'} \longrightarrow \hat{A} \quad ;$$

hence, $M^{\cdot\cdot}$ determines a point $\hat{\alpha}$ of G on $\hat{A}$ such that $b{\cdot}\hat{\alpha} = \hat{\alpha}$ and, according to Proposition 6.5, we know that:

6.6.5 *$M^{\cdot\cdot}$ defines a Morita stable equivalence between b and b' if and only if the structural map $A \to \hat{A}_{\hat{\alpha}}$ is an $\mathcal{O}G$-interior algebra stable isomorphism and b' has defect zero whenever b has defect zero.*

Notice that this statement remains true if we remove the word "stable" and the last condition.

6.7 Now we can apply results of Sections 4 and 5 in order to discuss that condition; as in 5.2, denote respectively by $\rho : P^{\cdot\cdot} \to G$ and $\rho' : P^{\cdot\cdot} \to G'$ the restrictions of the first and the second projection maps, by P and P' the images of ρ and ρ', and by $\sigma : P^{\cdot\cdot} \to P$ and $\sigma' : P^{\cdot\cdot} \to P'$ the corresponding group homomorphisms. Recall that, according to Theorem 4.4, we have the Hecke $\mathcal{O}G$-interior algebra isomorphism

6.7.1 $$H_{G,G'}(S^{\cdot\cdot}) : \hat{A} \cong \mathrm{Ind}_{\rho}\big(S^{\cdot\cdot} \underset{\mathcal{O}}{\otimes} \mathrm{Res}_{\rho'}(\mathcal{O}G')\big)$$

and we emphasize that the image of $(b')^\circ$ in $A^{\cdot\cdot}$ is contained in $z(\hat{A})$ and that

6.7.2 $$\begin{aligned} &H_{G,G'}(S^{\cdot\cdot})\Big((b')^\circ{\cdot}\mathrm{Tr}^{G\times G'}_{P^{\cdot\cdot}}\big((1,1)\otimes \mathrm{id}_{N^{\cdot\cdot}} \otimes (1,1)\big)\Big) \\ &= \mathrm{Tr}^G_P\Big(1 \otimes \big(1 \otimes (\mathrm{id}_{N^{\cdot\cdot}} \otimes b')\big) \otimes 1\Big) \end{aligned}$$

so that $H_{G,G'}(S^{\cdot\cdot})$ induces an $\mathcal{O}G$-interior algebra isomorphism

$$\hat{A}\cdot(b')^{\circ} \cong \mathrm{Ind}_{\rho}\big(S^{\cdot\cdot} \underset{\mathcal{O}}{\otimes} \mathrm{Res}_{\rho'}(A')\big) \quad ; \tag{6.7.3}$$

notice that, according to 4.9, the functor $\mathcal{F}_{M^{\cdot\cdot}}$ above coincides, up to isomorphism, with the restriction to $\mathrm{Mod}_{A'}$ of the functor $\mathcal{F}_{G,G'}(N^{\cdot\cdot})_{\hat{\imath}}$, where $\hat{\imath} \in \hat{\alpha}$. Moreover, it follows from Theorem 5.8 that the set of local points of P on $\mathrm{Ind}_{\sigma}\big(S^{\cdot\cdot} \otimes_{\mathcal{O}} \mathrm{Res}_{\rho'}(\mathcal{O}G')\big)$ is parametrized by the set of isomorphism classes of simple $k_*\hat{\bar{N}}_{P\times G'}(P^{\cdot\cdot}_{N^{\cdot\cdot}})$-modules; now, since the head $\mathrm{hd}(V^{\cdot\cdot})$ of $V^{\cdot\cdot}$ is a simple $k_*\hat{\bar{N}}_{G\times G'}(P^{\cdot\cdot}_{N^{\cdot\cdot}})$-module and $N_{G\times G'}(P^{\cdot\cdot}_{N^{\cdot\cdot}})$ normalizes $P \times G'$, the restriction of $\mathrm{hd}(V^{\cdot\cdot})$ to $k_*\hat{\bar{N}}_{P\times G'}(P^{\cdot\cdot}_{N^{\cdot\cdot}})$ is semisimple.

6.8 We choose an isotypic component $W^{\cdot\cdot}$ of the restriction of $\mathrm{hd}(V^{\cdot\cdot})$ to $k_*\hat{\bar{N}}_{P\times G'}(P^{\cdot\cdot}_{N^{\cdot\cdot}})$ and denote by $\hat{\gamma}$ the local point of P on $\mathrm{Ind}_{\sigma}\big(S^{\cdot\cdot} \otimes_{\mathcal{O}} \mathrm{Res}_{\rho'}(\mathcal{O}G')\big)$ determined by $W^{\cdot\cdot}$; in particular, the image of $\mathrm{Br}_P(\hat{\gamma})$ by isomorphism 5.8.2 acts nontrivially on $\mathrm{hd}(V^{\cdot\cdot})$ and moreover, denoting by $\hat{\bar{N}}$ the converse image of $\bar{N}_G(P_{\hat{\gamma}})$ in $\hat{\bar{N}}_{G\times G'}(P^{\cdot\cdot}_{N^{\cdot\cdot}})$, $\hat{\bar{N}}$ acts on $W^{\cdot\cdot}$ and we have a $k_*\hat{\bar{N}}_{G\times G'}(P^{\cdot\cdot}_{N^{\cdot\cdot}})$-module isomorphism

$$\mathrm{hd}(V^{\cdot\cdot}) \cong \mathrm{Ind}_{\hat{\bar{N}}}^{\hat{\bar{N}}_{G\times G'}(P^{\cdot\cdot}_{N^{\cdot\cdot}})}(W^{\cdot\cdot}) \quad , \tag{6.8.1}$$

so that, by Remark 5.10, $E_{G\times G'}(P^{\cdot\cdot}_{N^{\cdot\cdot}})$ acts transitively on the set of choices for $\hat{\gamma}$. More precisely, we denote by γ' the local point of P' on $\mathcal{O}G'$ determined by an isotypic component of the restriction of $W^{\cdot\cdot}$ to $kC_{G'}(P')$ (cf. 2.9.1), so that $\hat{\gamma}$ is still a local point of P on $\mathrm{Ind}_{\sigma}\Big(S^{\cdot\cdot} \otimes_{\mathcal{O}} \mathrm{Res}_{\sigma'}\big((\mathcal{O}G')_{\gamma'}\big)\Big)$ (cf. 2.11 and Remark 5.9) and, since $W^{\cdot\cdot}$ is a simple $k_*\hat{\bar{N}}$-module, $\hat{\bar{N}}$ acts transitively on the set of choices for γ', so that $E_{G\times G'}(P^{\cdot\cdot}_{N^{\cdot\cdot}})$ acts also transitively on the set of choices for the pairs $(\hat{\gamma}, \gamma')$. On the other hand, recall that $\hat{\gamma}$ is still a point of $P \times G'$ on $A^{\cdot\cdot}$ and that $\hat{A}(P_{\hat{\gamma}}) = A^{\cdot\cdot}\big((P \times G')_{\hat{\gamma}}\big)$, so that, up to identify $\bar{N}_G(P)$ to $\bar{N}_{G\times G'}(P \times G')$, we get the equalities (cf. 2.8)

$$\hat{\bar{N}}_G(P_{\hat{\gamma}}) = \hat{\bar{N}}_{G\times G'}\big((P \times G')_{\hat{\gamma}}\big) \text{ and } V_{\hat{A}}(P_{\hat{\gamma}}) = V_{A^{\cdot\cdot}}\big((P \times G')_{\hat{\gamma}}\big) \quad ; \tag{6.8.2}$$

actually, since $\mathrm{Br}_P(\hat{\gamma})$ acts nontrivially on $V^{\cdot\cdot}$, we have $(P \times G')_{\hat{\gamma}} \subset (G \times G')_{\hat{\alpha}}$ and therefore $(P \times G')_{\hat{\gamma}}$ is already a pointed group on $\mathrm{End}_{\mathcal{O}}(M^{\cdot\cdot})$ (cf. 2.11 and 6.6.3), so that $P'_{\gamma'}$ is a local pointed group on A' (cf. 6.7.3). Finally, for any $n \geq 0$, consider the $\mathcal{O}P$-interior algebra

$$T_n = \mathrm{End}_{\mathcal{O}}\big(\mathcal{O} \oplus (\mathcal{O}P)^n\big) \quad ; \tag{6.8.3}$$

notice that there is a unique $\mathcal{O}P$-interior algebra exoembedding $\mathcal{O} \to T_n$ and, by tensoring it by $\mathrm{Ind}_{\sigma}\big(S^{\cdot\cdot} \otimes_{\mathcal{O}} \mathrm{Res}_{\sigma'}(A'_{\gamma'})\big)_{\hat{\gamma}}$, we get the following one

$$\mathrm{Ind}_{\sigma}\big(S^{\cdot\cdot} \underset{\mathcal{O}}{\otimes} \mathrm{Res}_{\sigma'}(A'_{\gamma'})\big)_{\hat{\gamma}} \longrightarrow T_n \underset{\mathcal{O}}{\otimes} \mathrm{Ind}_{\sigma}\big(S^{\cdot\cdot} \underset{\mathcal{O}}{\otimes} \mathrm{Res}_{\sigma'}(A'_{\gamma'})\big)_{\hat{\gamma}} \quad . \tag{6.8.4}$$

Theorem 6.9 *With the notation above, $M^{\cdot\cdot}$ defines a Morita stable equivalence between b and b' if and only if P and P' are respectively defect groups of b and b', $P = 1$ implies $P' = 1$, and, for a suitable local point γ of P on A and a suitable $n \geq 0$, there is an $\mathcal{O}P$-interior algebra stable embedding*

6.9.1 $$e : A_\gamma \longrightarrow T_n \underset{\mathcal{O}}{\otimes} \mathrm{Ind}_\sigma\big(S^{\cdot\cdot} \underset{\mathcal{O}}{\otimes} \mathrm{Res}_{\sigma'}(A'_{\gamma'})\big)_{\hat{\gamma}}$$

such that we have a $k_*\hat{\bar{N}}_G(P_\gamma)$-module isomorphism

6.9.2 $$V_A(P_\gamma) \cong \mathrm{Res}_{\widehat{(\mathrm{Ind}_P^G(e))(P_\gamma)}}\Big(V_{M^{\cdot\cdot}}\big((P \times G')_{\hat{\gamma}}\big)\Big) \quad .$$

In that case, if n is minimal, $M^{\cdot\cdot}$ defines a Morita equivalence between b and b' if and only if $n = 0$ and e is an isomorphism

6.9.3 $$A_\gamma \cong \mathrm{Ind}_\sigma\big(S^{\cdot\cdot} \underset{\mathcal{O}}{\otimes} \mathrm{Res}_{\sigma'}(A'_{\gamma'})\big)_{\hat{\gamma}} \quad .$$

Remark 6.10 Notice that the existence of such an e determines the $\mathcal{O}G$-interior algebra stable embedding (cf. 6.4)

6.10.1 $$\mathrm{Ind}_P^G(e) : \mathrm{Ind}_P^G(A_\gamma) \longrightarrow \mathrm{Ind}_P^G\Big(T_n \underset{\mathcal{O}}{\otimes} \mathrm{Ind}_\sigma\big(S^{\cdot\cdot} \underset{\mathcal{O}}{\otimes} \mathrm{Res}_{\sigma'}(A'_{\gamma'})\big)_{\hat{\gamma}}\Big)$$

and therefore it implies $N_G(P_\gamma) = N_G(P_{\hat{\gamma}})$ (cf. 6.4.2) and induces a k^*-group isomorphism

6.10.2 $$\widehat{(\mathrm{Ind}_P^G(e))(P_\gamma)} : \hat{\bar{N}}_G(P_\gamma) \cong \hat{\bar{N}}_G(P_{\hat{\gamma}}) = \hat{\bar{N}}_{G\times G'}\big((P \times G')_{\hat{\gamma}}\big) \quad .$$

Remark 6.11 If $M^{\cdot\cdot}$ defines a Morita stable equivalence between b and b', it follows from 6.6.5 that, for a suitable projective $\mathcal{O}G$-bimodule U, we have $\hat{A}_{\hat{\alpha}} \cong A \oplus U$ as $\mathcal{O}G$-bimodules and, in particular, $\hat{A}_{\hat{\alpha}}$ has a $P \times P$-stable $\mathcal{O}$-basis by left and right multiplication; moreover, since we have $P_{\hat{\gamma}} \subset G_{\hat{\alpha}}$ on $\hat{A}$ up to suitable identifications, $\mathrm{Ind}_\sigma\big(S^{\cdot\cdot} \otimes_{\mathcal{O}} \mathrm{Res}_{\sigma'}(A'_{\gamma'})\big)_{\hat{\gamma}}$ still has a $P \times P$-stable $\mathcal{O}$-basis, which can be proved directly from the existence of e.

Proof: Assume first that $M^{\cdot\cdot}$ defines a Morita stable equivalence between b and b'. It follows from Proposition 5.3 that P is a defect group of $G_{\hat{\alpha}}$ and from 6.6.5 (or Proposition 6.5) that the structural map $A \longrightarrow \hat{A}_{\hat{\alpha}}$ is an $\mathcal{O}G$-interior algebra stable isomorphism, so that it induces a bijection between the sets of nonprojective pointed groups, which preserves inclusion and localness (cf. 6.4.2); hence, if $P = 1$ then there is no nontrivial local pointed groups on A, so that $P = 1$ is a defect group of b and then $P' = 1$ by 6.6.5 (or Proposition 6.5), whereas if $P \neq 1$ then $P_{\hat{\gamma}}$ is a defect pointed group of $G_{\hat{\alpha}}$ (cf. 2.9.5 and 6.8) and, denoting by γ the corresponding local point of P on A, P_γ is a defect pointed group of G_α where $\alpha = \{b\}$. In conclusion, P

is a defect group of b and therefore, by symmetry, P' is a defect group of b'; moreover, if $P = 1$ then $P' = 1$, so that $P^{\cdot\cdot} = 1 \times 1$, $N^{\cdot\cdot} \cong \mathcal{O}$, $A_\gamma \cong \mathcal{O} \cong A'_{\gamma'}$ and everything is obvious.

Consequently, by symmetry we may assume that $P \neq 1 \neq P'$; still denote by $P_{\hat{\gamma}}$ the corresponding local pointed group on $\hat{A}$, via isomorphism 6.7.1; according to our choices of $\hat{\gamma}$ and γ', isomorphism 6.7.1 determines an $\mathcal{O}P$-interior algebra isomorphism

6.9.4
$$\hat{A}_{\hat{\gamma}} \cong \operatorname{Ind}_\sigma\big(S^{\cdot\cdot} \underset{\mathcal{O}}{\otimes} \operatorname{Res}_{\sigma'}(A'_{\gamma'})\big)_{\hat{\gamma}} \qquad ;$$

moreover, since the structural map $f : A \longrightarrow \hat{A}_{\hat{\alpha}}$ is an $\mathcal{O}G$-interior algebra stable embedding, it induces both a k^*-group isomorphism (cf. 6.4.2)

6.9.5
$$\hat{f}(P_\gamma) : \hat{\bar{N}}_G(P_\gamma) \cong \hat{\bar{N}}_G(P_{\hat{\gamma}}) = \hat{\bar{N}}_{G\times G'}\big((P \times G')_{\hat{\gamma}}\big)$$

and a $k_*\hat{\bar{N}}_G(P_\gamma)$-module isomorphism

6.9.6
$$V_A(P_\gamma) \cong \operatorname{Res}_{\hat{f}(P_\gamma)}\big(V_{\hat{A}_{\hat{\alpha}}}(P_{\hat{\gamma}})\big) = \operatorname{Res}_{\hat{f}(P_\gamma)}\Big(V_{M^{\cdot\cdot}}\big((P \times G')_{\hat{\gamma}}\big)\Big) \qquad ,$$

and, choosing $j \in \gamma$ and $A_\gamma = j(\mathcal{O}G)j$, and setting $\hat{A}_\gamma = f(j)\hat{A}f(j)$, it determines an $\mathcal{O}P$-interior algebra stable isomorphism

6.9.7
$$f_\gamma : A_\gamma \longrightarrow \hat{A}_\gamma \qquad ;$$

furthermore, we have $f(j) = \hat{\jmath} + \hat{\imath}$, where $\hat{\jmath} \in \hat{\gamma}$, $\hat{\jmath}\hat{\imath} = 0 = \hat{\imath}\hat{\jmath}$ and $\hat{\imath} \in \hat{A}_1^P$, and, in particular, an $\mathcal{O}P$-interior algebra embedding

6.9.8
$$\hat{A}_{\hat{\gamma}} \longrightarrow \hat{A}_\gamma \qquad .$$

Let n be the maximal number in $\{\dim_k\big(V_{\hat{\imath}\hat{A}\hat{\imath}}(1_\varepsilon)\big)\}_\varepsilon$ where ε runs over the set of points of 1 on $\hat{\imath}\hat{A}\hat{\imath}$; we claim that there is an $\mathcal{O}P$-interior algebra exoembedding

6.9.9
$$\tilde{d} : \hat{A}_\gamma \longrightarrow T_n \underset{\mathcal{O}}{\otimes} \hat{A}_{\hat{\gamma}}$$

such that the following diagram, obtained by tensoring the unique $\mathcal{O}P$-interior algebra exoembedding $\mathcal{O} \longrightarrow T_n$ (cf. 6.8.3) and the exoembedding determined by embedding 6.9.8, is commutative

6.9.10
$$\begin{array}{ccc} \hat{A}_\gamma & \longrightarrow & T_n \underset{\mathcal{O}}{\otimes} \hat{A}_\gamma \\ \uparrow & \searrow^{\tilde{d}} & \uparrow \\ \hat{A}_{\hat{\gamma}} & \longrightarrow & T_n \underset{\mathcal{O}}{\otimes} \hat{A}_{\hat{\gamma}} \end{array} \qquad .$$

Indeed, since $(T_n \otimes_{\mathcal{O}} \hat{A}_\gamma)(P) = \hat{A}_\gamma(P)$ (cf. Lemma 7.10 below or [21, (5.6.1)], and 2.2.5) and $\mathrm{Br}_P(\hat{l}) = 0$, $\hat{\gamma}$ is the unique local point of P on all the $\mathcal{O}P$-interior algebras in diagram 6.9.10 and it has multiplicity one in each one; moreover all the other points of P on them are projective; but, since $\hat{A}_{\hat{\gamma}}$ and $\hat{A}_{\hat{\alpha}}$ are Morita equivalent $\mathcal{O}$-algebras (cf. 2.12), $\hat{A}_{\hat{\gamma}}$ and $\hat{A}_\gamma$ are Morita equivalent too and therefore any point $\hat{\varepsilon}$ of 1 on $\hat{A}_\gamma$ comes already from $\hat{A}_{\hat{\gamma}}$, so that we have (cf. 2.8)

6.9.11 $$m_{\hat{\varepsilon}}(\hat{l}\hat{A}\hat{l}) \leq n m_{\hat{\varepsilon}}(\hat{A}_{\hat{\gamma}}) = m_{\hat{\varepsilon}}(T_n \underset{\mathcal{O}}{\otimes} \hat{A}_{\hat{\gamma}}) \quad ;$$

consequently, assuming that $\hat{A}_{\hat{\gamma}} = \hat{j}\hat{A}\hat{j}$ and denoting by i an idempotent of $(T_n)^P$ such that $iT_n i = \mathcal{O}i$ (i.e. belonging to the local point of P on T_n), it follows from 2.8.4 that the image of $\hat{l}$ in $(T_n \otimes_{\mathcal{O}} \hat{A}_\gamma)^P$ is conjugate to the image of an idempotent of $(T_n \otimes_{\mathcal{O}} \hat{A}_{\hat{\gamma}})^P$ orthogonal to $i \otimes \hat{j}$. Finally, any element of $\big((T_n \otimes_{\mathcal{O}} \hat{A}_\gamma)^P\big)^*$ which realizes this conjugation and centralizes $i \otimes \hat{j}$ determines $\tilde{d}$.

If $M^{\cdot\cdot}$ defines a Morita equivalence between b and b' then, by Proposition 6.5, f is an isomorphism and, in particular, we have $\hat{l} = 0$, so that $n = 0$ and $\tilde{d}$ is an isomorphism too.

Conversely, assume that P and P' are respectively defect groups of b and b', and that we have a stable embedding as announced. If $P = 1 = P'$ then $M^{\cdot\cdot}$ certainly defines a Morita equivalence between b and b'. Consequently, we may assume that $P \neq 1$; as in Remark 6.10, we have an $\mathcal{O}G$-interior algebra stable embedding

6.9.12
$$g = \mathrm{Ind}_P^G(e) : \mathrm{Ind}_P^G(A_\gamma) \to \mathrm{Ind}_P^G\Big(T_n \underset{\mathcal{O}}{\otimes} \mathrm{Ind}_\sigma\big(S^{\cdot\cdot} \underset{\mathcal{O}}{\otimes} \mathrm{Res}_{\sigma'}(A'_{\gamma'})\big)_{\hat{\gamma}}\Big) = \hat{B}$$

and, since A_γ is a source algebra of $A_\alpha = A$, we consider a Higman embedding of $\mathcal{O}P$-interior algebras (cf. 2.12)

6.9.13 $$h : A \longrightarrow \mathrm{Ind}_P^G(A_\gamma) \quad ;$$

hence, G_α is the pointed group on $\mathrm{Ind}_P^G(A_\gamma)$ (cf. 2.11) having defect pointed group P_γ and being determined by the multiplicity $k_*\hat{\bar{N}}_G(P_\gamma)$-module $V_A(P_\gamma)$ (cf. 2.9.3). Now, it follows from 6.4.2 and 6.9.12 that there is a unique point $\hat{\beta}$ of G on $\hat{B}$ fulfilling

6.9.14 $$g(\alpha) \subset \hat{\beta} + \hat{B}_1^G \quad ;$$

moreover, embedding 6.8.4 shows that $P_{\hat{\gamma}}$ is still a local pointed group on $T_n \otimes_{\mathcal{O}} \mathrm{Ind}_\sigma\big(S^{\cdot\cdot} \otimes_{\mathcal{O}} \mathrm{Res}_{\sigma'}(A'_{\gamma'})\big)_{\hat{\gamma}}$ (cf. 2.11), thus on $\hat{B}$ (cf. 2.11), and it is quite clear that $g(\gamma) \subset \hat{\gamma} + \hat{B}_1^P$; consequently, by 6.4.2 again, $P_{\hat{\gamma}}$ is a defect pointed group of $G_{\hat{\beta}}$ and we have both a k^*-group and a $k_*\hat{\bar{N}}_G(P_\gamma)$-module isomorphisms

6.9.15 $$\hat{g}(P_\gamma) : \hat{\bar{N}}_G(P_\gamma) \cong \hat{\bar{N}}_G(P_{\hat{\gamma}}) \quad \text{and} \quad V_A(P_\gamma) \cong \mathrm{Res}_{\hat{g}(P_\gamma)}\big(V_{\hat{B}_{\hat{\beta}}}(P_{\hat{\gamma}})\big) \quad .$$

On the other hand, by inducing embedding 6.8.4 to G, we get an $\mathcal{O}G$-interior embedding

6.9.16 $$\mathrm{Ind}_P^G\Big(\mathrm{Ind}_\sigma\big(S^{\cdot\cdot}\underset{\mathcal{O}}{\otimes}\mathrm{Res}_{\sigma'}(A'_{\gamma'})\big)_{\hat\gamma}\Big)\longrightarrow \hat B$$

and, since $P_{\hat\gamma}$ is a defect pointed group of $G_{\hat\beta}$, it follows from 2.12.3 that $\hat\beta$ comes already from the left end of embedding 6.9.16; moreover, from the natural embedding

6.9.17 $$\mathrm{Ind}_\sigma\big(S^{\cdot\cdot}\underset{\mathcal{O}}{\otimes}\mathrm{Res}_{\sigma'}(A'_{\gamma'})\big)_{\hat\gamma}\longrightarrow \mathrm{Ind}_\sigma\big(S^{\cdot\cdot}\underset{\mathcal{O}}{\otimes}\mathrm{Res}_{\rho'}(\mathcal{O}G')\big)$$

and from 6.7.1 and Corollary 3.13, we get another $\mathcal{O}G$-interior algebra embedding

6.9.18 $$\mathrm{Ind}_P^G\Big(\mathrm{Ind}_\sigma\big(S^{\cdot\cdot}\underset{\mathcal{O}}{\otimes}\mathrm{Res}_{\sigma'}(A'_{\gamma'})\big)_{\hat\gamma}\Big)\longrightarrow \hat A \quad .$$

Consequently, $\hat\beta$ can be identified with a point of G on $\hat A$ (cf. 2.11), $P_{\hat\gamma}$ is still a defect pointed group of $G_{\hat\beta}$ on $\hat A$ and, from 6.9.16 and 6.9.18, we have a canonical $\hat{k}_*\hat N_G(P_{\hat\gamma})$-module isomorphism

6.9.19 $$V_{\hat B_{\hat\beta}}(P_{\hat\gamma})\cong V_{\hat A_{\hat\beta}}(P_{\hat\gamma})$$

(actually, we have an $\mathcal{O}G$-interior algebra isomorphism $\hat B_{\hat\beta}\cong \hat A_{\hat\beta}$); but we are assuming that isomorphism 6.9.2 holds; hence, by 6.8.2, 6.9.2, 6.9.15 and 6.9.19, we have $\hat{k}_*\hat N_G(P_\gamma)$-module isomorphisms

6.9.20 $$\mathrm{Res}_{\hat g(P_\gamma)}\big(V_{\hat A_{\hat\alpha}}(P_{\hat\gamma})\big)\cong V_{A_\alpha}(P_\gamma)\cong \mathrm{Res}_{\hat g(P_\gamma)}\big(V_{\hat A_{\hat\beta}}(P_{\hat\gamma})\big)$$

and therefore we get $\hat\alpha=\hat\beta$ (cf. 2.9.3); in particular, embeddings 6.9.16 and 6.9.18 induce $\mathcal{O}G$-interior algebra isomorphisms

6.9.21 $$\hat B_{\hat\beta}\cong \mathrm{Ind}_P^G\Big(\mathrm{Ind}_\sigma\big(S^{\cdot\cdot}\underset{\mathcal{O}}{\otimes}\mathrm{Res}_{\sigma'}(A'_{\gamma'})\big)_{\hat\gamma}\Big)_{\hat\beta}\cong \hat A_{\hat\alpha}$$

and we may identify f to the structural map $A\longrightarrow \hat B_{\hat\beta}$.

Since $g\circ h$ is a stable embedding and A is an indecomposable $\mathcal{O}G$-bimodule (cf. 2.14), there is an $\mathcal{O}G$-bimodule homomorphism

6.9.22 $$r: g\big(h(b)\big)\hat B g\big(h(b)\big)\longrightarrow A$$

such that we have $r\circ g\circ h=\mathrm{id}_A$ and $\mathrm{Ker}(r)$ is a projective $\mathcal{O}G$-bimodule; then, setting $g\big(h(b)\big)=\hat b+\hat c$ where $\hat b\in\hat\beta$ and $\hat c\in \hat B_1^G$ (cf. 6.9.14) and assuming that $\hat B_{\hat\beta}=\hat b\hat B\hat b$, $r(\hat c)$ belongs to $A_1^G\subset J\big(Z(A)\big)$ (cf. 2.9) and, for any $x\in G$, we have

6.9.23 $$xb=xr\Big(g\big(h(b)\big)\Big)=xr(\hat b)+xr(\hat c)=r\big(f(xb)\big)+xr(\hat c) \qquad ;$$

hence, we get $A = r\big(f(A)\big) + AJ\big(Z(A)\big)$ and therefore $r \circ f : A \longrightarrow A$ is an $\mathcal{O}G$-bimodule isomorphism. Consequently, f is injective and we have the $\mathcal{O}G$-bimodule isomorphisms

$$6.9.24 \quad \hat{b}\hat{B}\hat{b} \cong A \oplus \big(\hat{b}\hat{B}\hat{b} \cap \mathrm{Ker}(r)\big) \quad \text{and} \quad (\hat{b}+\hat{c})\hat{B}(\hat{b}+\hat{c}) \cong A \oplus \mathrm{Ker}(r) \quad ;$$

since $\hat{b}\hat{B}\hat{b}$ is a direct summand of $(\hat{b}+\hat{c})\hat{B}(\hat{b}+\hat{c})$ as $\mathcal{O}G$-bimodules, $\hat{b}\hat{B}\hat{b} \cap \mathrm{Ker}(r)$ is also a projective $\mathcal{O}G$-bimodule and therefore f is a stable embedding too. Finally, if $n = 0$ and e is an isomorphism then g is also an isomorphism and therefore we have $\hat{c} = 0$ and $f = g \circ h$ is an isomorphism too. In any case, Proposition 6.15 achieves the proof.

6.12 Now, let us explicitly show how Theorem 6.9 provides a classification of all the choices for $M^{\cdot\cdot}$ defining a Morita stable equivalence between b and b'; we may assume that b and b' are not projective (cf. 2.9). We fix defect pointed groups Q_δ of b and $Q'_{\delta'}$ of b', and consider all the subgroups $Q^{\cdot\cdot}$ of $Q \times Q'$ surjectively mapped onto both Q and Q' via the projection maps, noted $\tau : Q^{\cdot\cdot} \longrightarrow Q$ and $\tau' : Q^{\cdot\cdot} \longrightarrow Q'$; for such a subgroup $Q^{\cdot\cdot}$, we look for all the indecomposable $\mathcal{O}Q^{\cdot\cdot}$-modules $L^{\cdot\cdot}$ of vertex $Q^{\cdot\cdot}$ such that their restrictions to $\mathrm{Ker}(\tau) = Q^{\cdot\cdot} \cap (1 \times G')$ and to $\mathrm{Ker}(\tau') = Q^{\cdot\cdot} \cap (G \times 1)$ are both projective and, for some $n \geq 0$, there is an $\mathcal{O}Q$-interior algebra stable embedding

$$6.12.1 \quad f : A_\delta \longrightarrow \mathrm{End}_{\mathcal{O}}\big(\mathcal{O} \oplus (\mathcal{O}Q)^n\big) \underset{\mathcal{O}}{\otimes} \mathrm{Ind}_\tau\big(\mathrm{End}_{\mathcal{O}}(L^{\cdot\cdot}) \underset{\mathcal{O}}{\otimes} \mathrm{Res}_{\tau'}(A'_{\delta'})\big) \quad .$$

Then, we claim that:

6.12.2 *The triple $(Q^{\cdot\cdot}, L^{\cdot\cdot}, f)$ determines one choice for $M^{\cdot\cdot}$ defining a Morita stable equivalence between b and b'.*

6.13 Indeed, it is clear that f determines a local point $\hat{\delta}$ of Q on $\mathrm{Ind}_\tau\big(\mathrm{End}_{\mathcal{O}}(L^{\cdot\cdot}) \otimes_{\mathcal{O}} \mathrm{Res}_{\tau'}(A'_{\delta'})\big)$ and a k^*-group isomorphism (cf. 6.4.2)

$$6.13.1 \qquad \hat{f}(Q_\delta) : \hat{\bar{N}}_G(Q_\delta) \cong \hat{\bar{N}}_G(Q_{\hat{\delta}})$$

and, since we have the $\mathcal{O}$-algebra embeddings (cf. Proposition 3.7 and 4.4.1)

$$\mathrm{Ind}_\tau\big(\mathrm{End}_{\mathcal{O}}(L^{\cdot\cdot}) \underset{\mathcal{O}}{\otimes} \mathrm{Res}_{\tau'}(A'_{\delta'})\big)^Q \to \mathrm{Ind}_\tau\big(\mathrm{End}_{\mathcal{O}}(L^{\cdot\cdot}) \underset{\mathcal{O}}{\otimes} \mathrm{Res}_{\zeta'}(\mathcal{O}G')\big)^Q$$

$$6.13.2 \quad \mathrm{Ind}_\tau\big(\mathrm{End}_{\mathcal{O}}(L^{\cdot\cdot}) \underset{\mathcal{O}}{\otimes} \mathrm{Res}_{\zeta'}(\mathcal{O}G')\big)^Q \cong \mathrm{Ind}_{Q^{\cdot\cdot}}^{Q\times G'}\big(\mathrm{End}_{\mathcal{O}}(L^{\cdot\cdot})\big)^{Q\times G'}$$

$$\mathrm{Ind}_{Q^{\cdot\cdot}}^{Q\times G'}\big(\mathrm{End}_{\mathcal{O}}(L^{\cdot\cdot})\big)^{Q\times G'} \to \mathrm{End}_{\mathcal{O}}\big(\mathrm{Ind}_{Q^{\cdot\cdot}}^{G\times G'}(L^{\cdot\cdot})\big)^{Q\times G'} \quad ,$$

where $\zeta' : Q^{\cdot\cdot} \longrightarrow G'$ is the group homomorphism determined by π', the local point $\hat{\delta}$ can be identified with a point of $Q \times G'$ on $\mathrm{End}_{\mathcal{O}}\big(\mathrm{Ind}_{Q^{\cdot\cdot}}^{G\times G'}(L^{\cdot\cdot})\big)$; then, denoting by $Q^{\cdot\cdot}_{L^{\cdot\cdot}}$ the corresponding local pointed group on $\mathrm{End}_{\mathcal{O}}\big(\mathrm{Ind}_{Q^{\cdot\cdot}}^{G\times G'}(L^{\cdot\cdot})\big)$ (cf. 2.10), it follows from Theorem 5.8 that $Q^{\cdot\cdot}_{L^{\cdot\cdot}} \subset (Q \times G')_{\hat\delta}$. On the one hand, up to suitable identifications, we have the k^*-group equality

$$\hat{\bar{N}}_G(Q_{\hat\delta}) = \hat{\bar{N}}_{G\times G'}\big((Q \times G')_{\hat\delta}\big) \qquad ; \tag{6.13.3}$$

on the other hand, since $Q^{\cdot\cdot}_{L^{\cdot\cdot}}$ is a maximal local pointed group on $\mathrm{End}_{\mathcal{O}}\big(\mathrm{Ind}_{Q^{\cdot\cdot}}^{G\times G'}(L^{\cdot\cdot})\big)$ (cf. 2.11.3) and $N_{G\times G'}(Q^{\cdot\cdot}_{L^{\cdot\cdot}})$ normalizes $Q \times G'$, it follows from [2, Theorem 5.1] that, for any point $\beta^{\cdot\cdot}$ of $G \times G'$ on this $\mathcal{O}(G \times G')$-interior algebra such that

$$(Q \times G')_{\hat\delta} \subset (G \times G')_{\beta^{\cdot\cdot}} \qquad ,$$

the multiplicity module of $(Q \times G')_{\hat\delta}$ on $\mathrm{End}_{\mathcal{O}}\big(\mathrm{Ind}_{Q^{\cdot\cdot}}^{G\times G'}(L^{\cdot\cdot})\big)_{\beta^{\cdot\cdot}}$ is an indecomposable projective $k_*\hat{\bar{N}}_{G\times G'}\big((Q \times G')_{\hat\delta}\big)$-module, and that this correspondence defines a bijection between the set of such points and the set of isomorphism classes of such modules. Finally, notice that the restrictions of $\mathrm{Ind}_{Q^{\cdot\cdot}}^{G\times G'}(L^{\cdot\cdot})$ to $\mathcal{O}(G \times 1)$ and to $\mathcal{O}(1 \times G')$ are both projective. Consequently, it follows from Theorem 6.9 and 2.9.3 that:

6.13.4 *The indecomposable projective $k_*\hat{\bar{N}}_G(Q_\delta)$-module $V_A(Q_\delta)$ determines, via the k^*-group isomorphism $\hat{f}(Q_\delta)$, an indecomposable direct summand of the $\mathcal{O}(G \times G')$-module $\mathrm{Ind}_{Q^{\cdot\cdot}}^{G\times G'}(L^{\cdot\cdot})$ which defines a Morita stable equivalence between b and b'.*

7 Basic Morita stable equivalences between Brauer blocks

7.1 We keep the notation of Section 6; as in 6.6 and 6.7, $P^{\cdot\cdot}$ is a vertex and $N^{\cdot\cdot}$ an $\mathcal{O}P^{\cdot\cdot}$-source of $M^{\cdot\cdot}$, set $S^{\cdot\cdot} = \mathrm{End}_{\mathcal{O}}(N^{\cdot\cdot})$ and denote by $\sigma : P^{\cdot\cdot} \to P$ and $\sigma' : P^{\cdot\cdot} \to P'$ the surjective group homomorphisms determined by the projection maps on $G \times G'$. As we say in the introduction, in all the known situations where $\mathcal{O}$ has characteristic zero and $M^{\cdot\cdot}$ defines a Morita stable equivalence between b and b', it turns out that σ and σ' are both bijective and that $P^{\cdot\cdot}$ stabilizes by conjugation an $\mathcal{O}$-basis of $S^{\cdot\cdot}$ - that is to say, $S^{\cdot\cdot}$ is a so-called *Dade $P^{\cdot\cdot}$-algebra* (cf. 2.2) - which implies that p does not divide $\mathrm{rank}_{\mathcal{O}}(N^{\cdot\cdot})$ (cf. [38, (28.11)]). In this section we prove that all these statements are in fact equivalent to each other (see Corollary 7.4 below), and whenever they hold we say that the Morita stable equivalence between b and b' defined by $M^{\cdot\cdot}$ is *basic*. The most difficult part of the proof, namely the existence of a $P^{\cdot\cdot}$-stable $\mathcal{O}$-basis in $S^{\cdot\cdot}$, is a consequence of the following general (and surprising!) result. Recall that a *nondegenerate symmetric* $\mathcal{O}$-form $\mu : B \to \mathcal{O}$ of an $\mathcal{O}$-free $\mathcal{O}$-algebra B is an $\mathcal{O}$-linear map fulfilling $\mu(aa') = \mu(a'a)$ for any $a, a' \in B$ and inducing an $\mathcal{O}$-module isomorphism $B \cong \mathrm{Hom}_{\mathcal{O}}(B, \mathcal{O})$ which maps $a \in B$ on the $\mathcal{O}$-form mapping $a' \in B$ on $\mu(aa')$.

Theorem 7.2 *Let Q be a finite p-group, B and B' $\mathcal{O}$-free $\mathcal{O}Q$-interior algebras and N an indecomposable $\mathcal{O}$-free $\mathcal{O}Q$- module. Set $S = \mathrm{End}_{\mathcal{O}}(N)$ and assume that there is an $\mathcal{O}Q$-interior algebra embedding*

7.2.1 $$B \longrightarrow S \underset{\mathcal{O}}{\otimes} B' \quad .$$

If Q stabilizes by conjugation $\mathcal{O}$-bases of B and B', B admits a nondegenerate symmetric $\mathcal{O}$-form and we have $B(Q) \neq \{0\}$ then S is a Dade Q-algebra.

Remark 7.3 Since $(S \otimes_{\mathcal{O}} B')(Q) \cong S(Q) \otimes_k B'(Q)$ (cf. Lemma 7.10 below or [21, (5.6.1)]), we have $S(Q) \neq \{0\}$ and there are respectively local points δ and δ' of Q on B and B' such that 7.2.1 induces an $\mathcal{O}Q$-interior algebra exoembedding (cf. [21, (5.6.4)])

7.3.1 $$B_\delta \longrightarrow S \underset{\mathcal{O}}{\otimes} B'_{\delta'} \quad ;$$

consequently, it suffices to prove that S is a permutation $\mathcal{O}Q$-module by conjugation and we may assume that $B \cong B_\delta$ and $B' \cong B'_{\delta'}$. In that case, *a posteriori* the roles of B and B' are completely symmetric; indeed, once we know that S is a Dade Q-algebra, we have an $\mathcal{O}Q$-interior algebra embedding $\mathcal{O} \to S^\circ \otimes_{\mathcal{O}} S$ (cf. [21, 5.7]); then, by tensoring this embedding by B' and embedding 7.2.1 by S°, we get two $\mathcal{O}Q$-interior algebra embeddings

7.3.2 $$B' \longrightarrow S^\circ \underset{\mathcal{O}}{\otimes} S \underset{\mathcal{O}}{\otimes} B' \longleftarrow S^\circ \underset{\mathcal{O}}{\otimes} B$$

and, since $B' \cong B'_{\delta'}$ and $(S^\circ \otimes_{\mathcal{O}} S \otimes_{\mathcal{O}} B')(Q) \cong B'(Q)$ (cf. Lemma 7.10 or [21, (5.6.1)], and 2.2.7), the first exoembedding factorizes throughout the second one and an $\mathcal{O}Q$-interior algebra exoembedding (cf. [20, 1.6])

7.3.3
$$B' \longrightarrow S^\circ \underset{\mathcal{O}}{\otimes} B \quad ,$$

which proves that B' admits also a nondegenerate symmetric form.

Corollary 7.4 *Assume that $M^{\cdot\cdot}$ defines a Morita stable equivalence between b and b'. Then the following conditions are equivalent:*

7.4.1 *σ is a group isomorphism.*

7.4.2 *σ' is a group isomorphism.*

7.4.3 *p does not divide* $\mathrm{rank}_{\mathcal{O}}(N^{\cdot\cdot})$.

7.4.4 *$S^{\cdot\cdot}$ is a Dade $P^{\cdot\cdot}$-algebra.*

Remark 7.5 If $M^{\cdot\cdot}$ defines a Morita equivalence between b and b', and moreover $N^{\cdot\cdot} \cong \mathcal{O}$ or, equivalently, $\mathrm{Res}^{G\times G'}_{P^{\cdot\cdot}}(M^{\cdot\cdot})$ is a permutation $\mathcal{O}P^{\cdot\cdot}$-module then it follows from Theorem 6.9 and Corollary 7.4 that we have an $\mathcal{O}P$-interior algebra isomorphism

7.5.1
$$A_\gamma \cong \mathrm{Res}_{\sigma'\circ\sigma^{-1}}(A'_{\gamma'}) \quad ,$$

a fact proved by Leonard Scott [36] independently.

Proof: Since $N^{\cdot\cdot}$ is indecomposable, condition 7.4.4 implies $\mathrm{rank}_{\mathcal{O}}(S^{\cdot\cdot}) \equiv 1(p)$ (cf. [38, (28.11)]). On the other hand, since the restrictions of $N^{\cdot\cdot}$ to $\mathcal{O}(\mathrm{Ker}(\sigma))$ and to $\mathcal{O}(\mathrm{Ker}(\sigma'))$ are both projective (cf. 6.1), condition 7.4.3 implies that $\mathrm{Ker}(\sigma) = 1 \times 1 = \mathrm{Ker}(\sigma')$. Finally, assume that σ is a group isomorphism; then, since $\hat{A}_{\hat{\gamma}}$ has a P-stable $\mathcal{O}$-basis by Remark 6.11 and $A'_{\gamma'}$ is symmetric, Theorem 7.2 applies to the canonical $\mathcal{O}P$-interior algebra embedding (cf. 5.3.1)

7.4.5
$$\hat{A}_{\hat{\gamma}} \longrightarrow \mathrm{Res}_{\sigma^{-1}}(S^{\cdot\cdot}) \underset{\mathcal{O}}{\otimes} \mathrm{Res}_{\sigma'\circ\sigma^{-1}}(A'_{\gamma'})$$

(notice that the existence of this embedding forces $\hat{A}_{\hat{\gamma}}$ to be symmetric too).

7.6 Before proving Theorem 7.2, let us comment some consequences of the equivalent conditions above. So, assume that $M^{\cdot\cdot}$ defines a basic Morita stable equivalence between b and b'; then, for any nontrivial subgroup Q of P, the $\mathcal{O}P$-interior algebra stable embedding 6.9.1 induces an $\mathcal{O}C_P(Q)$-interior algebra embedding

7.6.1
$$A_\gamma(Q) \longrightarrow \mathrm{Res}_{\sigma_Q^{-1}}\big(S^{\cdot\cdot}(Q^{\cdot\cdot})\big) \underset{k}{\otimes} \mathrm{Res}_{\sigma'_Q\circ\sigma_Q^{-1}}\big(A'_{\gamma'}(Q')\big) \quad ,$$

where $Q^{\cdot\cdot} = \sigma^{-1}(Q)$, $Q' = \sigma'(Q^{\cdot\cdot})$ and we denote respectively by

$$\sigma_Q : C_{P^{\cdot\cdot}}(Q^{\cdot\cdot}) \cong C_P(Q) \quad \text{and} \quad \sigma'_Q : C_{P^{\cdot\cdot}}(Q^{\cdot\cdot}) \cong C_{P'}(Q')$$

the isomorphisms determined by σ and σ'; since $S^{\cdot\cdot}(Q^{\cdot\cdot})$ is a simple k-algebra (cf. 2.2.7), this embedding induces an injective map from $\mathcal{LP}_{A_\gamma}(Q)$ to $\mathcal{LP}_{A'_{\gamma'}}(Q')$, so that in particular we have $|\mathcal{LP}_{A_\gamma}(Q)| \leq |\mathcal{LP}_{A'_{\gamma'}}(Q')|$; consequently, by the symmetry of our hypothesis, embedding 7.6.1 induces a bijective map

7.6.2 $$\mathcal{LP}_{A_\gamma}(Q) \cong \mathcal{LP}_{A'_{\gamma'}}(Q') \quad .$$

Moreover, if this bijection maps $\delta \in \mathcal{LP}_{A_\gamma}(Q)$ on $\delta' \in \mathcal{LP}_{A'_{\gamma'}}(Q')$, it is well-known that (cf. [14, Lemma 1.17] and [20, Proposition 2.14])

7.6.3 $$\tilde{\sigma}^{-1} \circ F_A(Q_\delta, P_\gamma) \circ \tilde{\tau} = \tilde{\sigma}'^{-1} \circ F_{A'}(Q'_{\delta'}, P'_{\gamma'}) \circ \tilde{\tau}' \subset F_{S^{\cdot\cdot}}(Q^{\cdot\cdot}_{\delta^{\cdot\cdot}}, P^{\cdot\cdot}_{\gamma^{\cdot\cdot}}) \quad ,$$

where $\{\gamma^{\cdot\cdot}\} = \mathcal{LP}_{S^{\cdot\cdot}}(P^{\cdot\cdot}), \{\delta^{\cdot\cdot}\} = \mathcal{LP}_{S^{\cdot\cdot}}(Q^{\cdot\cdot})$ and we respectively denote by $\tau : Q^{\cdot\cdot} \cong Q$ and $\tau' : Q^{\cdot\cdot} \cong Q'$ the isomorphisms determined by σ and σ', that the stable embedding 6.9.1 determines an $\mathcal{O}Q$-interior algebra stable embedding (cf. [21, (5.6.4)])

7.6.4 $$A_\delta \longrightarrow T_n \underset{\mathcal{O}}{\otimes} \mathrm{Res}_{\tau^{-1}}(S^{\cdot\cdot}_{\delta^{\cdot\cdot}}) \underset{\mathcal{O}}{\otimes} \mathrm{Res}_{\tau' \circ \tau^{-1}}(A'_{\delta'})$$

and that there is a k^*-group isomorphism (cf. [14, 2.12.4] and [21, (5.11.1)])

7.6.5 $$\hat{F}_A(Q_\delta) \cong \hat{F}_{A'}(Q'_{\delta'}) \quad .$$

In particular, it follows from the main result of [20] that (cf. 2.15):

7.6.6 *The stable embedding 6.9.1 induces an equivalence between the stable local categories of b and b'.*

7.7 On the other hand, with $M^{\cdot\cdot}$, Q_δ and $Q'_{\delta'}$ as above, it is well-known that there are unique points β of $QC_G(Q)$ on A and β' of $Q'C_{G'}(Q')$ on A' such that (cf. 2.15.2)

7.7.1 $$Q_\delta \subset Q{\cdot}C_G(Q)_\beta \quad \text{and} \quad Q'_{\delta'} \subset Q'{\cdot}C_{G'}(Q')_{\beta'} \quad ;$$

then, up to replace Q_δ by a G-conjugate, we may assume that, for a suitable local point ε of $Q{\cdot}C_P(Q)$ on A, we have $Q{\cdot}C_P(Q)_\varepsilon \subset P_\gamma$ and $Q{\cdot}C_P(Q)_\varepsilon$ is a defect pointed group of $Q{\cdot}C_G(Q)_\beta$ (cf. 2.9.5 and 2.16.2); in that case, denoting by ε' the corresponding local point of $Q'{\cdot}C_{P'}(Q')$ on $A'_{\gamma'}$ (cf. 7.6.2), it is not difficult to see that $Q'{\cdot}C_{P'}(Q')_{\varepsilon'}$ is a defect pointed group of $Q'{\cdot}C_{G'}(Q')_{\beta'}$ too (cf. 2.16.2). Now, denoting respectively by $b(\delta)$ and $b(\delta')$ the blocks of $C_G(Q)$ and $C_{G'}(Q')$ determined by δ and δ', it is clear that we have (cf. 2.15.2)

7.7.2 $$\mathrm{Br}_Q(\beta) = \Big\{\mathrm{Br}_Q\big(b(\delta)\big)\Big\} \quad \text{and} \quad \mathrm{Br}_{Q'}(\beta') = \Big\{\mathrm{Br}_{Q'}\big(b(\delta')\big)\Big\}$$

and that $C_P(Q)_{\mathrm{Br}_Q(\varepsilon)}$ and $C_{P'}(Q')_{\mathrm{Br}_{Q'}(\varepsilon')}$ are respectively maximal local pointed groups on $\mathcal{O}C_G(Q)b(\delta)$ and on $\mathcal{O}C_{G'}(Q')b(\delta')$ (cf. 2.15.4); moreover, from embedding 7.6.1 we get the $kC_P(Q)$-interior algebra embedding

7.7.3 $$A_\varepsilon(Q) \longrightarrow \mathrm{Res}_{\sigma_Q^{-1}}\Big(\mathrm{End}_k\big(N^{\cdot\cdot}(Q_\delta)\big)\Big) \underset{k}{\otimes} \mathrm{Res}_{\sigma'_Q \circ\, \sigma_Q^{-1}}\big(A'_{\varepsilon'}(Q')\big) \quad ,$$

where, respectively denoting by $\delta^{\cdot\cdot}$ and $\varepsilon^{\cdot\cdot}$ the unique local points of $Q^{\cdot\cdot}$ and $Q^{\cdot\cdot}\cdot C_{P^{\cdot\cdot}}(Q^{\cdot\cdot})$ on $S^{\cdot\cdot}$, we set $N^{\cdot\cdot}(Q_\delta) = V_{S^{\cdot\cdot}_{\varepsilon^{\cdot\cdot}}}(Q^{\cdot\cdot}_{\delta^{\cdot\cdot}})$ (cf. 2.8). At that point, we are able to apply 6.11 and 6.12 to the blocks $\bar{b}(\delta)$ of $C_G(Q)$ and $\bar{b}(\delta')$ of $C_{G'}(Q')$ over k determined respectively by δ and δ', and to the triple $C_{P^{\cdot\cdot}}(Q^{\cdot\cdot})$, $N^{\cdot\cdot}(Q_\delta)$ and embedding 7.7.3, obtaining:

7.7.4 *The indecomposable projective $k_*\hat{\bar{N}}_{C_G(Q)}(C_P(Q)_{\mathrm{Br}_Q(\varepsilon)})$-module*

$$V_{kC_G(Q)\bar{b}(\delta)}(C_P(Q)_{\mathrm{Br}_Q(\varepsilon)})$$

determines, up to isomorphism, an indecomposable direct summand $M^{\cdot\cdot}(Q_\delta)$ of the $k(C_G(Q)\times C_{G'}(Q'))$-module $\mathrm{Ind}_{C_{P^{\cdot\cdot}}(Q^{\cdot\cdot})}^{C_{G\times G'}(Q^{\cdot\cdot})}(N^{\cdot\cdot}(Q_\delta))$, which defines a basic Morita equivalence between $\bar{b}(\delta)$ and $\bar{b}(\delta')$.

Remark 7.8 It is not difficult to prove, from 6.11 and 6.12, that $M^{\cdot\cdot}$ defines a basic Morita stable equivalence between b and b' if and only if $k\otimes_{\mathcal{O}} M^{\cdot\cdot}$ does it between the corresponding blocks over k and $N^{\cdot\cdot}$ is an endopermutation $\mathcal{O}P^{\cdot\cdot}$-module (for instance, see [21, 7.7]). In particular, the basic Morita equivalence between $\bar{b}(\delta)$ and $\bar{b}(\delta')$ in 7.7.4 can be lifted to a basic Morita equivalence between $b(\delta)$ and $b(\delta')$ whenever $N^{\cdot\cdot}(Q_\delta)$ can be lifted to an endopermutation $\mathcal{O}C_{P^{\cdot\cdot}}(Q^{\cdot\cdot})$-module.

7.9 In order to prove Theorem 7.2, we need the following canonical homomorphisms between Brauer sections. If H is a finite group, B an H-algebra over $\mathcal{O}$, Q a p-subgroup of H and R a normal subgroup of Q, setting $\bar{Q} = Q/R$ it is easily checked that there is a unique $\bar{N}_{\bar{N}_H(R)}(\bar{Q})$-algebra homomorphism

7.9.1 $$\tau_B(R,Q) : B(Q) \longrightarrow \big(B(R)\big)(\bar{Q})$$

which, for any $a\in B^Q$, maps $\mathrm{Br}_Q(a)$ on $\mathrm{Br}_{\bar{Q}}^{B(R)}\big(\mathrm{Br}_R(a)\big)$. Moreover, if C is a second H-algebra over $\mathcal{O}$, it is quite clear that there is a unique $\bar{N}_H(Q)$-algebra homomorphism

7.9.2 $$\alpha_{B,C}(Q) : B(Q)\underset{k}{\otimes} C(Q) \longrightarrow (B\underset{\mathcal{O}}{\otimes} C)(Q)$$

which, for any $a\in B^Q$ and any $c\in C^Q$, maps $\mathrm{Br}_Q(a)\otimes\mathrm{Br}_Q(c)$ on $\mathrm{Br}_Q(a\otimes c)$. The following lemmas collect the formal features of these homomorphisms that we need in the sequel.

Lemma 7.10 *With the notation above, if B has a Q-stable $\mathcal{O}$-basis then $\tau_B(R,Q)$ and $\alpha_{B,C}(Q)$ are both isomorphisms.*

Proof: If X is a Q-stable $\mathcal{O}$-basis of B, it is easily checked that $\mathrm{Br}_R(X^R)$ is a $\bar{Q}$-stable k-basis of $B(R)$ (cf. 2.2.5) and, similarly, that $\mathrm{Br}_{\bar{Q}}^{B(R)}\big(\mathrm{Br}_R(X^R)^{\bar{Q}}\big)$ is a k-basis of $\big(B(R)\big)(\bar{Q})$; but it is clear that we have $\mathrm{Br}_R(X^R)^{\bar{Q}} = \mathrm{Br}_R(X^Q)$ and therefore $\tau_B(R,Q)$ maps bijectively $\mathrm{Br}_Q(X^Q)$ onto $\mathrm{Br}_{\bar{Q}}^{B(R)}(\mathrm{Br}_R(X^R)^{\bar{Q}})$. On the other hand, we have the Q-stable decomposition

7.10.1 $$B \underset{\mathcal{O}}{\otimes} C = \bigoplus_{x\in X} x \otimes C \qquad ;$$

consequently, we get

7.10.2 $$\mathrm{Br}_Q\left(\left(\bigoplus_{x\in X-X^Q} x \otimes C\right)^Q\right) = \{0\} \quad ,$$

whereas, for any $x \in X^Q$, the evident map $C \to x \otimes C$ is an $\mathcal{O}Q$-module isomorphism, so that $\alpha_{B,C}(Q)$ induces a bijective map from $\mathrm{Br}_Q(x) \otimes C(Q)$ onto $\mathrm{Br}_Q(x \otimes C^Q)$; finally, since $\mathrm{Br}_Q(X^Q)$ is a k-basis of $B(Q)$ (cf. 2.2.5), $\alpha_{B,C}(Q)$ is indeed an isomorphism.

Lemma 7.11 *With the notation above, let B' be a third H-algebra over $\mathcal{O}$ and $g : B \to B'$ an H-algebra homomorphism. Then the following diagrams are commutative.*

7.11.1
$$\begin{array}{ccc} (B(R))(\bar{Q}) & \xrightarrow{(g(R))(\bar{Q})} & (B'(R))(\bar{Q}) \\ \uparrow{\scriptstyle \tau_B(R,Q)} & & \uparrow{\scriptstyle \tau_{B'}(R,Q)} \\ B(Q) & \xrightarrow{g(Q)} & B'(Q) \end{array} \qquad \begin{array}{ccc} (B \underset{\mathcal{O}}{\otimes} C)(Q) & \xrightarrow{(g\otimes \mathrm{id}_C)(Q)} & (B' \underset{\mathcal{O}}{\otimes} C)(Q) \\ \uparrow{\scriptstyle \alpha_{B,C}(Q)} & & \uparrow{\scriptstyle \alpha_{B',C}(Q)} \\ B(Q) \underset{k}{\otimes} C(Q) & \xrightarrow{g(Q)\otimes \mathrm{id}_{C(Q)}} & B'(Q) \underset{k}{\otimes} C(Q) \end{array} .$$

Proof: For any $a \in B^Q$ we have

$$\begin{aligned} (g(R))(\bar{Q})\Big(\tau_B(R,Q)(\mathrm{Br}_Q(a))\Big) &= (g(R))(\bar{Q})\Big(\mathrm{Br}_{\bar{Q}}^{B(R)}(\mathrm{Br}_R(a))\Big) \\ \text{7.11.2} \qquad = \mathrm{Br}_{\bar{Q}}^{B'(R)}\Big(g(R)(\mathrm{Br}_R(a))\Big) &= \mathrm{Br}_{\bar{Q}}^{B'(R)}\Big(\mathrm{Br}_R(g(a))\Big) \\ = \tau_{B'}(R,Q)\Big(\mathrm{Br}_Q(g(a))\Big) &= \tau_{B'}(R,Q)\Big(g(Q)(\mathrm{Br}_Q(a))\Big) \end{aligned}$$

and moreover, for any $c \in C^Q$, we get

7.11.3
$$\begin{aligned} (g \underset{\mathcal{O}}{\otimes} \mathrm{id}_C)(Q)\Big(\alpha_{B,C}(Q)(\mathrm{Br}_Q(a) \otimes \mathrm{Br}_Q(c))\Big) &= (g \underset{\mathcal{O}}{\otimes} \mathrm{id}_C)(Q)(\mathrm{Br}_Q(a \otimes c)) \\ = \mathrm{Br}_Q\big(g(a) \otimes c\big) &= \alpha_{B',C}(Q)\Big(\mathrm{Br}_Q(g(a)) \otimes \mathrm{Br}_Q(c)\Big) \\ = \alpha_{B',C}(Q)\Big(\big(g(Q) \underset{k}{\otimes} \mathrm{id}_{C(Q)}\big)&(\mathrm{Br}_Q(a) \otimes \mathrm{Br}_Q(c))\Big) \quad . \end{aligned}$$

Lemma 7.12 *With the notation above, the following diagram is commutative*

7.12.1
$$\begin{array}{ccc}
(B \underset{\mathcal{O}}{\otimes} C)(Q) & \xrightarrow{\tau_{B\otimes C}(R,Q)} & \big((B \underset{\mathcal{O}}{\otimes} C)(R)\big)(\bar{Q}) \\
 & & \uparrow {\scriptstyle (\alpha_{B,C}(R))(\bar{Q})} \\
{\scriptstyle \alpha_{B,C}(Q)} \uparrow & & \big(B(R) \underset{k}{\otimes} C(R)\big)(\bar{Q}) \\
 & & \uparrow {\scriptstyle \alpha_{B(R),C(R)}(\bar{Q})} \\
B(Q) \underset{k}{\otimes} C(Q) & \xrightarrow{\tau_B(R,Q)\otimes\tau_C(R,Q)} & \big(B(R)\big)(\bar{Q}) \underset{k}{\otimes} \big(C(R)\big)(\bar{Q})
\end{array} .$$

Proof: For any $a \in B^R$ and any $c \in C^R$, we have

7.12.2
$$\begin{aligned}
&\tau_{B\otimes_{\mathcal{O}}C}(R,Q)\Big(\alpha_{B,C}(Q)\big(\mathrm{Br}_Q(a)\otimes\mathrm{Br}_Q(c)\big)\Big) = \tau_{B\otimes_{\mathcal{O}}C}(R,Q)\big(\mathrm{Br}_Q(a\otimes c)\big) \\
&\quad = \mathrm{Br}_{\bar{Q}}^{(B\otimes_{\mathcal{O}}C)(R)}\big(\mathrm{Br}_R(a\otimes c)\big) = \mathrm{Br}_{\bar{Q}}^{(B\otimes_{\mathcal{O}}C)(R)}\Big(\alpha_{B,C}(R)\big(\mathrm{Br}_R(a)\otimes\mathrm{Br}_R(c)\big)\Big) \\
&\quad = \big(\alpha_{B,C}(R)\big)(\bar{Q})\Big(\mathrm{Br}_{\bar{Q}}^{B(R)\otimes_k C(R)}\big(\mathrm{Br}_R(a)\otimes\mathrm{Br}_R(c)\big)\Big) \\
&\quad = \big(\alpha_{B,C}(R)\big)(\bar{Q})\Big(\alpha_{B(R),C(R)}(\bar{Q})\Big(\mathrm{Br}_{\bar{Q}}^{B(R)}\big(\mathrm{Br}_R(a)\big)\otimes\mathrm{Br}_{\bar{Q}}^{C(R)}\big(\mathrm{Br}_R(c)\big)\Big)\Big) \\
&\quad = \big(\alpha_{B,C}(R)\big)(\bar{Q})\Big(\alpha_{B(R),C(R)}(\bar{Q})\Big(\tau_B(R,Q)\big(\mathrm{Br}_Q(a)\big)\otimes\tau_C(R,Q)\big(\mathrm{Br}_Q(c)\big)\Big)\Big) .
\end{aligned}$$

Lemma 7.13 *With the notation above, let T be a normal subgroup of Q contained in R and set $\bar{R} = R/T$ and $\bar{\bar{Q}} = Q/T$. Then the following diagram is commutative.*

7.13.1
$$\begin{array}{ccc}
\big(B(R)\big)(\bar{Q}) & \xrightarrow{(\tau_B(T,R))(\bar{Q})} & \Big(\big(B(T)\big)(\bar{R})\Big)(\bar{Q}) \\
{\scriptstyle \tau_B(R,Q)} \uparrow & & \uparrow {\scriptstyle \tau_{B(T)}(\bar{R},\bar{\bar{Q}})} \\
B(Q) & \xrightarrow{\tau_B(T,Q)} & \big(B(T)\big)(\bar{\bar{Q}})
\end{array} .$$

Proof: For any $a \in B^Q$ we have

7.13.2
$$\begin{aligned}
&\big(\tau_B(T,R)\big)(\bar{Q})\Big(\tau_B(R,Q)\big(\mathrm{Br}_Q(a)\big)\Big) = \big(\tau_B(T,R)\big)(\bar{Q})\Big(\mathrm{Br}_{\bar{Q}}^{B(R)}\big(\mathrm{Br}_R(a)\big)\Big) \\
&= \mathrm{Br}_{\bar{Q}}^{(B(T))(\bar{R})}\Big(\tau_B(T,R)\big(\mathrm{Br}_R(a)\big)\Big) = \mathrm{Br}_{\bar{Q}}^{(B(T))(\bar{R})}\Big(\mathrm{Br}_{\bar{R}}^{B(T)}\big(\mathrm{Br}_T(a)\big)\Big) \\
&= \tau_{B(T)}(\bar{R},\bar{\bar{Q}})\Big(\mathrm{Br}_{\bar{\bar{Q}}}^{B(T)}\big(\mathrm{Br}_T(a)\big)\Big) = \tau_{B(T)}(\bar{R},\bar{\bar{Q}})\Big(\tau_B(T,Q)\big(\mathrm{Br}_Q(a)\big)\Big) .
\end{aligned}$$

Lemma 7.14 *With the notation above, let C and D be H-algebras over $\mathcal{O}$. Then the following diagram is commutative.*

7.14.1

$$\begin{array}{ccc}
(B \underset{\mathcal{O}}{\otimes} C)(Q) \underset{k}{\otimes} D(Q) & \xrightarrow{\alpha_{B\otimes C,D}(Q)} & (B \underset{\mathcal{O}}{\otimes} C \underset{\mathcal{O}}{\otimes} D)(Q) \\
{\scriptstyle \alpha_{B,C}(Q)\otimes \mathrm{id}_{D(Q)}}\Big\uparrow & & \Big\uparrow{\scriptstyle \alpha_{B,C\otimes D}(Q)} \\
B(Q) \underset{k}{\otimes} C(Q) \underset{k}{\otimes} D(Q) & \xrightarrow{\mathrm{id}_{B(Q)}\otimes \alpha_{C,D}(Q)} & B(Q) \underset{k}{\otimes} (C \underset{\mathcal{O}}{\otimes} D)(Q)
\end{array} .$$

Proof: For any $a \in B^Q$, any $c \in C^Q$ and any $d \in D^Q$, we have

7.14.2

$$\begin{aligned}
\alpha_{B\otimes_{\mathcal{O}} C,D}(Q)\Big(\alpha_{B,C}(Q)\big(\mathrm{Br}_Q(a)\otimes \mathrm{Br}_Q(c)\big)\otimes \mathrm{Br}_Q(d)\Big) \\
&= \alpha_{B\otimes_{\mathcal{O}} C,D}(Q)\big(\mathrm{Br}_Q(a\otimes c)\otimes \mathrm{Br}_Q(d)\big) \\
&= \mathrm{Br}_Q(a\otimes c\otimes d) = \alpha_{B,C\otimes_{\mathcal{O}} D}(Q)\big(\mathrm{Br}_Q(a)\otimes \mathrm{Br}_Q(c\otimes d)\big) \\
&= \alpha_{B,C\otimes_{\mathcal{O}} D}(Q)\Big(\mathrm{Br}_Q(a)\otimes \alpha_{C,D}(Q)\big(\mathrm{Br}_Q(c)\otimes \mathrm{Br}_Q(d)\big)\Big) .
\end{aligned}$$

7.15 From now on, we prove Theorem 7.2; so, Q is a finite p-group, B and B' are $\mathcal{O}$-free $\mathcal{O}Q$-interior algebras having Q-stable $\mathcal{O}$-bases by conjugation and N is an indecomposable $\mathcal{O}$-free $\mathcal{O}Q$-module such that, setting $S = \mathrm{End}_{\mathcal{O}}(N)$, there is an $\mathcal{O}Q$-interior algebra embedding

7.15.1 $$g : B \longrightarrow S \underset{\mathcal{O}}{\otimes} B' \quad ;$$

moreover, we assume that B admits a nondegenerate symmetric $\mathcal{O}$-form $\mu : B \to \mathcal{O}$, that we have $B(Q) \neq \{0\}$ and, according to Remark 7.3, that the unity elements are primitive in both B^Q and $(B')^Q$; then the conclusion of Theorem 7.2 results from the following lemmas, which are actually easy consequences of that conclusion.

Lemma 7.16 *With the notation above, for any subgroup R of Q, the k-algebras $B(R)$, $B'(R)$ and $S(R)$ are nonzero and μ induces a nondegenerate symmetric k-form $\mu(R) : B(R) \to k$ which, for any $a \in B^R$, maps* $\mathrm{Br}_R(a)$ *on the image $\overline{\mu(a)}$ of $\mu(a)$ in k.*

Proof: Since μ is symmetric, for any nontrivial subgroup R of Q, any proper subgroup T of R and any $a \in B^T$, it is clear that $\mu\big(Tr_T^R(a)\big) \in p\mathcal{O}$ and therefore μ induces a symmetric k-form $\mu(R) : B(R) \to k$; then, if X is a Q-stable $\mathcal{O}$-basis of B, the dual $\mathcal{O}$-basis X^* with respect the nondegenerate scalar product determined by μ is still Q-stable and it is easily checked that the canonical bijection between X and X^* maps X^R onto $(X^*)^R$ for any subgroup R of Q, so that $\mathrm{Br}_R(X^R)$ and $\mathrm{Br}_R\big((X^*)^R\big)$ is a pair of dual k-bases of $B(R)$ with respect to the scalar product determined by $\mu(R)$; moreover, since $B(Q) \neq \{0\}$, the set X^Q is not empty, so that X^R is not empty too and therefore we have $B(R) \neq \{0\}$ (cf. 2.2.5); now, the existence of embedding 7.15.1 and Lemma 7.10 force

7.16.1 $$\{0\} \neq (S \underset{\mathcal{O}}{\otimes} B')(R) \cong S(R) \underset{k}{\otimes} B'(R) \quad .$$

Lemma 7.17 *With the notation above, for any subgroup R of Q and any subgroup $\bar{T}$ of $\bar{N}_Q(R)$, we have $(S(R))(\bar{T}) \neq \{0\}$. In particular, denoting by T the converse image $\bar{T}$ in $N_Q(R)$, there are local points ρ of R and τ of T on S such that $R_\rho \subset T_\tau$.*

Proof: It is clear that embedding 7.15.1 induces a k-algebra embedding

7.17.1 $$\big(B(R)\big)(\bar{T}) \longrightarrow \big((S \underset{\mathcal{O}}{\otimes} B')(R)\big)(\bar{T})$$

and successive applications of Lemma 7.10 transform this embedding in the following one

7.17.2 $$B(T) \longrightarrow \big(S(R)\big)(\bar{T}) \underset{k}{\otimes} B'(T)$$

which forces $\big(S(R)\big)(\bar{T}) \neq \{0\}$. In particular, since the homomorphism $\tau_S(R,T) : S(T) \to \big(S(R)\big)(\bar{T})$ is unitary, there is a local point τ of T on S such that $\tau_S(R,T)\ \big(\mathrm{Br}_T\ (\tau)\big) \neq \{0\}$; thus, if $j \in \tau$ then we have $\mathrm{Br}_R(j) \neq 0$ (cf. 7.9) and therefore there is a primitive idempotent i in S^R such that

7.17.3 $$0 \neq \mathrm{Br}_R(i) = \mathrm{Br}_R(ij) = \mathrm{Br}_R(ji) \quad ,$$

so that the point ρ of R on S containing i is local and we have $R_\rho \subset T_\tau$.

Lemma 7.18 *With the notation above, there is an embedding of $\mathcal{O}Q$-interior algebras*

7.18.1 $$f : \mathcal{O} \longrightarrow S^\circ \underset{\mathcal{O}}{\otimes} S \quad .$$

Proof: Since $S^\circ \otimes_{\mathcal{O}} S \cong \mathrm{End}_{\mathcal{O}}(S)$ as $\mathcal{O}Q$-interior algebras, it suffices to prove that $\mathcal{O}$ is a direct summand of S as $\mathcal{O}Q$-modules, by conjugation on S (cf. 2.10). On the one hand, according to our hypothesis, any indecomposable direct summand of the $\mathcal{O}Q$-module B' is isomorphic to $\mathrm{Ind}_R^Q(\mathcal{O})$ for a suitable subgroup R of Q, and therefore any indecomposable direct summand of the $\mathcal{O}Q$-module $S \otimes_{\mathcal{O}} B'$ is isomorphic to $\mathrm{Ind}_R^Q(W)$ for a suitable subgroup R of Q and a suitable indecomposable direct summand W of the $\mathcal{O}Q$-module S; on the other hand, always according to our hypothesis, $\mathcal{O}$ is a direct summand of B as $\mathcal{O}Q$-modules (cf. 2.2.5) and therefore, since embedding 7.15.1 is also a direct injection of $\mathcal{O}Q$-modules, $\mathcal{O}$ is still a direct summand of $S \otimes_{\mathcal{O}} B'$ as $\mathcal{O}Q$-modules. Consequently, at least once we have $R = Q$ and $W = \mathcal{O}$.

Lemma 7.19 *With the notation above, we have a k-algebra isomorphism*

7.19.1 $$(S^\circ \underset{\mathcal{O}}{\otimes} S \underset{\mathcal{O}}{\otimes} B')(Q) \cong S(Q)^\circ \underset{k}{\otimes} S(Q) \underset{k}{\otimes} B'(Q) \quad .$$

In particular, Q has a unique local point on $S^\circ \otimes_{\mathcal{O}} S \otimes_{\mathcal{O}} B'$, which has multiplicity one.

Proof: Applying Lemma 7.11 to embedding 7.15.1, we get the following commutative diagram

7.19.2 $$\begin{array}{ccc} (S^\circ \underset{\mathcal{O}}{\otimes} B)(Q) & \xrightarrow{(\mathrm{id}_{S^\circ}\otimes g)(Q)} & (S^\circ \underset{\mathcal{O}}{\otimes} S \underset{\mathcal{O}}{\otimes} B')(Q) \\ {\scriptstyle \alpha_{S^\circ,B}(Q)}\uparrow & & \uparrow{\scriptstyle \alpha_{S^\circ,S\otimes B'}(Q)} \\ S(Q)^\circ \underset{k}{\otimes} B(Q) & \xrightarrow{\mathrm{id}_{S(Q)^\circ}\otimes g(Q)} & S(Q)^\circ \underset{k}{\otimes} (S \underset{\mathcal{O}}{\otimes} B')(Q) \end{array} \quad ;$$

moreover, by Lemma 7.10, $\alpha_{S^\circ,B}(Q)$ is an isomorphism and $S(Q)^\circ \otimes_k (S \otimes_{\mathcal{O}} B')(Q)$ is isomorphic to $S(Q)^\circ \otimes_k S(Q) \otimes_k B'(Q)$; but, since $S(Q) \neq \{0\}$ and N is indecomposable, the unity element in $S(Q)^\circ \otimes_k S(Q) \otimes_{\mathcal{O}} B'(Q)$ is primitive; hence, the embedding $\mathrm{id}_{S(Q)^\circ} \otimes_k g(Q)$ is in fact an isomorphism and therefore, since the homomorphism $\alpha_{S^\circ,S\otimes_{\mathcal{O}}B'}(Q)$ is unitary, the embedding $(\mathrm{id}_{S^\circ} \otimes_{\mathcal{O}} g)(Q)$ is an isomorphism too, which forces $\alpha_{S^\circ,S\otimes_{\mathcal{O}}B'}(Q)$ to be an isomorphism.

Lemma 7.20 *With the notation above, there is an $\mathcal{O}Q$-interior algebra embedding $g' : B' \to S^\circ \otimes_{\mathcal{O}} B$ fulfilling*

7.20.1 $$(\widetilde{\mathrm{id}}_{S^\circ} \underset{\mathcal{O}}{\otimes} \tilde{g}) \circ \tilde{g}' = \tilde{f} \underset{\mathcal{O}}{\otimes} \widetilde{\mathrm{id}}_{B'} \quad \text{and} \quad (\widetilde{\mathrm{id}}_S \underset{\mathcal{O}}{\otimes} \tilde{g}') \circ \tilde{g} = \tilde{f} \underset{\mathcal{O}}{\otimes} \widetilde{\mathrm{id}}_B \quad .$$

Proof: According to Lemma 7.19, Q has a unique local point δ' on $S^\circ \otimes_{\mathcal{O}} S \otimes_{\mathcal{O}} B'$ and, according to Lemma 7.10, we have

7.20.2 $$(S^\circ \underset{\mathcal{O}}{\otimes} B)(Q) \cong S(Q)^\circ \underset{k}{\otimes} B(Q) \neq \{0\} \quad ;$$

hence δ' comes from $S^\circ \otimes_{\mathcal{O}} B$ throughout the $\mathcal{O}Q$-interior algebra embedding

7.20.3 $$\widetilde{\mathrm{id}}_{S^\circ} \underset{\mathcal{O}}{\otimes} \tilde{g} : S^\circ \underset{\mathcal{O}}{\otimes} B \longrightarrow S^\circ \underset{\mathcal{O}}{\otimes} S \underset{\mathcal{O}}{\otimes} B' \quad ;$$

but, according to our hypothesis, B' endowed with the $\mathcal{O}Q$-interior algebra embedding

7.20.4 $$\tilde{f} \underset{\mathcal{O}}{\otimes} \mathrm{id}_{B'} : B' \longrightarrow S^\circ \underset{\mathcal{O}}{\otimes} S \underset{\mathcal{O}}{\otimes} B'$$

is an *embedded algebra* associated with $Q_{\delta'}$ (cf. 2.7); consequently, the uniqueness of this embedded algebra (cf. [20, 1.6]) proves the existence of an $\mathcal{O}Q$-interior algebra embedding $g' : B' \to S^\circ \otimes_{\mathcal{O}} B$ such that

7.20.5 $$(\widetilde{\mathrm{id}}_{S^\circ} \underset{\mathcal{O}}{\otimes} \tilde{g}) \circ \tilde{g}' = \tilde{f} \underset{\mathcal{O}}{\otimes} \widetilde{\mathrm{id}}_{B'} \quad .$$

In order to prove the second equality in 7.20.1, we tensor the first one by $\widetilde{\mathrm{id}}_S$ getting

7.20.6 $$(\widetilde{\mathrm{id}}_{S \otimes_{\mathcal{O}} S^\circ} \underset{\mathcal{O}}{\otimes} \tilde{g}) \circ (\widetilde{\mathrm{id}}_S \underset{\mathcal{O}}{\otimes} \tilde{g}') = \widetilde{\mathrm{id}}_S \underset{\mathcal{O}}{\otimes} \tilde{f} \underset{\mathcal{O}}{\otimes} \widetilde{\mathrm{id}}_{B'} \quad ;$$

but, since the underlying $\mathcal{O}$-algebras of S and $S \otimes_{\mathcal{O}} S^\circ \otimes_{\mathcal{O}} S$ are full matrix algebras over $\mathcal{O}$, there is a unique $\mathcal{O}$-algebra exoembedding (cf. [19, Proposition 2.3])

7.20.7 $$\mathrm{Res}_1^Q(S) \longrightarrow \mathrm{Res}_1^Q(S \underset{\mathcal{O}}{\otimes} S^\circ \underset{\mathcal{O}}{\otimes} S)$$

and therefore it follows from 2.4.2 that

7.20.8 $$\widetilde{\mathrm{id}}_S \underset{\mathcal{O}}{\otimes} \tilde{f} = \tilde{f} \underset{\mathcal{O}}{\otimes} \widetilde{\mathrm{id}}_S \quad ;$$

consequently, by composing (on the right) both members of equality 7.20.6 with $\tilde{g}$, we get

7.20.9 $$\begin{aligned}(\widetilde{\mathrm{id}}_{S\otimes_{\mathcal{O}} S^\circ} \underset{\mathcal{O}}{\otimes} \tilde{g}) \circ (\widetilde{\mathrm{id}}_S \underset{\mathcal{O}}{\otimes} \tilde{g}') \circ \tilde{g} &= (\widetilde{\mathrm{id}}_S \underset{\mathcal{O}}{\otimes} \tilde{f} \underset{\mathcal{O}}{\otimes} \widetilde{\mathrm{id}}_{B'}) \circ \tilde{g} \\ &= (\tilde{f} \underset{\mathcal{O}}{\otimes} \widetilde{\mathrm{id}}_{S\otimes_{\mathcal{O}} B'}) \circ (\widetilde{\mathrm{id}}_{\mathcal{O}} \underset{\mathcal{O}}{\otimes} \tilde{g}) = \tilde{f} \underset{\mathcal{O}}{\otimes} \tilde{g} \\ &= (\widetilde{\mathrm{id}}_{S\otimes_{\mathcal{O}} S^\circ} \underset{\mathcal{O}}{\otimes} \tilde{g}) \circ (\tilde{f} \underset{\mathcal{O}}{\otimes} \widetilde{\mathrm{id}}_B)\end{aligned}$$

which implies the second equality in 7.20.1 (cf. 2.11.4).

Lemma 7.21 *With the notation above, let R be a subgroup of Q, set $\overline{S(R)} = S(R)/J\big(S(R)\big)$ and denote by $n(R) : S(R) \to \overline{S(R)}$ the natural map and by $\overline{g(R)}$ the composed $\bar{N}_Q(R)$-algebra homomorphism*

7.21.1 $$B(R) \xrightarrow{\alpha^{-1} \circ g(R)} S(R) \underset{k}{\otimes} B'(R) \xrightarrow{n(R) \otimes \mathrm{id}_{B'(R)}} \overline{S(Q)} \underset{k}{\otimes} B'(R)$$

where $\alpha = \alpha_{S,B'}(Q)$. Then $\overline{g(R)}$ is an $\bar{N}_Q(R)$-algebra embedding.

Proof: Since $g(R)$ is an embedding, it suffices to prove that $\overline{g(R)}$ is still injective. The embedding $f(R) : k \to (S \otimes_{\mathcal{O}} S^\circ)(R)$ determines a point of the k-algebra

$(S \otimes_{\mathcal{O}} S^\circ)(R)$ and then the trace map over the corresponding simple $(S \otimes_{\mathcal{O}} S^\circ)(R)$-module determines a symmetric k-form

7.21.2 $$\varphi_R : (S \underset{\mathcal{O}}{\otimes} S^\circ)(R) \longrightarrow k$$

which vanishes on the nilpotent elements and fulfills $\varphi_R \circ f(R) = \mathrm{id}_k$; in particular, considering the k-algebra homomorphism

7.21.3 $$\alpha_{S,S^\circ}(R) : S(R) \underset{k}{\otimes} S(R)^\circ \longrightarrow (S \underset{\mathcal{O}}{\otimes} S^\circ)(R)$$

we get

7.21.4 $$\varphi_R\Big(\alpha_{S,S^\circ}(R)\Big(J\big(S(R)\big) \underset{k}{\otimes} S(R)^\circ\Big)\Big) = \{0\}$$

and therefore φ_R factorizes throughout a symmetric k-form

7.21.5 $$\bar\varphi_R : \overline{S(R)} \underset{k}{\otimes} S(R)^\circ \longrightarrow k \quad .$$

Now, we claim that

7.21.6 $$(\bar\varphi_R \underset{k}{\otimes} \mu(R)) \circ \Big(\mathrm{id}_{\overline{S(R)}} \underset{k}{\otimes} \big(\alpha_{S^\circ,B}(R)^{-1} \circ g'(R)\big)\Big) \circ \overline{g(R)} = \mu(R)$$

which implies that $\mu(R)$ vanishes on $\mathrm{Ker}(\overline{g(R)})$, forcing $\overline{g(R)}$ to be injective (cf. Lemma 7.16). Indeed, first of all, by the commutativity of the tensor product we have

7.21.7
$$\begin{aligned}
&\Big(\mathrm{id}_{\overline{S(Q)}} \underset{k}{\otimes} \big(\alpha_{S^\circ,B}(R)^{-1} \circ g'(R)\big)\Big) \circ \overline{g(R)} \\
&= \Big(\mathrm{id}_{\overline{S(Q)}} \underset{k}{\otimes} \big(\alpha_{S^\circ,B}(R)^{-1} \circ g'(R)\big)\Big) \circ \big(n(R) \otimes \mathrm{id}_{B'(R)}\big) \circ \alpha^{-1} \circ g(R) \\
&= \Big(n(R) \underset{k}{\otimes} \big(\alpha_{S^\circ,B}(R)^{-1} \circ g'(R)\big)\Big) \circ \alpha^{-1} \circ g(R) \\
&= \big(n(R) \underset{k}{\otimes} \mathrm{id}_{S(R)^\circ} \underset{k}{\otimes} \mathrm{id}_{B(R)}\big) \circ \big(\mathrm{id}_{S(R)} \underset{k}{\otimes} \alpha_{S^\circ,B}(R)^{-1}\big) \circ h
\end{aligned}$$

where $h = \big(\mathrm{id}_{S(R)} \otimes g'(R)\big) \circ \alpha_{S,B'}(R)^{-1} \circ g(R)$; consequently, the left member of equality 7.21.6 is equal to

7.21.8
$$\begin{aligned}
&\Big(\big(\bar\varphi_R \circ (n(R) \underset{k}{\otimes} \mathrm{id}_{S(R)^\circ})\big) \underset{k}{\otimes} \mu(R)\Big) \circ \big(\mathrm{id}_{S(R)} \underset{k}{\otimes} \alpha_{S^\circ,B}(R)^{-1}\big) \circ h \\
&= \Big(\big(\varphi_R \circ \alpha_{S,S^\circ}(R)\big) \underset{k}{\otimes} \mu(R)\Big) \circ \big(\mathrm{id}_{S(R)} \underset{k}{\otimes} \alpha_{S^\circ,B}(R)^{-1}\big) \circ h \\
&= \big(\varphi_R \underset{k}{\otimes} \mu(R)\big) \circ \big(\alpha_{S,S^\circ}(R) \underset{k}{\otimes} \mathrm{id}_{B(R)}\big) \circ \big(\mathrm{id}_{S(R)} \underset{k}{\otimes} \alpha_{S^\circ,B}(R)^{-1}\big) \circ h \quad ;
\end{aligned}$$

but, according to Lemma 7.14, we have

7.21.9

$$\alpha_{S\otimes_{\mathcal{O}}S^\circ,B}(R)\circ\big(\alpha_{S,S^\circ}(R)\underset{k}{\otimes}\mathrm{id}_{B(R)}\big)=\alpha_{S,S^\circ\otimes_{\mathcal{O}}B}(R)\circ\big(\mathrm{id}_{S(R)}\underset{k}{\otimes}\alpha_{S^\circ,B}(R)\big)\quad;$$

consequently, by Lemma 7.10, that left member becomes

$$\begin{aligned}&\big(\varphi_R\underset{k}{\otimes}\mu(R)\big)\circ\alpha_{S\otimes_{\mathcal{O}}S^\circ,B}(R)^{-1}\circ\alpha_{S,S^\circ\otimes_{\mathcal{O}}B}(R)\circ h\\ &=l\circ\alpha_{S,S^\circ\otimes_{\mathcal{O}}B}(R)\circ\big(\mathrm{id}_{S(R)}\underset{k}{\otimes}g'(R)\big)\circ\alpha_{S,B'}(R)^{-1}\circ g(R)\\ &=l\circ(\mathrm{id}_S\underset{\mathcal{O}}{\otimes}g')(R)\circ g(R)\quad,\end{aligned}\tag{7.21.10}$$

where $l=\big(\varphi_R\otimes_k\mu(R)\big)\circ\alpha_{S\otimes_{\mathcal{O}}S^\circ,B}(R)^{-1}$, since it follows from Lemma 7.11 that we have

$$\alpha_{S,S^\circ\otimes_{\mathcal{O}}B}(R)\circ\big(\mathrm{id}_{S(R)}\underset{k}{\otimes}g'(R)\big)=(\mathrm{id}_S\underset{\mathcal{O}}{\otimes}g')(R)\circ\alpha_{S,B'}(R)\quad.\tag{7.21.11}$$

On the other hand, by Lemma 7.20, there is $a\in\big((S\otimes_{\mathcal{O}}S^\circ\otimes_{\mathcal{O}}B)^Q\big)^*$ such that we have

$$(f\underset{\mathcal{O}}{\otimes}\mathrm{id}_B)^a=(\mathrm{id}_S\underset{\mathcal{O}}{\otimes}g')\circ g\quad,\tag{7.21.12}$$

where $(f\otimes_{\mathcal{O}}\mathrm{id}_B)^a$ denotes the composition of $f\otimes_{\mathcal{O}}\mathrm{id}_B$ with the corresponding inner automorphism of $S\otimes_{\mathcal{O}}S^\circ\otimes_{\mathcal{O}}B$, and since $\alpha_{\mathcal{O},B}(R)=\mathrm{id}_{B(R)}$, by Lemma 7.11 we get

$$\alpha_{S\otimes_{\mathcal{O}}S^\circ,B}(R)\circ\big(f(R)\underset{k}{\otimes}\mathrm{id}_{B(R)}\big)=(f\underset{\mathcal{O}}{\otimes}\mathrm{id}_B)(R)\quad;\tag{7.21.13}$$

consequently, setting $a_R=\alpha_{S\otimes_{\mathcal{O}}S^\circ,B}(R)^{-1}\big(\mathrm{Br}_R(a)\big)$, we obtain

$$\begin{aligned}&l\circ(\mathrm{id}_S\underset{\mathcal{O}}{\otimes}g')(R)\circ g(R)=l\circ(f\underset{\mathcal{O}}{\otimes}\mathrm{id}_B)^a(R)\\ &=\big(\varphi_R\underset{k}{\otimes}\mu(R)\big)\circ\alpha_{S\underset{\mathcal{O}}{\otimes}S^\circ,B}(R)^{-1}\circ(f\underset{\mathcal{O}}{\otimes}\mathrm{id}_B)^a(R)\\ &=\big(\varphi_R\underset{k}{\otimes}\mu(R)\big)\circ\big(f(R)\underset{k}{\otimes}\mathrm{id}_{B(R)}\big)^{a_R}\quad;\end{aligned}\tag{7.21.14}$$

finally, since $\varphi_R\otimes_k\mu(R)$ is a symmetric k-form over $(S\otimes_{\mathcal{O}}S^\circ)(R)\otimes_k B(R)$ and we have $\varphi_R\circ f(R)=\mathrm{id}_k$, we get

$$\begin{aligned}&\big(\varphi_R\underset{k}{\otimes}\mu(R)\big)\circ\big(f(R)\underset{k}{\otimes}\mathrm{id}_{B(R)}\big)^{a_R}\\ &=\big(\varphi_R\underset{k}{\otimes}\mu(R)\big)\circ\big(f(R)\underset{k}{\otimes}\mathrm{id}_{B(R)}\big)=\mu(R)\quad.\end{aligned}\tag{7.21.15}$$

Lemma 7.22 *With the notation above, any subgroup R of Q has a unique local point on S.*

Proof: We borrow notation from Lemma 7.21 and argue by induction on $|Q : R|$; clearly, we may assume that $R \neq Q$, so let T be a subgroup of $N_Q(R)$ strictly containing R, set $\bar{T} = T/R \neq 1$ and denote by $\overline{\tau_S(R,T)}$ the composed homomorphism

7.22.1 $$S(T) \xrightarrow{\tau_S(R,T)} \big(S(R)\big)(\bar{T}) \xrightarrow{(n(R))(\bar{T})} \overline{S(R)}(\bar{T}) \quad ;$$

then, by the induction hypothesis, T has a unique local point τ on S and we have

7.22.2 $$S(T)/J\big(S(T)\big) \cong S(T_\tau) \quad ;$$

moreover, since T is subnormal in $N_Q(R)$, denoting by ν the unique local point of $N_Q(R)$ on S, it follows easily from Lemma 7.17 that

7.22.3 $$T_\tau \subset N_Q(R)_\nu \quad .$$

First of all, we claim that

7.22.4 $$\begin{aligned}&\overline{g(R)}(\bar{T}) \circ \tau_B(R,T)\\ &= \alpha_{\overline{S(R)},B'(R)}(\bar{T}) \circ \big(\overline{\tau_S(R,T)} \underset{k}{\otimes} \tau_{B'}(R,T)\big) \circ \alpha_{S,B'}(T)^{-1} \circ g(T) \quad ;\end{aligned}$$

indeed, according to our definition of $\overline{g(R)}$, we have

7.22.5 $$\overline{g(R)}(\bar{T}) = \big(n(R) \underset{k}{\otimes} \mathrm{id}_{B'(R)}\big)(\bar{T}) \circ \big(\alpha_{S,B'}(R)\big)(\bar{T})^{-1} \circ \big(g(R)\big)(\bar{T}) \quad ;$$

but, it follows from Lemma 7.11 that

7.22.6 $$\begin{gathered}\big(g(R)\big)(\bar{T}) \circ \tau_B(R,T) = \tau_{S\otimes_{\mathcal{O}}B'}(R,T) \circ g(T)\\ \big(n(R) \underset{k}{\otimes} \mathrm{id}_{B'(R)}\big)(\bar{T}) \circ \alpha_{S(R),B'(R)}(\bar{T}) = \alpha_{\overline{S(R)},B'(R)}(\bar{T}) \circ \Big(\big(n(R)\big)(\bar{T}) \underset{k}{\otimes} \mathrm{id}_{(B'(R))(\bar{T})}\Big)\end{gathered}$$

and, moreover, from Lemma 7.12 that

7.22.7 $$\begin{aligned}&\big(\alpha_{S,B'}(R)\big)(\bar{T}) \circ \alpha_{S(R),B'(R)}(\bar{T}) \circ \big(\tau_S(R,T) \underset{k}{\otimes} \tau_{B'}(R,T)\big)\\ &= \tau_{S\otimes_{\mathcal{O}}B'}(R,T) \circ \alpha_{S,B'}(T) \quad ;\end{aligned}$$

now, it is easily checked that equality 7.22.4 follows from Lemma 7.10 and equalities 7.22.5, 7.22.6 and 7.22.7.

In particular, from equality 7.22.4 and Lemma 7.10 we get

7.22.8 $$\begin{aligned}&\alpha_{\overline{S(R)},B'(R)}(\bar{T})^{-1} \circ \overline{g(R)}(\bar{T}) \circ \tau_B(R,T)\\ &= \big(\overline{\tau_S(R,T)} \underset{k}{\otimes} \tau_{B'}(R,T)\big) \circ \alpha_{S,B'}(T)^{-1} \circ g(T) \quad ;\end{aligned}$$

hence, since $\overline{g(R)}(\bar{T})$ is an embedding by Lemma 7.21, if i is a promitive idempotent of $B(T)$, it follows from equality 7.22.8 and Lemma 7.10 that the idempotent

7.22.9 $$\bar{\imath} = \left(\overline{\tau_S(R,T)} \underset{k}{\otimes} \tau_{B'}(R,T)\right)\left(\alpha_{S,B'}(T)^{-1}\left(g(T)(i)\right)\right)$$

is primitive in $\overline{S(R)}(\bar{T}) \otimes_k \left(B'(R)\right)(\bar{T})$; on the other hand, the idempotent $\alpha_{S,B'}(T)^{-1}\left(g(T)(i)\right)$ is already primitive in $S(T) \otimes_k B'(T)$ and therefore we have

7.22.10 $$\alpha_{S,B}(T)^{-1}\left(g(T)(i)\right) = (l \otimes i')^a$$

for suitable primitive idempotents l in $S(T)$ and i' in $B'(T)$, and a suitable $a \in \left(S(T) \otimes_k B'(T)\right)^*$; then, setting $\bar{l} = \overline{\tau_S(R,T)}(l)$, $\bar{\imath}' = \tau_{B'}(R,T)(i')$ and $\bar{a} = \left(\overline{\tau_S(R,T)} \otimes_k \tau_{B'}(R,T)\right)(a)$, we get

7.22.11 $$\bar{\imath} = (\bar{l} \otimes \bar{\imath}')^{\bar{a}} \quad ;$$

hence, $\bar{l} \otimes \bar{\imath}'$ is a primitive idempotent in $\overline{S(R)}(\bar{T}) \otimes_k \left(B'(R)\right)(\bar{T})$ which forces $\bar{l}$ to be primitive in $\overline{S(R)}(\bar{T})$.

In conclusion, the k-algebra homomorphism $\overline{\tau_S(R,T)}$ maps at least one primitive idempotent l of $S(T)$ on a primitive idempotent $\bar{l}$ of $\overline{S(R)}(\bar{T})$; but, by 7.22.2, all the primitive idempotents of $S(T)$ are conjugate to l; hence, $\overline{\tau_S(R,T)}$ maps *any* primitive idempotent of $S(T)$ on a primitive idempotent of $\overline{S(R)}(\bar{T})$ and, in particular, it maps bijectively a pairwise orthogonal primitive idempotent decomposition of the unity element in $S(T)$ onto such a decomposition in $\overline{S(R)}(\bar{T})$. Consequently, $\bar{T}$ has a unique local point $\bar{\tau}$ on $\overline{S(R)}$ and, considering S^R as an $\bar{N}_Q(R)$-algebra, so that τ is still a point of $\bar{T}$ on S^R, we have

7.22.12 $$S(T_\tau) = (S^R)(\bar{T}_\tau) \cong \overline{S(R)}(\bar{T}_{\bar{\tau}}) \quad .$$

In particular, since (cf. 2.9.1)

7.22.13 $$\overline{S(R)} = \prod_{\rho' \in \mathcal{LP}_S(R)} S(R_{\rho'}) \quad ,$$

there is a unique T-stable local point ρ of R on S such that $\left(S(R_\rho)\right)(\bar{T}) \neq \{0\}$ and then we have $\left(S(R_\rho)\right)(\bar{T}) \cong \overline{S(R)}(\bar{T})$; moreover, notice that, denoting by $\overline{\mathrm{Br}}_R$ the composed map

7.22.14 $$S^R \xrightarrow{\mathrm{Br}_R} S(R) \xrightarrow{n(R)} \overline{S(R)} \quad ,$$

$\overline{\mathrm{Br}}_R(\rho)$ is the unique point of 1 on $\overline{S(R)}$ such that $1_{\overline{\mathrm{Br}}_R(\rho)} \subset \bar{T}_{\bar{\tau}}$.

We claim that ρ does not depend on the choice of T; denoting by $\bar{\nu}$ the unique local point of $\bar{N}_Q(R)$ on $\overline{S(R)}$, it suffices to prove that $\bar{T}_{\bar{\tau}} \subset \bar{N}_Q(R)_{\bar{\nu}}$ since, in that case, $\overline{\mathrm{Br}}_R(\rho)$ is the unique point of 1 on $\overline{S(R)}$ such that $1_{\overline{\mathrm{Br}}_R(\rho)} \subset \bar{N}_Q(R)_{\bar{\nu}}$. But at this point, it is clear that, for *any* local pointed group $\bar{Z}_{\bar{\zeta}}$ on $\overline{S(R)}$, there is a point ζ of $\bar{Z}$ on S^R such that

$$7.22.15 \qquad \mathrm{Br}_{\bar{Z}}^{\overline{S(R)}}\big(\overline{\mathrm{Br}}_R(\zeta)\big) = \mathrm{Br}_{\bar{Z}}^{\overline{S(R)}}(\bar{\zeta}) \quad \text{and} \quad (S^R)(\bar{Z}) \cong \overline{S(R)}(\bar{Z}_{\bar{\zeta}}) \qquad ;$$

consequently, it follows from [38, (25.3) and (25.9)] (see also [21, Theorem 4.22]) that $\overline{\mathrm{Br}}_R$ is a *covering* $\bar{N}_Q(R)$*-algebra homomorphism*; in particular, inclusion 7.22.3 considered on S^R implies $\bar{T}_{\bar{\tau}} \subset \bar{N}_Q(R)_{\bar{\nu}}$ (actually, we have $\overline{\mathrm{Br}}_R(\tau) \subset \bar{\tau}$ and $\overline{\mathrm{Br}}_R\,(\nu) \subset \bar{\nu}$).

Finally, we are ready to prove that ρ is the unique local point of R on S. In 7.22.13, denote by e the idempotent of $\overline{S(R)}$ determined by the unity element of $S(R_\rho)$; since $\bar{T}$ fixes e and $\big(S(R_\rho)\big)(\bar{T}) \cong \overline{S(R)}(\bar{T})$, we have $\mathrm{Br}_{\bar{T}}^{\overline{S(R)}}(1-e) = 0$; consequently, when $\bar{T}$ runs over the set of all the nontrivial subgroups of $\bar{N}_Q(R)$ we get (cf. [8, Lemmas 1.11 and 1.12])

$$7.22.16 \qquad 1 - e \in \overline{S(R)}_1^{\bar{N}_Q(R)} = n(R)\big(S(R)_1^{\bar{N}_Q(R)}\big) = \overline{\mathrm{Br}}_Q(S_R^Q) \qquad ;$$

but, since N is indecomposable and $S(Q) \neq \{0\}$, zero is the unique idempotent in S_R^Q; hence, we have $e = 1$ and $\mathcal{LP}_S(R) = \{\rho\}$.

Lemma 7.23 *With the notation above, embedding 7.15.1 induces a bijection between the sets of local pointed groups on B and $S \otimes_{\mathcal{O}} B'$.*

Proof: Let R be a subgroup of Q; since g induces an injective map from $\mathcal{LP}_B(R)$ into $\mathcal{LP}_{S\otimes_{\mathcal{O}}B'}(R)$ (cf. 2.11.2), it suffices to prove that both sets have the same cardinal. But $\mathrm{Br}_R^{S\otimes_{\mathcal{O}}B'}$ induces a bijection from $\mathcal{LP}_{S\otimes_{\mathcal{O}}B'}(R)$ onto the set of points of the k-algebra $(S \otimes_{\mathcal{O}} B')(R)$ (cf. 2.9.1) which is isomorphic to $S(R) \otimes_k B'(R)$ by Lemma 7.10; moreover, by Lemma 7.22, the k-algebra $S(R)$ has a unique point and therefore we have a canonical bijection between the sets of points of the k-algebras $B'(R)$ and $S(R) \otimes_k B'(R)$, so that

$$7.23.1 \qquad \big|\mathcal{LP}_{S\otimes_{\mathcal{O}}B'}(R)\big| = |\mathcal{LP}_{B'}(R)| \qquad .$$

Similarly, we obtain

$$7.23.2 \qquad \big|\mathcal{LP}_{S^\circ\otimes_{\mathcal{O}}B}(R)\big| = |\mathcal{LP}_{B}(R)| \qquad .$$

Now, the lemma follows from the existence of the embedding g' in Lemma 7.20.

Lemma 7.24 *With the notation above, Q stabilizes an $\mathcal{O}$-basis of S.*

Proof: By Lemma 7.16 we have $B'(Q) \neq \{0\}$ and therefore, according to our hypothesis, $\mathcal{O}$ is a direct summand of B' as $\mathcal{O}Q$-modules, where Q acts by conjugation on B'; hence, S is also a direct summand of $S \otimes_{\mathcal{O}} B'$ as $\mathcal{O}Q$-modules and it suffices to prove that Q stabilizes an $\mathcal{O}$-basis of $S \otimes_{\mathcal{O}} B'$. Let D be a *maximal Q-stable commutative $\mathcal{O}$-semisimple $\mathcal{O}$-subalgebra* of $S \otimes_{\mathcal{O}} B'$ (cf. [22, 2.9.3]; see also [38, §24]), denote by I the set of primitive idempotents of D and consider the Q-stable direct sum decomposition

7.24.1 $$S \underset{\mathcal{O}}{\otimes} B' \cong \bigoplus_{i,j \in I} i(S \underset{\mathcal{O}}{\otimes} B')j \quad ;$$

now, it suffices to prove that the stabilizer in Q of any direct summand in 7.24.1 stabilizes an $\mathcal{O}$-basis of this summand. Let i and j be elements of I and denote respectively by R and T their stabilizers in Q; it is well-known that i and j belong respectively to local points of R and T on $S \otimes_{\mathcal{O}} B'$ (cf. [22, 2.9.3] or [38, (24.1)]) and therefore it follows from Lemma 7.23 that we have $i = g(\hat{\imath})^a$ and $j = g(\hat{\jmath})^c$ for suitable idempotents $\hat{\imath} \in B^R$ and $\hat{\jmath} \in B^T$, and suitable invertible elements $a \in (S \otimes_{\mathcal{O}} B')^R$ and $c \in (S \otimes_{\mathcal{O}} B')^T$; consequently, we get $\mathcal{O}(R \cap T)$-module isomorphisms

7.24.2 $$\hat{\imath} B \hat{\jmath} \cong a i(S \underset{\mathcal{O}}{\otimes} B')jc^{-1} \cong i(S \underset{\mathcal{O}}{\otimes} B')j$$

successively induced by g and by the multiplication on the left by a^{-1} and on the right by c, which proves that $i(S \otimes_{\mathcal{O}} B')j$ is a permutation $\mathcal{O}(R \cap T)$-module.

8 The Morita stable equivalent class of a nilpotent block

8.1 We keep again the notation of Section 6. Here we apply our analysis of Morita stable equivalences between blocks to give an affirmative answer - in characteristic zero - to the question we raise in [21, 1.8], namely whether the existence of an $\mathcal{O}$-algebra isomorphism between $\mathcal{O}Gb$ and a full matrix algebra over $\mathcal{O}P$ forces b to be a *nilpotent block*. Actually, once again we will discuss not only on Morita equivalences but on Morita *stable* equivalences between blocks since it does not modify essentially the proof. Our proof depends strongly on a criterion of Alfred Weiss (see our comment in 1.14 above): as a matter of fact, our analysis leads easily to the very formulation of Weiss' criterion for permutation modules over finite p-groups and complete valuation rings of characteristic zero; in Appendix 1 below, we give an independent (and short) proof of this criterion.

Theorem 8.2 *Assume that* $\mathrm{char}(\mathcal{O}) = 0$ *and that* $M^{\cdot\cdot}$ *defines a Morita stable equivalence between* b *and* b'. *Then* b *is a nilpotent block if and only if* b' *is so, and in that case the Morita stable equivalence is basic.*

Remark 8.3 Notice that, according to 7.6.3 and to the main result of [20], it suffices to prove that if one of the blocks b or b' is nilpotent then the Morita stable equivalence defined by $M^{\cdot\cdot}$ is *basic* or, equivalently, P stabilizes by conjugation an $\mathcal{O}$-basis of $S^{\cdot\cdot}$; however, Theorem 7.2 is useless in our proof.

Remark 8.4 As we recall in 6.3, when the *derived categories* of $\mathrm{Mod}_{\mathcal{O}Gb}$ and $\mathrm{Mod}_{\mathcal{O}G'b'}$ are equivalent (as *triangulated* categories), Jeremy Rickard exhibits in [31] an $\mathcal{O}(G \times G')$-module associated with $b \otimes (b')^\circ$ which defines a Morita stable equivalence between b and b', and therefore Theorem 8.2 applies.

Proof: If b' is a nilpotent block then, by the main theorem in [21], there is a Dade P'-algebra S' such that we have an $\mathcal{O}P'$-interior algebra isomorphism

8.2.1 $$A'_{\gamma'} \cong S'P' \quad ;$$

actually, the action of P' on S' can be lifted to a group homomorphism $P' \to (S')^*$ (cf. [21, (1.8.1)]) and then we get an $\mathcal{O}P'$-interior algebra isomorphism

8.2.2 $$A'_{\gamma'} \cong S' \underset{\mathcal{O}}{\otimes} \mathcal{O}P' \quad .$$

Now, from this isomorphism and Theorem 4.4, we obtain the following $\mathcal{O}P$-interior algebra isomorphisms

8.2.3 $$\begin{aligned}\mathrm{Ind}_\sigma\big(S^{\cdot\cdot} \underset{\mathcal{O}}{\otimes} \mathrm{Res}_{\sigma'}(A'_{\gamma'})\big) &\cong \mathrm{Ind}_\sigma\Big(\big(S^{\cdot\cdot} \underset{\mathcal{O}}{\otimes} \mathrm{Res}_{\sigma'}(S')\big) \underset{\mathcal{O}}{\otimes} \mathrm{Res}_{\sigma'}(\mathcal{O}P')\Big) \\ &\cong \mathrm{Ind}_{P^{\cdot\cdot}}^{P \times P'}\big(S^{\cdot\cdot} \underset{\mathcal{O}}{\otimes} \mathrm{Res}_{\sigma'}(S')\big)^{1 \times P'}\end{aligned}$$

and therefore the following $\mathcal{O}$-algebra isomorphism

8.2.4 $$\mathrm{Ind}_\sigma\big(S^{\cdot\cdot} \underset{\mathcal{O}}{\otimes} \mathrm{Res}_{\sigma'}(A'_{\gamma'})\big)^P \cong \mathrm{Ind}_{P^{\cdot\cdot}}^{P\times P'}\big(S^{\cdot\cdot} \underset{\mathcal{O}}{\otimes} \mathrm{Res}_{\sigma'}(S')\big)^{P\times P'} \quad ;$$

hence, by Green's Induction Theorem (cf. 2.12.2), there is a point $\delta^{\cdot\cdot}$ of $P^{\cdot\cdot}$ on $S^{\cdot\cdot} \otimes_{\mathcal{O}} \mathrm{Res}_{\sigma'}(S')$ such that, setting $T^{\cdot\cdot} = \big(S^{\cdot\cdot} \otimes_{\mathcal{O}} \mathrm{Res}_{\sigma'}(S')\big)_{\delta^{\cdot\cdot}}$, we have the $\mathcal{O}P$-interior algebra isomorphisms (cf. Theorem 4.4 and 6.8)

8.2.5 $$\hat{A}_{\hat{\gamma}} \cong \mathrm{Ind}_{P^{\cdot\cdot}}^{P\times P'}(T^{\cdot\cdot})^{1\times P'} \cong \mathrm{Ind}_\sigma\big(T^{\cdot\cdot} \underset{\mathcal{O}}{\otimes} \mathrm{Res}_{\sigma'}(\mathcal{O}P')\big) \quad .$$

We claim that $\delta^{\cdot\cdot}$ is a local point of $P^{\cdot\cdot}$; indeed, since $E_{G'}(P'_{\gamma'}) = 1$ (cf. [21, Theorem 3.8]), it follows from Corollary 5.14 that $\hat{\gamma}$ is the unique local point of P on $\mathrm{Ind}_\sigma\big(S^{\cdot\cdot} \otimes_{\mathcal{O}} \mathrm{Res}_{\sigma'}(A'_{\gamma'})\big)$ and has multiplicity one; but $P^{\cdot\cdot}$ has a (unique) local point $\varepsilon^{\cdot\cdot}$ on $S^{\cdot\cdot} \otimes_{\mathcal{O}} \mathrm{Res}_{\sigma'}(S')$ since by Lemma 7.10 we have

8.2.6 $$\big(S^{\cdot\cdot} \underset{\mathcal{O}}{\otimes} \mathrm{Res}_{\sigma'}(S')\big)(P^{\cdot\cdot}) \cong S^{\cdot\cdot}(P^{\cdot\cdot}) \neq \{0\}$$

and then Theorem 5.8 forces P to have one local point on the $\mathcal{O}P$-interior algebra $\mathrm{Ind}_\sigma\Big(\big(S^{\cdot\cdot} \otimes_{\mathcal{O}} \mathrm{Res}_{\sigma'}(S')\big)_{\varepsilon^{\cdot\cdot}} \otimes_{\mathcal{O}} \mathrm{Res}_{\sigma'}(\mathcal{O}P')\Big)$; consequently, we have $\delta^{\cdot\cdot} = \varepsilon^{\cdot\cdot}$.

Then, according to Remark 6.11, $\hat{A}_{\hat{\gamma}}$ has a P-stable $\mathcal{O}$-basis by conjugation and, according to isomorphism 8.2.5 and to our definition in 3.3, we have an $\mathcal{O}P$-module isomorphism

8.2.7 $$\hat{A}_{\hat{\gamma}} \cong \Big(\mathcal{O} \underset{\mathcal{O}\mathrm{Ker}(\sigma)}{\otimes} \big(T^{\cdot\cdot} \underset{\mathcal{O}}{\otimes} \mathrm{Res}_{\sigma'}(\mathcal{O}P')\big)\Big)^{\mathrm{Ker}(\sigma)} \quad ;$$

moreover, since $M^{\cdot\cdot}$ is a projective $\mathcal{O}(1 \times G')$-module and $\mathrm{Ker}(\sigma) = P^{\cdot\cdot} \cap (1 \times G')$, $N^{\cdot\cdot}$ is still a projective $\mathcal{O}\mathrm{Ker}(\sigma)$-module and therefore $S^{\cdot\cdot}$ is a projective $\mathcal{O}\mathrm{Ker}(\sigma)$-bimodule, so that successively $S^{\cdot\cdot} \otimes_{\mathcal{O}} \mathrm{Res}_{\sigma'}(S')$, $T^{\cdot\cdot}$ and $T^{\cdot\cdot} \otimes_{\mathcal{O}} \mathrm{Res}_{\sigma'}(\mathcal{O}P')$ are projective $\mathcal{O}\mathrm{Ker}(\sigma)$-bimodules too.

Consequently, considering $\mathcal{O} \otimes_{\mathcal{O}\mathrm{Ker}(\sigma)} \big(T^{\cdot\cdot} \otimes_{\mathcal{O}} \mathrm{Res}_{\sigma'}(\mathcal{O}P')\big)$ as an $\mathcal{O}P^{\cdot\cdot}$-module by conjugation, its restriction to $\mathcal{O}\mathrm{Ker}(\sigma)$ is projective and the submodule of $\mathrm{Ker}(\sigma)$-fixed elements is a permutation $\mathcal{O}P$-module (cf. 8.2.7); hence, it follows from Weiss' criterion (cf. Theorem A1.2 below) that $\mathcal{O} \otimes_{\mathcal{O}\mathrm{Ker}(\sigma)} \big(T^{\cdot\cdot} \otimes_{\mathcal{O}} \mathrm{Res}_{\sigma'}(\mathcal{O}P')\big)$ is a permutation $\mathcal{O}P^{\cdot\cdot}$-module; but, since

8.2.8 $$\mathrm{Ker}(\sigma) = P^{\cdot\cdot} \cap (1 \times G') = 1 \times U'$$

for a suitable subgroup U' of P' and since we have a canonical $\mathcal{O}P^{\cdot\cdot}$-module isomorphism

8.2.9 $$\mathcal{O} \underset{\mathcal{O}\mathrm{Ker}(\sigma)}{\otimes} \big(T^{\cdot\cdot} \underset{\mathcal{O}}{\otimes} \mathrm{Res}_{\sigma'}(\mathcal{O}U')\big) \cong T^{\cdot\cdot}$$

which, for any $t^{\cdot\cdot} \in T^{\cdot\cdot}$, maps $1 \otimes (t^{\cdot\cdot} \otimes 1)$ on $t^{\cdot\cdot}$, $T^{\cdot\cdot}$ is a direct summand of $\mathcal{O} \otimes_{\mathcal{O}\mathrm{Ker}(\sigma)} \big(T^{\cdot\cdot} \otimes_{\mathcal{O}} \mathrm{Res}_{\sigma'}(\mathcal{O}P')\big)$ as $\mathcal{O}P^{\cdot\cdot}$-modules. In conclusion, $T^{\cdot\cdot}$ admits a $P^{\cdot\cdot}$-stable $\mathcal{O}$-basis by conjugation and therefore, since $\delta^{\cdot\cdot}$ is local, $T^{\cdot\cdot}$ is a Dade $P^{\cdot\cdot}$-algebra (cf. 2.2).

In particular, $T^{\cdot\cdot} \otimes_{\mathcal{O}} \mathrm{Res}_{\sigma'}(S')^{\circ}$ is a Dade $P^{\cdot\cdot}$-algebra too and, since we have an $\mathcal{O}P'$-interior algebra embedding (cf. [21, 5.7])

8.2.10 $$\mathcal{O} \longrightarrow S' \otimes_{\mathcal{O}} (S')^{\circ} \quad ,$$

we have $\mathcal{O}P^{\cdot\cdot}$-interior algebra embeddings

8.2.11 $$S^{\cdot\cdot} \longrightarrow S^{\cdot\cdot} \otimes_{\mathcal{O}} \mathrm{Res}_{\sigma'}\big(S' \otimes_{\mathcal{O}} (S')^{\circ}\big) \longleftarrow T^{\cdot\cdot} \otimes_{\mathcal{O}} \mathrm{Res}_{\sigma'}(S')^{\circ} \quad .$$

But, by Lemma 7.10, we have

8.2.12 $$\Big(S^{\cdot\cdot} \otimes_{\mathcal{O}} \mathrm{Res}_{\sigma}\big(S' \otimes_{\mathcal{O}} (S')^{\circ}\big)\Big)(P^{\cdot\cdot}) \cong S^{\cdot\cdot}(P^{\cdot\cdot}) \neq \{0\}$$

and therefore $P^{\cdot\cdot}$ has a unique local point $\gamma^{\cdot\cdot}$ on $S^{\cdot\cdot} \otimes_{\mathcal{O}} \mathrm{Res}_{\sigma'}(S' \otimes_{\mathcal{O}} (S')^{\circ})$, so that, since the point of $P^{\cdot\cdot}$ on $S^{\cdot\cdot}$ is local, $S^{\cdot\cdot}$ is an embedded algebra associated with $\gamma^{\cdot\cdot}$ (cf. 2.7); hence, since $P^{\cdot\cdot}$ has a local point on $T^{\cdot\cdot} \otimes_{\mathcal{O}} \mathrm{Res}_{\sigma'}(S')^{\circ}$ too, the left exoembedding in 8.2.11 factorizes throughout the right one and an $\mathcal{O}P^{\cdot\cdot}$-interior algebra exoembedding (cf. [20, 1.6])

8.2.13 $$S^{\cdot\cdot} \longrightarrow T^{\cdot\cdot} \otimes_{\mathcal{O}} \mathrm{Res}_{\sigma'}(S')^{\circ} \quad .$$

The existence of this exoembedding proves that $S^{\cdot\cdot}$ is a Dade $P^{\cdot\cdot}$-algebra, so that the Morita stable equivalence defined by $M^{\cdot\cdot}$ is basic and, by 7.6.3 and the main result of [20], the block b is nilpotent too.

9 The differential Z-grading $\mathcal{O}$-algebra

9.1 As announced in the introduction, our purpose is to extend our analysis of the Morita equivalences between blocks to the so-called *Rickard equivalences* (see Section 18 below), where one considers the *derived categories* (cf. [30]) or, more generally, the *homotopy categories* of the corresponding categories of modules (see our comment in 1.11 above and 10.7 below) instead of the categories of modules themselves. In both cases, we have to extend our arguments to the differential $\mathbf{Z}$-graded $\mathcal{O}$-modules and, consequently, we have to deal with the so-called *differential $\mathbf{Z}$-graded $\mathcal{O}$-algebras*, obviously with some finite group involved. It is well-known by the specialists that one way of describing these objects is to consider both the differential action and the projectors of the $\mathbf{Z}$-graduation as *coefficients* in an extended coefficient ring $\mathcal{D}$ (no longuer commutative!), which is *a fortiori* an $\mathcal{O}$-algebra *not* finitely generated as $\mathcal{O}$-module. However, since Rickard equivalences depend on differential $\mathbf{Z}$-graded $\mathcal{O}$-modules which are actually finitely generated as $\mathcal{O}$-modules (see [31] or 18.2 below), we are in fact interested in the quotients of $\mathcal{D}$ which are so.

9.2 From this point of view, in order to define the tensor product of differential $\mathbf{Z}$-graded $\mathcal{O}$-modules, it would be helpful that $\mathcal{D}$ admitted a reasonable coproduct such that it became a *Hopf $\mathcal{O}$-algebra*; unfortunately, it is known that $\mathcal{D}$ behaves only as the *$\mathcal{O}$-dual* of a Hopf $\mathcal{O}$-algebra (see [18] or Remark 9.13 below); nevertheless, we observe that it contains, as an $\mathcal{O}$-subalgebra, a true Hopf $\mathcal{O}$-algebra $\mathcal{D}_\circ$ covering all the quotients of $\mathcal{D}$ which we are interested in, and this remark allows us to define that tensor product in spite of the lack of a suitable coproduct in $\mathcal{D}$. Although all these facts are more or less known, we have found no handy reference so that, in this section, we introduce $\mathcal{D}$ and $\mathcal{D}_\circ$ with some detail, fixing our notation and discussing the statements we need in the sequel.

9.3 For any $n \geq 1$, let $\mathcal{F}_n$ be the commutative $\mathcal{O}$-free $\mathcal{O}$-algebra of all the $\mathcal{O}$-valued functions on $\mathbf{Z}/2n\mathbf{Z}$ and identify $\mathcal{F}_n \otimes_{\mathcal{O}} \mathcal{F}_n$ with the $\mathcal{O}$-algebra of all the $\mathcal{O}$-valued functions on $\mathbf{Z}/2n\mathbf{Z} \times \mathbf{Z}/2n\mathbf{Z}$. Actually, $\mathcal{F}_n$ can be identified with the $\mathcal{O}$-dual $\mathrm{Hom}_{\mathcal{O}}\big(\mathcal{O}(\mathbf{Z}/2n\mathbf{Z}), \mathcal{O}\big)$ of the group $\mathcal{O}$-algebra of $\mathbf{Z}/2n\mathbf{Z}$, the product in $\mathcal{F}_n$ just being the dual map of the diagonal homomorphism

9.3.1 $$\mathcal{O}(\mathbf{Z}/2n\mathbf{Z}) \longrightarrow \mathcal{O}(\mathbf{Z}/2n\mathbf{Z} \times \mathbf{Z}/2n\mathbf{Z}) \quad ;$$

consequently, the dual map of the sum in $\mathbf{Z}/2n\mathbf{Z}$ defines a (co)unitary commutative associative coproduct

9.3.2 $$\Delta_n : \mathcal{F}_n \longrightarrow \mathcal{F}_n \underset{\mathcal{O}}{\otimes} \mathcal{F}_n$$

mapping $f \in \mathcal{F}_n$ on the element of $\mathcal{F}_n \otimes_{\mathcal{O}} \mathcal{F}_n$ which, as an $\mathcal{O}$-valued function on $\mathbf{Z}/2n\mathbf{Z} \times \mathbf{Z}/2n\mathbf{Z}$, maps $(x, y) \in \mathbf{Z}/2n\mathbf{Z} \times \mathbf{Z}/2n\mathbf{Z}$ on $f(x+y)$, and then $\mathcal{F}_n$ endowed with Δ_n is a *Hopf $\mathcal{O}$-algebra*, where the *antipodal isomorphism* $t_n : \mathcal{F}_n \cong \mathcal{F}_n$ maps $f \in \mathcal{F}_n$ on the $\mathcal{O}$-valued function mapping $x \in \mathbf{Z}/2n\mathbf{Z}$ on $f(-x)$.

9.4 Let $\mathcal{F}$ be the commutative $\mathcal{O}$-algebra of all the $\mathcal{O}$-valued functions on $\mathbf{Z}$ and denote by $t : \mathcal{F} \cong \mathcal{F}$ the automorphism maping $f \in \mathcal{F}$ on the $\mathcal{O}$-valued function which maps $z \in \mathbf{Z}$ on $f(-z)$; for any $n \geq 1$, the $\mathcal{O}$-algebra $\mathcal{F}_n$ can be identified with the $\mathcal{O}$-subalgebra of $\mathcal{F}$ formed by all the $2n$-periodic $\mathcal{O}$-valued functions on $\mathbf{Z}$ and then we set

9.4.1 $$\mathcal{F}_\circ = \bigcup_{n \geq 1} \mathcal{F}_n \quad .$$

Notice that $\mathcal{F}_\circ \otimes_\mathcal{O} \mathcal{F}_\circ$ can be still identified with the $\mathcal{O}$-algebra of $\mathcal{O}$-valued functions on $\mathbf{Z} \times \mathbf{Z}$ which are $(2n, 2m)$-periodic for some $n \geq 1$ and some $m \geq 1$, whereas the canonical map from $\mathcal{F} \otimes_\mathcal{O} \mathcal{F}$ into the $\mathcal{O}$-algebra of all the $\mathcal{O}$-valued functions on $\mathbf{Z} \times \mathbf{Z}$ is *not* surjective.

9.5 In other words, as above $\mathcal{F}$ can be identified with the $\mathcal{O}$-dual $\mathrm{Hom}_\mathcal{O}(\mathcal{O}\mathbf{Z}, \mathcal{O})$ of the group $\mathcal{O}$-algebra of $\mathbf{Z}$, the product being the dual map of the diagonal homomorphism

9.5.1 $$\mathcal{O}\mathbf{Z} \longrightarrow \mathcal{O}(\mathbf{Z} \times \mathbf{Z}) \cong \mathcal{O}\mathbf{Z} \underset{\mathcal{O}}{\otimes} \mathcal{O}\mathbf{Z} \quad ;$$

then the canonical map

9.5.2 $$\mathrm{Hom}_\mathcal{O}(\mathcal{O}\mathbf{Z}, \mathcal{O}) \underset{\mathcal{O}}{\otimes} \mathrm{Hom}_\mathcal{O}(\mathcal{O}\mathbf{Z}, \mathcal{O}) \longrightarrow \mathrm{Hom}_\mathcal{O}(\mathcal{O}\mathbf{Z} \underset{\mathcal{O}}{\otimes} \mathcal{O}\mathbf{Z}, \mathcal{O})$$

is injective but *not* surjective and, whereas the dual map of the sum in $\mathbf{Z}$ induces an $\mathcal{O}$-algebra homomorphism

9.5.3 $$\mathrm{Hom}_\mathcal{O}(\mathcal{O}\mathbf{Z}, \mathcal{O}) \longrightarrow \mathrm{Hom}_\mathcal{O}(\mathcal{O}\mathbf{Z} \underset{\mathcal{O}}{\otimes} \mathcal{O}\mathbf{Z}, \mathcal{O})$$

which maps $\mathcal{F}_\circ$ into the image of $\mathcal{F}_\circ \otimes_\mathcal{O} \mathcal{F}_\circ$ via homomorphism 9.5.2, the image of homomorphism 9.5.3 is *not* contained in the image of homomorphism 9.5.2. Now, as above, homomorphism 9.5.3 induces a (co)unitary commutative associative coproduct

9.5.4 $$\Delta_\circ : \mathcal{F}_\circ \longrightarrow \mathcal{F}_\circ \underset{\mathcal{O}}{\otimes} \mathcal{F}_\circ$$

which extends all the coproducts 9.3.2 and then $\mathcal{F}_\circ$ endowed with $\Delta_\circ$ is a Hopf $\mathcal{O}$-algebra, since $t(\mathcal{F}_\circ) = \mathcal{F}_\circ$ and the restriction of t provides the antipodal isomorphism; actually, $\mathcal{F}_\circ$ is the *direct limit* of the inductive family of Hopf $\mathcal{O}$-algebras $\mathcal{F}_n$ and Hopf $\mathcal{O}$-algebra homomorphisms $\mathcal{F}_n \subset \mathcal{F}_{nm}$ where $n, m \geq 1$.

9.6 For any subset X of $\mathbf{Z}$, denote by i_X the characteristic $\mathcal{O}$-valued function of X (and set $i_z = i_{\{z\}}$ for any $z \in \mathbf{Z}$); so i_X is an idempotent of $\mathcal{F}$ and, for any $n \geq 1$, the family $\{i_Y\}_{Y \in \mathbf{Z}/2n\mathbf{Z}}$ is both the set of primitive idempotents and an $\mathcal{O}$-basis of $\mathcal{F}_n$ where, for any $f \in \mathcal{F}_n$, we have

9.6.1 $$f = \sum_{Y \in \mathbf{Z}/2n\mathbf{Z}} f(Y) i_Y \quad ;$$

moreover, for any $X \in \mathbf{Z}/2n\mathbf{Z}$, we get

9.6.2
$$\Delta_n(\imath_X) = \sum_{Y \in \mathbf{Z}/2n\mathbf{Z}} \imath_Y \otimes \imath_{X-Y} \quad .$$

Proposition 9.7 *Let A be an $\mathcal{O}$-finite $\mathcal{O}$-algebra, $g : \mathcal{F} \to A$ an $\mathcal{O}$-algebra homomorphism and I the set of primitive idempotents of $g(\mathcal{F})$. For any $i \in I$ there is a unique $z_i \in \mathbf{Z}$ such that $g(\imath_{z_i}) = i$ and then, for any $f \in \mathcal{F}$, we have*

9.7.1
$$g(f) = \sum_{i \in I} f(z_i)i \quad .$$

Proof: Since $g(\mathcal{F})$ is a commutative $\mathcal{O}$-finite $\mathcal{O}$-algebra, for any $i \in I$, we have an $\mathcal{O}$-algebra homomorphism $\varphi_i : g(\mathcal{F}) \to k$ such that $\varphi_i(i) = 1$; set $g_i = \varphi_i \circ g$, consider $h \in \mathcal{F}$ such that $g_i(h) \neq 0$ and denote by X the set of $z \in \mathbf{Z}$ such that $h(z) \neq 0$; since $\imath_X\, h = h$, we get also $g_i(\imath_X) \neq 0$, so that $g_i(\imath_X) = 1$. Now consider a function $h_X : \mathbf{Z} \to \mathcal{O}$ vanishing on $\mathbf{Z}-X$ and mapping X injectivily into the canonical lifting of k^* in $\mathcal{O}^*$ (cf. [37, Chap. II, Proposition 8]); since $\imath_X = h_X\, h'$ for a suitable $h' \in \mathcal{F}$, we still have $g_i(h_X) \neq 0$ and, up to modify our choice of h_X (i.e. we multiply h_X by the canonical lifting of $g_i(h_X)^{-1}$), we may assume that $g_i(h_X) = 1$. Consequently, we have $g_i(h_X - \imath_X) = 0$; but, since h_X is injective on X, there is $z_i \in X$ such that, for any $x \in X - \{z_i\}$, $(h_X - \imath_X)(x)$ is inversible in $\mathcal{O}$ and therefore we have $\imath_X - \imath_{z_i} = (h_X - \imath_X)h'$ for a suitable $h' \in \mathcal{F}$, so that $g_i(\imath_{z_i}) = 1$ which forces $g(\imath_{z_i}) = i$.

Set $W = \{z_i\}_{i \in I}$; it is clear that $g(\imath_W)$ is the unity element in $g(\mathcal{F})$ and, in particular, $g(\imath_Y) = 0$ for any $Y \subset \mathbf{Z} - W$; moreover, if $n \geq 1$ fulfills $-2n < z_i \leq 2n$ for any $i \in I$ then, for any $f \in \mathcal{F}$, there is $f' \in \mathcal{F}_n$ such that $f'(z_i) = f(z_i)$ for any $i \in I$, so that $(f - f')\imath_W = 0$ which implies $g(f) = g(f')$; consequently, by 9.6.1, we get

9.7.2
$$g(f) = g(f') = \sum_{Y \in \mathbf{Z}/2n\mathbf{Z}} f'(Y)g(\imath_Y) = \sum_{i \in I} f(z_i)i$$

since, for any $Y \in \mathbf{Z}/2n\mathbf{Z}$, either $Y \cap W = \emptyset$ and $g(\imath_Y) = 0$, or $Y \cap W = \{z_i\}$ for some $i \in I$ and $0 = g(\imath_{Y-\{z_i\}}) = g(\imath_Y) - i$.

Proposition 9.8 *Let A be an $\mathcal{O}$-finite $\mathcal{O}$-algebra. Two $\mathcal{O}$-algebra homomorphisms from $\mathcal{F}$ to A having the same restriction to $\mathcal{F}_\circ$ coincide. Moreover, an $\mathcal{O}$-algebra homomorphism $g_\circ : \mathcal{F}_\circ \to A$ can be extended to $\mathcal{F}$ if and only if, for any primitive idempotent i of $g_\circ(\mathcal{F}_\circ)$, the intersection of all the subsets X of $\mathbf{Z}$ such that $\imath_X \in \mathcal{F}_\circ$ and $g_\circ(\imath_X) = i$ is not empty.*

Proof: Let I be the set of primitive idempotents of $g_\circ(\mathcal{F}_\circ)$; for any $i \in I$, we claim that the intersection W_i of all the subsets X of $\mathbf{Z}$ such that $\imath_X \in \mathcal{F}_\circ$ and $g_\circ(\imath_X) = i$ has at most one element; indeed, if $z, z' \in W_i$ then, for any $n \geq 1$, we have $z' - z \in 2n\mathbf{Z}$

since there is a unique $Y \in \mathbf{Z}/2n\mathbf{Z}$ such that $g_\circ(i_Y)i = i$. Now, assume that, for any $i \in I$, we have $W_i = \{z_i\}$; by 9.6.1, for any $n \geq 1$ and any $f \in \mathcal{F}_n$, we have

9.8.1 $$g_\circ(f) = \sum_{Y \in \mathbf{Z}/2n\mathbf{Z}} f(Y) g_\circ(i_Y) \quad ;$$

but, for any $i \in I$ and any $Y \in \mathbf{Z}/2n\mathbf{Z}$, $g_\circ(i_Y)i = i$ implies $z_i \in Y$ and in particular we have $f(Y) = f(z_i)$; consequently, we get

9.8.2 $$g_\circ(f) = \sum_{i \in I} f(z_i) i$$

which can be extended to the $\mathcal{O}$-algebra homomorphism $g : \mathcal{F} \to A$ mapping $f \in \mathcal{F}$ on $\Sigma_{i \in I} f(z_i)i$. Conversely, if an $\mathcal{O}$-algebra homomorphism $g' : \mathcal{F} \to A$ extends $g_\circ$ then it follows from Proposition 9.7 that, for any $i \in I$, there is $z'_i \in \mathbf{Z}$ such that $g'(i_{z'_i}) = i$ and that, for any $f \in \mathcal{F}$, we have

9.8.3 $$g'(f) = \sum_{i \in I} f(z'_i) i \quad ;$$

in particular, for any $i \in I$, any $X \subset \mathbf{Z}$ such that $i_X \in \mathcal{F}_\circ$ and $g_\circ(i_X) = i$ contains z'_i.

Corollary 9.9 *Let A be an $\mathcal{O}$-finite $\mathcal{O}$-algebra. For any $\mathcal{O}$-algebra homomorphism $g : \mathcal{F} \otimes_{\mathcal{O}} \mathcal{F} \to A$ there is a unique $\mathcal{O}$-algebra homomorphism $\Delta^*_A(g) : \mathcal{F} \to A$ such that, for any $f \in \mathcal{F}_\circ$, we have*

9.9.1 $$\big(\Delta^*_A(g)\big)(f) = g\big(\Delta_\circ(f)\big) \quad .$$

Proof: By Proposition 9.8, we may assume that $g\big(\Delta_\circ(\mathcal{F}_\circ)\big) \neq \{0\}$; in that case, let i be a primitive idempotent of $g\big(\Delta_\circ(\mathcal{F}_\circ)\big)$ and choose $f, h \in \mathcal{F}$ such that $g(f \otimes h)i \neq 0$; denoting respectively by X and Y the sets of x and y in $\mathbf{Z}$ such that $f(x) \neq 0$ and $h(y) \neq 0$, we still have $g(i_X \otimes i_Y)i \neq 0$ since $(i_X \otimes i_Y)(f \otimes h) = f \otimes h$; then it follows from Proposition 9.7, applied to the $\mathcal{O}$-algebra homomorphism from $\mathcal{F}$ to A mapping $\ell \in \mathcal{F}$ on $g(i_X \otimes \ell)i$, that there is $y \in Y$ such that $g(i_X \otimes i_y)i \neq 0$ and, applied to the $\mathcal{O}$-algebra homomorphism from $\mathcal{F}$ to A mapping $\ell \in \mathcal{F}$ on $g(\ell \otimes i_y)i$, that there is $x \in X$ such that $g(i_x \otimes i_y)i \neq 0$. Now, any $W \subset \mathbf{Z}$ such that $i_W \in \mathcal{F}_\circ$ and $g\big(\Delta_\circ(i_W)\big) = i$ contains $x + y$ since, by 9.4.1, i_W belongs to $\mathcal{F}_n$ for a suitable $n \geq 1$ and, by 9.6.2, there is $U \in \mathbf{Z}/2n\mathbf{Z}$ such that $x \in U$ and $y \in W - U = \{w - u | w \in W \text{ and } u \in U\}$; at that point, the corollary follows from Proposition 9.8.

9.10 In order to introduce the differential part, let us consider the *shift automorphism* sh of the $\mathcal{O}$-algebra $\mathcal{F}$, which maps $f \in \mathcal{F}$ on the $\mathcal{O}$-valued function mapping $z \in \mathbf{Z}$ on $\mathcal{F}(z+1)$, and the *sign function* $s \in \mathcal{F}_1$, which maps $z \in \mathbf{Z}$ on $(-1)^z$; notice that $\mathrm{sh}(s) = -s$, $\mathrm{sh}(\mathcal{F}_n) = \mathcal{F}_n$ for any $n \geq 0$ and

9.10.1 $$\Delta_\circ\big(\mathrm{sh}(f)\big) = (\mathrm{id}_F \underset{\mathcal{O}}{\otimes} \mathrm{sh})\big(\Delta_\circ(f)\big) = (\mathrm{sh} \underset{\mathcal{O}}{\otimes} \mathrm{id}_F)\big(\Delta_\circ(f)\big)$$

for any $\mathscr{f} \in \mathcal{F}_\circ$. Now, $\mathcal{D}$ is the $\mathcal{O}$-algebra containing $\mathcal{F}$ as a unitary $\mathcal{O}$-subalgebra and an element $\mathscr{d}$ such that

9.10.2 $\quad \mathcal{D} = \mathcal{F} \oplus \mathcal{F}\mathscr{d},\ \mathscr{d}^2 = 0$ *and* $\mathscr{d}\mathscr{f} = \mathrm{sh}(\mathscr{f})\mathscr{d} \neq 0$ *for any* $\mathscr{f} \in \mathcal{F} - \{0\}$.

For any $n \geq 0$, it is clear that $\mathcal{D}$ has the following $\mathcal{O}$-subalgebra

9.10.3
$$\mathcal{D}_n = \mathcal{F}_n \oplus \mathcal{F}_n \mathscr{d}$$

and it is easily checked that the coproduct in $\mathcal{F}_n$ can be extended to an $\mathcal{O}$-algebra homomorphism $\Delta_n : \mathcal{D}_n \to \mathcal{D}_n \otimes_{\mathcal{O}} \mathcal{D}_n$ by setting

9.10.4
$$\Delta_n(\mathscr{d}) = \mathscr{d} \otimes \mathscr{s} + 1 \otimes \mathscr{d} \qquad ;$$

this coproduct is clearly associative and has a counity, which extends to $\mathcal{D}_n$ the counity of $\mathcal{F}_n$ by mapping $\mathscr{d}$ on zero.

9.11 Denote by $s : \mathcal{D} \otimes_{\mathcal{O}} \mathcal{D} \to \mathcal{D} \otimes_{\mathcal{O}} \mathcal{D}$ the $\mathcal{O}$-algebra automorphism mapping $\mathscr{a} \otimes \mathscr{b} \in \mathcal{D} \otimes_{\mathcal{O}} \mathcal{D}$ on $\mathscr{b} \otimes \mathscr{a}$, and by $\mathscr{t} \in \mathcal{F}_1 \otimes_{\mathcal{O}} \mathcal{F}_1$ the (2, 2)-periodic $\mathcal{O}$-valued function mapping $(x, y) \in \mathbf{Z} \times \mathbf{Z}$ on $(-1)^{xy}$. For any $n \geq 0$, the coproduct Δ_n above is *not* commutative whenever $\mathrm{char}(\mathcal{O}) \neq 2$, but it is not difficult to check that

9.11.1
$$\mathscr{d} \otimes \mathscr{s} = (\mathscr{d} \otimes 1)^{\mathscr{t}}$$

and therefore, for any $\mathscr{a} \in \mathcal{D}_n$, we have

9.11.2
$$s\big(\Delta_n(\mathscr{a})\big) = \Delta_n(\mathscr{a})^{\mathscr{t}}$$

which will imply the commutativity of the tensor product of differential $\mathbf{Z}$-graded $\mathcal{O}$-modules (see 10.2.6 below). As usual, $\mathcal{D}^\circ$ denotes the opposite $\mathcal{O}$-algebra and, for any $n \geq 0$, $(\mathcal{D}_n)^\circ$ is endowed with the coproduct $s \circ \Delta_n$; it is clear that the automorphism $t : \mathcal{F} \cong \mathcal{F}$ above (cf. 9.4) can be extended to an $\mathcal{O}$-algebra isomorphism $t : \mathcal{D} \cong \mathcal{D}^\circ$ by setting

9.11.3
$$t(\mathscr{d}) = \mathscr{s}\mathscr{d} \qquad ;$$

for any $n \geq 0$, this isomorphism induces the antipodal isomorphism $t_n : \mathcal{D}_n \cong (\mathcal{D}_n)^\circ$, showing that $\mathcal{D}_n$ endowed with the coproduct Δ_n is a Hopf $\mathcal{O}$-algebra. As in 9.5, $\mathcal{D}_\circ$ is the direct limit of the inductive family of Hopf $\mathcal{O}$-algebras $\mathcal{D}_n$ and Hopf $\mathcal{O}$-algebra homomorphisms $\mathcal{D}_n \subset \mathcal{D}_{nm}$, $n, m \geq 1$.

Remark 9.12 For any $n \geq 0$, recall that the structure of Hopf $\mathcal{O}$-algebra on $\mathcal{D}_n$ supposes that the following diagram, where u_n, e_n and m_n are respectively the unity, the counity and the product map, is commutative (which can be easily checked)

9.12.1
$$\begin{array}{ccccc}
\mathcal{D}_n \underset{\mathcal{O}}{\otimes} \mathcal{D}_n & & \xrightarrow{\ \mathrm{id}\otimes t_n\ } & & \mathcal{D}_n \underset{\mathcal{O}}{\otimes} \mathcal{D}_n \\
{\scriptstyle \Delta_n}\uparrow & & & & \downarrow{\scriptstyle m_n} \\
\mathcal{D}_n & \xrightarrow{\ e_n\ } & \mathcal{O} & \xrightarrow{\ u_n\ } & \mathcal{D}_n
\end{array}$$

Setting $h_n = (\mathrm{id}_{\mathcal{D}_n} \otimes_{\mathcal{O}} m_n) \circ (\Delta_n \otimes_{\mathcal{O}} t_n)$, this commutativity implies that

9.12.2 $$(h_n)^2 = \mathrm{id}_{\mathcal{D}_n \otimes_{\mathcal{O}} \mathcal{D}_n}$$

which has the useful consequence

9.12.3 $$\Delta_n(\mathcal{D}_n)\cdot(1 \otimes \mathcal{D}_n) = \mathcal{D}_n \underset{\mathcal{O}}{\otimes} \mathcal{D}_n \quad ;$$

indeed, considering the evident right $\mathcal{D}_n$-module structure on $\mathcal{D}_n \otimes_{\mathcal{O}} \mathcal{D}_n$, it is quite clear that h_n is a $\mathcal{D}_n$-module homomorphism from $\mathcal{D}_n \otimes_{\mathcal{O}} \mathcal{D}_n$ to $\mathrm{Res}_{t_n}(\mathcal{D}_n \otimes_{\mathcal{O}} \mathcal{D}_n)$; moreover, for any $a \in \mathcal{D}_n$, it is not difficult, from the associtivity of the coproduct Δ_n and from the commutativity of diagram 9.12.1, to check the first of the following equalities, the second being clear

9.12.4 $$h_n\big(\Delta_n(a)\big) = a \otimes 1 \quad \text{and} \quad h_n(a \otimes 1) = \Delta_n(a) \quad ;$$

now, since $\mathcal{D}_n \otimes 1$ generates $\mathcal{D}_n \otimes_{\mathcal{O}} \mathcal{D}_n$ as $\mathcal{D}_n$-module, equalities 9.12.4 imply equality 9.12.2 and then imply that $\Delta_n(\mathcal{D}_n)$ generates $\mathcal{D}_n \otimes_{\mathcal{O}} \mathcal{D}_n$ too.

Remark 9.13 For any $n \geq 0$, consider the $\mathcal{O}$-algebra $\mathcal{C}_n$ containing the group $\mathcal{O}$-algebra $\mathcal{O}(\mathbf{Z}/2n\mathbf{Z})$ as a unitary $\mathcal{O}$-subalgebra and an element c such that

9.13.1 $\mathcal{C}_n = \mathcal{O}(\mathbf{Z}/2n\mathbf{Z}) \oplus \mathcal{O}(\mathbf{Z}/2n\mathbf{Z})c$, $c^2 = 0$ *and* $c\, e_X = \varepsilon_X\, e_X\, c$ *for any* $x \in \mathbf{Z}/2n\mathbf{Z}$, *where* e_X *is the corresponding element in the canonical* $\mathcal{O}$-*basis of* $\mathcal{O}(\mathbf{Z}/2n\mathbf{Z})$, *we set* $\varepsilon_X = (-1)^X$ *and* $\{e_X\, c\}_{X \in \mathbf{Z}/2n\mathbf{Z}}$ *is an* $\mathcal{O}$-*basis of* $\mathcal{O}(\mathbf{Z}/2n\mathbf{Z})c$.

Then, the (co)unitary associative coproduct on $\mathcal{O}(\mathbf{Z}/2n\mathbf{Z})$ given by the diagonal map can be extended to an $\mathcal{O}$-algebra homomorphism $\Gamma_n : \mathcal{C}_n \to \mathcal{C}_n \otimes_{\mathcal{O}} \mathcal{C}_n$ by setting

9.13.2 $$\Gamma_n(c) = c \otimes e_{1+2n\mathbf{Z}} + e_{2n\mathbf{Z}} \otimes c$$

and $\mathcal{C}_n$ endowed with Γ_n is a Hopf $\mathcal{O}$-algebra, the antipodal isomorphism mapping c on $e_{-1+2n\mathbf{Z}}c$ and e_X on e_{-X} for any $X \in \mathbf{Z}/2n\mathbf{Z}$. Now, it is easily checked that the $\mathcal{O}$-linear map $\mathcal{D} \to \mathrm{Hom}_{\mathcal{O}}(\mathcal{C}_\circ, \mathcal{O})$ defined by the duality relations

9.13.3 $$\begin{aligned} \langle f, e_z\rangle &= f(z), \quad \langle f, e_z\, c\rangle = 0 \\ \langle fd, e_z\rangle &= 0, \quad \langle fd, e_z\, c\rangle = f(z) \end{aligned}$$

where f runs on $\mathcal{F}$ and z on $\mathbf{Z}$, is an $\mathcal{O}$-algebra isomorphism between $\mathcal{D}$ and $\mathrm{Hom}_{\mathcal{O}}(\mathcal{C}_\circ, \mathcal{O})$ endowed with the product defined by the $\mathcal{O}$-dual map of $\Gamma_\circ$, and that, for any $n \geq 1$, this isomorphism induces a Hopf $\mathcal{O}$-algebra isomorphism between $\mathcal{D}_n$ and $\mathrm{Hom}_{\mathcal{O}}(\mathcal{C}_n, \mathcal{O})$ endowed with the product and the coproduct respectively defined by the $\mathcal{O}$-dual maps of the coproduct and the product of $\mathcal{C}_n$.

Proposition 9.14 *Let A be an* $\mathcal{O}$-*finite* $\mathcal{O}$-*algebra. Two* $\mathcal{O}$-*algebra homomorphisms* g *and* g' *from* $\mathcal{D}$ *to A having the same restriction to* $\mathcal{D}_\circ$ *coincide. Moreover, an* $\mathcal{O}$-*algebra homomorphism* $g_\circ : \mathcal{D}_\circ \to A$ *can be extended to* $\mathcal{D}$ *if and only if its restriction to* $\mathcal{F}_\circ$ *can be extended to* $\mathcal{F}$.

Proof: Since g and g' have the same restriction to $\mathcal{F}_\circ$, their restrictions to $\mathcal{F}$ coincide (cf. Proposition 9.8) and therefore, since $\mathcal{D}_\circ$ contains d, they coincide (cf. 9.10.2). Moreover, if the restriction of $g_\circ$ to $\mathcal{F}_\circ$ can be extended to an $\mathcal{O}$-algebra homomorphism $g'' : \mathcal{F} \to A$, we consider the map $\mathcal{D} \to A$ sending $f + hd$ to $g''(f) + g''(h)g_\circ(d)$ for any $f, h \in \mathcal{F}$, which clearly extends $g_\circ$; now, according to 9.10.2, it suffices to prove that, for any $f \in \mathcal{F}$, we have

9.14.1 $$g_\circ(d)g''(f) = g''\big(\mathrm{sh}(f)\big)g_\circ(d) \quad ;$$

but, denoting by I the set of primitive idempotents of $g''(\mathcal{F})$ and, for any $i \in I$, by z_i the element of $\mathbf{Z}$ such that $g''(i_{z_i}) = i$, and choosing $n \geq 1$ such that $-2n < z_i \leq 2n$ for any $i \in I$, it follows from Proposition 9.7 that, for any $f \in \mathcal{F}$, we have

9.14.2 $$g''(f) = \sum_{i\in I} f(z_i)i = \sum_{X\in\mathbf{Z}/2n\mathbf{Z}} f(X)g_\circ(i_X)$$

and therefore, for any $f \in \mathcal{F}$, we get

9.14.3 $$\begin{aligned} g_\circ(d)g''(f) &= \sum_{X\in\mathbf{Z}/2n\mathbf{Z}} f(X)g_\circ(di_X) = \sum_{X\in\mathbf{Z}/2n\mathbf{Z}} f(X)g_\circ(i_{X-1}d) \\ &= \sum_{X\in\mathbf{Z}/2n\mathbf{Z}} f(X+1)g_\circ(i_X)g_\circ(d) = g''\big(\mathrm{sh}(f)\big)g_\circ(d) \quad . \end{aligned}$$

Corollary 9.15 *Let A be an $\mathcal{O}$-finite $\mathcal{O}$-algebra. For any $\mathcal{O}$-algebra homomorphism $g : \mathcal{D} \otimes_{\mathcal{O}} \mathcal{D} \to A$ there is a unique $\mathcal{O}$-algebra homomorphism $\Delta_A^*(g) : \mathcal{D} \to A$ such that, for any $a \in \mathcal{D}_\circ$, we have*

9.15.1 $$\big(\Delta_A^*(g)\big)(a) = g\big(\Delta_\circ(a)\big) \quad .$$

Proof: By Corollary 9.9, the restriction to $\mathcal{F}_\circ$ of the composed $\mathcal{O}$-algebra homomorphism

9.15.2 $$\mathcal{D}_\circ \xrightarrow{\Delta_\circ} \mathcal{D}_\circ \underset{\mathcal{O}}{\otimes} \mathcal{D}_\circ \subset \mathcal{D} \underset{\mathcal{O}}{\otimes} \mathcal{D} \xrightarrow{g} A$$

can be extended to $\mathcal{F}$; then, it suffices to apply Proposition 9.14.

9.16 Finally, in order to discuss on the *tensor induction* of differential **Z**-graded structures in Appendix 2 below, we have to consider iterated coproducts and the corresponding iterated forms of equality 9.11.2. Let X be a nonempty finite set and, for any $n \geq 0$, denote by $T_X(\mathcal{D}_n)$ the *X-tensor power* of $\mathcal{D}_n$ (cf. 2.5); for any total order relation $R \subset X \times X$ on X, we define the *iterated coproduct*

9.16.1 $$\Delta_{n,R} : \mathcal{D}_n \longrightarrow T_X(\mathcal{D}_n)$$

in the following way: if $|X| = 1$ then $T_X(\mathcal{D}_n) = \mathcal{D}_n$, $R = X \times X$ and $\Delta_{n,R}$ is just the identity map; if $|X| \geq 2$ then, denoting by $X'' \subset X' \subset X$ the two last links in the maximal R-chain of X and by R' the total order relation on X' induced by R, and identifying $T_{X''}(\mathcal{D}_n) \otimes_{\mathcal{O}} \mathcal{D}_n$ to $T_{X'}(\mathcal{D}_n)$ and $T_{X'}(\mathcal{D}_n) \otimes_{\mathcal{O}} \mathcal{D}_n$ to $T_X(\mathcal{D}_n)$, we set

9.16.2 $$\Delta_{n,R} = (\mathrm{id}_{T_{X''}(\mathcal{D}_n)} \underset{\mathcal{O}}{\otimes} \Delta_n) \circ \Delta_{n,R'} \quad .$$

Notice that the restriction of $\Delta_{n,R}$ to $\mathcal{F}_n$ does not depend on the choice of R.

9.17 Now, if $\{A_x\}_{x\in X}$ is a family of $\mathcal{O}$-finite $\mathcal{O}$-algebras and, for any $x \in X$, $g_x : \mathcal{D}_n \to A_x$ is an $\mathcal{O}$-algebra homomorphism, setting $A = \otimes_{x\in X} A_X$ and $g = \otimes_{x\in X} g_x$, iterated applications of Corollary 9.15 imply, for any total order relation R on X, the existence of a unique $\mathcal{O}$-algebra homomorphism

9.17.1 $$\Delta^*_{A,R}(g) : \mathcal{D}_n \longrightarrow A$$

such that, for any $a \in \mathcal{D}_n$, we have

9.17.2 $$\big(\Delta^*_{A,R}(g)\big)(a) = g\big(\Delta_{n,R}(a)\big) \quad ;$$

once again, the restriction of $\Delta^*_{A,R}(g)$ to $\mathcal{F}$ does not depend on the choice of R. On the other hand, denote by $\mathcal{T}_1$ the subgroup of all the involutions in $T_X(\mathcal{F}_1)^*$; it is clear that, for any $n \geq 0$, $\mathcal{T}_1$ acts by conjugation on $T_X(\mathcal{D}_n)$ and, for any $t' \in \mathcal{T}_1$ and any $\mathcal{O}$-algebra homomorphism $\Lambda : \mathcal{D}_n \to T_X(\mathcal{D}_n)$, we denote by $\Lambda^{t'}$ the conjugate homomorphism.

Proposition 9.18 *With the notation above, for any $n \geq 0$, the quotient $\mathcal{T}_1/\{1,-1\}$ acts regularly on the set $\{(\Delta_{n,R})^{t'}\}_{R,t'}$, where R runs on the set of total order relations on X and t' on $\mathcal{T}_1$.*

Proof: If R and R' are total order relations on X, it is clear that we have a sequence $\{R_l\}_{0 \leq l \leq m}$ of total order relations on X such that $R_\circ = R$, $R_m = R'$ and for any $1 \leq l \leq m$ the symmetric sum $R_{l-1} + R_l$ is equal to $\{(x,y),(y,x)\}$ for suitable $x, y \in X$; then, it follows from 9.11.2 that there is $t_l \in \mathcal{T}_1$ such that $\Delta_{n,R_l} = (\Delta_{n,R_{l-1}})^{t_l}$ and therefore, considering $t_\circ = \Pi_{l=1}^m t_l$, we get $\Delta_{n,R'} = (\Delta_{n,R})^{t_\circ}$ which proves the transitivity part.

Now, if $\mathrm{char}(\mathcal{O}) = 2$, we have $\mathcal{T}_1 = \{1\}$; so, assume that $\mathrm{char}(\mathcal{O}) \neq 2$ and let t' be an element of the stabilizer of $\Delta_{n,R}$ in $\mathcal{T}_1$; since $\{i_{\bar z}\}_{\bar z\in \mathbf{Z}/2\mathbf{Z}}$ is an $\mathcal{O}$-basis of $\mathcal{F}_1$, it is clear that the family $\{\otimes_{x\in X} i_{\varphi(x)}\}_\varphi$, where φ runs on the set $(\mathbf{Z}/2\mathbf{Z})^X$ of $\mathbf{Z}/2\mathbf{Z}$-valued functions on X, is an $\mathcal{O}$-basis of $T_X(\mathcal{F}_1)$ and therefore, setting $\varepsilon_{z+2\mathbf{Z}} = (-1)^z$ in $\mathcal{O}$ for any $z \in \mathbf{Z}$, we have

9.18.1 $$t' = \sum_{\varphi\in(\mathbf{Z}/2\mathbf{Z})^X} \varepsilon_{\tau'(\varphi)}\big(\bigotimes_{x\in X} i_{\varphi(x)}\big)$$

for a suitable map $\tau' : (\mathbf{Z}/2\mathbf{Z})^X \to \mathbf{Z}/2\mathbf{Z}$. More generally, setting $d_{2\mathbf{Z}} = 1$, $d_{1+2\mathbf{Z}} = d$ and $b_{(\psi,\varphi)} = \otimes_{x\in X} d_{\psi(x)} i_{\varphi(x)}$ for any $\psi, \varphi \in (\mathbf{Z}/2\mathbf{Z})^X$, the family

$\{b_{(\psi,\varphi)}\}_{\psi,\varphi}$, where ψ and φ run on $(\mathbf{Z}/2\mathbf{Z})^X$, is an $\mathcal{O}$-basis of $T_X(\mathcal{D}_1)$ and it is easily checked that

9.18.2
$$\Delta_{n,R}(d) = \sum_{\varphi\in(\mathbf{Z}/2\mathbf{Z})^X} \left(\sum_{x\in X}\left(\prod_{x<x'} \varepsilon_{\varphi(x')}\right) b_{(\psi_x,\varphi)}\right) ,$$

where, for any $x \in X$, ψ_x is the characteristic $\mathbf{Z}/2\mathbf{Z}$-valued function of $\{x\}$ in X. Then, since t' stabilizes $\Delta_{n,R}$, it commutes with $\Delta_{n,R}(d)$; but, from 9.18.1 and 9.18.2, it is easily checked that

9.18.3
$$\Delta_{n,R}(d)t' = \sum_{\varphi\in(\mathbf{Z}/2\mathbf{Z})^X} \left(\sum_{x\in X} \varepsilon_{\tau'(\varphi)}\left(\prod_{x<x'} \varepsilon_{\varphi(x')}\right) b_{(\psi_x,\varphi)}\right)$$

whereas, since $i_{\bar z}\, d = d i_{\bar z+1}$ for any $\bar z \in \mathbf{Z}/2\mathbf{Z}$, we get

9.18.4
$$t'\Delta_{n,R}(d) = \sum_{\varphi\in(\mathbf{Z}/2\mathbf{Z})^X} \left(\sum_{x\in X} \varepsilon_{\tau'(\varphi+\psi_x)}\left(\prod_{x<x'} \varepsilon_{\varphi(x')}\right) b_{(\psi_x,\varphi)}\right) .$$

Consequently, since $\mathrm{char}(\mathcal{O}) \neq 2$, we get $\varepsilon_{\tau'(\varphi)} = \varepsilon_{\tau'(\varphi+\psi_x)}$ for any $\varphi \in (\mathbf{Z}/2\mathbf{Z})^X$ and any $x \in X$, which proves that τ' is a constant map.

10 $\mathcal{D}G$-modules

10.1 Let G be a finite group; in this section we introduce the *differential* $\mathbf{Z}$*-graded* $\mathcal{O}G$*-modules* (or *complexes* of $\mathcal{O}G$-modules) and develop the usual terminology in terms of $\mathcal{D}G$*-modules*. As explained in 9.1, here we are interested only on the differential $\mathbf{Z}$-graded $\mathcal{O}G$-modules which are finitely generated as $\mathcal{O}$-modules (in short, $\mathcal{O}$*-finite*) and it is now clear that they are just the $\mathcal{O}$*-finite* $\mathcal{D}G$*-modules*. In other words, an $\mathcal{O}$-finite complexe of $\mathcal{O}G$-modules M is an $\mathcal{O}$-finite $\mathcal{O}G$-module endowed with a unitary $\mathcal{O}$-algebra homomorphism

10.1.1 $$\mathcal{D} \longrightarrow \mathrm{End}_{\mathcal{O}G}(M) \quad ;$$

indeed, notice that for all $z \in \mathbf{Z}$ but a finite set we have $i_z(M) = \{0\}$ and, by Proposition 9.7, the family of inclusions $i_z(M) \subset M$, where z runs on $\mathbf{Z}$, determines an $\mathcal{O}G$-module isomorphism

10.1.2 $$M \cong \bigoplus_{z \in Z} i_z(M)$$

(we denote by $a(m)$ the image of $m \in M$ by the action of $a \in \mathcal{D}$ to avoid confusion in the $\mathcal{O}$*-finite* $\mathcal{D}$*-interior algebras*: see 11.2 below). If M is an $\mathcal{O}$-finite $\mathcal{D}G$-module, we endow the $\mathcal{O}$-dual $\mathcal{O}G$-module $M^* = \mathrm{Hom}_{\mathcal{O}}(M, \mathcal{O})$ with the $\mathcal{D}G$-module structure given by the composed $\mathcal{O}$-algebra homomorphism (cf. 9.4 and 9.11.3)

10.1.3 $$\mathcal{D} \overset{t}{\cong} \mathcal{D}^\circ \longrightarrow \mathrm{End}_{\mathcal{O}G}(M)^\circ \longrightarrow \mathrm{End}_{\mathcal{O}G}(M^*) \quad ,$$

where the last homomorphism maps $f \in \mathrm{End}_{\mathcal{O}G}(M)$ on the $\mathcal{O}$-dual map $f^* : M^* \to M^*$. We consider any ($\mathcal{O}$-finite) $\mathcal{O}G$-module M as a $\mathcal{D}G$-module by mapping $f \in \mathcal{F}$ on $f(0)\mathrm{id}_M$ and d on the zero map.

10.2 Let M and M' be $\mathcal{O}$-finite $\mathcal{D}G$-modules; coherently, a differential $\mathbf{Z}$-graded $\mathcal{O}G$-module homomorphism $f : M \to M'$ is just a $\mathcal{D}G$-linear map and we denote by $\mathrm{Hom}_{\mathcal{D}G}(M, M')$ or $\mathrm{Hom}_{\mathcal{D}}(M, M')^G$ the set of all the $\mathcal{D}G$-module homomorphisms from M to M'; it is clear that the $\mathcal{O}$-dual map $f^* : (M')^* \to M^*$ is still $\mathcal{D}G$-linear. On the other hand, it follows from Corollary 9.15 that the composed $\mathcal{O}$-algebra homomorphism

10.2.1 $$\begin{gathered} \mathcal{D}_\circ \overset{\Delta_\circ}{\longrightarrow} \mathcal{D}_\circ \otimes_{\mathcal{O}} \mathcal{D}_\circ \subset \mathcal{D} \otimes_{\mathcal{O}} \mathcal{D} \longrightarrow \mathrm{End}_{\mathcal{O}G}(M) \otimes_{\mathcal{O}} \mathrm{End}_{\mathcal{O}G}(M') \\ \longrightarrow \mathrm{End}_{\mathcal{O}G}(M \otimes_{\mathcal{O}} M') \end{gathered}$$

can be extended to a unique $\mathcal{O}$-algebra homomorphism

10.2.2 $$\mathcal{D} \longrightarrow \mathrm{End}_{\mathcal{O}G}(M \otimes_{\mathcal{O}} M') \quad ,$$

so that $M \otimes_{\mathcal{O}} M'$ becomes a $\mathcal{D}G$-module, and it is easily checked that, for any $m \in M$, any $m' \in M'$ and any $\mathscr{f} \in \mathcal{F}$, we have

10.2.3
$$\begin{aligned} \mathscr{f}(m \otimes m') &= \sum_{z,z' \in \mathbf{Z}} \mathscr{f}(z) \imath_{z'}(m) \otimes \imath_{z-z'}(m') \\ d(m \otimes m') &= d(m) \otimes \mathscr{s}(m') + m \otimes d(m') \end{aligned} \quad .$$

In particular, if $f : M \to N$ and $f' : M' \to N'$ are $\mathcal{D}G$-module homomorphisms, N and N' being $\mathcal{O}$-finite $\mathcal{D}G$-modules, the tensor product $f \otimes_{\mathcal{O}} f'$ is a $\mathcal{D}G$-module homomorphism too. More generally, if H is a normal subgroup of G then the kernel of the canonical map from $M \otimes_{\mathcal{O}} M'$ to $M \otimes_{\mathcal{O}H} M'$ (cf. 2.5)

10.2.4
$$\sum_{y \in H} \Big(\sum_{m \in M} \Big(\sum_{m' \in M'} \mathcal{O}\big(y(m) \otimes m' - m \otimes y^{-1}(m')\big)\Big)\Big)$$

is a $\mathcal{D}G$-submodule of $M \otimes_{\mathcal{O}} M'$ and therefore $M \otimes_{\mathcal{O}H} M'$ becomes a $\mathcal{D}(G/H)$-module; notice that, denoting by $\mathscr{t}^H_{M,M'}$ the image of $\mathscr{t} \in (\mathcal{F}_1 \otimes_{\mathcal{O}} \mathcal{F}_1)^*$ in $\mathrm{End}_{\mathcal{O}(G/H)}(M \otimes_{\mathcal{O}H} M')$ and by

10.2.5
$$s^H_{M,M'} : M \underset{\mathcal{O}H}{\otimes} M' \cong M' \underset{\mathcal{O}H}{\otimes} M$$

the canonical $\mathcal{O}(G/H)$-module isomorphism, we get from 9.11.2 and from the uniqueness of homomorphism 10.2.2 the canonical $\mathcal{D}(G/H)$-module isomorphism

10.2.6
$$s^H_{M,M'} \circ \mathscr{t}^H_{M,M'} : M \underset{\mathcal{O}H}{\otimes} M' \cong M' \underset{\mathcal{O}H}{\otimes} M \quad .$$

Moreover, if M'' is a third $\mathcal{O}$-finite $\mathcal{D}G$-module and H' a second normal subgroup of G, the associativity of the coproduct $\Delta_\circ$ and the uniqueness of homomorphism 10.2.2 force the following canonical $\mathcal{O}(G/H{\cdot}H')$-module isomorphism to be $\mathcal{D}(G/H{\cdot}H')$-linear

10.2.7
$$(M \underset{\mathcal{O}H}{\otimes} M') \underset{\mathcal{O}H'}{\otimes} M'' \cong M \underset{\mathcal{O}H}{\otimes} (M' \underset{\mathcal{O}H'}{\otimes} M'') \quad .$$

Remark 10.3 For any $n \geq 0$, the corresponding tensor product $M \otimes_{\mathcal{O}H} M'$ of $\mathcal{D}_n G$-modules M and M' can be alternatively defined in the following way: first we consider the extension of M to $(\mathcal{D}_n \otimes_{\mathcal{O}} \mathcal{D}_n)G$ via the $\mathcal{O}$-algebra homomorphism determined by Δ_n

10.3.1
$$\mathcal{D}_n G \longrightarrow (\mathcal{D}_n \underset{\mathcal{O}}{\otimes} \mathcal{D}_n)G \quad ;$$

then, denoting by $(\mathcal{D}_n \otimes_{\mathcal{O}} \mathcal{D}_n) \otimes_{\mathcal{D}_n} M$ this extension and considering its right $\mathcal{D}_n H$-module structure given by the $\mathcal{O}$-algebra homomorphism

10.3.2 $$(\mathcal{D}_n H)^{\circ} \longrightarrow (\mathcal{D}_n \underset{\mathcal{O}}{\otimes} \mathcal{D}_n) G$$

mapping a_y on $\big(1 \otimes t_n(a)\big) y^{-1}$ for any $a \in \mathcal{D}_n$ and any $y \in H$, the tensor product $\big((\mathcal{D}_n \otimes_{\mathcal{O}} \mathcal{D}_n) \otimes_{\mathcal{D}_n} M\big) \otimes_{\mathcal{D}_n H} M'$ has a $\mathcal{D}_n(G/H)$-module structure given by

10.3.3 $$a\bar{x}\Big(\big((a' \otimes a'') \otimes m\big) \otimes m'\Big) = \big((aa' \otimes a'') \otimes x(m)\big) \otimes x(m')$$

for any a, a', $a'' \in \mathcal{D}_n$ and any $x \in G$, where $\bar{x}$ denotes the image of x in G/H, and we claim that:

10.3.4 *We have a $\mathcal{D}_n(G/H)$-module isomorphism*

$$M \underset{\mathcal{O}H}{\otimes} M' \cong \big((\mathcal{D}_n \underset{\mathcal{O}}{\otimes} \mathcal{D}_n) \underset{\mathcal{D}_n}{\otimes} M\big) \underset{\mathcal{D}_n H}{\otimes} M'$$

mapping $m \otimes m'$ on $\big((1 \otimes 1) \otimes m\big) \otimes m'$ for any $m \in M$ and any $m' \in M'$.

Indeed, the existence of such a canonical $\mathcal{O}$-linear map is clear and it is easily checked that it is actually $\mathcal{D}_n(G/H)$-linear; moreover, it follows from equality 9.12.3 that it is a surjective map; on the other hand, considering the evident extension of this natural map to the pairs of finitely generated $\mathcal{D}_n H$ modules, it is quite clear that it suffices to prove the injectivity when $M = M' = \mathcal{D}_n H$; but, in that case, both members of the isomorphism in 10.3.4 become $(\mathcal{D}_n \otimes_{\mathcal{O}} \mathcal{D}_n) H$ and the map becomes the identity.

10.4 Let M and M' be again $\mathcal{O}$-finite $\mathcal{D}G$-modules; similarly, the composed $\mathcal{O}$-algebra homomorphism

10.4.1 $$\begin{gathered} \mathcal{D}_{\circ} \xrightarrow{\Delta_{\circ}} \mathcal{D}_{\circ} \underset{\mathcal{O}}{\otimes} \mathcal{D}_{\circ} \xrightarrow{t_{\circ} \otimes \mathrm{id}} (\mathcal{D}_{\circ})^{\circ} \underset{\mathcal{O}}{\otimes} \mathcal{D}_{\circ} \subset \mathcal{D}^{\circ} \underset{\mathcal{O}}{\otimes} \mathcal{D} \\ \longrightarrow \mathrm{End}_{\mathcal{O}G}(M)^{\circ} \underset{\mathcal{O}}{\otimes} \mathrm{End}_{\mathcal{O}G}(M') \longrightarrow \mathrm{End}_{\mathcal{O}G}\big(\mathrm{Hom}_{\mathcal{O}}(M, M')\big) \end{gathered}$$

can be extended to a unique $\mathcal{D}G$-module structure on $\mathrm{Hom}_{\mathcal{O}}(M, M')$ and it is easily checked that, respectively denoting by a_M and $a_{M'}$ the images of $a \in \mathcal{D}$ in $\mathrm{End}_{\mathcal{O}G}(M)$ and $\mathrm{End}_{\mathcal{O}G}(M')$, for any $h \in \mathrm{Hom}_{\mathcal{O}}(M, M')$ and any $f \in \mathcal{F}$ we have

10.4.2 $$\begin{aligned} f(h) &= \sum_{z,z' \in Z} f(z)(i_{z'})_{M'} \circ h \circ (i_{z+z'})_M \\ d(h) &= (d_{M'} \circ h - h \circ d_M) \circ s_M \qquad ; \end{aligned}$$

notice that if H is a normal subgroup of G then $\mathrm{Hom}_{\mathcal{O}H}(M, M')$ is a $\mathcal{D}G$-submodule of $\mathrm{Hom}_{\mathcal{O}}(M, M')$, so that it becomes a $\mathcal{D}(G/H)$-module. Finally, it is easy to check that the canonical $\mathcal{O}G$-module homomorphism

10.4.3 $$\mathrm{Hom}_{\mathcal{O}}(M, M') \underset{\mathcal{O}}{\otimes} M \longrightarrow M'$$

which, for any $h \in \mathrm{Hom}_{\mathcal{O}}(M, M')$ and any $m \in M$, maps $h\otimes m$ on $h(m)$ is $\mathcal{D}G$-linear and that if M'' is a third $\mathcal{O}$-finite $\mathcal{D}G$-module then:

10.4.4 *The composition determines a $\mathcal{D}G$-module homomorphism*

$$\mathrm{Hom}_{\mathcal{O}}(M', M'') \underset{\mathcal{O}}{\otimes} \mathrm{Hom}_{\mathcal{O}}(M, M') \longrightarrow \mathrm{Hom}_{\mathcal{O}}(M, M'') \quad .$$

Remark 10.5 If we remove the $\mathcal{O}$-finiteness condition, a *differential* **Z**-*graded* $\mathcal{O}G$-*module* is a $\mathcal{D}G$-module *fulfilling condition* 10.1.2; then, if M and M' are two differential **Z**-graded $\mathcal{O}G$-modules, homomorphism 10.2.1 can be still extended to a $\mathcal{D}G$-module structure on $M \otimes_{\mathcal{O}} M'$ by formulae 10.2.3. On the other hand, condition 10.1.2 forces to consider only the $\mathcal{O}$-linear maps $f : M \to M'$ such that $f\big(i_z(M)\big) = \{0\}$ for all $z \in \mathbf{Z}$ but a finite set and, denoting by $\mathrm{BHom}_{\mathcal{O}}(M, M')$ (B as "bounded") the set of them, the corresponding $\mathcal{O}$-algebra homomorphism

10.5.1 $$\mathcal{D}_\circ \longrightarrow \mathrm{End}_{\mathcal{O}G}\big(\mathrm{BHom}_{\mathcal{O}}(M, M')\big)$$

can be similarly extended to a $\mathcal{D}G$-module structure on $\mathrm{BHom}_{\mathcal{O}}(M, M')$ fulfilling condition 10.1.2.

10.6 As usual, for any $\mathcal{O}$-finite $\mathcal{D}G$-module M, denoting by d_M the image of d in $\mathrm{End}_{\mathcal{O}G}(M)$, we consider the $\mathcal{F}G$-modules

10.6.1 $\mathsf{B}(M) = \mathrm{Im}(d_M) \subset \mathsf{C}(M) = \mathrm{Ker}(d_M)$ and $\mathsf{H}(M) = \mathsf{C}(M)/\mathsf{B}(M)$;

actually, they are $\mathcal{D}G$-modules where the action of d is the zero map; moreover, for any $z \in \mathbf{Z}$, we set

10.6.2 $$\mathsf{B}_z(M) = i_z\big(\mathsf{B}(M)\big), \quad \mathsf{C}_z(M) = i_z\big(\mathsf{C}(M)\big) \quad \text{and} \\ \mathsf{H}_z(M) = i_z\big(\mathsf{H}(M)\big) = \mathsf{C}_z(M)/\mathsf{B}_z(M)$$

which are $\mathcal{O}G$-modules. If M' is a second $\mathcal{O}$-finite $\mathcal{D}G$-module and $f : M \to M'$ a $\mathcal{D}G$-module homomorphism, it is clear that, for any $z \in \mathbf{Z}$, we have the $\mathcal{O}G$-module homomorphisms

10.6.3 $$\begin{aligned} \mathsf{B}_z(f) &: \mathsf{B}_z(M) \longrightarrow \mathsf{B}_z(M') \\ \mathsf{C}_z(f) &: \mathsf{C}_z(M) \longrightarrow \mathsf{C}_z(M') \\ \mathsf{H}_z(f) &: \mathsf{H}_z(M) \longrightarrow \mathsf{H}_z(M') \quad . \end{aligned}$$

It is easily checked from 10.4.4 that $\mathsf{C}(\mathrm{End}_{\mathcal{O}}(M))$ and $\mathsf{C}_\circ(\mathrm{End}_{\mathcal{O}}(M))$ are G-stable $\mathcal{O}$-subalgebras of $\mathrm{End}_{\mathcal{O}}(M)$, whereas $\mathsf{B}(\mathrm{End}_{\mathcal{O}}(M))$ and $\mathsf{B}_\circ(\mathrm{End}_{\mathcal{O}}(M))$ are respectively G-stable two-sided ideals of these $\mathcal{O}$-subalgebras, so that $\mathsf{H}(\mathrm{End}_{\mathcal{O}}(M))$ and $\mathsf{H}_\circ(\mathrm{End}_{\mathcal{O}}(M))$ are G-algebras over $\mathcal{O}$; actually, these four G-algebras are $\mathcal{O}G$-interior algebras since it follows from 10.4.2 applied to id_M that, for any $f \in \mathcal{F}$, we have

10.6.4 $$f(\mathrm{id}_M) = f(0)\mathrm{id}_M \quad \text{and} \quad d(\mathrm{id}_M) = 0 \quad .$$

Moreover, it is not difficult to check from 10.4.3 that, for any $x, y \in \mathbf{Z}$, we have

10.6.5 $$\begin{aligned} \mathsf{C}_x(\mathrm{End}_{\mathcal{O}}(M))(\mathsf{C}_y(M)) &\subset \mathsf{C}_{x+y}(M) \\ \mathsf{C}_x(\mathrm{End}_{\mathcal{O}}(M))(\mathsf{B}_y(M)) &\subset \mathsf{B}_{x+y}(M) \\ \mathsf{B}_x(\mathrm{End}_{\mathcal{O}}(M))(\mathsf{C}_y(M)) &\subset \mathsf{B}_{x+y}(M) \end{aligned}$$

and therefore homomorphism 10.4.3 induces unitary $\mathcal{O}G$-interior algebra homomorphisms

10.6.6
$$\mathsf{H}(\mathrm{End}_{\mathcal{O}}(M)) \longrightarrow \mathrm{End}_{\mathcal{O}}(\mathsf{H}(M)) \quad \text{and} \quad \mathsf{H}_\circ(\mathrm{End}_{\mathcal{O}}(M)) \longrightarrow \mathrm{End}_{\mathcal{O}}(\mathsf{H}_\circ(M)) \quad ;$$

notice that $\mathsf{H}(\mathrm{End}_{\mathcal{O}}(M)) = \{0\}$ is equivalent to $\mathsf{H}_\circ(\mathrm{End}_{\mathcal{O}}(M)) = \{0\}$ and implies that $\mathsf{H}(M) = \{0\}$. Finally, notice that (cf. 10.4.2)

10.6.7 $$\mathsf{C}_\circ(\mathrm{Hom}_{\mathcal{O}G}(M, M')) = \mathrm{Hom}_{\mathcal{D}G}(M, M') \quad .$$

10.7 Recall that a $\mathcal{D}G$-module homomorphism $f : M \to M'$ between two $\mathcal{O}$-finite $\mathcal{D}G$-modules M and M' is *homotopically null* or *contractile* if there is $h \in \mathrm{Hom}_{\mathcal{O}G}(M, M')$ such that (cf. [10, Chap. IV, §3])

10.7.1 $$f = d_{M'} \circ h + h \circ d_M \quad ;$$

where d_M and $d_{M'}$ are respectively the images of d in $\mathrm{End}_{\mathcal{O}G}(M)$ and $\mathrm{End}_{\mathcal{O}G}(M')$; in that case, it is clear that the compositions of f, both on the right and on the left, with $\mathcal{D}G$-module homomorphisms are also contractile. We denote by $\overline{\mathrm{Mod}}_{\mathcal{D}G}$ the category where the objects are the $\mathcal{O}$-finite $\mathcal{D}G$-modules and the morphisms are the classes of $\mathcal{D}G$-module homomorphisms *modulo* the contractile ones, and we call it the *homotopy category* of $\mathcal{O}G$. If M'' is a third $\mathcal{O}$-finite $\mathcal{D}G$-module, notice that, by 10.2.3 and 10.6.4, equality 10.7.1 implies

10.7.2 $$f \underset{\mathcal{O}}{\otimes} \mathrm{id}_{M''} = d_{M' \otimes_{\mathcal{O}} M''} \circ (h \underset{\mathcal{O}}{\otimes} \mathrm{id}_{M''}) + (h \underset{\mathcal{O}}{\otimes} \mathrm{id}_{M''}) \circ d_{M \otimes_{\mathcal{O}} M''}$$

and, from 10.4.2 and 10.6.4, we get analogous equalities for $\mathrm{Hom}_{\mathcal{O}}(f, \mathrm{id}_{M''})$ and $\mathrm{Hom}_{\mathcal{O}}(\mathrm{id}_{M''}, f)$, so that:

10.7.3 *If f is contractile then $f \otimes_{\mathcal{O}} \mathrm{id}_{M''}$, $\mathrm{Hom}_{\mathcal{O}}(f, \mathrm{id}_{M''})$ and $\mathrm{Hom}_{\mathcal{O}}(\mathrm{id}_{M''}, f)$ are contractile too.*

Moreover, recall that an $\mathcal{O}$-finite $\mathcal{D}G$-module M is *contractile* if id_M is contractile or, equivalently, if M is an initial object in $\overline{\mathrm{Mod}}_{\mathcal{D}G}$. It is well-known (cf. [10, Chap. IV, Proposition 2.3]) that contractility is just *relative projectivity* with respect to $\mathcal{O}G$; we state it explicitely in the next proposition for reader's convenience. We employ the notation $\mathrm{Res}^{\mathcal{D}}_{\mathcal{F}}$, $\mathrm{Res}^{\mathcal{F}}_{\mathcal{O}}$ and $\mathrm{Ind}^{\mathcal{D}}_{\mathcal{F}}$ for the corresponding restrictions and extension of the coefficient ring to avoid confusion with the tensor products above. First of all, notice that:

10.7.4 *Any $\mathcal{O}$-finite $\mathcal{F}G$-module M is a direct summand of* $\mathrm{Ind}^{\mathcal{F}}_{\mathcal{O}}\big(\mathrm{Res}^{\mathcal{F}}_{\mathcal{O}}(M)\big)$.

Indeed, the $\mathcal{F}G$-module homomorphism $M \to \mathrm{Inf}^{\mathcal{F}}_{\mathcal{O}}\big(\mathrm{Res}^{\mathcal{F}}_{\mathcal{O}}(M)\big)$ mapping $m \in M$ on $\Sigma_{z\in Z}\, \imath_z \otimes \imath_z(m)$ is a section of the $\mathcal{F}G$-module homomorphism from $\mathrm{Ind}^{\mathcal{F}}_{\mathcal{O}}\big(\mathrm{Res}^{\mathcal{F}}_{\mathcal{O}}(M)\big)$ to M mapping $1 \otimes m$ on $m \in M$.

Proposition 10.8 *Let M and M' be $\mathcal{O}$-finite $\mathcal{D}G$-modules and $f : M \to M'$ a $\mathcal{D}G$-module homomorphism. The following conditions on f are equivalent*

10.8.1 *It factorizes in* $\mathrm{Mod}_{\mathcal{D}G}$ *via a contractile $\mathcal{O}$-finite $\mathcal{D}G$-module.*

10.8.2 *It factorizes in* $\mathrm{Mod}_{\mathcal{D}G}$ *throughout the $\mathcal{D}G$-module homomorphism* $\mathrm{Ind}^{\mathcal{D}}_{\mathcal{F}}\big(\mathrm{Res}^{\mathcal{D}}_{\mathcal{F}}(M')\big) \to M'$ *mapping* $1 \otimes m'$ *on* $m' \in M'$.

10.8.3 *It is contractible.*

10.8.4 *It belongs to* $\mathsf{B}_\circ\big(\mathrm{Hom}_{\mathcal{O}G}(M, M')\big)$.

Proof: Set $N' = \mathrm{Ind}^{\mathcal{D}}_{\mathcal{F}}\big(\mathrm{Res}^{\mathcal{D}}_{\mathcal{F}}(M')\big)$ and notice that $N' = 1 \otimes M' \oplus d \otimes M'$; defining h by $h(1 \otimes m') = d \otimes m'$ and $h(d \otimes m') = 1 \otimes m'$ for any $m' \in M'$, we have

10.8.5
$$\begin{aligned} d\big(h(1 \otimes m')\big) \;&+\; h\big(d(1 \otimes m')\big) = 1 \otimes m' \\ d\big(h(d \otimes m')\big) \;&+\; h\big(d(d \otimes m')\big) = d \otimes m' \quad ; \end{aligned}$$

hence, N' is contractile and therefore condition 10.8.2 implies condition 10.8.1. Secondly, if M'' is a contractile $\mathcal{O}$-finite $\mathcal{D}G$-module and we have $f = f' \circ f''$ for suitable $\mathcal{D}G$-module homomorphisms $f'' : M \to M''$ and $f' : M'' \to M'$ then, since we have (cf. 10.2.3, 10.4.4 and 10.6.7)

10.8.6
$$f' \circ \mathsf{B}_\circ\big(\mathrm{End}_{\mathcal{O}G}(M'')\big) \circ f'' \subset \mathsf{B}_\circ\big(\mathrm{Hom}_{\mathcal{O}G}(M, M')\big) \quad ,$$

$f = f' \circ \mathrm{id}_{M''} \circ f''$ belongs to $\mathsf{B}_\circ\big(\mathrm{Hom}_{\mathcal{O}G}(M, M')\big)$. Thirdly, if $f = d(h)$ for some $h \in \mathrm{Hom}_{\mathcal{O}G}(M, M')$ then, by 10.4.2, we get

10.8.7
$$f = d_{M'} \circ (h \circ \imath_M) + (h \circ \imath_M) \circ d_M \quad .$$

Finally, assume that $f = d_{M'} \circ h + h \circ d_M$ for some $h \in \mathrm{Hom}_{\mathcal{O}G}(M, M')$; since, for any $z \in \mathbf{Z}$ and any $m \in M$, we have $f\big(\imath_z(m)\big) = \imath_z\big(f(m)\big)$, up to modify our choice of h, we may assume that $h\big(i_z(m)\big) = i_{z+1}\big(h(m)\big)$ which, by Proposition 9.7, implies that, for any $f \in \mathcal{F}$, we have

10.8.8 $$h(f(m)) = \big(\mathrm{sh}(f)\big)\big(h(m)\big) \quad ;$$

then, it suffices to consider the $\mathcal{O}G$-module homomorphism $M \to N'$ mapping $m \in M$ on $d \otimes h(M) + 1 \otimes h\big(d(m)\big)$; indeed, for any $f \in \mathcal{F}$ and any $m \in M$, we have

10.8.9
$$\begin{aligned} d \otimes h(f(m)) + 1 \otimes h\Big(d\big(f(m)\big)\Big) &= d \otimes \big(\mathrm{sh}^{-1}(f)\big)\big(h(m)\big) + 1 \otimes f\Big(h\big(d(m)\big)\Big) \\ &= d\mathrm{sh}^{-1}(f) \otimes h(m) + f \otimes h\big(d(m)\big) \\ &= f\Big(d \otimes h(m) + 1 \otimes h\big(d(m)\big)\Big) \end{aligned}$$

and obviously

10.8.10 $$d \otimes h\big(d(m)\big) + 1 \otimes h\Big(d\big(d(m)\big)\Big) = d\Big(d \otimes h(m) + 1 \otimes h\big(d(m)\big)\Big) \ ,$$

so that this map is actually $\mathcal{D}G$-linear; moreover, the composition of it with the $\mathcal{D}G$-module homomorphism $N' \to M'$ mapping $1 \otimes m'$ on $m' \in M'$ coincides clearly with f.

Corollary 10.9 *The following conditions on an $\mathcal{O}$-finite $\mathcal{D}G$-module M are equivalent.*

10.9.1 *It is a direct summand of* $\mathrm{Ind}_{\mathcal{F}}^{\mathcal{D}}\big(\mathrm{Res}_{\mathcal{F}}^{\mathcal{D}}(M)\big)$.

10.9.2 *It is contractile.*

10.9.3 *We have* $\mathsf{H}_\circ\big(\mathrm{End}_{\mathcal{O}G}(M)\big) = \{0\}$.

Proof: If $\mathsf{H}_\circ\big(\mathrm{End}_{\mathcal{O}G}(M)\big) = \{0\}$ then id_M belongs to $\mathsf{B}_\circ\big(\mathrm{End}_{\mathcal{O}G}(M)\big)$ and therefore id_M is contractile by Proposition 10.8. Secondly, if id_M is contractile then it factorizes throughout the $\mathcal{D}G$-module homomorphism $\mathrm{Ind}_{\mathcal{F}}^{\mathcal{D}}\big(\mathrm{Res}_{\mathcal{F}}^{\mathcal{D}}(M)\big) \to M$ mapping $1 \otimes m$ on $m \in M$, so that such a factorization is a section of this homomorphism. Finally, it follows from Proposition 10.8 that the $\mathcal{D}G$-module $\mathrm{Ind}_{\mathcal{F}}^{\mathcal{D}}\big(\mathrm{Res}_{\mathcal{F}}^{\mathcal{D}}(M)\big)$ is contractile, so that we have

10.9.4 $$\mathsf{H}_\circ\Big(\mathrm{End}_{\mathcal{O}}\Big(\mathrm{Ind}_{\mathcal{F}}^{\mathcal{D}}\big(\mathrm{Res}_{\mathcal{F}}^{\mathcal{D}}(M)\big)\Big)\Big) = \{0\} \quad .$$

Corollary 10.10 *Two $\mathcal{O}$-finite $\mathcal{D}G$-modules M and M' are contractile if and only if the direct sum $M \oplus M'$ is contractile.*

Proof: If id_M and $\mathrm{id}_{M'}$ are both contractile then it is easily checked that $\mathrm{id}_{M\oplus M'}$ is contractile too. Conversely, if we have

10.10.1 $$\mathsf{H}_\circ\big(\mathrm{End}_{\mathcal{O}G}(M \oplus M')\big) = \{0\} \quad ,$$

it is clear that both $\mathsf{H}_\circ\big(\mathrm{End}_{\mathcal{O}G}(M)\big)$ and $\mathsf{H}_\circ\big(\mathrm{End}_{\mathcal{O}G}(M')\big)$ are zero.

Corollary 10.11 *Two $\mathcal{O}$-finite $\mathcal{D}G$-modules M and M' having no nonzero contractile direct summands are isomorphic in the homotopy category if and only if they are isomorphic.*

Proof: If M and M' are isomorphic in the homotopy category then it is clear that the images of id_M and $\mathrm{id}_{M'}$ in $\mathsf{H}_\circ(\mathrm{End}_{\mathcal{O}G}(M \oplus M'))$ are conjugate (cf. 10.6.7 and Proposition 10.8); but, since $M \oplus M'$ has no nonzero contractile direct summands, $\mathsf{B}_\circ(\mathrm{End}_{\mathcal{O}G}(M \oplus M'))$ contains no nonzero idempotents (cf. Proposition 10.8), so that we have

10.11.1 $$\mathsf{B}_\circ(\mathrm{End}_{\mathcal{O}G}(M \oplus M')) \subset J\Big(\mathsf{C}_\circ(\mathrm{End}_{\mathcal{O}G}(M \oplus M'))\Big) \quad ;$$

consequently, id_M and $\mathrm{id}_{M'}$ are still conjugate in (cf. 10.6.7)

10.11.2 $$\mathsf{C}_\circ(\mathrm{End}_{\mathcal{O}G}(M \oplus M')) = \mathrm{End}_{\mathcal{D}G}(M \oplus M') \quad ,$$

so that we have $M \cong M'$.

10.12 As we explain in Section 18 below, the projective $\mathcal{O}(G \times G)$-module U appearing in condition 6.3.1 above for Morita stable equivalences is replaced by a *contractile* $\mathcal{D}(G \times G)$-module in the analogous condition for *Rickard equivalences* and, coherently, the $\mathcal{O}(G \times G)$-module $\mathcal{O}G$ is considered there as a $\mathcal{D}(G \times G)$-module (cf. 10.1). Consequently, we are interested in $\mathcal{O}$-finite $\mathcal{D}G$-modules which are the direct sum of a contractile one and an $\mathcal{O}$-finite $\mathcal{D}G$-module M such that $M = \mathsf{C}(M) \cong \mathsf{H}(M)$ or, more precisely, such that $M = \mathsf{C}_\circ(M) \cong \mathsf{H}_\circ(M)$; following Rickard [33], we call them *split* $\mathcal{D}G$-modules in the first case and 0-*split* $\mathcal{D}G$-*modules* in the second; that is to say, an $\mathcal{O}$-finite $\mathcal{D}G$-module is 0-*split* if and only if it is isomorphic to an $\mathcal{O}$-finite $\mathcal{O}G$-module in the homotopy category.

Proposition 10.13 *Let M be an $\mathcal{O}$-finite $\mathcal{D}G$-module and denote by d_M the image of d in* $\mathrm{End}_{\mathcal{O}G}(M)$. *The following conditions on M are equivalent.*

10.13.1 *It is split.*

10.13.2 *The following exact sequences split*

$$0 \to \mathsf{C}(M) \longrightarrow M \xrightarrow{d_M} \mathsf{B}(M) \to 0 \quad \textit{and}$$
$$0 \to \mathsf{B}(M) \longrightarrow \mathsf{C}(M) \longrightarrow \mathsf{H}(M) \to 0 \quad .$$

10.13.3 *There is $h \in \mathrm{End}_{\mathcal{O}G}(M)$ such that $d_M = d_M \circ h \circ d_M$.*
Moreover, in that case, homomorphisms 10.6.6 *are bijective.*

Proof: If $h \in \mathrm{End}_{\mathcal{O}G}(M)$ fulfills $d_M = d_M \circ h \circ d_M$ then we have

10.13.4 $$\begin{aligned}&(d_M \circ h + h \circ d_M - d_M \circ h^2 \circ d_M)^2\\ &= d_M \circ h + d_M \circ h^2 \circ d_M - d_M \circ h^2 \circ d_M + h \circ d_M - d_M \circ h^2 \circ d_M\end{aligned}$$

so that $i = d_M \circ h + h \circ d_M - d_M \circ h^2 \circ d_M$ is an idempotent in $\mathrm{End}_{\mathcal{O}G}(M)$ and therefore we get the $\mathcal{O}G$-module direct sum decomposition

10.13.5 $$M = i(M) \oplus (\mathrm{id}_M - i)(M) \quad ;$$

moreover, notice that

10.13.6 $$d_M \circ i = d_M = (h \circ d_M) \circ d_M + d_M \circ (h \circ d_M)$$

which proves that $i(M)$ is contractile and that $(\mathrm{id}_M - i)(M)$ coincides with $\mathsf{C}\big((\mathrm{id}_M - i)(M)\big)$; consequently, M is split. Secondly, if $M = \mathsf{C}(M) \cong \mathsf{H}(M)$ then the exact sequences in 10.13.2 obviously split and we have also (cf. 10.4.2)

10.13.7 $$\mathrm{End}_{\mathcal{O}}\big(\mathsf{H}(M)\big) \cong \mathrm{End}_{\mathcal{O}}(M) = \mathsf{C}\big(\mathrm{End}_{\mathcal{O}}(M)\big) \cong \mathsf{H}\big(\mathrm{End}_{\mathcal{O}}(M)\big) \quad ;$$

whereas if M is contractile then we have

10.13.8 $$\mathsf{H}\big(\mathrm{End}_{\mathcal{O}}(M)\big) = \{0\} = \mathrm{End}_{\mathcal{O}}\big(\mathsf{H}(M)\big)$$

and, according to Proposition 10.8, M is a direct summand of $N = \mathrm{Ind}_{\mathcal{F}}^{\mathcal{D}}\big(\mathrm{Res}_{\mathcal{F}}^{\mathcal{D}}(M)\big)$ as $\mathcal{D}G$-modules; but it is easily checked that $\mathsf{C}(N) = d \otimes M = \mathsf{B}(N)$ and that $d_N : N \to d \otimes M$ splits; consequently, condition 10.13.1 implies condition 10.13.2 and the last statement. Finally, assume that 10.13.2 holds and let $h : \mathsf{B}(M) \to M$ be a section of $d_M : M \to \mathsf{B}(M)$ as $\mathcal{O}G$-modules; on the one hand, since $d_M \circ h = \mathrm{id}_{\mathsf{B}(M)}$, for any $m \in M$ we have $d_M\Big(h\big(d_M(m)\big)\Big) = d_M(m)$; on the other hand, since $\mathsf{B}(M)$ and $\mathsf{C}(M)$ are respectively direct summands of $\mathsf{C}(M)$ and M as $\mathcal{O}G$-modules, h can be extended to an $\mathcal{O}G$-module endomorphism of M.

11 $\mathcal{D}$-algebras and $\mathcal{D}G$-interior algebras

11.1 Following Bernhard Keller [13], a *differential* $\mathbf{Z}$*-graded* $\mathcal{O}$*-algebra* or, in short, a $\mathcal{D}$*-algebra* is an $\mathcal{O}$-algebra A endowed with a differential $\mathbf{Z}$-graded $\mathcal{O}$-module structure (see Remark 10.5 above) such that the product map $A \otimes_{\mathcal{O}} A \to A$ is $\mathcal{D}$-linear or, equivalently, such that, for any $a, a' \in A$ and any $f \in \mathcal{F}$, we have (cf. 10.2.3)

11.1.1
$$\begin{aligned} f(aa') &= \sum_{z,z' \in \mathbf{Z}} f(z) i_{z'}(a) i_{z-z'}(a') \\ d(aa') &= d(a) s(a') + a d(a') \quad . \end{aligned}$$

But, the $\mathcal{D}$-algebras A we have to consider are finitely generated as $\mathcal{O}$-modules (in short, $\mathcal{O}$*-finite*), as we point out in 9.1 above, and on the other hand their $\mathcal{D}$-module structure comes - in the sense that we explain in 11.2 below - from a unitary $\mathcal{O}$-algebra homomorphism $\mathcal{D} \to A$. As it happens when the actions of a group G on A comes from a group homomorphism $G \to A^*$, we lose information if we forget that $\mathcal{O}$-algebra homomorphism and consider only the $\mathcal{D}$-module structure; for instance, notice that a unitary $\mathcal{O}$-algebra homomorphism $\mathcal{D} \to A$ induces on A a $\mathcal{D}$*-bimodule structure.*

11.2 Consequently, we put the following definition: an $\mathcal{O}$*-finite* $\mathcal{D}$*-interior algebra* is an $\mathcal{O}$-finite $\mathcal{O}$-algebra A endowed with a unitary $\mathcal{O}$-algebra homomorphism

11.2.1
$$f : \mathcal{D} \longrightarrow A \quad ;$$

as usual, we omit f writing $g{\cdot}a{\cdot}g'$ instead of $f(g)af(g')$ for any $g, g' \in \mathcal{D}$ and any $a \in A$. Thus, if M is an $\mathcal{O}$-finite $\mathcal{D}$-module then $\mathrm{End}_{\mathcal{O}}(M)$ is an $\mathcal{O}$-finite $\mathcal{D}$-interior algebra. By Corollary 9.15, if A is an $\mathcal{O}$-finite $\mathcal{D}$-interior algebra then the composed $\mathcal{O}$-algebra homomorphism

11.2.2
$$\mathcal{D}_\circ \xrightarrow{\Delta_\circ} \mathcal{D}_\circ \underset{\mathcal{O}}{\otimes} \mathcal{D}_\circ \xrightarrow{\mathrm{id} \otimes t_\circ} \mathcal{D}_\circ \underset{\mathcal{O}}{\otimes} (\mathcal{D}_\circ)^\circ \subset \mathcal{D} \underset{\mathcal{O}}{\otimes} \mathcal{D}^\circ \to A \underset{\mathcal{O}}{\otimes} A^\circ \to \mathrm{End}_{\mathcal{O}}(A) \quad ,$$

where the last homomorphism is defined by multiplication on the left and on the right, can be extended to a unique unitary $\mathcal{O}$-algebra homomorphism

11.2.3
$$\mathcal{D} \longrightarrow \mathrm{End}_{\mathcal{O}}(A)$$

and we claim that A endowed with this $\mathcal{D}$-module structure is a $\mathcal{D}$-algebra; indeed, it is easily checked from proposition 9.7 that, for any $a \in A$ and any $f \in \mathcal{F}$, we have

11.2.4
$$\begin{aligned} f(a) &= \sum_{z,z' \in \mathbf{Z}} f(z) i_{z'}{\cdot}a{\cdot}i_{z'-z} \\ d(a) &= (d{\cdot}a - a{\cdot}d){\cdot}s \end{aligned}$$

and therefore, for any $a, a' \in A$ and any $f \in \mathcal{F}$, we get

11.2.5
$$\begin{aligned} f(aa') &= \sum_{z,z' \in \mathbf{Z}} f(z) i_{z'} \cdot aa' \cdot i_{z-z'} = \sum_{z,z'' \in \mathbf{Z}} f(z) i_{z''}(a) i_{z-z''}(a') \\ d(aa') &= d \cdot aa' \cdot s - aa' \cdot ds = d(a)s(a') + a d(a') \end{aligned}$$

since we have $s(a) = s \cdot a \cdot s$ (the fact that A holds both a $\mathcal{D}$-module and a $\mathcal{D}$-bimodule structures motivates the choice of our notation).

11.3 In particular, for any $\mathcal{O}$-finite $\mathcal{D}$-interior algebra A, we have (cf. 10.6.1 and 11.2.4)

11.3.1 $$\mathsf{C}(A) = \{a \in A | d \cdot a = a \cdot d\} \quad \text{and} \quad \mathsf{B}(A) = \{d \cdot a + a \cdot d | a \in A\}$$

and therefore $\mathsf{C}(A)$ is a unitary $\mathcal{D}$-subalgebra of A which contains $\mathsf{B}(A)$ as both a two-sided ideal and a $\mathcal{D}$-submodule, so that $\mathsf{H}(A)$ is a $\mathcal{D}$-algebra; actually, d annihilates both $\mathsf{C}(A)$ and $\mathsf{H}(A)$, so that they have to be considered simply as $\mathcal{F}$-algebras, according to the obvious definition. We are specially interested in $\mathsf{C}_\circ(A)$ and $\mathsf{H}_\circ(A)$; since $i_\circ(a) = \Sigma_{z \in \mathbf{Z}} i_z \cdot a \cdot i_z$ for any $a \in A$, it is easily checked that

11.3.2 $$\mathsf{C}_\circ(A) = \{a \in A | g \cdot a = a \cdot g \text{ for any } g \in \mathcal{D}\} \quad ,$$

so that $\mathsf{C}_\circ(A)$ is a unitary $\mathcal{O}$-subalgebra of A, $\mathsf{B}_\circ(A)$ a two-sided ideal of $\mathsf{C}_\circ(A)$ and $\mathsf{H}_\circ(A)$ an $\mathcal{O}$-algebra; in particular, for any idempotent i in $\mathsf{C}_\circ(A)$, iAi has a $\mathcal{D}$-interior algebra structure given by the homomorphism $\mathcal{D} \to iAi$ mapping $g \in \mathcal{D}$ on $g \cdot i = i \cdot g$. Notice that $\mathsf{H}_\circ(A) = \{0\}$ is equivalent to $\mathsf{H}(A) = \{0\}$; in that case, we say that A is *contractile* and actually, as it is easily checked, A endowed with the $\mathcal{D}$-module structure given by 11.2.4 is a contractile module; notice that if M is an $\mathcal{O}$-finite $\mathcal{D}$-module then $\mathrm{End}_\mathcal{O}(M)$ is a contractile $\mathcal{D}$-interior algebra if and only if M is a contractile $\mathcal{D}$-module (cf. 10.7).

11.4 Let G be a finite group and H a normal subgroup of G; recall that an $\mathcal{O}$-finite $\mathcal{O}H$-interior G-algebra is an $\mathcal{O}$-finite G-algebra endowed with a unitary G-algebra homomorphism from $\mathcal{O}H$ which lifts the action of H (cf. 2.3). Now, an *$\mathcal{O}$-finite $\mathcal{D}H$-interior G-algebra* is an $\mathcal{O}$-finite $\mathcal{O}H$-interior G-algebra A endowed with a unitary $\mathcal{O}$-algebra homomorphism

11.4.1 $$\mathcal{D} \longrightarrow A^G \quad ;$$

in particular, according to 2.3 and 11.2, A has an evident $\mathcal{D}(H \times H) \cdot \Delta(G)$-module structure, where $\Delta(G)$ is the diagonal subgroup of $G \times G$; for any subgroup F of G, we denote by $\mathrm{Res}^G_F(A)$ the corresponding $\mathcal{D}(H \cap F)$-interior F-algebra. As it happens over $\mathcal{O}$ (cf. 2.3), we are mainly interested in the case where $H = G$ and then we call it *$\mathcal{D}G$-interior algebra*; thus, if M is an $\mathcal{O}$-finite $\mathcal{D}G$-module then $\mathrm{End}_\mathcal{O}(M)$ is a $\mathcal{D}G$-interior algebra. Let A be an $\mathcal{O}$-finite $\mathcal{D}H$-interior G-algebra; we say that

A is *contractile* if A^G is a contractile $\mathcal{D}$-interior algebra, namely if $\mathsf{H}(A^G) = \{0\}$. Notice that, on the one hand, A itself is a $\mathcal{D}$-interior algebra and then $\mathsf{C}(A)$ and $\mathsf{H}(A)$ become *H-interior $\mathcal{D}G$-algebras* (see 11.6 below) or, actually, *H-interior $\mathcal{F}G$-algebras*, whereas $\mathsf{C}_\circ(A)$ and $\mathsf{H}_\circ(A)$ are simply $\mathcal{O}H$-interior G-algebras. On the other hand, for any subgroup F of G, A^F and $A(F)$ are $\mathcal{D}C_H(F)$-interior $N_G(F)$-algebras; moreover, for any $z \in \mathbf{Z}$, we get

11.4.2 $$\mathsf{C}_z(A^F) = \mathsf{C}_z(A)^F \quad \text{and} \quad \mathsf{B}_z(A^F) \subset \mathsf{B}_z(A)^F \quad .$$

and therefore we have unitary $C_H(F)$-interior $\mathcal{F}N_G(F)$-algebra homomorphisms (see 11.6 below)

11.4.3 $$\mathsf{H}(A^F) \longrightarrow \mathsf{H}(A)^F \quad \text{and} \quad (\mathsf{C}(A))(F) \longrightarrow \mathsf{C}(A(F)) \quad .$$

Remark 11.5 Although homomorphisms 11.4.3 need not be injective nor surjective, notice that

11.5.1 *If A is a split $\mathcal{D}F$-module then homomorphisms 11.4.3 are bijective.*

Indeed, if $M \cong M \oplus M'$ as $\mathcal{D}F$-modules, where M is contractile and $d(M') = \{0\}$ (cf. 10.12), on the one hand we have also $d\big((M')^F\big) = \{0\} = d\big(M'(F)\big)$, so that

11.5.2
$$(M')^F = \mathsf{C}\big((M')^F\big) \cong \mathsf{H}\big((M')^F\big) \quad \text{and} \quad (\mathsf{C}(M'))(F) = M'(F) = \mathsf{C}\big(M'(F)\big) \quad ;$$

on the other hand we have $\mathsf{H}(M^F) = \{0\}$, M is a direct summand of $\mathrm{Ind}_{\mathcal{F}}^{\mathcal{D}}\big(\mathrm{Res}_{\mathcal{F}}^{\mathcal{D}}(M)\big) = \mathcal{D} \otimes_{\mathcal{F}} M$ (cf. Corollary 10.9) and it is quite clear that

11.5.3 $$\big(\mathsf{C}(\mathcal{D} \underset{\mathcal{F}}{\otimes} M)\big)(F) = (d \otimes M)(F) \cong d \otimes M(F) = \mathsf{C}\big(\mathcal{D} \underset{\mathcal{F}}{\otimes} M(F)\big) \quad .$$

11.6 Let A and A' be $\mathcal{O}$-finite $\mathcal{D}H$-interior G-algebras; in particular, they are $\mathcal{O}H$-interior G-algebras and $\mathcal{D}$-algebras (cf. 11.2) where the action of $\mathrm{Ker}(e)$, $e : \mathcal{D} \to \mathcal{O}$ being the augmentation map, centralizes the action of G and annihilates the image of H (cf. 11.2.4): let us call this kind of structure *H-interior $\mathcal{D}G$-algebra*; we are occasionally interested in the corresponding kind of homomorphisms: we say that a map $f : A \to A'$ is an *H-interior $\mathcal{D}G$-algebra homomorphism* if it is a *$\mathcal{D}$-linear* $\mathcal{O}H$-interior G-algebra homomorphism, whereas we say that f is a *$\mathcal{D}H$-interior G-algebra homomorphism* if moreover it is a $\mathcal{D}$-bimodule homomorphism or, equivalently, if f is both a G-algebra homomorphism and a $\mathcal{D}H$-bimodule homomorphism; as above, when $H = G$ we do not repeat G. Notice that, considering $\mathcal{O}H$ endowed with the $\mathcal{O}$-algebra homomorphism $\mathcal{D} \to \mathcal{O}H$ mapping $f \in F$ on $f(0)$ and d on zero as a $\mathcal{D}H$-interior G-algebra, the structural map $\mathcal{O}H \to A$ is *always* an H-interior $\mathcal{D}G$-algebra homomorphism since the images of $\mathcal{D}$ and H in A centralize each other, whereas if the image of d in A is not zero then that map is *not* a $\mathcal{D}H$-interior G-algebra homomorphism. Let $f : A \to A'$ be an H-interior

$\mathfrak{D}G$-algebra homomorphism; then, it is a $\mathfrak{D}(H \times H)\cdot\Delta(G)$-module homomorphism (cf. 11.4) and, for any subgroup F of G and any $z \in \mathbf{Z}$, we clearly have

11.6.1 $$f\big(\mathsf{C}_z(A^F)\big) \subset \mathsf{C}_z(A')^F \quad \text{and} \quad f\big(\mathsf{B}_z(A^F)\big) \subset \mathsf{B}_z\big((A')^F\big) \quad ,$$

so that f induces a $C_H(F)$-interior $\mathcal{F}N_G(F)$-algebra homomorphism (according to the obvious definition)

11.6.2 $$\mathsf{H}(f^F) : \mathsf{H}(A^F) \longrightarrow \mathsf{H}\big((A')^F\big) \quad .$$

As usual, we say that f is an *embedding* when moreover we have

11.6.3 $$\mathrm{Ker}(f) = \{0\} \quad \text{and} \quad \mathrm{Im}(f) = f(1)A'f(1) \quad .$$

11.7 But, since the *Rickard equivalences* between blocks (see Section 18 below) are concerned with the *homotopy categories* of the blocks (cf. 10.7), in the same way that in Section 6 above it is useful to consider the $\mathcal{O}G$-interior algebra *stable* embeddings (cf. 6.4), it is convenient to put here the following definition: we say that an H-interior $\mathfrak{D}G$-algebra homomorphism $f : A \to A'$ is a *homotopy embedding* if f induces an isomorphism in the homotopy category of $\mathcal{O}(H \times H)\cdot\Delta(G)$ between the $\mathfrak{D}(H \times H)\cdot\Delta(G)$-modules A and $f(1)A'f(1)$ (cf. 11.4); in that case, it is quite clear that, for any subgroup F of G, the $C_H(F)$-interior $\mathfrak{D}N_G(F)$-algebra homomorphisms

11.7.1 $$f^F : A^F \longrightarrow (A')^F \quad \text{and} \quad f(F) : A(F) \longrightarrow A'(F)$$

are homotopy embeddings too, $\mathsf{H}(f^F)$ and $\mathsf{H}\big(f(F)\big)$ being true embeddings. If moreover f is unitary, we say that f is a *homotopy isomorphism* and then $\mathsf{H}(f^F)$ and $\mathsf{H}\big(f(F)\big)$ are true isomorphisms.

11.8 By 11.3.2, any element a of $\mathsf{C}_\circ(A^G)^*$ determines by conjugation a $\mathfrak{D}H$-interior G-algebra automorphism of A, that we denote by $\mathrm{int}(a)$; then, as usual, the *exterior homomorphism* or *exomorphism* $\tilde{f}$ determined by an H-interior $\mathfrak{D}G$-algebra homomorphism $f : A \to A'$ is the set

11.8.1 $$\tilde{f} = \big\{\mathrm{int}(a') \circ f \circ \mathrm{int}(a)\big\}_{a,a'} = \big\{\mathrm{int}(a') \circ f\big\}_{a'}$$

where a and a' respectively run on $\mathsf{C}_\circ(A^G)^*$ and $\mathsf{C}_\circ\big((A')^G\big)^*$; notice that if f is actually a $\mathfrak{D}H$-interior G-algebra homomorphism then all the homomorphisms in $\tilde{f}$ are so. If F is a subgroup of G, we denote by $\mathrm{Res}^G_F(\tilde{f})$ the $(H \cap F)$-interior $\mathfrak{D}F$-algebra exomorphism from $\mathrm{Res}^G_F(A)$ to $\mathrm{Res}^G_F(A')$ containing $\tilde{f}$; moreover, notice that $\tilde{f}$ determines unique $C_H(F)$-interior $\mathcal{F}N_G(F)$-algebra exomorphisms

11.8.2 $$\mathsf{C}(\tilde{f}^F) : \mathsf{C}(A^F) \longrightarrow \mathsf{C}\big((A')^F\big) \quad \text{and} \quad \mathsf{H}(\tilde{f}^F) : \mathsf{H}(A^F) \longrightarrow \mathsf{H}\big((A')^F\big)$$

and also unique $\mathcal{O}C_H(F)$-interior $N_G(F)$-algebra exomorphisms

11.8.3
$$\mathsf{C}_\circ(\tilde{f}^F) : \mathsf{C}_\circ(A^F) \longrightarrow \mathsf{C}_\circ\big((A')^F\big) \text{ and } \mathsf{H}_\circ(\tilde{f}^F) : \mathsf{H}_\circ(A^F) \longrightarrow \mathsf{H}_\circ\big((A')^F\big) \quad .$$

Finally, if A'' is a third $\mathcal{O}$-finite $\mathcal{D}H$-interior G-algebra and $f' : A' \to A''$ a second H-interior $\mathcal{D}G$-algebra homomorphism then the composition $f' \circ f$ is an H-interior $\mathcal{D}G$-algebra homomorphism too and we have

11.8.4
$$\widetilde{f' \circ f} = \tilde{f}' \circ \tilde{f} \quad ;$$

obviously, if f and f' are actually $\mathcal{D}H$-interior G-algebras, so is $f' \circ f$. Next proposition generalizes to the $\mathcal{D}$-structures the well-known good behavior by restriction of $\mathcal{O}G$-interior exomorphisms (cf. 2.4.2); notice that even for $\mathcal{D}$ the restriction of exomorphisms is sensitive to the "interiority" condition.

Proposition 11.9 *Let A and A' be $\mathcal{O}$-finite $\mathcal{D}G$-interior algebras and consider two G-interior $\mathcal{D}$-algebra homomorphisms $f : A \to A'$ and $g : A \to A'$. Then, we have $\tilde{f} = \tilde{g}$ if one of the following conditions holds.*

11.9.1 *f and g determine the same $\mathcal{D}$-algebra exomorphism.*

11.9.2 *f and g are $\mathcal{D}G$-interior algebra homomorphisms and determine the same $\mathcal{O}$-algebra exomorphism.*

Proof: Assume that one condition holds and set either $B = \mathcal{O}G$ and $C' = \mathsf{C}_\circ(A')$ if 11.9.1 holds or $B = \mathcal{D}G$ and $C' = A'$ if 11.9.2 holds; respectively denote by i' and j' the images by f and g of the unity element of A; since i' and j' belong to $\mathsf{C}_\circ(A')^G$, $C'i'$ and $C'j'$ become $C' \otimes_\mathcal{O} B^\circ$-modules by left and right multiplication. On the other hand, according to our hypothesis, there is $c' \in (C')^*$ such that $g(a) = f(a)^{c'}$ for any $a \in A$. Consequently, we have $C'i'c' = C'j'$ and, for any $b \in B$, we get

11.9.3
$$(i'{\cdot}b)c' = f(1{\cdot}b)c' = c'g(1{\cdot}b) = c'j'{\cdot}b = i'c'{\cdot}b \quad ;$$

that is to say, the multiplication by c' on the right defines a $C' \otimes_\mathcal{O} B^\circ$-module isomorphism $m_{c'}$ between $C'i'$ and $C'j'$; hence, considering also C' as a $C' \otimes_\mathcal{O} B^\circ$-module, the $C' \otimes_\mathcal{O} B^\circ$-submodules $C'(1-i')$ and $C'(1-j')$ are isomorphic too and therefore $m_{c'}$ can be extended to a $C' \otimes_\mathcal{O} B^\circ$-module automorphism h' of C' such that $h'\big(C'(1-i')\big) = C'(1-j')$. But, setting $h'(1) = u'$, it is clear that $h'(a') = a'u'$ for any $a' \in C'$ and, in particular, u' is invertible in C'; moreover, for any $b \in B$, we have

11.9.4
$$u'{\cdot}b = h'(1{\cdot}b) = (1{\cdot}b)u' = b{\cdot}u'$$

and therefore u' belongs to $\big(\mathsf{C}_\circ(A')^G\big)^*$. Then, we have $i'u' = h'(i') = i'c'$ and $(1-i')u'j' = h'(1-i')j' = 0$, so that $u'j' = i'u'j' = i'c'$ which implies that $(c')^{-1}i' = (u')^{-1}i'$, and therefore, for any $a \in A$, we get

11.9.5
$$f(a)^{u'} = (u')^{-1}i'f(a)i'u' = f(a)^{c'} = g(a) \quad .$$

Proposition 11.10 *Let A, A', B and B' be $\mathcal{O}$-finite $\mathcal{D}H$-interior G-algebras, $f : A \to A'$ and $g : A \to A'$ H-interior $\mathcal{D}G$-algebra homomorphisms and moreover $e : B \to A$ and $e' : A' \to B'$ H-interior $\mathcal{D}G$-algebra embeddings. Assume that the number of points of G on $\mathsf{C}_\circ(A)$ and on $\mathsf{C}_\circ(B)$ coincide. Then $\tilde{e}' \circ \tilde{f} \circ \tilde{e} = \tilde{e}' \circ \tilde{g} \circ \tilde{e}$ is equivalent to $\tilde{f} = \tilde{g}$.*

Proof: Assume that there is $b' \in \left(\mathsf{C}_\circ(B')^G\right)^*$ such that $e'\Big(g\big(e(b)\big)\Big) = e'\Big(f\big(e(b)\big)\Big)^{b'}$ for any $b \in B$; since the matrices of multiplicities of $\mathsf{C}_\circ(\tilde{e})^G$ and $\mathsf{C}_\circ(\tilde{e}')^G$ are respectively bijective and injective (cf. 2.11.2), for any point α of G on $\mathsf{C}_\circ(A)$ and any point α' of G on $\mathsf{C}_\circ(A')$, we have

11.10.1 $$m(\tilde{f})_\alpha^{\alpha'} = m(\tilde{g})_\alpha^{\alpha'}$$

and, in particular, $f(1)$ and $g(1)$ are idempotents of $\mathsf{C}_\circ(A')^G$ having the same multiplicity at each point of this $\mathcal{O}$-algebra; hence, there is $a' \in \left(\mathsf{C}_\circ(A')^G\right)^*$ such that $g(1) = f(1)^{a'}$ (cf. 2.8.4). Similarly, $f\big(e(1)\big)$ and $g\big(e(1)\big)$ are conjugate in $\mathsf{C}_\circ(A')^G$ and therefore, up to modify our choice of g in $\tilde{g}$, we may assume that $f\big(e(1)\big) = g\big(e(1)\big)$; in that case, we have

11.10.2 $$e'\Big(f\big(e(1)\big)\Big)^{b'} = e'\Big(g\big(e(1)\big)\Big) = e'\Big(f\big(e(1)\big)\Big)$$

and therefore there is $c' \in \left(\mathsf{C}_\circ(A')^G\right)^*$ such that (cf. 2.4)

11.10.3 $$e'(c') = e'\Big(f\big(e(1)\big)\Big)b' = b'e'\Big(g\big(e(1)\big)\Big) \qquad ;$$

hence, for any $b \in B$, we get

11.10.4 $$\begin{aligned} e'\Big(c'g\big(e(b)\big)\Big) &= b'e'\Big(g\big(e(1)\big)\Big)e'\Big(g\big(e(b)\big)\Big) = e'\Big(f\big(e(b)\big)\Big)b' \\ &= e'\Big(f\big(e(b)\big)c'\Big) \quad , \end{aligned}$$

so that $g\big(e(b)\big) = f\big(e(b)\big)^{c'}$, and therefore, up to modify again our choice of g in $\tilde{g}$, we may assume that $f \circ e = g \circ e$.

Now, for any point α of G on $\mathsf{C}_\circ(A)$, according to our hypothesis, the set $e^{-1}(\alpha) \cap \mathsf{C}_\circ(B)$ is not empty; choose $j_\alpha \in e^{-1}(\alpha) \cap \mathsf{C}_\circ(B)$ and consider $U_\alpha \subset \left(\mathsf{C}_\circ(A)^G\right)^*$ such that $\left\{e(j_\alpha)^u\right\}_{\alpha,u}$ where α runs on $\mathcal{P}_{\mathsf{C}_\circ(A)}(G)$ and u on U_α is a pairwise orthogonal idempotent decomposition of the unity element in $\mathsf{C}_\circ(A)^G$; then, setting $i'_\alpha = f\big(e(j_\alpha)\big) = g\big(e(j_\alpha)\big)$ for any $\alpha \in \mathcal{P}_{\mathsf{C}_\circ(A)}(G)$ and considering in $\mathsf{C}_\circ(A')^G$ the elements

11.10.5 $$\begin{aligned} u' &= \sum_{\alpha \in \mathcal{P}_{\mathsf{C}_\circ(A)}(G)} \Big(\sum_{u \in U_\alpha} f(u^{-1})i'_\alpha g(u)\Big) + (1 - f(1))a'(1 - g(1)) \\ v' &= \sum_{\alpha \in \mathcal{P}_{\mathsf{C}_\circ(A)}(G)} \Big(\sum_{u \in U_\alpha} g(u^{-1})i'_\alpha f(u)\Big) + (1 - g(1))a'^{-1}(1 - f(1)) \end{aligned}$$

(up to modify our choice of a' according to our new choice of g), it is easily checked that $v'u' = 1 = u'v'$ and that, for any $a \in A$, we have

11.10.6
$$\begin{aligned} v'f(a)u' &= \sum_{\alpha,\beta\in\mathcal{P}_{\mathrm{C}_\circ(A)}(G)} \Big(\sum_{u\in U_\alpha}\Big(\sum_{v\in U_\beta} g(u^{-1})f\big(e(j_\alpha)uav^{-1}e(j_\beta)\big)g(v)\Big)\Big) \\ &= \sum_{\alpha,\beta\in\mathcal{P}_{\mathrm{C}_\circ(A)}(G)} \Big(\sum_{u\in U_\alpha}\Big(\sum_{v\in U_\beta} g(u^{-1})g\big(e(j_\alpha)uav^{-1}e(j_\beta)\big)g(v)\Big)\Big) \\ &= g(a) \end{aligned}$$

since $e(j_\alpha)uav^{-1}e(j_\beta)$ belongs to $\mathrm{Im}(e)$ for any $\alpha, \beta \in \mathcal{P}_{\mathrm{C}_\circ(A)}(G)$, any $u \in U_\alpha$ and any $v \in U_\beta$.

11.11 Let A be an $\mathcal{O}$-finite $\mathcal{D}H$-interior G-algebra; it is clear that the opposite $\mathcal{O}$-algebra A° is an $\mathcal{O}H$-interior G-algebra and it becomes a $\mathcal{D}H$-interior G-algebra once endowed with the composed $\mathcal{O}$-algebra homomorphism

11.11.1
$$\mathcal{D} \overset{t}{\cong} \mathcal{D}^\circ \longrightarrow (A^G)^\circ = (A^\circ)^G \quad .$$

Let A' be a second $\mathcal{O}$-finite $\mathcal{D}H$-interior G-algebra; first of all, notice that the direct product $A \times A'$ has a unique $\mathcal{D}H$-interior G-algebra structure such that both projection maps are $\mathcal{D}H$-interior G-algebra homomorphisms. As in 10.2, it follows from Corollary 9.15 that the composed $\mathcal{O}$-algebra homomorphism

11.11.2
$$\mathcal{D}_\circ \xrightarrow{\Delta_\circ} \mathcal{D}_\circ \underset{\mathcal{O}}{\otimes} \mathcal{D}_\circ \subset \mathcal{D} \underset{\mathcal{O}}{\otimes} \mathcal{D} \longrightarrow A^G \underset{\mathcal{O}}{\otimes} (A')^G \longrightarrow (A \underset{\mathcal{O}}{\otimes} A')^G$$

can be extended to a unique $\mathcal{D}$-interior algebra structure on $(A \otimes A')^G$, so that $A \otimes_{\mathcal{O}} A'$ becomes an $\mathcal{O}$-finite $\mathcal{D}H$-interior G-algebra. Then, if $f : A \to B$ and $f' : A' \to B'$ are H-interior $\mathcal{D}G$-algebra homomorphisms, B and B' being $\mathcal{O}$-finite $\mathcal{D}H$-interior G-algebras, by 10.2.3 the tensor product $f \otimes_{\mathcal{O}} f'$ is clearly a $\mathcal{D}$-linear $\mathcal{O}H$-interior G-algebra homomorphism; similarly, if f and f' are $\mathcal{D}H$-interior G-algebra homomorphisms, so is $f \otimes_{\mathcal{O}} f'$. Notice that, for any subgroup F of G and any $x, y \in \mathbf{Z}$, in $A \otimes_{\mathcal{O}} A'$ we have

11.11.3
$$\begin{gathered} \mathsf{C}_x(A)^F \underset{\mathcal{O}}{\otimes} \mathsf{C}_y(A')^F \subset \mathsf{C}_{x+y}(A \underset{\mathcal{O}}{\otimes} A')^F \\ \mathsf{C}_x(A)^F \underset{\mathcal{O}}{\otimes} \mathsf{B}_y\big((A')^F\big) + \mathsf{B}_x(A^F) \underset{\mathcal{O}}{\otimes} \mathsf{C}_y(A')^F \subset \mathsf{B}_{x+y}\big((A \underset{\mathcal{O}}{\otimes} A')^F\big) \end{gathered}$$

and therefore we get a unitary $\mathcal{F}C_H(F)$-interior $N_G(F)$-algebra homomorphism

11.11.4
$$\mathsf{H}(A^F) \underset{\mathcal{O}}{\otimes} \mathsf{H}\big((A')^F\big) \to \mathsf{H}\big((A \underset{\mathcal{O}}{\otimes} A')^F\big) \quad .$$

As in 10.2 again, denoting by $s_{A,A'} : A \otimes_{\mathcal{O}} A' \cong A' \otimes_{\mathcal{O}} A$ the canonical $\mathcal{O}H$-interior G-algebra isomorphism and by $\imath_{A,A'}$ the image of $\imath \in (\mathcal{F}_1 \otimes_{\mathcal{O}} \mathcal{F}_1)^*$ in $((A \otimes_{\mathcal{O}} A')^G)^*$, we get from 9.11.2 and propositions 9.8 and 9.14, the canonical $\mathcal{D}H$-interior G-algebra isomorphism

11.11.5 $$\mathrm{int}(\imath_{A,A'}) \circ s_{A,A'} : A \underset{\mathcal{O}}{\otimes} A' \cong A' \underset{\mathcal{O}}{\otimes} A \quad ;$$

moreover, if A'' is a third $\mathcal{O}$-finite $\mathcal{D}H$-interior G-algebra then the associativity of the coproduct $\Delta_\circ$ and Propositions 9.8 and 9.14 force the canonical bijection

11.11.6 $$(A \underset{\mathcal{O}}{\otimes} A') \underset{\mathcal{O}}{\otimes} A'' \cong A \underset{\mathcal{O}}{\otimes} (A' \underset{\mathcal{O}}{\otimes} A'')$$

to be a $\mathcal{D}H$-interior G-algebra isomorphism.

Remark 11.12 We can remove the $\mathcal{O}$-finiteness condition provided we work with *nonunitary* $\mathcal{O}$-algebras; precisely, a *$\mathcal{D}H$-interior G-algebra* (or a *$\mathcal{D}G$-interior algebra* if $H = G$) is an $\mathcal{O}$-algebra A, not necessarily unitary, endowed with both a G-action and a $\mathcal{D}H$-bimodule structure fulfilling the following three conditions:

11.12.1 The action of G is $\mathcal{D}H$-semilinear on both sides, with respect the action by conjugation of G on H, and extends the action of H on A mapping $a \in A$ on $y{\cdot}a{\cdot}y^{-1}$ for any $y \in H$.

11.12.2 For any $a, a' \in A$ and any $c \in \mathcal{D}H$, we have $(a{\cdot}c)a' = a(c{\cdot}a')$.

11.12.3 The inclusions $\imath_z{\cdot}A{\cdot}\imath_{z'} \subset A$, where z and z' run on Z, induce an $\mathcal{O}$-module isomorphism $A \cong \oplus_{z,z' \in \mathbf{Z}} \imath_z{\cdot}A{\cdot}\imath_{z'}$.

For instance, if M is a $\mathcal{D}G$-module fulfilling condition 10.1.2 then the composition on the left and on the right defines a $\mathcal{D}G$-interior algebra structure on the "bounded" endomorphism $\mathcal{O}$-algebra $\mathrm{BEnd}_{\mathcal{O}}(M)$ (cf. Remark 10.5). Notice that if A has a unity element then we have $\imath_z{\cdot}A{\cdot}\imath_{z'} = \{0\}$ for all $z, z' \in \mathbf{Z}$ but a finite set; moreover, if A has a unity element and is a finitely generated $\mathcal{O}$-module then A is an $\mathcal{O}$-finite $\mathcal{D}H$-interior G-algebra as defined in 11.4 above. Finally, if A' is a second $\mathcal{D}H$-interior G-algebra, $A \otimes_{\mathcal{O}} A'$ has an evident $\mathcal{D}H$-interior G-algebra structure (cf. Remark 10.5).

11.13 As in the $\mathcal{D}G$-module case (cf. Corollary 10.9), *contractility* in $\mathcal{D}H$-interior G-algebras can be characterized in terms of a suitable *induction from $\mathcal{F}$ to $\mathcal{D}$* that we describe here. Let B be an *$\mathcal{O}$-finite $\mathcal{F}H$-interior G-algebra*, namely an $\mathcal{O}$-finite $\mathcal{O}H$-interior G-algebra endowed with a unitary $\mathcal{O}$-algebra homomorphism

11.13.1 $$\mathcal{F} \longrightarrow B^G \quad ;$$

thus, any $\mathcal{O}$-finite $\mathcal{D}H$-interior G-algebra A determines an evident $\mathcal{F}H$-interior G-algebra, that we denote by $\mathrm{Res}^{\mathcal{D}}_{\mathcal{F}}(A)$. Consider $\mathcal{D}$ as an $\mathcal{F}$-bimodule by multiplication on the left after shifting and on the right, and denote by

11.13.2 $$\pi : \mathcal{D} \longrightarrow \mathcal{F}$$

the $\mathcal{F}$-bimodule homomorphism mapping 1 on zero and d on 1; then, we set

11.13.3 $$\mathrm{Ind}_{\mathcal{F}}^{\mathcal{D}}(B) = \mathcal{D} \underset{\mathcal{F}}{\otimes} B \underset{\mathcal{F}}{\otimes} \mathcal{D}$$

endowed with the distributive product defined by

11.13.4 $$(u \otimes b \otimes v)(u' \otimes b' \otimes v') = u \otimes b{\cdot}\pi(vu'){\cdot}b' \otimes v' \quad ,$$

where $u, v, u', v' \in \mathcal{D}$ and $b, b' \in B$, with the action of G defined by

11.13.5 $$(u \otimes b \otimes v)^x = u \otimes b^x \otimes v \quad ,$$

where $u, v \in \mathcal{D}$, $b \in B$ and $x \in G$, and with the G-algebra homomorphism

11.13.6 $$\mathcal{D}H \longrightarrow \mathrm{Ind}_{\mathcal{F}}^{\mathcal{D}}(B)$$

mapping $f \in \mathcal{F}$ on $1 \otimes \mathrm{sh}(f){\cdot}1 \otimes d + d \otimes f{\cdot}1 \otimes 1$, d on $d \otimes 1 \otimes d$ and $y \in H$ on $1 \otimes y{\cdot}1 \otimes d + d \otimes y{\cdot}1 \otimes 1$. It is easily checked that $\mathrm{Ind}_{\mathcal{F}}^{\mathcal{D}}(B)$ is an $\mathcal{O}$-finite $\mathcal{D}H$-interior G-algebra and the next proposition shows that, as in the usual case, the induction of algebras is compatible with the induction of modules.

Proposition 11.14 *Let M be an $\mathcal{O}$-finite $\mathcal{F}G$-module. There is a $\mathcal{D}G$-interior algebra isomorphism*

11.14.1 $$\mathrm{Ind}_{\mathcal{F}}^{\mathcal{D}}\big(\mathrm{End}_{\mathcal{O}}(M)\big) \cong \mathrm{End}_{\mathcal{O}}\big(\mathrm{Ind}_{\mathcal{F}}^{\mathcal{D}}(M)\big)$$

which, for any $u, v \in \mathcal{D}$ and any $f \in \mathrm{End}_{\mathcal{O}}(M)$, maps $u \otimes f \otimes v$ on the $\mathcal{O}$-module endomorphism of $\mathrm{Ind}_{\mathcal{F}}^{\mathcal{D}}(M)$ mapping $w \otimes m$ on $u \otimes f\big(\pi(vw)(m)\big)$ for any $w \in \mathcal{D}$ and any $m \in M$.

Proof: It is easily checked that this statement defines indeed a $\mathcal{D}G$-interior algebra homomorphism

11.14.2 $$\mathrm{Ind}_{\mathcal{F}}^{\mathcal{D}}\big(\mathrm{End}_{\mathcal{O}}(M)\big) \longrightarrow \mathrm{End}_{\mathcal{O}}\big(\mathrm{Ind}_{\mathcal{F}}^{\mathcal{D}}(M)\big) \quad ;$$

but, since $\mathrm{Ind}_{\mathcal{F}}^{\mathcal{D}}(M) \cong 1 \otimes M \oplus d \otimes M$ as $\mathcal{O}$-modules, we have an evident $\mathcal{O}$-module isomorphism

11.14.3 $$\begin{aligned}\mathrm{End}_{\mathcal{O}}\big(\mathrm{Ind}_{\mathcal{F}}^{\mathcal{D}}(M)\big) &\cong \mathrm{End}_{\mathcal{O}}(1 \otimes M) \oplus \mathrm{Hom}_{\mathcal{O}}(1 \otimes M, d \otimes M)\\ &\oplus \mathrm{Hom}_{\mathcal{O}}(d \otimes M, 1 \otimes M) \oplus \mathrm{End}_{\mathcal{O}}(d \otimes M) \quad ;\end{aligned}$$

on the other hand, we have also a canonical $\mathcal{O}$-module isomorphism

11.14.4 $$\begin{aligned}\mathrm{Ind}_{\mathcal{F}}^{\mathcal{D}}\big(\mathrm{End}_{\mathcal{O}}(M)\big) &\cong \big(1 \otimes \mathrm{End}_{\mathcal{O}}(M) \otimes d\big) \oplus \big(d \otimes \mathrm{End}_{\mathcal{O}}(M) \otimes d\big)\\ &\oplus \big(1 \otimes \mathrm{End}_{\mathcal{O}}(M) \otimes 1\big) \oplus \big(d \otimes \mathrm{End}_{\mathcal{O}}(M) \otimes 1\big) \quad ;\end{aligned}$$

then, it is quite clear that homomorphism 11.14.2 maps bijectively each term of the right member in 11.14.4 onto the corresponding term of the right member in 11.14.3.

11.15 It is easily checked that, as in the usual induction (cf. 2.6), we have a $\mathcal{F}H$-interior G-algebra embedding

11.15.1 $$d_{\mathcal{F}}^{\mathcal{D}}(B) : B \longrightarrow \mathrm{Res}_{\mathcal{F}}^{\mathcal{D}}\big(\mathrm{Ind}_{\mathcal{F}}^{\mathcal{D}}(B)\big)$$

mapping $b \in B$ on $d \otimes b \otimes 1$. On the other hand, since the unity element in $\mathrm{Ind}_{\mathcal{F}}^{\mathcal{D}}(B)$ is equal to

11.15.2 $$1 \otimes 1 \otimes d + d \otimes 1 \otimes 1 = d\cdot(1 \otimes 1 \otimes 1) + (1 \otimes 1 \otimes 1)\cdot d \quad ,$$

it belongs to $\mathsf{B}\big(\mathrm{Ind}_{\mathcal{F}}^{\mathcal{D}}(B)\big)$ (cf. 11.3.1) and therefore we get

11.15.3 $$\mathsf{H}\big(\mathrm{Ind}_{\mathcal{F}}^{\mathcal{D}}(B)^G\big) = \{0\} \quad ;$$

that is to say, the induced $\mathcal{D}H$-interior G-algebras are contractile and the next proposition shows that the converse is "almost" true.

Proposition 11.16 *An $\mathcal{O}$-finite $\mathcal{D}H$-interior G-algebra A is contractile if and only if there is a $\mathcal{D}H$-interior G-algebra embedding*

11.16.1 $$h : A \longrightarrow \mathrm{Ind}_{\mathcal{F}}^{\mathcal{D}}\big(\mathrm{Res}_{\mathcal{F}}^{\mathcal{D}}(A)\big)$$

contained in the same $\mathcal{O}$-algebra exomorphism than $d_{\mathcal{F}}^{\mathcal{D}}\big(\mathrm{Res}_{\mathcal{F}}^{\mathcal{D}}(A)\big)$ and then all such embeddings form a unique $\mathcal{D}H$-interior G-algebra exomorphism.

Remark 11.17 Actually, Corollary 10.9 is now a particular case of this proposition, according to proposition 11.14 above.

Proof: By 11.15.3, the existence of embedding 11.16.1 implies that $\mathsf{H}(A^G) = \{0\}$. Conversely, assume that there is $c \in A^G$ such that $d\cdot c + c\cdot d = 1$; notice that, up to modify our choice of c, we may assume that $c = \Sigma_{z\in\mathbf{Z}} i_z\cdot c\cdot i_{z+1}$; then, setting

11.16.2 $$c' = d \otimes 1 \otimes d + d \otimes d\cdot 1 \otimes 1 \text{ and } c'' = d \otimes c \otimes 1 + 1 \otimes c\cdot d \otimes 1\ ,$$

we consider the map

11.16.3 $$h : A \longrightarrow \mathcal{D} \underset{\mathcal{F}}{\otimes} A \underset{\mathcal{F}}{\otimes} \mathcal{D}$$

sending $a \in A$ to $c''(d \otimes a \otimes 1)c'$; since G fixes c' and c'', and we have

11.16.4 $$c'c'' = d \otimes 1 \otimes 1 \quad ,$$

it is easily checked that h is a $\mathcal{D}H$-interior G-algebra homomorphism. On the other hand, denoting by g the $\mathcal{F}H$-interior G-algebra embedding

11.16.5 $$d_{\mathcal{F}}^{\mathcal{D}}\big(\mathrm{Res}_{\mathcal{F}}^{\mathcal{D}}(A)\big) : \mathrm{Res}_{\mathcal{F}}^{\mathcal{D}}(A) \longrightarrow \mathrm{Res}_{\mathcal{F}}^{\mathcal{D}}\Big(\mathrm{Ind}_{\mathcal{F}}^{\mathcal{D}}\big(\mathrm{Res}_{\mathcal{F}}^{\mathcal{D}}(A)\big)\Big) \quad ,$$

for any $a \in A$ we get

11.16.6 $$h(a) = c''g(a)c' \text{ and } g(a)c'c'' = g(a) = c'c''g(a) \quad ;$$

hence, it follows from Lemma 3.7 in [19] that g and h are contained in the same $\mathcal{O}$-algebra exomorphism and, in particular, h is an embedding. The uniqueness of $\tilde{h}$ is an immediate consequence of proposition 11.9.

12 Induction of $\mathcal{D}G$-interior algebras

12.1 In this section we extend to $\mathcal{D}G$-interior algebras - from now on, all the $\mathcal{D}G$- or $\mathcal{O}G$-interior algebras we consider are $\mathcal{O}$-finite and we omit to write it - what has been done for $\mathcal{O}G$-interior algebras in Section 3 above. That is to say, let G and G' be finite groups, $\varphi : G \to G'$ a group homomorphism and A a $\mathcal{D}G$-interior algebra; here we construct the *induced $\mathcal{D}G'$-interior algebra* $\mathrm{Ind}_\varphi(A)$ just by giving an $\mathcal{O}$-algebra homomorphism from $\mathcal{D}$ to the $\mathcal{O}$-subalgebra of G'-fixed elements in the corresponding induced $\mathcal{O}G'$-interior algebra. *Mutatis mutandis* we will apply this construction in Section 15 to give an alternative description of the so-called *Hecke $\mathcal{D}G$-interior algebras* which appear naturally in Section 18 when dealing with *Rickard equivalences* between blocks. On the other hand, in order to state the analogous result of Theorem 6.9 on these equivalences (see Theorem 18.8 below), the functor $T_n \otimes_\mathcal{O}$ on $\mathcal{O}P$-interior algebras that we employ there - which can be considered as a "direct sum" of the identity functor and n times the "restriction-induction" throughout the trivial subgroup - has to be replaced by a more sophisticated functor on $\mathcal{D}P$-interior algebras consisting of the "direct sum" of the identity functor and n times a "direct sum of restriction-inductions" involving the family of all the proper subgroups of P with their isomorphisms; in the last part of this section, we introduce these "direct sums of restriction-inductions".

12.2 Set $K = \mathrm{Ker}(\varphi)$, denote by $\mathrm{st}_A : \mathcal{D} \to A^G$ the structural map and consider the induced $\mathcal{O}G'$-interior algebra that we still denote by $\mathrm{Ind}_\varphi(A)$. Recall that A^G is contained in $N_A(K)$ (cf. 2.3.2) and that we have an $\mathcal{O}G$-interior algebra homomorphism $d_\varphi(A)$ from $N_A(K)$ to $\mathrm{Res}_\varphi\big(\mathrm{Ind}_\varphi(A)\big)$ (cf. 3.4.2) which has its image contained in $1 \otimes (\mathcal{O} \otimes_{\mathcal{O}K} A)^K \otimes 1$, so that

12.2.1 $$d_\varphi(A)(A^G) \subset 1 \otimes (\mathcal{O} \underset{\mathcal{O}K}{\otimes} A)^G \otimes 1 \subset \mathrm{Ind}_\varphi(A)^{\varphi(G)} \quad ;$$

moreover, notice that the relative trace map

12.2.2 $$\mathrm{Tr}_{\varphi(G)}^{G'} : \mathrm{Ind}_\varphi(A)^{\varphi(G)} \longrightarrow \mathrm{Ind}_\varphi(A)^{G'}$$

induces an $\mathcal{O}$-algebra homomorphism when restricted to $1 \otimes (\mathcal{O} \otimes_{\mathcal{O}K} A)^G \otimes 1$. Consequently, the composed map

12.2.3 $$\mathcal{D} \xrightarrow{\mathrm{st}_A} A^G \xrightarrow{d_\varphi(A)} 1 \otimes (\mathcal{O} \underset{\mathcal{O}K}{\otimes} A)^G \otimes 1 \xrightarrow{\mathrm{Tr}_{\varphi(G)}^{G'}} \mathrm{Ind}_\varphi(A)^{G'}$$

is an $\mathcal{O}$-algebra homomorphism and $\mathrm{Ind}_\varphi(A)$ becomes a $\mathcal{D}G'$-interior algebra; explicitely, the image of $g \in \mathcal{D}$ in $\mathrm{Ind}_\varphi(A)^{G'}$ is $\Sigma_{x'} x' \otimes (1 \otimes g\cdot 1) \otimes x'^{-1}$, where x' runs on a set of representatives for $G'/\varphi(G)$ in G' and therefore, for any $g, h \in \mathcal{D}$, any $x', y' \in G'$ and any $a \in A$ such that K fixes $1 \otimes a$ in $\mathcal{O} \otimes_{\mathcal{O}K} A$, we have

12.2.4 $$g\cdot\big(x' \otimes (1 \otimes a) \otimes y'\big)\cdot h = x' \otimes (1 \otimes g\cdot a\cdot h) \otimes y' \quad .$$

In particular, considering the case where $G' = G/K$ and φ is the canonical map, $(\mathcal{O} \otimes_{\mathcal{O}K} A)^K$ becomes always a $\mathcal{D}(G/K)$-interior algebra. On the other hand, when G is a subgroup of G' and φ is the inclusion map, we write $\mathrm{Ind}_G^{G'}(A)$ instead of $\mathrm{Ind}_\varphi(A)$.

12.3 Actually, $N_A(K)$ is a $\mathcal{D}G$-interior subalgebra of A and the map (cf. 3.4.2)

12.3.1 $$d_\varphi(A) : N_A(K) \longrightarrow \mathrm{Res}_\varphi\big(\mathrm{Ind}_\varphi(A)\big)$$

becomes now a $\mathcal{D}G$-interior algebra homomorphism (written $d_G^{G'}(A)$ in the subgroup case above). Moreover, if B is a second $\mathcal{D}G$-interior algebra and $f : A \to B$ a G-interior $\mathcal{D}$-algebra or a $\mathcal{D}G$-interior algebra homomorphism, it follows immediately from 12.2.4 that the induced map $\mathrm{Ind}_\varphi(f)$ in 3.5.3 is now respectively a G'-interior $\mathcal{D}$-algebra or a $\mathcal{D}G'$-interior algebra homomorphism. Notice that in the case where $A = \mathrm{End}_\mathcal{O}(M)$ for some $\mathcal{D}G$-module M, homomorphism 3.7.1 induces a $\mathcal{D}G'$-module structure on $\mathrm{Ind}_\varphi(M)$; explicitely, for any $\mathcal{g} \in \mathcal{D}$, any $x' \in G'$ and any $m \in M$, we have

12.3.2 $$\mathcal{g}(x' \otimes m) = x' \otimes \mathcal{g}(m) \quad .$$

12.4 Since $\mathsf{C}(A)$ is actually a G-interior $\mathcal{F}$-subalgebra of A, it makes sense to consider the corresponding induced $\mathcal{O}G'$-interior algebra $\mathrm{Ind}_\varphi\big(\mathsf{C}(A)\big)$, which is clearly a $\mathcal{F}$-algebra, and the induced homomorphism of the inclusion map $\mathsf{C}(A) \subset A$, which is $\mathcal{F}$-linear; then, it is quite clear that we get a unitary G'-interior $\mathcal{F}$-algebra homomorphism

12.4.1 $$\mathrm{Ind}_\varphi\big(\mathsf{C}(A)\big) \longrightarrow \mathsf{C}\big(\mathrm{Ind}_\varphi(A)\big) \quad ;$$

moreover, it is easily checked from 12.2.4 that this homomorphism maps the sum

12.4.2 $$\sum_{x',y' \in G'} x' \otimes \big(1 \otimes \mathsf{B}(A)\big)^K \otimes y'$$

onto $\mathsf{B}\big(\mathrm{Ind}_\varphi(A)\big)$ and therefore we still have a unitary G'-interior $\mathcal{F}$-algebra homomorphism

12.4.3 $$\mathrm{Ind}_\varphi\big(\mathsf{H}(A)\big) \longrightarrow \mathsf{H}\big(\mathrm{Ind}_\varphi(A)\big) \quad ;$$

in particular, if A is contractile then $\mathrm{Ind}_\varphi(A)$ is contractile too. Moreover, notice that

12.4.4 *If φ is injective then homomorphisms 12.4.1 and 12.4.3 are bijective.*

12.5 As in 3.10 above, let G'' be a third finite group and $\varphi' : G' \to G''$ a second group homomorphism, and set $K' = \mathrm{Ker}(\varphi')$ and $L = \mathrm{Ker}(\varphi' \circ \varphi) = \varphi^{-1}(K')$; it is quite clear from 12.2.4 that the $\mathcal{O}$-linear map 3.10.2 becomes now a $\mathcal{D}$-linear map

12.5.1 $$r_{\varphi,K}(A) : \big(\mathcal{O} \underset{\mathcal{O}(L/K)}{\otimes} (\mathcal{O} \underset{\mathcal{O}K}{\otimes} A)^K\big)^{L/K} \longrightarrow \big(\mathcal{O} \underset{\mathcal{O}K'}{\otimes} \mathrm{Ind}_\varphi(A)\big)^{\varphi(L)}$$

sending $1 \otimes (1 \otimes a)$ to $1 \otimes \big(1 \otimes (1 \otimes a) \otimes 1\big)$ for any $a \in A$ such that K fixes $1 \otimes a$ in $\mathcal{O} \otimes_{\mathcal{O}K} A$ and L/K fixes $1 \otimes (1 \otimes a)$ in $\mathcal{O} \otimes_{\mathcal{O}(L/K)} (\mathcal{O} \otimes_{\mathcal{O}K} A)^K$. Moreover, as in 3.10, we consider the composed map

12.5.2

$$\big(\mathcal{O} \underset{\mathcal{O}(L/K)}{\otimes} (\mathcal{O} \underset{\mathcal{O}K}{\otimes} A)^K\big)^{L/K} \xrightarrow{r_{\varphi,K}(A)} (\mathcal{O} \underset{\mathcal{O}K'}{\otimes} \mathrm{Ind}_\varphi(A))^{\varphi(L)} \xrightarrow{\mathrm{Tr}^{K'}_{\varphi(L)}} \big(\mathcal{O} \underset{\mathcal{O}K'}{\otimes} \mathrm{Ind}_\varphi(A)\big)^{K'}$$

and then, as in 3.12, the restricted map

12.5.3 $$s_{K,L}(A) : \big(\mathcal{O} \underset{\mathcal{O}(L/K)}{\otimes} (\mathcal{O} \underset{\mathcal{O}K}{\otimes} A)^K\big)^{L/K} \longrightarrow (\mathcal{O} \underset{\mathcal{O}L}{\otimes} A)^L$$

which are now $\mathcal{D}\big(\varphi'(\varphi(G))\big)$-interior algebra homomorphisms. The following proposition and corollary respectively extend Proposition 3.11 and Corollary 3.13 to the present situation.

Proposition 12.6 *With the notation above, the $\mathcal{D}\big(\varphi'(\varphi(G))\big)$-interior algebra homomorphism*

12.6.1 $$\big(\mathcal{O} \underset{\mathcal{O}(L/K)}{\otimes} (\mathcal{O} \underset{\mathcal{O}K}{\otimes} A)^K\big)^{L/K} \longrightarrow \mathrm{Res}^{\varphi'(G')}_{\varphi'(\varphi(G))}\Big(\big(\mathcal{O} \underset{\mathcal{O}K'}{\otimes} \mathrm{Ind}_\varphi(A)\big)^{K'}\Big)$$

mapping $1 \otimes (1 \otimes a)$ on $\mathrm{Tr}^{K'}_{\varphi(L)}\Big(1 \otimes \big(1 \otimes (1 \otimes a) \otimes 1\big)\Big)$, for any $a \in A$ such that K fixes $1 \otimes a$ in $\mathcal{O} \otimes_{\mathcal{O}K} A$ and L/K fixes $1 \otimes (1 \otimes a)$ in $\mathcal{O} \otimes_{\mathcal{O}(L/K)} (\mathcal{O} \otimes_{\mathcal{O}K} A)^K$ is an embedding which induces a $\mathcal{D}G''$-interior algebra isomorphism

12.6.2 $$\mathrm{Ind}^{G''}_{\varphi'(\varphi(G))}\Big(\big(\mathcal{O} \underset{\mathcal{O}(L/K)}{\otimes} (\mathcal{O} \underset{\mathcal{O}K}{\otimes} A)^K\big)^{L/K}\Big) \cong \mathrm{Ind}_{\varphi'}\big(\mathrm{Ind}_\varphi(A)\big) \quad .$$

Proof: According to Proposition 3.11, it suffices to prove the $\mathcal{D}$-linearity on the left and on the right of the above maps, which follows immediately from 12.2.4.

Corollary 12.7 *With the notation above, there is a unique $\mathcal{D}G''$-interior algebra homomorphism*

12.7.1 $$t_{\varphi',\varphi}(A) : \mathrm{Ind}_{\varphi'}\big(\mathrm{Ind}_\varphi(A)\big) \longrightarrow \mathrm{Ind}_{\varphi'\circ\varphi}(A)$$

mapping $1 \otimes \mathrm{Tr}^{K'}_{\varphi(L)}\Big(1 \otimes \big(1 \otimes (1 \otimes a) \otimes 1\big)\Big) \otimes 1$ on $1 \otimes (1 \otimes a) \otimes 1$ for any $a \in A$ such that K fixes $1 \otimes a$ in $\mathcal{O} \otimes_{\mathcal{O}K} A$ and L/K fixes $1 \otimes (1 \otimes a)$ in $\mathcal{O} \otimes_{\mathcal{O}(L/K)} (\mathcal{O} \otimes_{\mathcal{O}K} A)^K$. Moreover, $t_{\varphi',\varphi}(A)$ is an isomorphism if and only if the inclusion, $(\mathcal{O} \otimes_{\mathcal{O}K} A)^K \subset \mathcal{O} \otimes_{\mathcal{O}K} A$ induces an isomorphism.

12.7.2 $$\big(\mathcal{O} \underset{\mathcal{O}(L/K)}{\otimes} (\mathcal{O} \underset{\mathcal{O}K}{\otimes} A)^K\big)^{L/K} \cong (\mathcal{O} \underset{\mathcal{O}L}{\otimes} A)^L \quad .$$

In particular, if either φ or φ' is injective, or A is a $1 \times L$-projective $\mathcal{O}(L \times L)$-module by left and right multiplication then $t_{\varphi',\varphi}(A)$ is an isomorphism.

Proof: As above, according to Corollary 3.13, it suffices to prove the $\mathcal{D} \otimes_{\mathcal{O}} \mathcal{D}^{\circ}$-linearity of the $\mathcal{O}G''$-interior algebra homomorphism $t_{\varphi',\varphi}(A)$ obtained in that corollary, which follows immediately from 12.2.4.

12.8 Now, it is clear that the diagrams obtained when replacing $\mathcal{O}$-algebras by $\mathcal{D}$-interior algebras and $\mathcal{O}$-algebra homomorphisms by $\mathcal{D}$-algebra homomorphisms in diagrams 3.14.1, 3.15.1 and 3.15.2 make sense and are commutative, since so are their restrictions to $\mathcal{O}$. Similarly, notice that the criterion in Remark 3.14 can be extended to the $\mathcal{D}G$-interior algebra A by assuming that there are two families $\{X_i\}_{i\in I}$ and $\{Y_i\}_{i\in I}$ of $\mathcal{D}G$-modules such that A is a direct summand of $\oplus_{i\in I} X_i \otimes_{\mathcal{O}} Y_i^{\circ}$ as $\mathcal{D}G$-bimodules and that, for any $i \in I$, $\mathcal{O} \otimes_{\mathcal{O}K} X_i$ and $\mathcal{O} \otimes_{\mathcal{O}L} X_i$ are $\mathcal{O}$-free. Next proposition extends Proposition 3.16 to the present situation.

Proposition 12.9 *For any $\mathcal{D}G'$-interior algebra A' there is a unique $\mathcal{D}G'$-interior algebra homomorphism*

12.9.1 $$F_{\varphi}(A,A') : \mathrm{Ind}_{\varphi}(A) \underset{\mathcal{O}}{\otimes} A' \longrightarrow \mathrm{Ind}_{\varphi}\big(A \underset{\mathcal{O}}{\otimes} \mathrm{Res}_{\varphi}(A')\big)$$

mapping $\big(1\otimes(1\otimes a)\otimes 1\big)\otimes a'$ on $1\otimes\big(1\otimes(a\otimes a')\big)\otimes 1$ for any $a \in A$ such that K fixes $1\otimes a$ in $\mathcal{O}\otimes_{\mathcal{O}K} A$, and any $a' \in A'$. Moreover, if either A' is $\mathcal{O}$-free or A is a 1×1-projective $\mathcal{O}(K\times K)$-module by left and right multiplication then $F_{\varphi}(A,A')$ is an isomorphism.

Proof: Once again, it suffices to prove that the $\mathcal{O}G'$-interior algebra homomorphism $F_{\varphi}(A,A')$ obtained in Proposition 3.16 makes commutative the following diagram with the structural maps

12.9.2 $$\begin{array}{ccc} \mathrm{Ind}_{\varphi}(A) \underset{\mathcal{O}}{\otimes} A' & \xrightarrow{F_{\varphi}(A,A')} & \mathrm{Ind}_{\varphi}\big(A \underset{\mathcal{O}}{\otimes} \mathrm{Res}_{\varphi}(A')\big) \\ & \nwarrow \quad \nearrow & \\ & \mathcal{D} & \end{array}$$

but this time we feel excessive to claim immediateness. By Propositions 9.8 and 9.14, in order to prove this commutativity, it suffices to consider $\mathcal{D}_{\circ}$ or, by 9.4.1 and 9.10.3, $\mathcal{D}_n$ for any $n \geq 1$. Now, respectively considering $\mathcal{D}_n G$ and $\mathcal{D}_n G'$ as $\mathcal{O}G$- and $\mathcal{O}G'$-interior algebras, and the structural maps $\mathcal{D}_n G \to A$ and $\mathcal{D}_n G' \to A'$, from the naturality of $F_{\varphi}(A,A')$ in Proposition 3.16 (cf. 3.19.1), we get the following commutative diagram of $\mathcal{O}G'$-interior algebra homomorphisms

12.9.3 $$\begin{array}{ccc} \mathrm{Ind}_{\varphi}(A) \underset{\mathcal{O}}{\otimes} A' & \xrightarrow{F_{\varphi}(A,A')} & \mathrm{Ind}_{\varphi}\big(A \underset{\mathcal{O}}{\otimes} \mathrm{Res}_{\varphi}(A')\big) \\ \uparrow & & \uparrow \\ \mathrm{Ind}_{\varphi}(\mathcal{D}_n G) \underset{\mathcal{O}}{\otimes} \mathcal{D}_n G' & \xrightarrow{F_{\varphi}(\mathcal{D}_n G,\mathcal{D}_n G')} & \mathrm{Ind}_{\varphi}\big(\mathcal{D}_n G \underset{\mathcal{O}}{\otimes} \mathrm{Res}_{\varphi}(\mathcal{D}_n G')\big) \end{array}$$

Consequently, according to the $\mathcal{D}$-interior algebra structures of the tensor product (cf. 11.11.2) and the induction (cf. 12.2.3), it suffices to prove the commutativity of the following diagram of $\mathcal{O}$-algebra homomorphisms

12.9.4
$$\begin{array}{ccc}
\big(\mathrm{Ind}_\varphi(\mathcal{D}_nG)\underset{\mathcal{O}}{\otimes}\mathcal{D}_nG'\big)^{G'} & \xrightarrow{F_\varphi(\mathcal{D}_nG,\,\mathcal{D}_nG')} & \mathrm{Ind}_\varphi\big(\mathcal{D}_nG\underset{\mathcal{O}}{\otimes}\mathrm{Res}_\varphi(\mathcal{D}_nG')\big)^{G'} \\
\Big\uparrow {\scriptstyle \mathrm{Tr}^{G'}_{\varphi(G)}\circ(d\otimes\mathrm{id})} & & \Big\uparrow {\scriptstyle \mathrm{Ind}_\varphi(\Delta_{\mathrm{id},\varphi})} \\
\big(\mathcal{D}_nG\underset{\mathcal{O}}{\otimes}\mathrm{Res}_\varphi(\mathcal{D}_nG')\big)^{G} & & \mathrm{Ind}_\varphi(\mathcal{D}_nG)^{G'} \\
{\scriptstyle \Delta_{\mathrm{id},\varphi}}\nwarrow & & \nearrow{\scriptstyle \mathrm{Tr}^{G'}_{\varphi(G)}\circ d} \\
& (\mathcal{D}_nG)^G &
\end{array} ,$$

where $d = d_\varphi(\mathcal{D}_nG)$ and $\Delta_{\mathrm{id},\varphi} : \mathcal{D}_nG \to \mathcal{D}_nG \otimes_{\mathcal{O}} \mathrm{Res}_\varphi(\mathcal{D}_nG')$ maps $g \in \mathcal{D}_n$ on $\Delta_n(g)$ and $x \in G$ on $x \otimes \varphi(x)$. Consider the composition of the $\mathcal{O}G$-interior algebra homomorphism $d_\varphi\big(\mathcal{D}_nG \otimes_{\mathcal{O}} \mathrm{Res}_\varphi(\mathcal{D}_nG')\big)$ restricted to $\big(\mathcal{D}_nG \otimes_{\mathcal{O}} \mathrm{Res}_\varphi(\mathcal{D}_nG')\big)^G$ with the relative trace map

12.9.5
$$\mathrm{Tr}^{G'}_{\varphi(G)} : \mathrm{Ind}_\varphi\big(\mathcal{D}_nG\underset{\mathcal{O}}{\otimes}\mathrm{Res}_\varphi(\mathcal{D}_nG')\big)^{\varphi(G)} \longrightarrow \mathrm{Ind}_\varphi\big(\mathcal{D}_nG\underset{\mathcal{O}}{\otimes}\mathrm{Res}_\varphi(\mathcal{D}_nG')\big)^{G'} \quad ;$$

on the one hand, since $F_\varphi(\mathcal{D}_nG, \mathcal{D}_nG')$ and $\mathrm{Ind}_\varphi(\Delta_{\mathrm{id},\varphi})$ are $\mathcal{O}G'$-interior algebra homomorphisms, they commute with the corresponding relative trace maps $\mathrm{Tr}^{G'}_{\varphi(G)}$; on the other hand, it follows from the naturality of $d_\varphi(A)$ that we have (cf. 3.5.4)

12.9.6
$$\mathrm{Ind}_\varphi(\Delta_{\mathrm{id},\varphi}) \circ d_\varphi(\mathcal{D}_nG) = d_\varphi\big(\mathcal{D}_nG\underset{\mathcal{O}}{\otimes}\mathrm{Res}_\varphi(\mathcal{D}_nG')\big) \circ \Delta_{\mathrm{id},\varphi}$$

and from Proposition 3.16 that

12.9.7
$$F_\varphi(\mathcal{D}_nG, \mathcal{D}_nG') \circ \big(d_\varphi(\mathcal{D}_nG)\underset{\mathcal{O}}{\otimes}\mathrm{id}_{\mathcal{D}_nG'}\big) = d_\varphi\big(\mathcal{D}_nG\underset{\mathcal{O}}{\otimes}\mathrm{Res}_\varphi(\mathcal{D}_nG')\big) \quad ;$$

hence, diagram 12.9.4 is commutative.

Remark 12.10 In that proof we *never* employ the precise form of the coproduct Δ_n.

12.11 Once again, it is clear that diagrams 3.19.1 and 3.19.2 make sense here, up to change the nature of the homomorphisms, and still are commutative. Moreover, Remark 3.17 remains true without change, whereas in Remark 3.18 it suffices to consider a $\mathcal{D}G$-module M such that its restriction to $\mathcal{O}K$ is projective and a $\mathcal{D}G'$-module M' to get a canonical $\mathcal{D}G'$-module isomorphism

12.11.1
$$\mathrm{Ind}_\varphi(M)\underset{\mathcal{O}}{\otimes} M' \cong \mathrm{Ind}_\varphi\big(M\underset{\mathcal{O}}{\otimes}\mathrm{Res}_\varphi(M')\big) \quad .$$

As in 3.20, in order to extend Proposition 3.21, let H' be a third finite group and $\eta' : H' \to G'$ a second group homomorphism, and for any $x' \in G'$ consider the group homomorphism $\eta'_{x'} : H' \to G'$ mapping $y' \in H'$ on $\eta'(y')^{x'}$ and the pull-back

12.11.2
$$\begin{array}{ccc} G & \xrightarrow{\ \varphi\ } & G' \\ {\scriptstyle \eta_{x'}}\uparrow & & \uparrow{\scriptstyle \eta'_{x'}} \\ H_{x'} & \xrightarrow{\ \psi_{x'}\ } & H' \end{array},$$

setting $H_1 = H$, $\eta_1 = \eta$ and $\psi_1 = \psi$.

Proposition 12.12 *With the notation above, there is a unique $\mathcal{D}H'$-interior algebra embedding*

12.12.1
$$d_{\eta',\varphi}(A) : \mathrm{Ind}_\psi\big(\mathrm{Res}_\eta(A)\big) \longrightarrow \mathrm{Res}_{\eta'}\big(\mathrm{Ind}_\varphi(A)\big)$$

mapping $1 \otimes (1 \otimes a) \otimes 1 \in \mathrm{Ind}_\psi\big(\mathrm{Res}_\eta(A)\big)$ *on* $1 \otimes (1 \otimes a) \otimes 1 \in \mathrm{Ind}_\varphi(A)$ *for any* $a \in A$ *such that K fixes* $1 \otimes a$ *in* $\mathcal{O} \otimes_{\mathcal{O}K} A$. *Moreover, the set*

12.12.2
$$\Big\{\big(d_{\eta'_{x'},\varphi}(A)\big)\Big(\mathrm{Tr}^{H'}_{\psi_{x'}(H_{x'})}\big(1 \otimes (1 \otimes 1) \otimes 1\big)\Big)^{x'^{-1}}\Big\}_{x'},$$

where x' runs over a set of representatives X' for $\eta'(H')\backslash G'/\varphi(G)$ in G', is a pairwise orthogonal idempotent decomposition of the unity element in $\mathrm{C}_\circ\big(\mathrm{Ind}_\varphi(A)\big)$ *which does not depend on the choice of X'.*

Proof: By Proposition 3.21, we have already a unique $\mathcal{O}H'$-interior algebra embedding as described above, together with the announced decomposition of the unity element in $\mathrm{Ind}_\varphi(A)$; but, since the set of elements $1 \otimes (1 \otimes a) \otimes 1$ in $\mathrm{Ind}_\psi\big(\mathrm{Res}_\eta(A)\big)$, where a runs on the set of elements of A such that K fixes $1 \otimes a$ in $\mathcal{O} \otimes_{\mathcal{O}K} A$ (cf. 3.3.1), generates $\mathrm{Ind}_\psi\big(\mathrm{Res}_\eta(A)\big)$ as an $\mathcal{O}G'$-bimodule, it is easily checked that $d_{\eta',\varphi}(A)$ is also $\mathcal{D} \otimes_{\mathcal{O}} \mathcal{D}^\circ$-linear; moreover, it is clear that the image of $\mathcal{D}$ in $\mathrm{Ind}_\psi\big(\mathrm{Res}_\eta(A)\big)$ centralizes the set 12.12.2.

12.13 Again, diagrams 3.22.1 and 3.22.4 translated to our present situation make sense and remain commutative. The commutativity of the second one has the following consequence which will be useful in Propositions 14.18 and 16.6 below; assume that η' is injective (in 12.11), so that we can respectively identify H and H' to subgroups of G and G', and consider the evident iterated pull-backs of groups

12.13.1
$$\begin{array}{ccccc} H & \xrightarrow{\ \eta\ } & G & \xrightarrow{\ \varphi\ } & G' \\ {\scriptstyle \mathrm{id}}\uparrow & & {\scriptstyle \eta}\uparrow & & {\scriptstyle \eta'}\uparrow \\ H & \xrightarrow{\ \mathrm{id}\ } & H & \xrightarrow{\ \psi\ } & H' \end{array} \quad \text{and} \quad \begin{array}{ccccc} H & \xrightarrow{\ \psi\ } & H' & \xrightarrow{\ \eta'\ } & G' \\ {\scriptstyle \mathrm{id}}\uparrow & & {\scriptstyle \mathrm{id}}\uparrow & & {\scriptstyle \eta'}\uparrow \\ H & \xrightarrow{\ \psi\ } & H' & \xrightarrow{\ \mathrm{id}\ } & H' \end{array};$$

then, for any $\mathcal{D}H$-interior algebra B, it is easily checked that the commutativity of the corresponding diagrams 3.22.4 provides the following commutative diagram of

$\mathcal{D}H'$-interior algebra embeddings (cf. Corollary 12.7)

12.13.2
$$\begin{array}{ccc}
\mathrm{Res}^{G'}_{H'}\Big(\mathrm{Ind}^{G'}_{H'}\big(\mathrm{Ind}_{\psi}(B)\big)\Big) & \cong & \mathrm{Res}^{G'}_{H'}\Big(\mathrm{Ind}_{\varphi}\big(\mathrm{Ind}^{G}_{H}(B)\big)\Big) \\
{\scriptstyle d^{G'}_{H'}(\mathrm{Ind}_{\psi}(B))}\Big\uparrow & & \Big\uparrow{\scriptstyle d^{G'}_{H',\varphi}(\mathrm{Ind}^{G}_{H}(B))} \\
\mathrm{Ind}_{\psi}(B) & \xrightarrow{\mathrm{Ind}_{\psi}(d^{G}_{H}(B))} & \mathrm{Ind}_{\psi}\Big(\mathrm{Res}^{G}_{H}\big(\mathrm{Ind}^{G}_{H}(B)\big)\Big)
\end{array}$$

(actually, a direct checking is possible and even shorter).

12.14 In particular, Proposition 12.12 is useful to include group actions which are only *partially* interior (cf. 11.4); explicitely, let F be a finite group and $\alpha : F \to \mathrm{Aut}(G)$, $\alpha' : F \to \mathrm{Aut}(G')$ and $\omega : F \to \mathrm{End}_{\mathcal{O}}(A)^*$ group homomorphisms such that, for any $x \in F$, we have $\alpha'(x) \circ \varphi = \varphi \circ \alpha(x)$ and $\omega(x)$ is a $\mathcal{D}G$-interior algebra isomorphism between A and $\mathrm{Res}_{\alpha(x)}(A)$; then, by 12.3 and Proposition 12.12, we have the $\mathcal{D}G'$-interior algebra isomorphisms

12.14.1
$$\begin{aligned}
\mathrm{Ind}_{\varphi}\big(\omega(x)\big) \quad &: \quad \mathrm{Ind}_{\varphi}(A) \cong \mathrm{Ind}_{\varphi}\big(\mathrm{Res}_{\alpha(x)}(A)\big) \\
d_{\alpha'(x),\varphi}(A) \quad &: \quad \mathrm{Ind}_{\varphi}\big(\mathrm{Res}_{\alpha(x)}(A)\big) \cong \mathrm{Res}_{\alpha'(x)}\big(\mathrm{Ind}_{\varphi}(A)\big)
\end{aligned}$$

where x runs over F, and therefore we get a group homomorphism

$$\omega' : F \to \mathrm{End}_{\mathcal{O}}\big(\mathrm{Ind}_{\varphi}(A)\big)^*$$

mapping $x \in F$ on the composed $\mathcal{D}G'$-interior algebra isomorphism

12.14.2
$$d_{\alpha'(x),\varphi}(A) \circ \mathrm{Ind}_{\varphi}\big(\omega(x)\big) : \mathrm{Ind}_{\varphi}(A) \cong \mathrm{Res}_{\alpha'(x)}\big(\mathrm{Ind}_{\varphi}(A)\big) \quad .$$

Moreover, if x is an element of F such that, for a suitable $y \in G$, we have $\alpha(x) = \mathrm{int}(y)$, $\alpha'(x) = \mathrm{int}\big(\varphi(y)\big)$ and $\big(\omega(x)\big)(a) = a^y$ for any $a \in A$ then we have $\big(\omega'(x)\big)(a') = a'^{\varphi(y)}$ for any $a' \in \mathrm{Ind}_{\varphi}(A)$.

12.15 On the other hand, let $\mathfrak{H}'$ be a finite family $\{(H'_i, \eta_i)\}_{i\in I}$ of pairs (H'_i, η_i) formed by a subgroup H'_i of G' and a group homomorphism η_i from H'_i to G; as announced in 12.1 above, we define here the $\mathfrak{H}'$*-induced* $\mathcal{D}G'$*-interior algebra* of A, noted $\mathrm{Ind}^{G'}_{\mathfrak{H}'}\big(\mathrm{Res}^{G}_{\mathfrak{H}'}(A)\big)$. For any $i, j \in I$ denote by ${}_iA_j$ the $\mathcal{O}(H'_i \times H'_j)$-module obtained from the $\mathcal{O}G$-bimodule A by restriction throughout the group homomorphism $\eta_i \times \eta_j$ and consider the $\mathcal{O}G'$-bimodule $\mathcal{O}G' \otimes_{\mathcal{O}H'_i} ({}_iA_j) \otimes_{\mathcal{O}H'_j} \mathcal{O}G'$; to avoid confusion, for any $x', y' \in G'$ and any $a \in A$, we denote by $x' \otimes_i a \otimes_j y'$ the corresponding element of that $\mathcal{O}G'$-bimodule. Then, we set

12.15.1
$$\mathrm{Ind}^{G'}_{\mathfrak{H}'}\big(\mathrm{Res}^{G}_{\mathfrak{H}'}(A)\big) = \bigoplus_{i,j\in I} \big(\mathcal{O}G' \underset{\mathcal{O}H'_i}{\otimes} ({}_iA_j) \underset{\mathcal{O}H'_j}{\otimes} \mathcal{O}G'\big)$$

endowed with the distributive product that, for any $x', s', y', t' \in G'$, any $i, l, j, k \in I$ and any $a, b \in A$ maps $(x' \otimes_i a \otimes_l s', y' \otimes_j b \otimes_k t')$ either on zero or on $x' \otimes_i a{\cdot}\eta_j(s'y'){\cdot}b \otimes_k t'$ whenever $l = j$ and $s'y' \in H'_j$, and with the $\mathcal{O}$-algebra homomorphism

12.15.2 $$\mathcal{D}G' \longrightarrow \mathrm{Ind}_{\mathfrak{H}'}^{G'}\big(\mathrm{Res}_{\mathfrak{H}'}^{G}(A)\big)$$

mapping $x' \in G'$ on $\Sigma_{i\in I}(\Sigma_{y'\in Y'_i} x'y' \otimes_i 1 \otimes_i y'^{-1})$ and $g \in \mathcal{D}$ on

$$\Sigma_{i\in I}\big(\Sigma_{y'\in Y'_i} y' \otimes_i \mathrm{st}_A(g) \otimes_i y'^{-1}\big) \quad ,$$

where Y'_i is a set of representatives for G'/H'_i in G' for any $i \in I$, and $\mathrm{st}_A : \mathcal{D} \to A^G$ is the structural map.

12.16 It is quite clear that, as in 12.4, we have canonical G'-interior $\mathcal{F}$-algebra isomorphisms

12.16.1 $$\begin{aligned}\mathrm{Ind}_{\mathfrak{H}'}^{G'}\Big(\mathrm{Res}_{\mathfrak{H}'}^{G}(\mathsf{C}(A))\Big) &\cong \mathsf{C}\Big(\mathrm{Ind}_{\mathfrak{H}'}^{G'}\big(\mathrm{Res}_{\mathfrak{H}'}^{G}(A)\big)\Big)\\ \mathrm{Ind}_{\mathfrak{H}'}^{G'}\Big(\mathrm{Res}_{\mathfrak{H}'}^{G}(\mathsf{H}(A))\Big) &\cong \mathsf{H}\Big(\mathrm{Ind}_{\mathfrak{H}'}^{G'}\big(\mathrm{Res}_{\mathfrak{H}'}^{G}(A)\big)\Big) \quad .\end{aligned}$$

Notice that, for any $i \in I$, we have an evident $\mathcal{D}G'$-interior algebra embedding

12.16.2 $$\tilde{f}_i : \mathrm{Ind}_{H'_i}^{G'}(\mathrm{Res}_{\eta_i}(A)) \longrightarrow \mathrm{Ind}_{\mathfrak{H}'}^{G'}\big(\mathrm{Res}_{\mathfrak{H}'}^{G}(A)\big)$$

whereas, setting $\tilde{d}_i = \mathrm{Res}_1^{H'_i}\Big(\tilde{d}_{H'_i}^{G'}(\mathrm{Res}_{\eta_i}(A))\Big)$, the corresponding composed $\mathcal{D}$-interior algebra exoembedding

12.16.3 $$\mathrm{Res}_1^G(A) \xrightarrow{\tilde{d}_i} \mathrm{Res}_1^{G'}\Big(\mathrm{Ind}_{H'_i}^{G'}(\mathrm{Res}_{\eta_i}(A))\Big) \longrightarrow \mathrm{Res}_1^{G'}\Big(\mathrm{Ind}_{\mathfrak{H}'}^{G'}\big(\mathrm{Res}_{\mathfrak{H}'}^{G}(A)\big)\Big)$$

does not depend on the choice of i; in particular, it follows from Propositions 11.9 and 11.10 that:

12.16.4 *For any $i, j \in I$ such that $(H'_i, \eta_i) = (H'_j, \eta_j)$, we have $\tilde{f}_i = \tilde{f}_j$.*

On the other hand, if $A = \mathrm{End}_{\mathcal{O}}(M)$ for some $\mathcal{D}G$-module M then, by 2.6.5 and 12.3.2, there is a canonical $\mathcal{D}G'$-interior algebra isomorphism

12.16.5 $$\mathrm{Ind}_{\mathfrak{H}'}^{G'}\Big(\mathrm{Res}_{\mathfrak{H}'}^{G}(\mathrm{End}_{\mathcal{O}}(M))\Big) \cong \mathrm{End}_{\mathcal{O}}\Big(\bigoplus_{i\in I}\mathrm{Ind}_{H'_i}^{G'}(\mathrm{Res}_{\eta_i}(M))\Big) \quad .$$

Moreover, when $G' = G$ and, for any $i \in I$, η_i is the inclusion map, setting $H'_i = H_i$ and $\mathfrak{H}' = \mathfrak{H}$, as in Proposition 12.9 we easily get a $\mathcal{D}G$-interior algebra isomorphism

12.16.6 $$\mathrm{Ind}_{\mathfrak{H}}^{G}(\mathrm{Res}_{\mathfrak{H}}^{G}(A)) \cong \mathrm{End}_{\mathcal{O}}\left(\bigoplus_{i\in I}\mathrm{Ind}_{H_i}^{G}(\mathcal{O}_{H_i})\right) \underset{\mathcal{O}}{\otimes} A \quad ,$$

where $\mathcal{O}_{H_i}$ is the trivial $\mathcal{O}H_i$-module for any $i \in I$. Finally, if B is a second $\mathcal{O}G$-interior algebra and $f : A \to B$ a $\mathcal{O}G$-interior algebra homomorphism, it is clear that we have a canonical $\mathcal{O}G'$-interior algebra homomorphism

12.16.7 $$\mathrm{Ind}_{\mathfrak{H}'}^{G'}\big(\mathrm{Res}_{\mathfrak{H}'}^{G}(f)\big) : \mathrm{Ind}_{\mathfrak{H}'}^{G'}\big(\mathrm{Res}_{\mathfrak{H}'}^{G}(A)\big) \longrightarrow \mathrm{Ind}_{\mathfrak{H}'}^{G'}\big(\mathrm{Res}_{\mathfrak{H}'}^{G}(B)\big)$$

and that, for any $i \in I$, the following diagram is commutative

12.16.8 $$\begin{array}{ccc} \mathrm{Ind}_{H_i'}^{G'}\big(\mathrm{Res}_{\eta_i}(A)\big) & \longrightarrow & \mathrm{Ind}_{\mathfrak{H}'}^{G'}\big(\mathrm{Res}_{\mathfrak{H}'}^{G}(A)\big) \\ {\scriptstyle \mathrm{Ind}_{H_i'}^{G'}(\mathrm{Res}_{\eta_i}(f))}\big\downarrow & & \big\downarrow{\scriptstyle \mathrm{Ind}_{\mathfrak{H}'}^{G'}(\mathrm{Res}_{\mathfrak{H}'}^{G}(f))} \\ \mathrm{Ind}_{H_i'}^{G'}\big(\mathrm{Res}_{\eta_i}(B)\big) & \longrightarrow & \mathrm{Ind}_{\mathfrak{H}'}^{G'}\big(\mathrm{Res}_{\mathfrak{H}'}^{G}(B)\big) \end{array} .$$

13 Braüer sections in basic induced $\mathcal{D}G$-interior algebras

13.1 In Section 7 we have shown that the existence of a suitable stable $\mathcal{O}$-basis in a Morita equivalence between two blocks has strong consequences on the relationship between these blocks. As we will show in Section 19, a generalization of this phenomenon appears in *Rickard equivalences* between blocks. In order to study this generalization, the starting point is a general result on the Brauer sections of the induced $\mathcal{D}G$-interior algebras introduced in Section 12, under the hypothesis of existence of a suitable stable $\mathcal{O}$-basis (see Theorem 13.9 below); this section is devoted to introduce this hypothesis and to prove that result. Recall that the Brauer section of a $\mathcal{D}G$-interior algebra corresponding to a p-subgroup P of G (cf. 2.2.1) has a natural structure of $\mathcal{D}C_G(P)$-interior $N_G(P)$-algebra; hence, for inductive purposes, we will start already with a $\mathcal{D}H$-interior G-algebra, where H is a normal subgroup of G. Notice that, although any $\mathscr{k}$-structure is also an $\mathcal{O}$-structure, we have to be more accurate when considering $\mathscr{k}$-bases, replacing $\mathcal{D}$ by the differential $\mathbf{Z}$-grading $\mathscr{k}$-algebra $\mathscr{k} \otimes_{\mathcal{O}} \mathcal{D}$.

13.2 Let G be a finite group, H a normal subgroup of G and A a $\mathcal{D}H$-interior G-algebra; if J is a normal subgroup of G contained in H, we say that A is *J-basic* if it is $\mathcal{O}$-free and a Sylow p-subgroup P of G stabilizes by conjugation an $\mathcal{O}$-basis of A where the direct product $(J \cap P) \times (J \cap P)$ acts freely by left and right multiplication; that is to say, denoting by $\Delta(P)$ the diagonal subgroup of $P \times P$ and setting $U = J \cap P$, A is J-basic if and only if A is a permutation $\mathcal{O}(U \times U)\cdot\Delta(P)$-module which restricted to $\mathcal{O}(U \times U)$ becomes projective (cf. 2.2). In that case, for any p-subgroup Q of G, the Brauer section $A(Q)$ is a $C_J(Q)$-basic $\mathscr{k} \otimes_{\mathcal{O}} \mathcal{D}C_H(Q)$-interior $N_G(Q)$-algebra. Notice that, by Weiss' criterion (cf. Theorem A1.2), we have that:

13.2.1 *If $\mathcal{O}$ has characteristic zero, A is J-basic if and only if A is a projective $\mathcal{O}(U \times U)$-module by left and right multiplication and $(\Sigma_{u\in U} u)\cdot A\cdot(\Sigma_{u\in U} u)$ is a permutation $\mathcal{O}(P/U)$-module.*

When $J = 1$ we say that A is *basic* instead of 1-basic.

13.3 Let K be a normal subgroup of G contained in J, set $G' = G/K$, $H' = H/K$ and $J' = J/K$, and denote by $\varphi : G \to G'$ and $\psi : H \to H'$ the corresponding canonical maps. According to 12.14, $\mathrm{Ind}_\psi(A)$ is actually a *$\mathcal{D}H'$-interior G'-algebra*, and not only a $\mathcal{D}H'$-interior one; precisely, for any $x' \in G'$ and any $a \in A$ such that K fixes $1 \otimes a$ in $\mathcal{O} \otimes_{\mathcal{O}K} A$, from 12.14 we get the action

13.3.1 $$(1 \otimes a)^{x'} = 1 \otimes a^x \quad ,$$

where x lifts x' to G. In particular, if P' is a p-subgroup of G' such that K has a complement P in $\varphi^{-1}(P')$ then, denoting by $\psi_P : C_H(P) \to \varphi(C_H(P))$ the group homomorphism determined by ψ, the induced algebra $\mathrm{Ind}_{\psi_P}(A(P))$ is a

$\mathcal{D}\varphi\big(C_H(P)\big)$-interior $\varphi\big(N_G(P)\big)$-algebra, since $N_K(P) = C_K(P)$. On the other hand, next lemma shows that the "basic" condition above behaves well with respect to induction.

Lemma 13.4 *With the notation above, if A is J-basic then* $\mathrm{Ind}_\psi(A)$ *is a J'-basic $\mathcal{D}H'$-interior G'-algebra and we have*

13.4.1 $$\big(d_\psi(A)\big)(A^J) = \mathrm{Ind}_\psi(A)^{J'} \quad .$$

Proof: Let P be a Sylow p-subgroup of G, set $P' = K{\cdot}P/K \subset G'$ and $U' = J' \cap P' = K{\cdot}U/K$, denote by

$$\zeta : J{\cdot}P \to J'{\cdot}P' \quad \text{and} \quad \xi : J{\cdot}(P \cap H) \to J'{\cdot}(P' \cap H')$$

the canonical maps, and consider the $\mathcal{D}J{\cdot}(P \cap H)$-interior $J{\cdot}P$-algebra $B = \mathrm{Ind}_{P\cap H}^{J\cdot(P\cap H)}\big(\mathrm{Res}_P^G(A)\big)$ (cf. 2.6); then, since $P \cap H$ is a Sylow p-subgroup of $J{\cdot}(P \cap H)$, A is a direct summand of B as $\mathcal{O}(J \times J){\cdot}\Delta(P)$-modules (cf. [19, Lemma 3.8]) and therefore $\mathcal{O} \otimes_{\mathcal{O}K} A$ is a direct summand of $\mathcal{O} \otimes_{\mathcal{O}K} B$ as $\mathcal{O}(J' \times J){\cdot}\overline{\Delta(P)}$-modules, where $\overline{\Delta(P)}$ is the image of $\Delta(P)$ in $P' \times P$. Now, if W is a $(U \times U){\cdot}\Delta(P)$-stable $\mathcal{O}$-basis of A (so that A is $\mathcal{O}$-free) where $U \times U$ acts freely, it is quite clear that the set $J \otimes W \otimes J$ in B is a $(J \times J){\cdot}\Delta(P)$-stable $\mathcal{O}$-basis of B where $J \times J$ acts freely; hence, the set $1 \otimes (J \otimes W \otimes J)$ in $\mathcal{O} \otimes_{\mathcal{O}K} B$ is a $(J' \times J){\cdot}\overline{\Delta(P)}$-stable $\mathcal{O}$-basis of $\mathcal{O} \otimes_{\mathcal{O}K} B$ (so that it is $\mathcal{O}$-free) where $J' \times J$ acts freely and therefore the set $\mathrm{Tr}_1^K\big(1 \otimes (J \otimes W \otimes J)\big)$ is a $(J' \times J'){\cdot}\Delta(P')$-stable $\mathcal{O}$-basis of the $\mathcal{D}J'{\cdot}(P' \cap H')$-interior $J'{\cdot}P'$-algebra $\mathrm{Ind}_\xi(B)$ where $J' \times J'$ acts freely; moreover, it is easily checked in these $\mathcal{O}$-bases that $1 \otimes B^J = (\mathcal{O} \otimes_{\mathcal{O}K} B)^J$. Consequently, $\mathrm{Ind}_\psi(A)$ is a permutation $\mathcal{O}$-free $\mathcal{O}(U' \times U'){\cdot}\Delta(P')$-module which restricted to $\mathcal{O}(U' \times U')$ is projective, and equality 13.4.1 holds.

13.5 Now, if P' is a p-subgroup of G', in order to determine $\big(\mathrm{Ind}_\psi(A)\big)(P')$ we consider the relationship between $(\mathcal{O} \otimes_{\mathcal{O}K} A)^K(P')$ and $\big(\prod_P A(P)\big)^K$ where P runs on the set of p-subgroups of G lifting P'; we have no general answer, but when A is K-basic the following lemma on direct summands of permutation modules - which generalizes Lemma 4.2 in [33] - allow us to clarify that relationship. In order to state it, notice that the group $\varphi^{-1}\big(N_{G'}(P')\big) = N_G\big(\varphi^{-1}(P')\big)$ acts on the (eventually empty) set $\mathfrak{C}$ of all the complements of K in $\varphi^{-1}(P')$ and therefore, if M is an $\mathcal{O}G$-module, it acts on the direct product $\prod_{P\in\mathfrak{C}} M(P)$, so that $N_{G'}(P')$ acts on $\big(\prod_{P\in\mathfrak{C}} M(P)\big)^K$.

Lemma 13.6 *Let M be a direct summand of a permutation $\mathcal{O}G$-module and assume that the restriction of M to $\mathcal{O}K$ is projective. Then, for any p-subgroup P' of G', the inclusion $M^K \subset M$ induces a $kN_{G'}(P')$-module isomorphism*

13.6.1 $$(M^K)(P') \cong \Big(\prod_P M(P)\Big)^K \quad ,$$

where P runs over the set of complements of K in $\varphi^{-1}(P')$.

Remark 13.7 Notice that, for such a complement P, we can identify canonically $M(P)^{C_K(P)}$ with a direct summand in the second member of isomorphism 13.6.1, namely with $\big(\prod_y M(P^y)\big)^K$ where y runs on a set of representatives for $K/C_K(P)$ in K; thus, this member becomes $\prod_P M(P)^{C_K(P)}$, where P runs on a set of representatives for the orbits of K in the set of such complements.

Proof: It is clear that the inclusion $M^K \subset M$ induces a k-linear map $(M^K)(P) \to M(P)$ for any p-subgroup P of G and it is easily checked that $(M^K)(P) = (M^K)\big(\varphi(P)\big)$; hence, denoting by $\mathfrak{C}$ the set of all the complements of K in $\varphi^{-1}(P')$, we have a $kN_{G'}(P')$-module homomorphism

$$13.6.2 \qquad (M^K)(P') \longrightarrow \Big(\prod_{P\in\mathfrak{C}} M(P)\Big)^K$$

and it suffices to prove that this map is bijective. On the other hand, if S is a Sylow p-subgroup of $\varphi^{-1}(P')$, it follows from our hypothesis that M has a S-stable $\mathcal{O}$-basis W where $K \cap S$ acts freely (cf. [38, (27.2)]); then, the set $K \otimes W$ in $\mathrm{Ind}_S^{\varphi^{-1}(P')}\big(\mathrm{Res}_S^G(M)\big) = N$ is a S-stable $\mathcal{O}$-basis of N where K acts freely; consequently, since homomorphism 13.6.2 is natural on M, up to replace M by N we may assume that W is $\varphi^{-1}(P')$-stable and K acts freely on it. In that case, the set $\mathrm{Tr}_1^K(W)$ is a P'-stable $\mathcal{O}$-basis of M^K and therefore the image by $\mathrm{Br}_{P'}$ of the set of fixed points $\mathrm{Tr}_1^K(W)^{P'}$ is a k-basis of $(M^K)(P')$ (cf. 2.2.5); but, if P' fixes $\mathrm{Tr}_1^K(w)$ for some $w \in W$, $\varphi^{-1}(P')$ stabilizes the set $\{y.w\}_{y\in K}$, where K acts regularly, and therefore the stabilizer P of w in $\varphi^{-1}(P')$ belongs to $\mathfrak{C}$; conversely, for any $P \in \mathfrak{C}$ and any $w \in W$ fixed by P, P' fixes $\mathrm{Tr}_1^K(w)$; hence, the set of $w \in W$ such that P' fixes $\mathrm{Tr}_1^K(w)$ is the disjoint union $\bigcup_{P\in\mathfrak{C}} W^P$; consequently, since $\mathrm{Br}_P(W^P)$ is a k-basis of $M(P)$ for any $P \in \mathfrak{C}$ (cf. 2.2.5), homomorphism 13.6.2 maps bijectively $\mathrm{Tr}_1^K(W)^{P'}$ onto a k-basis of the right end.

13.8 If A is K-basic, it follows from Lemma 13.6 above that, for any p-subgroup P' of G', the inclusion $A^K \subset A$ induces a canonical $kN_{G'}(P')$-module isomorphism

$$13.8.1 \qquad (A^K)(P') \cong \Big(\prod_P A(P)\Big)^K \quad ,$$

where P runs over the set of complements of K in $\varphi^{-1}(P')$; actually, since that inclusion is a $\mathfrak{D}C_H(K)$-interior G-algebra homomorphism, isomorphism 13.8.1 is a $\mathfrak{D}C_H\big(\varphi^{-1}(P')\big)$-interior $N_G\big(\varphi^{-1}(P')\big)$-algebra isomorphism and, in particular, for any complement P of K in $\varphi^{-1}(P')$, we get a canonical $\mathfrak{D}C_H\big(\varphi^{-1}(P')\big)$-interior $N_G(P)$-algebra embedding (cf. Remark 13.7)

$$13.8.2 \qquad c_{\varphi,P}(A) : A(P)^{C_K(P)} \longrightarrow (A^K)(P') \quad .$$

Moreover, recall that we have a canonical unitary $\mathfrak{D}C_H(K)$-interior G-algebra homomorphism (cf. 3.4.2)

13.8.3 $$d_\psi(A) : A^K \longrightarrow \mathrm{Res}_\varphi\big(\mathrm{Ind}_\psi(A)\big)$$

and similarly, for any p-subgroup P' of G' and any complement P of K in $\varphi^{-1}(P')$, denoting by $\varphi_P : N_G(P) \to \varphi\big(N_G(P)\big)$ and $\psi_P : C_H(P) \to \varphi\big(C_H(P)\big)$ the group homomorphisms determinated by φ, we have the canonical unitary $\mathfrak{D}C_H\big(\varphi^{-1}(P')\big)$-interior $N_G(P)$-algebra homomorphism

13.8.4 $$d_{\psi_P}\big(A(P)\big) : A(P)^{C_K(P)} \longrightarrow \mathrm{Res}_{\varphi_P}\Big(\mathrm{Ind}_{\psi_P}\big(A(P)\big)\Big) \quad .$$

Finally, for any p-subgroup P' of G', we choose a set of representatives $\mathfrak{C}_{\varphi,P'}$ for the set of orbits of K in the set of complements of K in $\varphi^{-1}(P')$.

Theorem 13.9 *With the notation above, assume that A is K-basic and let P' be a p-subgroup of G'. For any complement P of K in $\varphi^{-1}(P')$, there is a unique $\mathfrak{D}\varphi\big(C_H(P)\big)$-interior $\varphi\big(N_G(P)\big)$-algebra embedding*

13.9.1 $$e_{\varphi,P}(A) : \mathrm{Ind}_{\psi_P}\big(A(P)\big) \longrightarrow \big(\mathrm{Ind}_\psi(A)\big)(P')$$

such that we have

13.9.2 $$e_{\varphi,P}(A) \circ d_{\psi_P}\big(A(P)\big) = \big(d_\psi(A)\big)(P') \circ c_{\varphi,P}(A) \quad .$$

Moreover, the set $\Big\{\big(e_{\varphi,Q}(A)\big)(1)\Big\}_{Q\in\mathfrak{C}_{\varphi,P'}}$ is a pairwise orthogonal idempotent decomposition of the unity element in $\big(\mathrm{Ind}_\psi(A)\big)(P')$, which does not depend on the choice of $\mathfrak{C}_{\varphi,P'}$.

Remark 13.10 For any $x \in N_G\big(\varphi^{-1}(P')\big)$, according to Remark 13.7 and to the uniqueness of $e_{\varphi,P}(A)$, and denoting by $\sigma_x : N_G(P) \to N_G(P^x)$ and by $f_x : A(P) \to \mathrm{Res}_{\sigma_x}\big(A(P^x)\big)$ the corresponding isomorphisms induced by conjugation by x, we have

13.10.1 $$e_{\varphi,P^x}(A) \circ \mathrm{Ind}_{\psi_P}(f_x) = e_{\varphi,P}(A) \quad .$$

In particular, this proves that the set $\Big\{\big(e_{\varphi,Q}(A)\big)(1)\Big\}_{Q\in\mathfrak{C}_{\varphi,P'}}$ does not depend on the choice of $\mathfrak{C}_{\varphi,P'}$.

Proof: By Remark 13.7, isomorphism 13.8.1 becomes

13.9.3 $$(A^K)(P') \cong \prod_{P\in\mathfrak{C}_{\varphi,P'}} A(P)^{C_K(P)}$$

and, for any $P \in \mathfrak{C}_{\varphi,P'}$, we denote by $\bar{\imath}_P$ the unity element of $\mathrm{Im}(c_{\varphi,P})$; moreover, from homomorphism 13.8.3 we get the $\mathfrak{D}C_H(\varphi^{-1}(P'))$-interior $N_G(\varphi^{-1}(P'))$-algebra homomorphism

13.9.4 $$\big(d_\psi(A)\big)(P') : (A^K)(P') \longrightarrow \mathrm{Res}_{\varphi_{P'}}\Big(\big(\mathrm{Ind}_\psi(A)\big)(P')\Big) \quad ,$$

where $\varphi_{P'} : N_G(\varphi^{-1}(P')) \to N_{G'}(P')$ is the group homomorphism determined by φ, and, for any $P \in \mathfrak{C}_{\varphi,P'}$, we set $\bar{\jmath}_P = \Big(\big(d_\psi(A)\big)(P')\Big)(\bar{\imath}_P)$. At this point, we claim that it suffices to prove the existence of an injective $\mathcal{O}$-algebra homomorphism

13.9.5 $$f : \prod_{P \in \mathfrak{C}_{\varphi,P'}} \mathrm{Ind}_{\psi_P}\big(A(P)\big) \longrightarrow \big(\mathrm{Ind}_\psi(A)\big)(P')$$

such that the following diagram is commutative

13.9.6 $$\begin{array}{ccc} \prod_{P \in \mathfrak{C}_{\varphi,P'}} A(P)^{C_K(P)} & \cong & (A^K)(P') \\ \Big\downarrow {\scriptstyle \prod_{P \in \mathfrak{C}_{\varphi,P'}} d_\psi(A(P))} & & \Big\downarrow {\scriptstyle (d_\psi(A))(P')} \\ \prod_{P \in \mathfrak{C}_{\varphi,P'}} \mathrm{Ind}_{\psi_P}\big(A(P)\big) & \xrightarrow{\ f\ } & \big(\mathrm{Ind}_\psi(A)\big)(P') \end{array}$$

and, for any $P \in \mathfrak{C}_{\varphi,P'}$, we have

13.9.7 $$f\Big(\mathrm{Ind}_{\psi_P}\big(A(P)\big)\Big) = \bar{\jmath}_P\big(\mathrm{Ind}_\psi(A)\big)(P')\bar{\jmath}_P \quad ;$$

indeed, in that case, since $d_{\psi_P}\big(A(P)\big)$ is surjective for any $P \in \mathfrak{C}_{\varphi,P'}$ (cf. 13.2 and 13.4.1), f is completely determined by $\big(d_\psi(A)\big)(P')$ and isomorphism 13.9.3, which implies that f is actually an injective $\mathfrak{D}C_H(\varphi^{-1}(P'))$-interior $N_G(\varphi^{-1}(P'))$-algebra homomorphism and, by equality 13.9.7, its restriction to the factor $\mathrm{Ind}_{\psi_P}\big(A(P)\big)$ is an embedding for any $P \in \mathfrak{C}_{\varphi,P'}$.

Consequently, in order to prove the existence of f, we may forget the $\mathfrak{D}$-interior algebra structures. First of all notice that, since a Sylow p-subgroup S of $\varphi^{-1}(P')$ stabilizes by conjugation an $\mathcal{O}$-basis of A where $K \cap S$ acts freely (cf. [38, (27.2)]), for any $P \in \mathfrak{C}_{\varphi,P'}$ and any p-subgroup Q of $\varphi^{-1}(P') = K.P$ containing strictly P, by Lemma 7.10 we have $A(Q) = \{0\}$ and therefore we get easily (cf. 2.2.6)

13.9.8 $$A(P)^{C_K(P)} = A(P)_1^{C_K(P)} \quad .$$

Now, consider a pairwise orthogonal primitive idempotent decomposition I of the unity element in $A^{\varphi^{-1}(P')}$ and, for any $P \in \mathfrak{C}_{\varphi,P'}$, denote by i_P the sum of all i in I belonging to points of $\varphi^{-1}(P') = K.P$ on A which admit P as a defect group (cf. 2.9); then, by equality 13.9.8, $\mathrm{Br}_P(i_P)$ is the unity element in A(P) (cf. 2.9.2), whereas $\mathrm{Br}_Q(i_P) = 0$ for any $Q \in \mathfrak{C}_{\varphi,P'} - \{P\}$ (cf. 2.9.5), and therefore, setting $A_P = i_P A i_P$ considered as an $\mathcal{O}K \cdot P$-interior algebra, we have

13.9.9 $$\prod_{Q \in \mathfrak{C}_{\varphi,P'}} A_P(Q) \cong A(P) \quad \text{and} \quad (A_P)^K(P') \cong \bar{\imath}_P(A^K)(P')\bar{\imath}_P \quad .$$

Hence, it suffices to prove that, for any $P \in \mathfrak{C}_{\varphi,P'}$, there is an $\mathcal{O}$-algebra isomorphism

13.9.10 $$f_P : (\mathcal{O} \underset{\mathcal{O}C_K(P)}{\otimes} A_P(P))^{C_K(P)} \cong ((\mathcal{O} \underset{\mathcal{O}K}{\otimes} A_P)^K)(P')$$

such that the following diagram is commutative

13.9.11 $$\begin{array}{ccc} A_P(P)^{C_K(P)} & \cong & (A_P)^K(P') \\ \downarrow & & \downarrow \\ (\mathcal{O} \underset{\mathcal{O}C_K(P)}{\otimes} A_P(P))^{C_K(P)} & \overset{f_P}{\cong} & ((\mathcal{O} \underset{\mathcal{O}K}{\otimes} A_P)^K)(P') \end{array}$$

From now on, we fix $P \in \mathfrak{C}_{\varphi,P'}$. Since i_P belongs to $A_P^{K \cdot P}$ (cf. 2.9.4), we have a Higman embedding of $\mathcal{O}K \cdot P$-interior algebras (cf. [19, Lemma 3.8])

13.9.12 $$A_P \longrightarrow \mathrm{Ind}_P^{K \cdot P}(\mathrm{Res}_P^{K \cdot P}(A_P)) = B_P$$

which induces the following commutative diagram (cf. 3.5.4)

13.9.13
$$\begin{array}{ccccccc} (\mathcal{O} \underset{\mathcal{O}C_K(P)}{\otimes} A_P(P))^{C_K(P)} & \leftarrow & A_P(P)^{C_K(P)} & \cong & (A_P)^K(P') & \rightarrow & ((\mathcal{O} \underset{\mathcal{O}K}{\otimes} A_P)^K)(P') \\ \downarrow & & \downarrow & & \downarrow & & \downarrow \\ (\mathcal{O} \underset{\mathcal{O}C_K(P)}{\otimes} B_P(P))^{C_K(P)} & \leftarrow & B_P(P)^{C_K(P)} & \cong & (B_P)^K(P') & \rightarrow & ((\mathcal{O} \underset{\mathcal{O}K}{\otimes} B_P)^K)(P') \end{array}$$

since again $B_P(Q) = \{0\}$ for any $Q \in \mathfrak{C}_{\varphi,P'} - \{P\}$ (cf. 2.9.5). Consequently, it suffices to prove the existence of an $\mathcal{O}$-algebra isomorphism

13.9.14 $$g_P : (\mathcal{O} \underset{\mathcal{O}C_K(P)}{\otimes} B_P(P))^{C_K(P)} \cong ((\mathcal{O} \underset{\mathcal{O}K}{\otimes} B_P)^K)(P')$$

such that the following diagram is commutative

13.9.15 $$\begin{array}{ccc} B_P(P)^{C_K(P)} & \cong & (B_P)^K(P') \\ \downarrow & & \downarrow \\ (\mathcal{O} \underset{\mathcal{O}C_K(P)}{\otimes} B_P(P))^{C_K(P)} & \overset{g_P}{\cong} & (\mathcal{O} \underset{\mathcal{O}K}{\otimes} B_P)^K(P') \end{array} \quad .$$

But, setting $Z = Z(P)$, the canonical $\ell C_K(P)\cdot Z$-interior algebra embedding (cf. 2.6.4 and 2.6.6)

13.9.16 $$\mathrm{Ind}_Z^{C_K(P)\cdot Z}\big(A_P(P)\big) \longrightarrow B_P(P)$$

which, for any $a \in (A_P)^P$ and any $x, y \in C_K(P)$, maps $x \otimes \mathrm{Br}_P(a) \otimes y$ on $\mathrm{Br}_P(x \otimes a \otimes y)$ is an isomorphism since

13.9.17 $$\mathrm{Br}_P\big(\mathrm{Tr}_P^{K\cdot P}(1 \otimes 1 \otimes 1)\big) = \mathrm{Tr}_P^{N_{K\cdot P}(P)}\big(\mathrm{Br}_P(1 \otimes 1 \otimes 1)\big)$$

and clearly $N_{K\cdot P}(P) = C_K(P)\cdot P$. Moreover, by Lemma 13.11 below applied to A_P and to $A_P(P)$, we have the following commutative diagram (cf. Lemma 7.10)

13.9.18
$$\begin{array}{ccccc}
(B_P)^K(P') \cong (\mathcal{O}K \underset{\mathcal{O}}{\otimes} A_P)(P) & \cong & kC_K(P) \underset{\mathcal{O}}{\otimes} A_P(P) & \cong & \mathrm{Ind}_Z^{C_K(P).Z}\big(A_P(P)\big)^{C_K(P)} \\
\downarrow \quad (\varepsilon_K \otimes \mathrm{id})(P) \searrow & & & \swarrow \bar{\varepsilon}_{C_K(P)} \otimes \mathrm{id} & \downarrow \\
(\mathcal{O} \underset{\mathcal{O}}{\otimes} B_P)^K(P') & \cong & A_P(P) & \cong & \Big(\mathcal{O} \underset{\mathcal{O}C_K(P)}{\otimes} \mathrm{Ind}_Z^{C_K(P).Z}\big(A_P(P)\big)\Big)^{C_K(P)}
\end{array}$$

where ε_K and $\bar{\varepsilon}_{C_K(P)}$ are the corresponding augmentation maps. Finally, since the composition of the isomorphisms

13.9.19 $$(B_P)^K(P') \cong \mathrm{Ind}_Z^{C_K(P)\cdot Z}\big(A_P(P)\big)^{C_K(P)} \cong B_P(P)^{C_K(P)} \quad ,$$

respectively obtained from the composition of all those in the top of diagram 13.9.18 and from isomorphism 13.9.16, maps $\mathrm{Br}_{P'}\big(\mathrm{Tr}_1^K(x \otimes a \otimes y)\big)$ on $\mathrm{Tr}_1^{C_K(P)}\big(x \otimes \mathrm{Br}_P\,(a) \otimes y\big)$ for any $a \in (A_P)^P$ and any $x, y \in C_K(P)$, so that it coincides with isomorphism 13.9.3 for B_P, we get the commutativity of diagram 13.9.15 taking g_P equal to the composition of the isomorphisms

13.9.20
$$\begin{aligned}
\Big(\mathcal{O} \underset{\mathcal{O}C_K(P)}{\otimes} B_P(P)\Big)^{C_K(P)} &\cong \Big(\mathcal{O} \underset{\mathcal{O}C_K(P)}{\otimes} \mathrm{Ind}_Z^{C_K(P).Z}\big(A_P(P)\big)\Big)^{C_K(P)} \\
&\cong \big((\mathcal{O} \underset{\mathcal{O}K}{\otimes} B_P)^K\big)(P')
\end{aligned}$$

respectively obtained from the inverses of isomorphism 13.9.16 and of the composition of those in the bottom of diagram 13.9.18.

Lemma 13.11 *Assume that K has a complement L in G and let B be an $\mathcal{O}L$-interior algebra. Then we have an L-algebra isomorphism*

13.11.1 $$\mathcal{O}K \underset{\mathcal{O}}{\otimes} B \cong \mathrm{Ind}_L^G(B)^K$$

which, for any $y \in K$ and any $b \in B$, maps $y \otimes b$ on $\mathrm{Tr}_1^K(y^{-1} \otimes b \otimes 1)$ and determines the following commutative diagram of L-algebra homomorphisms

13.11.2
$$\begin{array}{ccc} B & \cong & \mathrm{Ind}_\varphi\big(\mathrm{Ind}_L^G(B)\big) \\ \Big\uparrow{\scriptstyle \varepsilon\otimes\mathrm{id}} & & \Big\uparrow{\scriptstyle d_\varphi(\mathrm{Ind}_L^G(B))} \\ \mathcal{O}K \underset{\mathcal{O}}{\otimes} B & \cong & \mathrm{Ind}_L^G(B)^K \end{array}$$

where $\varepsilon : \mathcal{O}K \to \mathcal{O}$ is the augmentation map and the isomorphism in the top is the corresponding homomorphism 3.13.1.

Proof: It is easily checked that the announced map in 13.11.1 is an injective L-algebra homomorphism; moreover, it is surjective since $\mathrm{Ind}_L^G(B)^K = \mathrm{Ind}_L^G(B)_1^K$ and we have

13.11.3 $$\mathrm{Tr}_1^K(x \otimes b \otimes y) = \mathrm{Tr}_1^K(yx \otimes b \otimes 1)$$

for any $x, y \in K$ and any $b \in B$. On the other hand, it follows from Corollary 3.13 that, up to identify L to G' via φ, we have an $\mathcal{O}L$-interior algebra isomorphism

13.11.4 $$\mathrm{Ind}_\varphi\big(\mathrm{Ind}_L^G(B)\big) \cong B$$

mapping $\mathrm{Tr}_1^K\big(1 \otimes (1 \otimes b \otimes 1)\big)$ on b for any $b \in B$; now, it is clear that, with this isomorphism, diagram 13.11.2 is commutative.

13.12 It is quite clear that embedding 13.9.1 depends basically on K, P' and $\varphi^{-1}(P')$. In order to be more precise, let F' be a subgroup of G', set $F = \varphi^{-1}(F')$, $L' = F' \cap H'$ and $L = \varphi^{-1}(L') = F \cap H$, and denote by $\zeta : F \to F'$, by $\xi : L \to L'$ and, for any p-subgroup P of F, by $\xi_P : C_L(P) \to \varphi\big(C_L(P)\big)$ the group homomorphisms determined by φ. Recall that, by Proposition 12.12, we have a canonical isomorphism of $\mathcal{D}L'$-interior F'-algebras

13.12.1 $$d_{L',\psi}^{H'}(A) : \mathrm{Ind}_\xi\big(\mathrm{Res}_L^H(A)\big) \cong \mathrm{Res}_{L'}^{H'}\big(\mathrm{Ind}_\psi(A)\big)$$

and, for any p-subgroup P of F, one of $\mathcal{D}\varphi\big(C_L(P)\big)$-interior $\varphi\big(N_F(P)\big)$-algebras

13.12.2
$$d_{\varphi(C_L(P)),\psi_P}^{\varphi(C_H(P))}\big(A(P)\big) : \mathrm{Ind}_{\xi_P}\Big(\mathrm{Res}_{C_L(P)}^{C_H(P)}\big(A(P)\big)\Big) \cong \mathrm{Res}_{\varphi(C_L(P))}^{\varphi(C_H(P))}\Big(\mathrm{Ind}_{\psi_P}\big(A(P)\big)\Big)\ .$$

Corollary 13.13 *With the notation above, assume that A is K-basic. For any p-subgroup P' of F' and any complement P of K in $\varphi^{-1}(P')$, we have*

13.13.1 $$e_{\varphi,P}(A) \circ d_{\varphi(C_L(P)),\psi_P}^{\varphi(C_H(P))}\big(A(P)\big) = \big(d_{L',\psi}^{H'}(A)\big)(P') \circ e_{\zeta,P}\big(\mathrm{Res}_F^G(A)\big)\ .$$

Proof: From 13.9.2 and Proposition 12.12 we get

$$e_{\varphi,P}(A)\circ d^{\varphi(C_H(P))}_{\varphi(C_L(P)),\psi_P}\big(A(P)\big)\circ d_{\xi_P}\Big(\mathrm{Res}^{C_H(P)}_{C_L(P)}\big(A(P)\big)\Big)$$

$$= e_{\varphi,P}(A)\circ d_{\psi_P}\big(A(P)\big) = \big(d_\psi(A)\big)(P)\circ c_{\varphi,P}(A)$$

13.13.2
$$= \big(d^{H'}_{L',\psi}(A)\big)(P')\circ d_\xi\big(\mathrm{Res}^H_L(A)\big)\circ c_{\zeta,P}\big(\mathrm{Res}^G_F(A)\big)$$

$$= \big(d^{H'}_{L',\psi}(A)\big)(P')\circ e_{\zeta,P}\big(\mathrm{Res}^G_F(A)\big)\circ d_{\xi_P}\Big(\mathrm{Res}^{C_H(P)}_{C_L(P)}\big(A(P)\big)\Big)$$

and, by equality 13.4.1, $d_{\xi_P}\Big(\mathrm{Res}^{C_H(P)}_{C_L(P)}\big(A(P)\big)\Big)$ is surjective.

13.14 We discuss now on iterated Brauer sections of $\mathrm{Ind}_\psi(A)$; this analysis will be useful for inductive arguments in Section 17 below. If P' is a p-subgroup of G' and Q' a normal subgroup of P', it is clear that, for any complement P of K in $\varphi^{-1}(P')$, the intersection $P\cap\varphi^{-1}(Q')$ is a complement of K in $\varphi^{-1}(Q')$ so that, up to modify our choice of $\mathfrak{C}_{\varphi,P'}$, we may assume that the intersection with $\varphi^{-1}(Q')$ determines a map

13.14.1
$$\omega^{Q'}_{\varphi,P'} : \mathfrak{C}_{\varphi,P'} \longrightarrow \mathfrak{C}_{\varphi,Q'} \qquad ;$$

moreover, it is clear that P' acts by conjugation on the set of orbits of K in the set of complements of K in $\varphi^{-1}(Q')$, which determines an action of P' on $\mathfrak{C}_{\varphi,Q'}$ and P' fixes $Q\in\mathfrak{C}_{\varphi,Q'}$ if and only if $P'\subset\varphi\big(N_G(Q)\big)$; in particular, we get easily that:

13.14.2 $\mathrm{Im}(\omega^{Q'}_{\varphi,P'})\subset(\mathfrak{C}_{\varphi,Q'})^{P'}$ *and, for any* $Q\in(\mathfrak{C}_{\varphi,Q'})^{P'}-\mathrm{Im}(\omega^{Q'}_{\varphi,P'})$ *and any* $P\in\mathfrak{C}_{\varphi_Q,P'}$*, we have* $Q\not\subset P$.

(Recall that $\varphi_Q : N_G(Q)\to\varphi\big(N_G(Q)\big)$ is the group homomorphism determined by φ). If $Q\in\mathrm{Im}(\omega^{Q'}_{\varphi,P'})$, any $P\in(\omega^{Q'}_{\varphi,P'})^{-1}(Q)$ normalizes Q so that, up to modify our choice of $\mathfrak{C}_{\varphi_Q,P'}$, we may assume that $(\omega^{Q'}_{\varphi,P'})^{-1}(Q)\subset\mathfrak{C}_{\varphi_Q,P'}$ and then we have that:

13.14.3 $P\in(\omega^{Q'}_{\varphi,P'})^{-1}(Q)$ *if and only if* $P\in\mathfrak{C}_{\varphi_Q,P'}$ *and* $Q\subset P$.

Indeed, if $P\in\mathfrak{C}_{\varphi_Q,P'}$ then P is still a complement of K in $\varphi^{-1}(P')$ since $[K\cap P,Q]\subset K\cap Q=1$, and therefore we have $P^y\in\mathfrak{C}_{\varphi,P'}$ for a suitable $y\in K$; hence, if $Q\subset P$, the existence of $\omega^{Q'}_{\varphi,P'}$ forces $Q^y\in\mathfrak{C}_{\varphi,Q'}$ and therefore $Q^y=Q$, which implies that y belongs to $C_K(Q)$ and P^y to $(\omega^{Q'}_{\varphi,P'})^{-1}(Q)\subset\mathfrak{C}_{\varphi_Q,P'}$, so that $P^y=P$.

13.15 Assume that A is K-basic; then, by Lemma 13.4, $\mathrm{Ind}_\psi(A)$ is basic and, by Lemma 7.10, we have a canonical isomorphism

13.15.1
$$\Big(\big(\mathrm{Ind}_\psi(A)\big)(Q')\Big)(P')\cong\big(\mathrm{Ind}_\psi(A)\big)(P') \qquad ;$$

moreover, if P and Q are respectively complements of K in $\varphi^{-1}(P')$ and in $\varphi^{-1}(Q')$, and we have $Q \subset P$ then P normalizes Q and, by Lemma 7.10 again, we have also a canonical isomorphism

13.15.2 $$\big(A(Q)\big)(P) \cong A(P) \qquad ;$$

notice that if P normalizes Q but does not contain it, always by Lemma 7.10, we get

13.15.3 $$\big(A(Q)\big)(P) \cong \big(A(Q)\big)(Q{\cdot}P) \cong A(Q{\cdot}P) \cong \big(A(K \cap Q{\cdot}P)\big)(Q{\cdot}P) = \{0\}$$

since $K \cap Q{\cdot}P \neq 1$ and A is a projective $\mathcal{O}K$-module. In the following result we identify to each other both members in isomorphisms 13.15.1 and 13.15.2.

Proposition 13.16 *With the notation above, assume that A is K-basic. For any p-subgroup P' of G', any normal subgroup Q' of P' and any complement P of K in $\varphi^{-1}(P')$, setting $Q = P \cap \varphi^{-1}(Q')$, $A(Q)$ is a $C_K(Q)$-basic $k \otimes_{\mathcal{O}} \mathcal{D}C_H(Q)$-interior $N_G(Q)$-algebra and we have the following commutative diagram*

13.16.1 $$\begin{array}{ccc} \mathrm{Ind}_{\psi_P}\big(A(P)\big) & \xrightarrow{\;e_{\varphi,P}(A)\;} & \big(\mathrm{Ind}_{\psi}(A)\big)(P') \\ {\scriptstyle e_{\varphi_Q,P}(A(Q))}\searrow & & \nearrow{\scriptstyle (e_{\varphi,Q}(A))(P')} \\ & \Big(\mathrm{Ind}_{\psi_Q}\big(A(Q)\big)\Big)(P') & \end{array}$$

Proof: Let P' be a p-subgroup of G' and Q' a normal subgroup of P'; it follows from Lemma 13.6 and Remark 13.7 that we have a canonical P'-algebra isomorphism

13.16.2 $$(A^K)(Q') \cong \prod_{Q \in \mathfrak{C}_{\varphi,Q'}} A(Q)^{C_K(Q)} \qquad ;$$

then, since P' permutes the direct factors in the right member of this isomorphism, it is quite clear that this isomorphism induces, up to suitable identifications, a k-algebra isomorphism

13.16.3 $$(A^K)(P') \cong \prod_{Q \in (\mathfrak{C}_{\varphi,Q'})^{P'}} \big(A(Q)^{C_K(Q)}\big)(P') \qquad .$$

But, for any $Q \in (\mathfrak{C}_{\varphi,Q'})^{P'}$, $A(Q)$ is a $C_K(Q)$-basic $k \otimes_{\mathcal{O}} \mathcal{D}C_H(Q)$-interior $N_G(Q)$-algebra (cf. 13.2) and therefore, again by Lemma 13.6 and Remark 13.7, we have, up to suitable identifications, a canonical k-algebra isomorphism

13.16.4 $$\big(A(Q)^{C_K(Q)}\big)(P') \cong \prod_{P \in \mathfrak{C}_{\varphi_Q,P'}} A(Q{\cdot}P)^{C_K(Q \cdot P)} \qquad ;$$

hence, according to 13.14.2 and 13.15.3, we have $\big(A(Q)^{C_K(Q)}\big)(P') = \{0\}$ unless Q belongs to $\mathrm{Im}(\omega^{Q'}_{\varphi,P'})$ and then, by 13.14.3 and 13.15.3, isomorphism 13.16.4 becomes

13.16.5 $$\big(A(Q)^{C_K(Q)}\big)(P') \cong \prod_{P\in(\omega^{Q'}_{\varphi,P'})^{-1}(Q)} A(P)^{C_K(P)} \quad .$$

Consequently, from isomorphisms 13.16.3 and 13.16.5, we get the $\mathcal{k}$-algebra isomorphism

13.16.6 $$(A^K)(P') \cong \prod_{Q\in\mathrm{Im}(\omega^{Q'}_{\varphi,P'})} \Bigg(\prod_{P\in(\omega^{Q'}_{\varphi,P'})^{-1}(Q)} A(P)^{C_K(P)} \Bigg)$$

and it is easily checked that it coincides with the corresponding isomorphism obtained from Lemma 13.6 and Remark 13.7. In particular, if $P \in \mathfrak{C}_{\varphi,P'}$ and $Q = P\cap\varphi^{-1}(Q')$ then we have (cf. 13.8.2)

13.16.7 $$c_{\varphi,P}(A) = \big(c_{\varphi,Q}(A)\big)(P') \circ c_{\varphi_Q,P}\big(A(Q)\big) \quad ;$$

hence, according to 13.9.2, we get

13.16.8 $$\begin{aligned} e_{\varphi,P}(A) \circ d_{\psi_P}\big(A(P)\big) &= \big(d_\psi(A)\big)(P') \circ c_{\varphi,P}(A) \\ &= \Big(\big(d_\psi(A)\big)(Q')\Big)(P') \circ \big(c_{\varphi,Q}(A)\big)(P') \circ c_{\varphi_Q,P}\big(A(Q)\big) \\ &= \big(e_{\varphi,Q}(A)\big)(P') \circ \Big(d_{\psi_Q}\big(A(Q)\big)\Big)(P') \circ c_{\varphi_Q,P}\big(A(Q)\big) \\ &= \big(e_{\varphi,Q}(A)\big)(P') \circ e_{\varphi_Q,P}\big(A(Q)\big) \circ d_{\psi_P}\big(A(P)\big) \end{aligned}$$

and the commutativity of diagram 13.16.1 follows from the surjectivity of the map $d_{\psi_P}\big(A(P)\big)$ (cf. 13.4.1).

Proposition 13.17 *Let A and B be K-basic $\mathfrak{D}H$-interior G-algebras and $f : A \to B$ an H-interior $\mathfrak{D}G$-algebra homomorphism. For any p-subgroup P' of G' and any complement P of K in $\varphi^{-1}(P')$, we have the following commutative diagram*

13.17.1 $$\begin{array}{ccc} \mathrm{Ind}_{\psi_P}\big(A(P)\big) & \xrightarrow{e_{\varphi,P}(A)} & \big(\mathrm{Ind}_\psi(A)\big)(P') \\ {\scriptstyle \mathrm{Ind}_{\psi_P}(f(P))}\big\downarrow & & \big\downarrow{\scriptstyle (\mathrm{Ind}_\psi(f))(P')} \\ \mathrm{Ind}_{\psi_P}\big(B(P)\big) & \xrightarrow{e_{\varphi,P}(B)} & \big(\mathrm{Ind}_\psi(B)\big)(P') \end{array} \quad .$$

Proof: Let P' be a p-subgroup of G'; since $f(A^K) \subset B^K$, for any $P \in \mathfrak{C}_{\varphi,P'}$ we have the following commutative diagram

13.17.2
$$\begin{array}{ccc} (A^K)(P') & \longrightarrow & A(P) \\ {\scriptstyle (f^K)(P')}\big\downarrow & & \big\downarrow{\scriptstyle f(P)} \\ (B^K)(P') & \longrightarrow & B(P) \end{array} \quad ;$$

then, according to the isomorphisms obtained from Lemma 13.6 and Remark 13.7, for any $P \in \mathfrak{C}_{\varphi,P'}$ we get easily (cf. 13.8.2)

13.17.3
$$(f^K)(P') \circ c_{\varphi,P}(A) = c_{\varphi,P}(B) \circ f(P) \quad ;$$

consequently, according to 3.5.4 and 13.9.2, for any $P \in \mathfrak{C}_{\varphi,P'}$, we have

13.17.4
$$\begin{aligned} e_{\varphi,P}(B) \circ \mathrm{Ind}_{\psi_P}\big(f(P)\big) \circ d_{\psi_P}\big(A(P)\big) &= e_{\varphi,P}(B) \circ d_{\psi_P}\big(B(P)\big) \circ f(P) \\ &= \big(d_\psi(B)\big)(P') \circ c_{\varphi,P}(B) \circ f(P) \\ &= \big(d_\psi(B)\big)(P') \circ (f^K)(P') \circ c_{\varphi,P}(A) \\ &= \big(\mathrm{Ind}_\psi(f)\big)(P') \circ \big(d_\psi(A)\big)(P') \circ c_{\varphi,P}(A) \\ &= \big(\mathrm{Ind}_\psi(f)\big)(P') \circ e_{\varphi,P}(A) \circ d_{\psi_P}\big(A(P)\big) \end{aligned}$$

and the commutativity of diagram 13.17.1 follows from equality 13.4.1.

13.18 Actually, all the results above can be extended, under suitable "basic" hypothesis, to nonsurjective group homomorphisms and, on the other hand, the composition of group homomorphisms induces the composition of the corresponding embeddings 13.9.1. We need it only in Section 17 below, in the particular case where a surjective group homomorphism factorizes throughout an injective one. Thus, let F be a subgroup of G such that $G = K{\cdot}F$, set $L = H \cap F$ and $I = K \cap F$, and denote by $\zeta : F \to G'$ and $\xi : L \to H'$ the group homomorphisms determined by φ. Let B be a $\mathfrak{D}L$-interior F-algebra; it follows from 12.14 that the induced $\mathfrak{D}H$-interior algebra $\mathrm{Ind}_L^H(B)$ has a canonical F-algebra structure, so that it becomes a $\mathfrak{D}H$-interior G-algebra (see also 2.6), and we denote by g the following composed map

13.18.1
$$B^I \xrightarrow{d_L^H(B)} \mathrm{Ind}_L^H(B)^I \xrightarrow{\mathrm{Tr}_I^K} \mathrm{Ind}_L^H(B)^K$$

which, as it is easily checked, is a G'-algebra homomorphism. Similarly, if P is a p-subgroup of F, the induced $\mathfrak{D}C_H(P)$-interior algebra $\mathrm{Ind}_{C_L(P)}^{C_H(P)}\big(B(P)\big)$ admits a canonical $N_F(P)$-algebra structure, becoming a $\mathfrak{D}C_H(P)$-interior $C_H(P){\cdot}N_F(P)$-algebra and we denote by g_P the following composed map

13.18.2
$$B(P)^{C_I(P)} \xrightarrow{d_{C_L(P)}^{C_H(P)}(B(P))} \mathrm{Ind}_{C_L(P)}^{C_H(P)}\big(B(P)\big)^{C_I(P)} \xrightarrow{\mathrm{Tr}_{C_L(P)}^{C_H(P)}} \mathrm{Ind}_{C_L(P)}^{C_H(P)}\big(B(P)\big)^{C_K(P)}$$

which is actually *a* $\varphi\big(C_H(P)\cdot N_F(P)\big)$-algebra homomorphism; moreover, according to Proposition 12.12, we have a canonical $\mathfrak{D}C_H(P)$-interior algebra embedding

13.18.3 $$\mathrm{Ind}_{C_L(P)}^{C_H(P)}\big(\mathrm{Res}^L_{C_L(P)}(B)\big) \longrightarrow \mathrm{Res}^H_{C_H(P)}\big(\mathrm{Ind}_L^H(B)\big)$$

which is clearly compatible with the actions of $N_F(P)$ in both ends, so that we get a $\mathfrak{D}C_H(P)$-interior $C_H(P)\cdot N_F(P)$-algebra embedding

13.18.4 $$d_P : \mathrm{Ind}_{C_L(P)}^{C_H(P)}\big(B(P)\big) \longrightarrow \big(\mathrm{Ind}_L^H(B)\big)(P) \quad .$$

Notice that, denoting by $\xi_P : C_L(P) \to \varphi\big(C_L(P)\big)$ the group homomorphism determined by φ, by Corollary 12.7 we have canonical isomorphisms

13.18.5 $$\begin{aligned} t^* &: \mathrm{Ind}_\xi(B) \cong \mathrm{Ind}_\psi\big(\mathrm{Ind}_L^H(B)\big) \\ t_P^* &: \mathrm{Ind}_{\xi_P}\big(B(P)\big) \cong \mathrm{Ind}_{\psi_P}\Big(\mathrm{Ind}_{C_L(P)}^{C_H(P)}\big(B(P)\big)\Big) \quad . \end{aligned}$$

Proposition 13.19 *With the notation above, B is I-basic if and only if* $\mathrm{Ind}_L^H(B)$ *is K-basic. In that case, for any p-subgroup P' of G' and any complement P of I in $\zeta^{-1}(P')$, the following diagram is commutative*

13.19.1
$$\begin{array}{ccccc} (B^I)(P') & & \xrightarrow{\quad g(P') \quad} & & \big(\mathrm{Ind}_L^H(B)^K\big)(P') \\ {\scriptstyle c_{\zeta,P}(B)}\uparrow & & & & \uparrow{\scriptstyle c_{\varphi,P}(\mathrm{Ind}_L^H(B))} \\ B(P)^{C_I(P)} & \xrightarrow{g_P} & \mathrm{Ind}_{C_L(P)}^{C_H(P)}\big(B(P)\big)^{C_K(P)} & \xrightarrow{d_P} & \big(\mathrm{Ind}_L^H(B)\big)(P)^{C_K(P)} \end{array} \quad .$$

Proof: Let S be a Sylow p-subgroup of G containing a Sylow p-subgroup T of F and set $U = K \cap S$, $V = I \cap T$ and $C = \mathrm{Ind}_L^H(B)$. Since the map

13.19.2 $$d_L^H(B) : B \longrightarrow \mathrm{Res}_F^G(C)$$

is a direct injection of $\mathcal{O}(L \times L)\cdot\Delta(F)$-modules, if C is a permutation $\mathcal{O}(U \times U)\cdot\Delta(S)$-module which restricted to $\mathcal{O}(U \times U)$ is projective then B is a permutation $\mathcal{O}(V \times V)\cdot\Delta(T)$-module which restricted to $\mathcal{O}(V \times V)$ is projective too.

Conversely, assume that B has a $(V \times V)\cdot\Delta(T)$-stable $\mathcal{O}$-basis W where $V \times V$ acts freely and consider

13.19.3 $$\dot{B} = \mathrm{Ind}_{T\cap L}^{I\cdot(T\cap L)}\big(\mathrm{Res}_T^F(B)\big)$$

which, by 12.14, has a $I\cdot T$-algebra structure (see also 2.6); then, the set $\dot{W} = I \otimes W \otimes I$ in $\dot{B}$ is a $(I \times I)\cdot\Delta(T)$-stable $\mathcal{O}$-basis of $\dot{B}$ where $I \times I$ acts freely. Similarly, setting

13.19.4 $$\dot{C} = \mathrm{Ind}_{I\cdot(T\cap L)}^{K\cdot(S\cap H)}(\dot{B}) \quad ,$$

$\dot{C}$ is a $K{\cdot}S$-algebra (cf. 2.6) and, since $K{\cdot}S = K{\cdot}T$, the set $K \otimes \dot{W} \otimes K$ in $\dot{C}$ is a $(K \times K){\cdot}\Delta(S)$-stable $\mathcal{O}$-basis of $\dot{C}$ where $K \times K$ acts freely. On the other hand, notice that

13.19.5
$$\begin{aligned}\dot{C} &\cong \mathrm{Ind}_{T\cap L}^{K\cdot(S\cap H)}\big(\mathrm{Res}_T^F(B)\big) \cong \mathrm{Ind}_{S\cap H}^{K\cdot(S\cap H)}\Big(\mathrm{Ind}_{T\cap L}^{S\cap H}\big(\mathrm{Res}_T^F(B)\big)\Big)\\ &\cong \mathrm{Ind}_{S\cap H}^{K\cdot(S\cap H)}\Big(\mathrm{Res}_S^G\big(\mathrm{Ind}_L^H(B)\big)\Big) \cong \mathrm{Ind}_{S\cap H}^{K\cdot(S\cap H)}\big(\mathrm{Res}_S^G(C)\big) \quad ;\end{aligned}$$

in particular, considering C and $\dot{C}$ as $\mathcal{O}(K \times K){\cdot}\Delta(S)$-modules, $\dot{C}$ is just the induced module, from $\mathcal{O}(U \times U){\cdot}\Delta(S)$ to $\mathcal{O}(K \times K){\cdot}\Delta(S)$, of the corresponding restriction of C; thus, since $(U \times U){\cdot}\Delta(S)$ is a Sylow p-subgroup of $(K \times K){\cdot}\Delta(S)$, C is a direct summand of $\dot{C}$ (cf. [19, Lemma 3.8]). Consequently, C is a permutation $\mathcal{O}(U \times U){\cdot}\Delta(S)$-module which restricted to $\mathcal{O}(U \times U)$ is projective.

Let P' be a subgroup of $\varphi(S)$; for any complement P of I in $\zeta^{-1}(P')$, choose a set of representatives for $C_K(P)\backslash K/I$ in K and consider the subset Y_P of elements y such that $[y, P] \subset I$; then, for any $y \in Y_P$, denote by $c_P^y : C(P^y) \cong C(P)$ the isomorphism determined by conjugation by y, by

13.19.6
$$r_P^y : \prod_{x\in Y_P} B(P^x) \longrightarrow B(P^y)$$

the corresponding projection map, and by f_P^y the composed $\Bbbk$-algebra homomorphism

13.19.7
$$\prod_{x\in Y_P} B(P^x)^{C_I(P^x)} \xrightarrow{r_P^y} B(P^y)^{C_I(P^y)} \xrightarrow{d_{P^y}\circ g_{P^y}} C(P^y)^{C_K(P^y)} \xrightarrow{c_P^y} C(P)^{C_K(P)} \quad ;$$

it is not difficult to check that the idempotents $f_P^y(1)$, where y runs on Y_P, are pairwise orthogonal and therefore the sum

13.19.8
$$f_P = \sum_{y\in Y_P} f_P^y : \prod_{y\in Y_P} B(P^y)^{C_I(P^y)} \longrightarrow C(P)^{C_K(P)}$$

is a $\Bbbk$-algebra homomorphism too. Now, we claim that the following diagram determined by the inclusions $B^I \subset B$ and $C^K \subset C$ is commutative

13.19.9
$$\begin{array}{ccc} (B^I)(P') & \xrightarrow{g(P')} & (C^K)(P') \\ \big\downarrow & & \big\downarrow \\ \prod_{y\in Y_P} B(P^y)^{C_I(P^y)} & \xrightarrow{f_P} & C(P)^{C_K(P)} \end{array} \quad .$$

Indeed, respectively denote by $\dot{g}$, $\dot{g}_P$, $\dot{d}_P$, $\dot{c}_P^y$, $\dot{r}_P^y$, $\dot{f}_P^y$ and $\dot{f}_P$ the maps obtained by replacing B, C, H and L by $\dot{B}, \dot{C}, K{\cdot}(S \cap H) = \dot{H}$ and $I{\cdot}(T \cap L) = \dot{L}$ in

13.18.1, 13.18.2, 13.18.3, 13.19.6, 13.19.7 and 13.19.8, and consider the canonical $\mathcal{O}(I \times I)\cdot\Delta(T)$- and $\mathcal{O}(K \times K)\cdot\Delta(S)$-module homomorphisms

13.19.10
$$m : \dot{B} \longrightarrow B \quad \text{and} \quad n : \dot{C} \longrightarrow C$$

mapping $1 \otimes b \otimes 1$ on $b \in B$ and $1 \otimes c \otimes 1$ on $c \in C$; similarly, setting

13.19.11
$$C_P = \mathrm{Ind}_{C_L(P)}^{C_H(P)}\big(B(P)\big) \quad \text{and} \quad \dot{C}_P = \mathrm{Ind}_{C_{\dot{L}}(P)}^{C_{\dot{H}}(P)}\big(\dot{B}(P)\big) \quad ,$$

$m(P)$ induces a $\mathcal{O}\big(C_K(P) \times C_K(P)\big)\cdot\Delta\big(N_S(P)\big)$-module homomorphism

13.19.12
$$n_P : \dot{C}_P \longrightarrow C_P$$

mapping $1 \otimes \mathrm{Br}_P^{\dot{B}}(\dot{b}) \otimes 1$ on $1 \otimes \mathrm{Br}_P^{B}\big(m(\dot{b})\big) \otimes 1$ for any $\dot{b} \in \dot{B}^P$. Now, it is not difficult to check that the following diagrams are commutative

13.19.13
$$\begin{array}{ccccccccc} B & \xrightarrow{d_L^H(B)} & C & \quad & B(P) & \xrightarrow{d_{C_L(P)}^{C_H(P)}(B(P))} & C_P & \xrightarrow{d_P} & C(P) \\ {\scriptstyle m}\uparrow & & \uparrow{\scriptstyle n} & & \uparrow{\scriptstyle m(P)} & & \uparrow{\scriptstyle n_P} & & \uparrow{\scriptstyle n(P)} \\ \dot{B} & \xrightarrow{d_{\dot{L}}^{\dot{H}}(\dot{B})} & \dot{C} & & \dot{B}(P) & \xrightarrow{d_{C_{\dot{L}}(P)}^{C_{\dot{H}}(P)}(\dot{B}(P))} & \dot{C}_P & \xrightarrow{\dot{d}_P} & \dot{C}(P) \end{array}$$

and from them we get easily the following new commutative diagrams

13.19.14
$$\begin{array}{ccc} B^I & \xrightarrow{g} & C^K \\ {\scriptstyle m}\uparrow & & \uparrow{\scriptstyle n} \\ \dot{B}^I & \xrightarrow{\dot{g}} & \dot{C}^K \end{array} \quad \text{and} \quad \begin{array}{ccc} \prod_{y\in Y_P} B(P^y)^{C_I(P^y)} & \xrightarrow{f_P} & C(P)^{C_K(P)} \\ {\scriptstyle \prod_{y\in Y_P} m(P^y)}\uparrow & & \uparrow{\scriptstyle n(P)} \\ \prod_{y\in Y_P} \dot{B}(P^y)^{C_I(P^y)} & \xrightarrow{\dot{f}_P} & \dot{C}(P)^{C_K(P)} \end{array} \quad .$$

Moreover, employing again the fact that $(U \times U)\cdot\Delta(S)$ is a Sylow p-subgroup of $(K \times K)\cdot\Delta(S)$, we get that n admits an $\mathcal{O}(K \times K)\cdot\Delta(S)$-linear section; similarly, m admits an $\mathcal{O}(I \times I)\cdot\Delta(T)$-linear section (see, for instance, the proof of Lemma 3.8 in [19]). It follows that all the vertical arrows in diagrams 13.19.14 are surjective maps; hence, as it is easily checked, in order to prove the commutativity of diagram 13.19.9, we may respectively replace B, C, g and f_P by $\dot{B}$, $\dot{C}$, $\dot{g}$ and $\dot{f}_P$; in that case, since $\dot{W}$ is a $(I \times I)\cdot\Delta(T)$-stable $\mathcal{O}$-basis of $\dot{B}$ where $I \times I$ acts freely, $\mathrm{Tr}_1^I(\dot{W})$ is a P'-stable $\mathcal{O}$-basis of $\dot{B}^I$ and it is quite clear that the disjoint union

13.19.15
$$\bigcup_{Q\in\mathfrak{C}_{\zeta,P'}} \mathrm{Br}_{P'}^{\dot{B}^I}\big(\mathrm{Tr}_1^I(\dot{W}^Q)\big)$$

is a $\mathscr{k}$-basis of $(\dot{B}^I)(P')$; moreover, we may assume that $\{P^y\}_{y\in Y_P} \subset \mathfrak{C}_{\zeta,P'}$ and then, for any $Q \in \mathfrak{C}_{\zeta,P'}$, any $\dot{w} \in \dot{W}^Q$ and any $y \in Y_P$, we have

$$\text{13.19.16} \qquad \mathrm{Br}^{\dot{B}}_{P^y}\big(\mathrm{Tr}^I_1(\dot{w})\big) = \begin{cases} \mathrm{Tr}_1^{C_I(P^y)}\big(Br^{\dot{B}}_{P^y}(\dot{w})\big) & \text{if } Q = P^y \\ 0 & \text{otherwise} \end{cases}$$

since P^y fixes $\dot{w}^x$, where $x \in I$, if and only if P^y fixes $\dot{w}$ and x; on the other hand, we have (cf. 13.18.1)

$$\text{13.19.17} \qquad \big(\dot{g}(P')\big)\Big(\mathrm{Br}^{\dot{B}^I}_{P'}\big(\mathrm{Tr}^I_1(\dot{w})\big)\Big) = \mathrm{Br}^{\dot{C}^K}_{P'}\big(\mathrm{Tr}^K_1(1 \otimes \dot{w} \otimes 1)\big)$$

and, as above, if $Q = P^y$ for some $y \in Y_P$ then we get (see [8, Lemma 1.12])

$$\text{13.19.18} \qquad \mathrm{Br}^{\dot{C}}_P\big(\mathrm{Tr}^K_1(1 \otimes \dot{w} \otimes 1)\big) = \mathrm{Tr}_1^{C_K(P)}\big(\mathrm{Br}^{\dot{C}}_P(y \otimes \dot{w} \otimes y^{-1})\big)$$

whereas $\mathrm{Br}^{\dot{C}}_P\big(\mathrm{Tr}^K_1(1 \otimes \dot{w} \otimes 1)\big) = 0$ otherwise; finally, for any $y \in Y_P$ and any $\dot{w} \in \dot{W}^{P^Y}$, it is clear that

$$\text{13.19.19} \qquad \dot{c}^y_P\bigg(\dot{d}_{P^y}\Big(\dot{g}_{P^y}\Big(\mathrm{Tr}_1^{C_I(P^y)}\big(\mathrm{Br}^{\dot{B}}_{P^y}(\dot{w})\big)\Big)\Big)\bigg) = \mathrm{Tr}_1^{C_K(P)}\big(\mathrm{Br}^{\dot{C}}_P(y \otimes \dot{w} \otimes y^{-1})\big)$$

which achieves the proof of our claim.

Furthermore, any complement Q of K in $\varphi^{-1}(P')$ such that $\dot{C}(Q) \neq \{0\}$ has a K-conjugate which is contained in $I.T$ (cf. 13.19.4 and 2.11.3) and therefore is a complement of I in $\zeta^{-1}(P')$; consequently, we may assume that for a suitable $\mathfrak{C} \subset \mathfrak{C}_{\varphi,P'}$ we have $\dot{C}(Q) = \{0\}$, for any $Q \in \mathfrak{C}_{\varphi,P'} - \mathfrak{C}$, and the equality (where the union in the second member is disjoint)

$$\text{13.19.20} \qquad \mathfrak{C}_{\zeta,P'} = \bigcup_{P\in\mathfrak{C}} \{P^y\}_{y\in Y_P} \quad ;$$

then, according to Lemma 13.6 and to the commutativity of diagram 13.19.9, we get the following commutative diagram

$$\text{13.19.21} \qquad \begin{array}{ccc} (B^I)(P') & \xrightarrow{g(P')} & (C^K)(P') \\ \wr\| & & \wr\| \\ \prod_{P\in\mathfrak{C}_{\zeta,P'}} B(P)^{C_I(P)} & \xrightarrow{f_{P'}} & \prod_{P\in\mathfrak{C}} C(P)^{C_K(P)} \end{array}$$

where $f_{P'} = \prod_{P\in\mathfrak{C}} f_P$, which immediately implies the announced commutativity of diagram 13.19.1.

Corollary 13.20 *With the notation above, assume that B is I-basic. For any p-subgroup P' of G' and any complement P of I in $\zeta^{-1}(P')$, the following diagram is commutative*

13.20.1

$$\begin{array}{ccc} \big(\mathrm{Ind}_\zeta(B)\big)(P') & \overset{t^*(P')}{\cong} & \Big(\mathrm{Ind}_\psi\big(\mathrm{Ind}_L^H(B)\big)\Big)(P') \\ \uparrow{\scriptstyle e_{\zeta,P}(B)} & & \uparrow{\scriptstyle e_{\varphi,P}(\mathrm{Ind}_L^H(B))} \\ \mathrm{Ind}_{\xi_P}\big(B(P)\big) \overset{t_P^*}{\cong} \mathrm{Ind}_{\psi_P}\Big(\mathrm{Ind}_{C_L(P)}^{C_H(P)}\big(B(P)\big)\Big) & \xrightarrow{\mathrm{Ind}_{\psi_P}(d_P)} & \mathrm{Ind}_{\psi_P}\Big(\big(\mathrm{Ind}_L^H(B)\big)(P)\Big) \end{array} .$$

Proof: By Proposition 13.19, $\mathrm{Ind}_L^H(B)$ is K-basic and we may apply Theorem 13.9 to both B and $\mathrm{Ind}_L^H(B)$. Thus, by 13.9.2, we have

13.20.2 $$t^*(P') \circ e_{\zeta,P}(B) \circ d_{\xi_P}\big(B(P)\big) = \big(t^* \circ d_\xi(B)\big)(P') \circ c_{\zeta,P}(B) \quad ;$$

but, it follows from Corollary 12.7 and 13.18.1 that

13.20.3 $$t^* \circ d_\xi(B) = d_\psi\big(\mathrm{Ind}_L^H(B)\big) \circ g \quad ;$$

then according to the commutative diagram 13.19.1 and successively applying 13.9.2 to $\mathrm{Ind}_L^H(B)$ and 3.5.4 to d_P, we get

13.20.4 $$\begin{aligned} t^*(P') \circ e_{\zeta,P}(B) \circ d_{\xi_P}\big(B(P)\big) & \\ &= \Big(d_\psi\big(\mathrm{Ind}_L^H(B)\big)\Big)(P') \circ c_{\varphi,P}\big(\mathrm{Ind}_L^H(B)\big) \circ d_P \circ g_P \\ &= e_{\varphi,P}\big(\mathrm{Ind}_L^H(B)\big) \circ d_{\psi_P}\Big(\big(\mathrm{Ind}_L^H(B)\big)(P)\Big) \circ d_P \circ g_P \\ &= e_{\varphi,P}\big(\mathrm{Ind}_L^H(B)\big) \circ \mathrm{Ind}_{\psi_P}(d_P) \circ d_{\psi_P}\Big(\mathrm{Ind}_{C_L(P)}^{C_H(P)}\big(B(P)\big)\Big) \circ g_P \quad ; \end{aligned}$$

moreover, *mutatis mutandis* equality 13.20.3 becomes

13.20.5 $$t_P^* \circ d_{\xi_P}\big(B(P)\big) = d_{\psi_P}\Big(\mathrm{Ind}_{C_L(P)}^{C_H(P)}\big(B(P)\big)\Big) \circ g_P \quad ;$$

finally, we get

13.20.6
$$t^*(P') \circ e_{\zeta,P}(B) \circ d_{\xi_P}\big(B(P)\big) = e_{\varphi,P}\big(\mathrm{Ind}_L^H(B)\big) \circ \mathrm{Ind}_{\psi_P}(d_P) \circ t_P^* \circ d_{\xi_P}\big(B(P)\big)$$

and it suffices to apply equality 13.4.1 to obtain the announced commutativity of diagram 13.20.1.

14 Pointed groups on $\mathcal{D}G$-interior algebras and Higman embeddings

14.1 In order to develop for the *Rickard equivalences* between blocks the same analysis that we do for the Morita ones in Sections 6 and 7 above, we need to extend to $\mathcal{D}G$-interior algebras the notion of *pointed group* on an $\mathcal{O}G$-interior algebra (cf. 2.7). Although this extension can be expressed in a few words - the pointed groups on a $\mathcal{D}G$-interior algebra A are just the usual pointed groups on the $\mathcal{O}G$-interior algebra $\mathsf{C}_\circ(A)$ - in this section we develop the relationship between pointed groups and induction further than it has been done for $\mathcal{O}G$-interior algebras, in particular by giving a description of the *local structure* of induced $\mathcal{D}G$-interior algebras; moreover, since Rickard equivalences between blocks concern the homotopy categories of the blocks (cf. 10.7), we have to introduce the *contractile pointed groups* which are a specific notion for $\mathcal{D}G$-interior algebras. Actually this extension can be more generally carried out for $\mathcal{D}H$-interior G-algebras where H is a normal subgroup of G (cf. 11.4), except that it requires harder notation and, as a matter of fact, most of the definitions concern only $\mathcal{D}G$-interior algebras; the interested reader will easily fill in this lack of generality.

14.2 Let G be a finite group and A a $\mathcal{D}G$-interior algebra; a *pointed group* H_β on A is an ordinary pointed group on the $\mathcal{O}G$-interior algebra $\mathsf{C}_\circ(A)$; that is to say, H is a subgroup of G and β a conjugacy class of primitive idempotents of the $\mathcal{O}$-algebra $\mathsf{C}_\circ(A)^H = \mathsf{C}_\circ(A^H)$, called *point* of H on $\mathsf{C}_\circ(A)$ or A; then, for any $j \in \beta$, the $\mathcal{O}$-subalgebra jAj endowed with the $\mathcal{O}$-algebra homomorphism $\mathcal{D}H \to jAj$ mapping $c \in \mathcal{D}H$ on $c\cdot j = j\cdot c$ (cf. 11.3.2) is a $\mathcal{D}H$-interior algebra and it is quite clear that all these $\mathcal{D}H$-interior algebras, when j runs over β, determine a *unique* $\mathcal{D}H$-interior subalgebra of $\mathrm{Res}_H^G(A)$ in the category of $\mathcal{D}H$-interior algebras and $\mathcal{D}H$-interior algebra *exomorphisms* (cf. 11.8 and Proposition 11.9); that is to say, setting $A_\beta = jAj$ and denoting by

14.2.1 $$\tilde{f}_\beta : A_\beta \longrightarrow \mathrm{Res}_H^G(A)$$

the $\mathcal{D}H$-interior algebra exoembedding determined by the inclusion $jAj \subset A$, the pair $(A_\beta, \tilde{f}_\beta)$ is unique up to unique $\mathcal{D}H$-interior algebra exoisomorphisms and we call it the *embedded algebra* of H_β; notice that $\mathsf{C}_\circ(A_\beta) \cong \mathsf{C}_\circ(A)_\beta$ (cf. 2.7). On the other hand, we say that H_β or β is *contractile* if $\beta \subset \mathsf{B}_\circ(A^H)$ or, equivalently, if A_β is a contractile $\mathcal{D}H$-interior algebra (cf. 11.4), namely if

14.2.2 $$\mathsf{H}_\circ(A_\beta^H) = \{0\} \quad ;$$

notice that:

14.2.3 *The set of noncontractile points of H on A corresponds bijectively with the set of points of the $\mathcal{O}$-algebra* $\mathsf{H}_\circ(A^H)$.

14.3 If H_β and L_ε are pointed groups on A, we say that L_ε *is contained in* H_β, and write $L_\varepsilon \subset H_\beta$, if this is true as pointed groups on $\mathsf{C}_\circ(A)$ or, equivalently, if $L \subset H$ and there is a $\mathcal{D}G$-interior algebra exoembedding

14.3.1
$$\tilde{f}_\varepsilon^\beta : A_\varepsilon \longrightarrow \mathrm{Res}_L^H(A_\beta)$$

such that we have

14.3.2
$$\mathrm{Res}_L^H(\tilde{f}_\beta) \circ \tilde{f}_\varepsilon^\beta = \tilde{f}_\varepsilon \quad ;$$

in this case, by Proposition 11.10, $\tilde{f}_\varepsilon^\beta$ is uniquely determined by equality 14.3.2. More generally, we say that a group exomorphism $\tilde{\varphi} : L \to H$ is an *A-fusion from* L_ε *to* H_β if it is an ordinary $\mathsf{C}_\circ(A)$-fusion from L_ε to H_β on $\mathsf{C}_\circ(A)$ (cf. 2.13), namely if, choosing $\varphi \in \tilde{\varphi}$, $j \in \beta$ and $l \in \varepsilon$, there is $c \in \mathsf{C}_\circ(A)^*$ such that $j^c l = l = l j^c$ and $cl{\cdot}y = \varphi(y){\cdot}cl$ for any $y \in L$; in that case, it is clear that $\mathrm{int}_A(c)$ induces a unique $\mathcal{D}L$-interior algebra exoembedding

14.3.3
$$\tilde{f}_\varphi : A_\varepsilon \longrightarrow \mathrm{Res}_\varphi(A_\beta)$$

such that we have

14.3.4
$$\mathrm{Res}_1^H(\tilde{f}_\beta) \circ \mathrm{Res}_1^L(\tilde{f}_\varphi) = \mathrm{Res}_1^L(\tilde{f}_\varepsilon)$$

(the uniqueness follows from Propositions 11.9 and 11.10); notice that:

14.3.5 *If* $F_{\mathsf{C}_\circ(A)}(L_\varepsilon, H_\beta) \neq \emptyset$ *and* H_β *is contractile then* L_ε *is contractile too.*

14.4 Let A' be a second $\mathcal{D}G$-interior algebra and $\tilde{f} : A \to A'$ a G-interior $\mathcal{D}$-algebra exomorphism (cf. 11.6); by 11.8.3, $\tilde{f}$ induces by restriction an $\mathcal{O}G$-interior algebra exomorphism

14.4.1
$$\mathsf{C}_\circ(\tilde{f}) : \mathsf{C}_\circ(A) \longrightarrow \mathsf{C}_\circ(A')$$

and, in particular, if H_β and $H_{\beta'}$ are respectively pointed groups on A and on A' then it makes sense to consider the usual *multiplicity* of $\mathsf{C}_\circ(\tilde{f})$ at (β, β') (cf. 2.11.1), namely the number of elements of β' in a pairwise orthogonal primitive idempotent decomposition of $f(j)$ in $\mathsf{C}_\circ(A')^H$ for any f in $\tilde{f}$ and any j in β, which we denote here by $m(\tilde{f})_\beta^{\beta'}$; notice that:

14.4.2 *If* β *is contractile and* $m(\tilde{f})_\beta^{\beta'} \neq 0$ *then* β' *is contractile too.*

On the other hand, if $\tilde{f}$ is an exoembedding then $\mathsf{C}_\circ(\tilde{f})$ is still an exoembedding of $\mathcal{O}G$-interior algebras, so that, for any pointed group H_β on A, $\tilde{f}(\beta)$ is contained in a point β' of H on A' (cf. 2.11.2) and this correspondence is a bijective map, which preserves inclusion and contractility, between the set of pointed groups on A and the set of pointed groups $H_{\beta'}$ on A' such that $\beta' \cap f(1)A'f(1) \neq \emptyset$ for some f in $\tilde{f}$ (cf. 2.11.2). But, if $\tilde{f}$ is only a *homotopy exoembedding* (cf. 11.7) then, by 11.7.2, $\mathsf{H}_\circ(\tilde{f}^H)$ is still an exoembedding of $\mathcal{O}G$-interior algebras for any subgroup H of G; hence, for any *noncontractile* pointed group H_β on A, there is a unique *noncontractile* point β' of H on A' such that $m(\tilde{f})_\beta^{\beta'} \neq 0$ and this correspondence is also a bijective map, which preserves inclusion, between the set of noncontractile pointed groups on A and the set of noncontractile pointed groups $H_{\beta'}$ on A' such that $(\beta' + \mathsf{B}_\circ(A'^H)) \cap f(1)A'f(1) \neq \emptyset$ for some f in $\tilde{f}$.

14.5 On the same spirit, a pointed group P_γ on A is *local* if it is local as a pointed group on $\mathsf{C}_\circ(A)$(cf. 2.9), namely if $\mathrm{Br}_P^{\mathsf{C}_\circ(A)}(\gamma) \neq |0|$; that is to say, the set of *local points* of P on A corresponds bijectively with the set of points of the k-algebra $(\mathsf{C}_\circ(A))(P)$ which *need not coincide* with $\mathsf{C}_\circ(A(P))$ (see Remark 11.5 above); now, it follows easily from 14.2.3 above that the set of *noncontractile local points* of P on A corresponds bijectively with the set of points of the k-algebra

14.5.1 $$k \underset{\mathcal{O}}{\otimes} \Big(\mathsf{H}_\circ(A^P)/\sum_Q \mathrm{Im}(\mathsf{H}_\circ(\mathrm{Tr}_Q^P))\Big)$$

where Q runs on the set of proper subgroups of P and, for such a Q, the trace map $\mathrm{Tr}_Q^P : A^Q \to A^P$ is $\mathcal{D}$-linear, so that $\mathsf{H}_\circ(\mathrm{Tr}_Q^P)$ makes sense (cf. 10.6.3); in particular, notice that (cf. 11.7):

14.5.2 *The G-interior D-algebra homotopy exoembeddings preserve localness on noncontractile pointed groups.*

As usual, we call the local pointed groups which are maximal contained in a pointed group H_β on A *defect pointed groups* of H_β, and H acts transitively on the set of them (cf. 2.9); moreover, a local pointed group P_γ on A contained in H_β is maximal if and only if it fulfills (cf. 2.9.4 and 2.9.5)

14.5.3 $$\beta \subset \mathrm{Tr}_P^H(\mathsf{C}_\circ(A^P)\cdot\gamma\cdot\mathsf{C}_\circ(A^P)) \qquad ;$$

in that case, notice that:

14.5.4 *P_γ is contractile if and only if H_β is contractile.*

14.6 Recall that condition 14.5.3 is just Higman's criterion on relative projectivity for pointed groups on the $\mathcal{O}G$-interior algebra $\mathsf{C}_\circ(A)$; explicitely, if H_β and L_ε are pointed groups on A, so on $\mathsf{C}_\circ(A)$, such that

14.6.1 $$L_\varepsilon \subset H_\beta \quad \text{and} \quad \beta \subset \mathrm{Tr}_L^H(\mathsf{C}_\circ(A^L)\cdot\varepsilon\cdot\mathsf{C}_\circ(A^L))$$

then there is a unique $\mathcal{O}H$-interior algebra exoembedding (cf. 2.12.1)

14.6.2 $$\tilde{h}_\beta^\varepsilon : \mathsf{C}_\circ(A)_\beta \longrightarrow \mathrm{Ind}_L^H(\mathsf{C}_\circ(A)_\varepsilon)$$

such that we have

14.6.3 $$\mathrm{Res}_L^H(\tilde{h}_\beta^\varepsilon) \circ \tilde{f}_\varepsilon^\beta(\mathsf{C}_\circ(A)) = \tilde{d}_L^H(\mathsf{C}_\circ(A)_\varepsilon)$$

(we write $\tilde{f}_\varepsilon^\beta(\mathrm{C}_\circ(A))$ to emphasize that it is exoembedding 2.7.2 for $\mathrm{C}_\circ(A)$, which coincides actually with $\mathrm{C}_\circ(\tilde{f}_\varepsilon^\beta)$ when $\tilde{f}_\varepsilon^\beta$ concerns A); as it could be expected, in our present situation $\tilde{h}_\beta^\varepsilon$ can be extended to a unique $\mathcal{D}H$-interior algebra exoembedding from A_β to $\mathrm{Ind}_L^H(A_\varepsilon)$ fulfilling the analogous condition. However, in that form, this statement has the inconvenience that, when H has more than one point on A, there is no evident connection between $\mathrm{Ind}_L^H(A_\varepsilon)$ and A; but, as a matter of fact, *there is* such a connection, namely a new $\mathcal{D}G$-interior algebra - we call it the *Higman envelope* of A - which contains all these situations. As we show below, the Higman envelope plays an important role in the description of the local structure of induced and subsequent Hecke $\mathcal{D}G$-interior algebras.

Proposition 14.7 *There are a $\mathcal{D}G$-interior algebra $\mathsf{E}(A)$ and a $\mathcal{D}G$-interior algebra exoembedding $\tilde{e}(A) : A \to \mathsf{E}(A)$, unique up to unique $\mathcal{D}G$-interior algebra exoisomorphisms, fulfilling the following conditions*

14.7.1 For any pointed group H_β on A there is a G-interior $\mathcal{D}$-algebra exoembedding

$$\tilde{h}_\beta : \mathrm{Ind}_H^G(A_\beta) \longrightarrow \mathsf{E}(A)$$

such that we have $\mathrm{Res}_H^G(\tilde{h}_\beta) \circ \tilde{d}_H^G(A_\beta) = \mathrm{Res}_H^G\big(\tilde{e}(A)\big) \circ \tilde{f}_\beta$.

14.7.2 If A' is a $\mathcal{D}G$-interior algebra and $\tilde{f} : A \to A'$ a G-interior $\mathcal{D}$-algebra exoembedding fulfilling condition 14.7.1 then there is a unique G-interior $\mathcal{D}$-algebra exoembedding $\tilde{g} : \mathsf{E}(A) \to A'$ such that $\tilde{g} \circ \tilde{e}(A) = \tilde{f}$.

Moreover $\tilde{h}_\beta$ is a $\mathcal{D}G$-interior algebra exoembedding which is unique fulfilling that equality, $\tilde{g}$ is a $\mathcal{D}G$-interior algebra exoembedding whenever $\tilde{f}$ is so, any local pointed group on $\mathsf{E}(A)$ comes from A, and if A is K-basic for some normal subgroup K of G then $\mathsf{E}(A)$ is K-basic too.

Proof: For any subgroup H of G denote by $\mathcal{O}_H$ the trivial $\mathcal{D}H$-module where $\mathfrak{f} \in \mathcal{F}$ acts by multiplication by $\mathfrak{f}(0)$ and d annihilates it, and set

$$E = \mathrm{End}_{\mathcal{O}}\left(\bigoplus_H \mathrm{Ind}_H^G(\mathcal{O}_H)\right) \qquad , \tag{14.7.3}$$

where H runs on the set of subgroups of G; in particular, we have an evident $\mathcal{D}G$-interior algebra exoembedding

$$\mathcal{O} \longrightarrow E \tag{14.7.4}$$

which is unique by Proposition 11.9 above and Proposition 2.3 in [19], and determines by tensor product a $\mathcal{D}G$-interior algebra exoembedding

$$\tilde{e} : A \longrightarrow E \underset{\mathcal{O}}{\otimes} A \qquad ; \tag{14.7.5}$$

more generally, if H_β is a pointed group on A, denoting by $\tilde{h}_\beta$ the composed $\mathcal{D}G$-interior algebra exoembedding (cf. Proposition 12.9)

14.7.6 $$\mathrm{Ind}_H^G(A_\beta) \xrightarrow{\mathrm{Ind}_H^G(\tilde{f}_\beta)} \mathrm{Ind}_H^G\big(\mathrm{Res}_H^G(A)\big) \cong \mathrm{Ind}_H^G(\mathcal{O}) \underset{\mathcal{O}}{\otimes} A \longrightarrow E \underset{\mathcal{O}}{\otimes} A \quad ,$$

the uniqueness of exoembedding 14.7.4 restricted to H and 3.5.4 imply that

14.7.7 $$\mathrm{Res}_H^G(\tilde{h}_\beta) \circ \tilde{d}_H^G(A_\beta) = \mathrm{Res}_H^G(\tilde{e}) \circ \tilde{f}_\beta \quad .$$

Now, for any point α of G on $E \otimes_{\mathcal{O}} A$, denote by n_α the smallest nonnegative integer such that $n_\alpha \geq m_\alpha(A)$ if $\tilde{e}^{-1}(\alpha) \neq \emptyset$ and $n_\alpha \geq m_\alpha\big(\mathrm{Ind}_H^G(A_\beta)\big)$ for any pointed group H_β on A such that $\tilde{h}_\beta^{-1}(\alpha) \neq \emptyset$ (cf. 2.8 and 2.11), and consider an idempotent i in $\mathsf{C}_\circ(E \otimes_{\mathcal{O}} A)^G$ having multiplicity n_α at any point α of G on $E \otimes_{\mathcal{O}} A$ (which is possible since the existence of $\tilde{e}$ and all the $\tilde{h}_\beta$ implies that we have $m_\alpha(E \otimes_{\mathcal{O}} A) \geq n_\alpha$ for such an α). Then we set

14.7.8 $$\mathsf{E}(A) = i(E \underset{\mathcal{O}}{\otimes} A)i$$

and it is clear that $\tilde{e}$ and all the $\tilde{h}_\beta$ factorize throughout the exoembedding determined by the inclusion $i(E \otimes_{\mathcal{O}} A)i \subset E \otimes_{\mathcal{O}} A$; by Proposition 11.10 these factorizations are uniquely determined and, denoting them by the same letters, fulfill equalities 14.7.7.

On the other hand, if P is a p-subgroup of G, it is quite clear that the trivial $\mathcal{O}P$-module $\mathcal{O}_P$ determines the unique local point of P on E (cf. 2.10) and actually $E(P)$ is a matrix algebra over k (cf. 2.2.5); moreover, since E admits a P-stable $\mathcal{O}$-basis by conjugation, by Lemma 7.10 we have k-algebra isomorphisms (cf. 11.3.2)

14.7.9 $$\big(\mathsf{C}_\circ(E \underset{\mathcal{O}}{\otimes} A)\big)(P) \cong \big(E \underset{\mathcal{O}}{\otimes} \mathsf{C}_\circ(A)\big)(P) \cong E(P) \underset{k}{\otimes} \big(\mathsf{C}_\circ(A)\big)(P) \quad ;$$

consequently, any local point of P on $E \otimes_{\mathcal{O}} A$, and a *fortiori* on $\mathsf{E}(A)$, comes from A throughout $\tilde{e}$ (cf. 2.11.2).

Finally, let A' be a $\mathcal{D}G$-interior algebra and $\tilde{f} : A \to A'$ a G-interior $\mathcal{D}$-algebra exoembedding fulfilling condition 14.7.1; by tensoring $\tilde{f}$ and embedding 14.7.4, we get the following commutative diagram of G-interior $\mathcal{D}$-algebra exoembeddings which are $\mathcal{D}G$-interior algebra exoembeddings if $\tilde{f}$ is so

14.7.10 $$\begin{array}{ccc} A & \xrightarrow{\tilde{e}} & E \underset{\mathcal{O}}{\otimes} A \\ {\scriptstyle \tilde{f}}\big\downarrow & & \big\downarrow{\scriptstyle \widetilde{\mathrm{id}}_E \otimes \tilde{f}} \\ A' & \xrightarrow{\tilde{e}'} & E \underset{\mathcal{O}}{\otimes} A' \end{array} \quad ;$$

then, for any pointed group H_β on A, denoting by

14.7.11 $$\tilde{h}'_\beta : \mathrm{Ind}_H^G(A_\beta) \longrightarrow A'$$

a G-interior $\mathcal{D}$-algebra such that

14.7.12 $$\mathrm{Res}^G_H(\tilde{h}'_\beta) \circ \tilde{d}^G_H(A_\beta) = \mathrm{Res}^G_H(\tilde{f}) \circ \tilde{f}_\beta \quad ,$$

we claim that

14.7.13 $$(\widetilde{\mathrm{id}}_E \underset{\mathcal{O}}{\otimes} \tilde{f}) \circ \tilde{h}_\beta = \tilde{e}' \circ \tilde{h}'_\beta \quad .$$

Indeed, from 14.7.7 and 14.7.13, we get

14.7.14
$$\begin{aligned}\mathrm{Res}^G_H(\widetilde{\mathrm{id}}_E \underset{\mathcal{O}}{\otimes} \tilde{f}) \circ \mathrm{Res}^G_H(\tilde{h}_\beta) \circ \tilde{d}^G_H(A_\beta) &= \mathrm{Res}^G_H(\widetilde{\mathrm{id}}_E \underset{\mathcal{O}}{\otimes} \tilde{f}) \circ \mathrm{Res}^G_H(\tilde{e}) \circ \tilde{f}_\beta \\ &= \mathrm{Res}^G_H(\tilde{e}') \circ \mathrm{Res}^G_H(\tilde{f}) \circ \tilde{f}_\beta \\ &= \mathrm{Res}^G_H(\tilde{e}') \circ \mathrm{Res}^G_H(\tilde{h}'_\beta) \circ \tilde{d}^G_H(A_\beta) ;\end{aligned}$$

but, by 12.4.4, we have

14.7.15 $$\mathsf{C}_\circ(\mathrm{Ind}^G_H(A_\beta)) \cong \mathrm{Ind}^G_H(\mathsf{C}_\circ(A_\beta))$$

and therefore the number of points of the trivial subgroup 1 on A_β and on $\mathrm{Ind}^G_H(A_\beta)$ coincide; consequently, by Proposition 11.10, equality 14.7.14 implies that the restriction of equality 14.7.13 to the trivial subgroup 1 holds and then equality 14.7.13 follows from 11.9.1.

In particular, for any point α of G on $E \otimes_{\mathcal{O}} A$, if the image of α in $E \otimes_{\mathcal{O}} A'$ comes from A' throughout $\tilde{e}'$ then, by equality 14.7.13 and the commutativity of diagram 14.7.10, we have $m_\alpha(A') \geq n_\alpha$, and otherwise we have $n_\alpha = 0$; consequently, there is also an idempotent i' in $\mathsf{C}_\circ(A')^G$ having multiplicity n_α at any point α of G on $E \otimes_{\mathcal{O}} A$ such that its image in $E \otimes_{\mathcal{O}} A'$ comes from A' throughout $\tilde{e}'$, and zero elsewhere; now, the images of i and i' in $E \otimes_{\mathcal{O}} A'$ by respective representatives of $\widetilde{\mathrm{id}}_E \otimes_{\mathcal{O}} \tilde{f}$ and $\tilde{e}'$ have the same multiplicity at any point of G on $E \otimes_{\mathcal{O}} A'$ and therefore they are conjugate in $\mathsf{C}_\circ(E \otimes_{\mathcal{O}} A')^G$ (cf. 2.8.4), showing the existence of a G-interior $\mathcal{D}$-algebra exoembedding $\tilde{g} : \mathsf{E}(A) \to A'$ such that $\tilde{g} \circ \tilde{e} = \tilde{f}$ (by Proposition 11.10), which is a $\mathcal{D}G$-interior algebra exoembedding if $\widetilde{\mathrm{id}}_E \otimes_{\mathcal{O}} \tilde{f}$ is so. Since the number of points of 1 on A, on $E \otimes_{\mathcal{O}} A$ and on $\mathsf{E}(A)$ coincide (recall that $\mathsf{C}_\circ(E \otimes_{\mathcal{O}} A) \cong E \otimes_{\mathcal{O}} \mathsf{C}_\circ(A)$), the uniqueness of $\tilde{g}$ follows once again from Propositions 11.9 and 11.10.

14.8 If A' is a second $\mathcal{D}G$-interior algebra and $\tilde{f} : A \to A'$ a G-interior $\mathcal{D}$-algebra exoembedding, the composed exoembedding

14.8.1 $$A \xrightarrow{\tilde{f}} A' \xrightarrow{\tilde{e}(A')} \mathsf{E}(A')$$

fulfills clearly condition 14.7.1 and therefore we get a unique G-interior $\mathcal{D}$-algebra exoembedding

14.8.2 $$\mathsf{E}(\tilde{f}) : \mathsf{E}(A) \longrightarrow \mathsf{E}(A')$$

such that $\tilde{e}(A') \circ \tilde{f} = \mathsf{E}(\tilde{f}) \circ \tilde{e}(A)$; notice that, always by Proposition 14.7, $\mathsf{E}(\tilde{f})$ is a $\mathcal{D}G$-interior algebra exoembedding whenever $\tilde{f}$ is so. On the other hand, if H is a subgroup of G, it is clear that the $\mathcal{D}H$-interior algebra exoembedding

14.8.3 $$\mathrm{Res}^G_H\big(\tilde{e}(A)\big) : \mathrm{Res}^G_H(A) \longrightarrow \mathrm{Res}^G_H\big(\mathsf{E}(A)\big)$$

fulfills condition 14.7.1 with respect to $\mathrm{Res}^G_H(A)$ and consequently we obtain a $\mathcal{D}H$-interior algebra exoembedding

14.8.4 $$\tilde{r}^G_H(A) : \mathsf{E}\big(\mathrm{Res}^G_H(A)\big) \longrightarrow \mathrm{Res}^G_H\big(\mathsf{E}(A)\big)$$

such that we have

14.8.5 $$\tilde{r}^G_H(A) \circ \tilde{e}\big(\mathrm{Res}^G_H(A)\big) = \mathrm{Res}^G_H\big(\tilde{e}(A)\big) \quad ;$$

notice that if $A \cong \mathsf{E}(A)$ then $\mathrm{Res}^G_H(A) \cong \mathsf{E}\big(\mathrm{Res}^G_H(A)\big)$ and that, in general, if L_ε is a pointed group on A such that $L \subset H$ then, denoting respectively by $\tilde{h}_\varepsilon(A)$ and $\tilde{h}_\varepsilon\big(\mathrm{Res}^G_H(A)\big)$ the exoembeddings coming from condition 14.7.1 for A and $\mathrm{Res}^G_H(\mathrm{A})$, Propositions 11.9 and 11.10 force the following equality

14.8.6 $$\mathrm{Res}^G_H\big(\tilde{h}_\varepsilon(A)\big) \circ \tilde{d}^G_H\big(\mathrm{Ind}^H_L(A_\varepsilon)\big) = \tilde{r}^G_H(A) \circ \tilde{h}_\varepsilon\big(\mathrm{Res}^G_H(A)\big) \quad ;$$

indeed, it is easily checked from the equalities in condition 14.7.1 and in 14.8.5 that the compositions with $\tilde{d}^H_L(A_\varepsilon)$ of both members restricted to L coincide.

Theorem 14.9 *Let H_β and L_ε be pointed groups on A such that $L \subset H$. We have $\beta \subset \mathrm{Tr}^H_L\big(\mathrm{C}_\circ(A^L){\cdot}\varepsilon{\cdot}\mathrm{C}_\circ(A^L)\big)$ if and only if there is a $\mathcal{D}H$-interior algebra exoembedding*

14.9.1 $$\tilde{h}^\varepsilon_\beta : A_\beta \longrightarrow \mathrm{Ind}^H_L(A_\varepsilon)$$

such that we have

14.9.2 $$\tilde{h}_\varepsilon \circ \mathrm{Ind}^G_H(\tilde{h}^\varepsilon_\beta) = \tilde{h}_\beta$$

and then $\tilde{h}^\varepsilon_\beta$ is unique.

Remark 14.10 Exoembedding 14.9.1 is the announced extension of exoembedding 14.6.2; indeed, it is not difficult to prove from 11.9.1, Proposition 11.10 and condition 14.7.1 that when $L_\varepsilon \subset H_\beta$, equality 14.9.2 is equivalent to the following one

14.10.1 $$\mathrm{Res}^H_L(\tilde{h}^\varepsilon_\beta) \circ \tilde{f}^\beta_\varepsilon = \tilde{d}^H_L(A_\varepsilon) \quad .$$

But notice that in Theorem 14.9 there is no inclusion hypothesis between the pointed groups: the fact that condition $L_\varepsilon \subset H_\beta$ can be removed has already been noticed by Laurence Barker [1, Theorem 6.4] and Jacques Thévenaz [38, (17.11)].

Proof: Since $\mathrm{Tr}_L^H(1 \otimes 1 \otimes 1)$ is the unity element in $\mathrm{Ind}_L^H(A_\varepsilon)$, the existence of $\tilde{h}_\beta^\varepsilon$ fulfilling equality 14.9.2 forces, by 3.5.4, the inclusion

14.9.3 $$\beta \subset \mathrm{Tr}_L^H\Big(\mathrm{C}_\circ\big(\mathsf{E}(A)^L\big)\cdot\varepsilon\cdot\mathrm{C}_\circ\big(\mathsf{E}(A)^L\big)\Big)$$

in $\mathsf{E}(A)$, up to suitable identifications; but, since β and ε come from A throughout $\tilde{e}(A)$ (cf. 2.11.2), the analogous inclusion in A holds. Always by 3.5.4, in that case we have

14.9.4
$$\begin{aligned}\mathrm{Res}_H^G(\tilde{h}_\varepsilon)\circ \tilde{d}_H^G\big(\mathrm{Ind}_L^H(A_\varepsilon)\big)\circ \tilde{h}_\beta^\varepsilon &= \mathrm{Res}_H^G(\tilde{h}_\varepsilon)\circ \mathrm{Res}_H^G\big(\mathrm{Ind}_H^G(\tilde{h}_\beta^\varepsilon)\big)\circ \tilde{d}_H^G(A_\beta)\\ &= \mathrm{Res}_H^G(\tilde{h}_\beta)\circ \tilde{d}_H^G(A_\beta)\end{aligned}$$

and therefore the uniqueness of $\tilde{h}_\beta^\varepsilon$ follows from this equality and Proposition 11.10.

Conversely, choose $i \in \beta$ and $j \in \varepsilon$, and assume that i belongs to $\mathrm{Tr}_L^H\big(\mathrm{C}_\circ(A^L)j\mathrm{C}_\circ(A^L)\big)$ or, equivalently, that

14.9.5 $$i\mathrm{C}_\circ(A^H)i = \mathrm{Tr}_L^H\big(i\mathrm{C}_\circ(A^L)j\mathrm{C}_\circ(A^L)i\big) \quad ;$$

since the semisimple quotient of $i\mathrm{C}_\circ(A^H)i$ is just k, there are $a' \in i\mathrm{C}_\circ(A^L)j$ and $a'' \in j\mathrm{C}_\circ(A^L)i$ such that

14.9.6 $$i = \mathrm{Tr}_L^H(a'a'')$$

and we claim that the $\mathcal{O}$-module homomorphism

14.9.7 $$h : iAi \longrightarrow \mathrm{Ind}_L^H(jAj)$$

mapping $a \in iAi$ on $\Sigma_{x\in X}(\Sigma_{y\in Y} x \otimes a''x^{-1}aya' \otimes y^{-1})$, where X and Y are sets of representatives for H/L in H, is actually a $\mathfrak{D}H$-interior algebra embedding fulfilling equality 14.9.2. First of all notice that h does not depend on the choice of X and Y, and therefore, for any $a \in iAi$ and any $x, y \in H$, we have

14.9.8 $$h(x\cdot a\cdot y) = x\cdot h(a)\cdot y \quad .$$

Consider the following elements of $\mathrm{C}_\circ\Big(\mathrm{Ind}_L^H\big(\mathrm{Res}_L^G(A)\big)\Big)^L$

14.9.9 $$c' = \sum_{y\in Y} 1 \otimes ya' \otimes y^{-1} \quad \text{and} \quad c'' = \sum_{x\in X} x \otimes a''x^{-1} \otimes 1$$

and notice that, by 14.9.6, we have $c'c'' = 1 \otimes i \otimes 1$; hence, it is easily checked that $l = c''c'$ is an idempotent of $\mathrm{C}_\circ\Big(\mathrm{Ind}_L^H\big(\mathrm{Res}_L^G(A)\big)\Big)^L$ which has the same multiplicity than $1 \otimes i \otimes 1$ at any point of this $\mathcal{O}$-algebra; consequently, there is $u \in \bigg(\mathrm{C}_\circ\Big(\mathrm{Ind}_L^H\big(\mathrm{Res}_L^G(A)\big)\Big)^L\bigg)^*$ such that $l = (1 \otimes i \otimes 1)^u$ (cf. 2.8.4). Then, setting

14.9.10 $$c = c' + \big(1 - (1 \otimes i \otimes 1)\big)u(1 - l)$$

and denoting by

14.9.11 $$f : iAi \longrightarrow \mathrm{Ind}_L^H\big(\mathrm{Res}_L^G(A)\big) \text{ and } g : iAi \longrightarrow \mathrm{Ind}_L^H\big(\mathrm{Res}_L^G(A)\big) \quad ,$$

the maps defined, up to suitable identifications, by $f(a) = 1 \otimes a \otimes 1$ and $g(a) = h(a)$ for any $a \in iAi$, it is easily checked that c is an invertible element of $\mathrm{C}_\circ\Big(\mathrm{Ind}_L^H\big(\mathrm{Res}_L^G(A)\big)\Big)^L$ with

$$c^{-1} = c'' + (1-l)u^{-1}\big(1 - (1 \otimes i \otimes 1)\big)$$

and that we have $g(a) = f(a)^c$ for any $a \in iAi$; this proves that h is a $\mathcal{D}L$-interior algebra embedding since so is f and therefore, by 14.9.8, h is actually a $\mathcal{D}H$-interior algebra embedding; moreover, this proves that the following diagram is commutative

14.9.12 $$\begin{array}{ccc} \mathrm{Res}_L^G(A) & \xrightarrow{\tilde d_L^H(\mathrm{Res}_L^G(A))} & \mathrm{Res}_L^H\Big(\mathrm{Ind}_L^H\big(\mathrm{Res}_L^G(A)\big)\Big) \\ {\scriptstyle \mathrm{Res}_L^H(\tilde f_\beta)}\uparrow & & \uparrow{\scriptstyle \mathrm{Res}_L^H(\mathrm{Ind}_L^H(\tilde f_\varepsilon))} \\ \mathrm{Res}_L^H(A_\beta) & \xrightarrow{\mathrm{Res}_L^H(\tilde h)} & \mathrm{Res}_L^H\big(\mathrm{Ind}_L^H(A_\varepsilon)\big) \end{array} \quad .$$

Now, it suffices to prove that

14.9.13 $$\tilde h_\varepsilon \circ \mathrm{Ind}_H^G(\tilde h) = \tilde h_\beta$$

and we claim that this equality follows from the commutativity of diagram 14.9.12. First of all notice that, according to condition 14.7.1, we have

14.9.14 $$\mathrm{Res}_L^G\big(\tilde e(A)\big) \circ \mathrm{Res}_L^H(\tilde f_\beta) = \mathrm{Res}_L^G(\tilde h_\beta) \circ \mathrm{Res}_L^H\big(\tilde d_H^G(A_\beta)\big)$$

and similarly

14.9.15 $$\begin{aligned} &\mathrm{Res}_L^H\Big(\mathrm{Ind}_L^H\Big(\mathrm{Res}_L^G\big(\tilde e(A)\big)\Big)\Big) \circ \mathrm{Res}_L^H\big(\mathrm{Ind}_L^H(\tilde f_\varepsilon)\big) \\ &\qquad = \mathrm{Res}_L^H\Big(\mathrm{Ind}_L^H\big(\mathrm{Res}_L^G(\tilde h_\varepsilon)\big)\Big) \circ \mathrm{Res}_L^H\Big(\mathrm{Ind}_L^H\big(\tilde d_L^G(A_\varepsilon)\big)\Big) \quad . \end{aligned}$$

On the other hand, it follows from Propositions 11.9 and 11.10 that the following diagram of $\mathcal{D}L$-interior algebra exoembeddings is commutative

14.9.16 $$\begin{array}{ccc} & \mathrm{Res}_L^H\bigg(\mathrm{Ind}_L^H\Big(\mathrm{Res}_L^G\big(\mathrm{Ind}_L^G(A_\varepsilon)\big)\Big)\bigg) & \\ {\scriptstyle \mathrm{Res}_L^H(\mathrm{Ind}_L^H(\tilde d_L^G(A_\varepsilon)))}\nearrow & & \nwarrow{\scriptstyle \tilde d_L^H(\mathrm{Res}_L^G(\mathrm{Ind}_L^G(A_\varepsilon)))} \\ \mathrm{Res}_L^H\big(\mathrm{Ind}_L^H(A_\varepsilon)\big) & \xrightarrow{\mathrm{Res}_L^H(\tilde d_H^G(\mathrm{Ind}_L^H(A_\varepsilon)))} & \mathrm{Res}_L^G\big(\mathrm{Ind}_L^G(A_\varepsilon)\big) \end{array} \quad ;$$

indeed, the composition with $\tilde{d}_L^H(A_\varepsilon)$ of both exoembeddings from $\mathrm{Res}_L^H(\mathrm{Ind}_L^H(A_\varepsilon))$ to $\mathrm{Res}_L^H\Big(\mathrm{Ind}_L^H\big(\mathrm{Res}_L^G(\mathrm{Ind}_L^G(A_\varepsilon))\big)\Big)$ coincide since, for any $a \in A_\varepsilon$, the images of $1 \otimes a \otimes 1 \in \mathrm{Res}_L^H(\mathrm{Ind}_L^H(A_\varepsilon))$ in $\mathrm{Res}_L^H\Big(\mathrm{Ind}_L^H\big(\mathrm{Res}_L^G(\mathrm{Ind}_L^G(A_\varepsilon))\big)\Big)$ are both equal to $1 \otimes (1 \otimes a \otimes 1) \otimes 1$. Consequently, from the naturality of d_H^G and d_L^H, from the equalities 14.9.14 and 14.9.15, and from the commutativity of diagrams 14.9.12 and 14.9.16, we get

14.9.17

$$\begin{aligned}
&\tilde{d}_L^H\big(\mathrm{Res}_L^G(\mathsf{E}(A))\big) \circ \mathrm{Res}_L^G(\tilde{h}_\beta) \circ \mathrm{Res}_L^H(\tilde{d}_H^G(A_\beta))\\
&= \tilde{d}_L^H\big(\mathrm{Res}_L^G(\mathsf{E}(A))\big) \circ \mathrm{Res}_L^G(\tilde{e}(A)) \circ \mathrm{Res}_L^H(\tilde{f}_\beta)\\
&= \mathrm{Res}_L^H\Big(\mathrm{Ind}_L^H\big(\mathrm{Res}_L^G(\tilde{e}(A))\big)\Big) \circ \tilde{d}_L^H(\mathrm{Res}_L^G(A)) \circ \mathrm{Res}_L^H(\tilde{f}_\beta)\\
&= \mathrm{Res}_L^H\Big(\mathrm{Ind}_L^H\big(\mathrm{Res}_L^G(\tilde{e}(A))\big)\Big) \circ \mathrm{Res}_L^H(\mathrm{Ind}_L^H(\tilde{f}_\varepsilon)) \circ \mathrm{Res}_L^H(\tilde{h})\\
&= \mathrm{Res}_L^H\big(\mathrm{Ind}_L^H(\mathrm{Res}_L^G(\tilde{h}_\varepsilon))\big) \circ \mathrm{Res}_L^H\big(\mathrm{Ind}_L^H(\tilde{d}_L^G(A_\varepsilon))\big) \circ \mathrm{Res}_L^H(\tilde{h})\\
&= \mathrm{Res}_L^H\big(\mathrm{Ind}_L^H(\mathrm{Res}_L^G(\tilde{h}_\varepsilon))\big) \circ \tilde{d}_L^H\big(\mathrm{Res}_L^G(\mathrm{Ind}_L^G(A_\varepsilon))\big) \circ \mathrm{Res}_L^H\big(\tilde{d}_H^G(\mathrm{Ind}_L^H(A_\varepsilon)) \circ \tilde{h}\big)\\
&= \tilde{d}_L^H\big(\mathrm{Res}_L^G(\mathsf{E}(A))\big) \circ \mathrm{Res}_L^G(\tilde{h}_\varepsilon) \circ \mathrm{Res}_L^G(\mathrm{Ind}_H^G(\tilde{h})) \circ \mathrm{Res}_L^H(\tilde{d}_H^G(A_\beta))
\end{aligned}$$

and equality 14.9.13 follows from Propositions 11.9 and 11.10.

Corollary 14.11 *Let H_β be a pointed group on A and P_γ a defect pointed group of H_β. There is a unique $\mathcal{D}H$-interior algebra exoembedding*

14.11.1 $$\tilde{h}_\beta^\gamma : A_\beta \longrightarrow \mathrm{Ind}_P^H(A_\gamma)$$

such that we have

14.11.2 $$\mathrm{Res}_P^H(\tilde{h}_\beta^\gamma) \circ \tilde{f}_\gamma^\beta = \tilde{d}_P^H(A_\gamma) \quad \textit{and} \quad \tilde{h}_\gamma \circ \mathrm{Ind}_H^G(\tilde{h}_\beta^\gamma) = \tilde{h}_\beta \quad .$$

Proof: Since $\beta \subset \mathrm{Tr}_P^H(\mathsf{C}_\circ(A^P)\cdot\gamma\cdot\mathsf{C}_\circ(A^P))$ (cf. 14.5.3), by Theorem 14.9 there is a $\mathcal{D}H$-interior algebra exoembedding

14.11.3 $$\tilde{h} : A_\beta \longrightarrow \mathrm{Ind}_P^H(A_\gamma)$$

such that we have $\tilde{h}_\gamma \circ \mathrm{Ind}_H^G(\tilde{h}) = \tilde{h}_\beta$ and therefore, by 3.5.4 and 14.7.1, we get

$$\mathrm{Res}_P^G(\tilde{h}_\gamma)\circ \mathrm{Res}_P^H\Big(\tilde{d}_H^G(\mathrm{Ind}_P^H(A_\gamma))\Big)\circ \mathrm{Res}_P^H(\tilde{h})\circ \tilde{f}_\gamma^\beta$$

$$= \mathrm{Res}_P^G(\tilde{h}_\gamma)\circ \mathrm{Res}_P^G(\mathrm{Ind}_H^G(\tilde{h}))\circ \mathrm{Res}_P^H(\tilde{d}_H^G(A_\beta))\circ \tilde{f}_\gamma^\beta$$

14.11.4 $$= \mathrm{Res}_P^G(\tilde{h}_\beta)\circ \mathrm{Res}_P^H(\tilde{d}_H^G(A_\beta))\circ \tilde{f}_\gamma^\beta$$

$$= \mathrm{Res}_P^G(\tilde{e}(A))\circ \tilde{f}_\gamma = \mathrm{Res}_P^G(\tilde{h}_\gamma)\circ \tilde{d}_P^G(A_\gamma)$$

$$= \mathrm{Res}_P^G(\tilde{h}_\gamma)\circ \mathrm{Res}_P^H\Big(\tilde{d}_H^G(\mathrm{Ind}_P^H(A_\gamma))\Big)\circ \tilde{d}_P^H(A_\gamma)$$

which, by Proposition 11.10, implies the first equality in 14.11.2. Then, since any point of 1 on A_β comes from A_γ throughout $\tilde{f}_\gamma^\beta$ (cf. 2.9.5), the uniqueness of $\tilde{h}$ follows from Propositions 11.9 and 11.10 applied to the first equality in 14.11.2.

Corollary 14.12 *We have* $\mathsf{E}(A)\cong \mathsf{E}(\mathsf{E}(A))$ *and* $\tilde{e}(\mathsf{E}(A)) = \mathsf{E}(\tilde{e}(A))$.

Remark 14.13 Since A and $\mathsf{E}(A)$ have the same local structure by Proposition 14.7 and [20, Proposition 2.14], up to replace A by $\mathsf{E}(A)$ we may assume without loss of generality that $A\cong \mathsf{E}(A)$ whenever we are only concerned by local pointed groups.

Proof: It suffices to prove that $\mathsf{E}(A)$ and $\widetilde{\mathrm{id}}_{\mathsf{E}(A)}$ fulfill condition 14.7.1 with respect to $\mathsf{E}(A)$. Let H_β be a pointed group on $\mathsf{E}(A)$ and P_γ a defect pointed group of H_β; by Proposition 14.7, P_γ comes from A throughout $\tilde{e}(A)$ (cf. 2.11.2) and therefore we have a $\mathfrak{D}G$-interior algebra exoembedding

14.12.1 $$\tilde{h}_\gamma : \mathrm{Ind}_P^G(A_\gamma)\longrightarrow \mathsf{E}(A)$$

such that

14.12.2 $$\mathrm{Res}_P^G(\tilde{h}_\gamma)\circ d_P^G(A_\gamma) = \mathrm{Res}_P^G(\tilde{e}(A))\circ \tilde{f}_\gamma \quad ;$$

but, by Corollary 14.11, there is a $\mathfrak{D}H$-interior algebra exoembedding

14.12.3 $$\tilde{h}_\beta^\gamma : \mathsf{E}(A)_\beta \longrightarrow \mathrm{Ind}_P^H(A_\gamma)$$

such that we have

14.12.4 $$\mathrm{Res}_P^H(\tilde{h}_\beta^\gamma)\circ \tilde{f}_\gamma^\beta = \tilde{d}_P^H(A_\gamma)$$

(where $\tilde{f}_\gamma^\beta$ is an exoembedding from $A_\gamma\cong E(A)_\gamma$ to $\mathsf{E}(A)_\beta$); consequently, we get the composed $\mathfrak{D}G$-interior algebra exoembedding

14.12.5 $$\mathrm{Ind}_H^G(\mathsf{E}(A)_\beta)\xrightarrow{\mathrm{Ind}_H^G(\tilde{h}_\beta^\gamma)} \mathrm{Ind}_P^G(A_\gamma)\xrightarrow{\tilde{h}_\gamma} \mathsf{E}(A)$$

that, by 3.5.4, 14.7.1 and 14.12.4, fulfills (up to suitable identifications)

$$\mathrm{Res}_P^G(\tilde{h}_\gamma)\circ \mathrm{Res}_P^G(\mathrm{Ind}_H^G(\tilde{h}_\beta^\gamma))\circ \mathrm{Res}_P^H\Big(\tilde{d}_H^G(\mathsf{E}(A)_\beta)\Big)\circ \tilde{f}_\gamma^\beta$$

$$= \mathrm{Res}_P^G(\tilde{h}_\gamma)\circ \mathrm{Res}_P^H\Big(\tilde{d}_H^G(\mathrm{Ind}_P^H(A_\gamma))\Big)\circ \mathrm{Res}_P^H(\tilde{h}_\beta^\gamma)\circ \tilde{f}_\gamma^\beta$$

14.12.6 $$= \mathrm{Res}_P^G(\tilde{h}_\gamma)\circ \tilde{d}_P^G(A_\gamma) = \mathrm{Res}_P^G(\tilde{e}(A))\circ \tilde{f}_\gamma = \tilde{f}_\gamma(\mathsf{E}(A))$$

$$= \mathrm{Res}_P^H\Big(\tilde{f}_\beta(\mathsf{E}(A))\Big)\circ \tilde{f}_\gamma^\beta$$

and therefore, since any point of 1 on $\mathsf{E}(A)_\beta$ comes from A_γ throughout $\tilde{f}^\beta_\gamma$ (cf. 2.9.5), it follows from Propositions 11.9 and 11.10 that

$$\text{14.12.7} \qquad \mathrm{Res}^G_H(\tilde{h}_\gamma \circ \mathrm{Ind}^G_H(\tilde{h}^\gamma_\beta)) \circ \tilde{d}^G_H(\mathsf{E}(A)_\beta) = \tilde{f}_\beta(\mathsf{E}(A))$$

which agree with condition 14.7.1.

Proposition 14.14 *For any $\mathcal{D}G$-interior algebra A' there is a unique $\mathcal{D}G$-interior algebra exoembedding*

$$\text{14.14.1} \qquad \tilde{e}(A, A') : \mathsf{E}(A \underset{\mathcal{O}}{\otimes} A') \longrightarrow \mathsf{E}(A) \underset{\mathcal{O}}{\otimes} A'$$

such that we have

$$\text{14.14.2} \qquad \tilde{e}(A, A') \circ \tilde{e}(A \underset{\mathcal{O}}{\otimes} A') = \tilde{e}(A) \underset{\mathcal{O}}{\otimes} \widetilde{\mathrm{id}}_{A'} \quad .$$

In particular $A \cong \mathsf{E}(A)$ implies that $A \otimes_{\mathcal{O}} A' \cong \mathsf{E}(A \otimes_{\mathcal{O}} A')$.

Proof: Let $H_{\beta''}$ be a pointed group on $A \otimes_{\mathcal{O}} A'$; it is quite clear that there is a point β of H on A and a $\mathcal{D}H$-interior algebra exoembedding

$$\text{14.14.3} \qquad \tilde{f}^\beta_{\beta''} : (A \underset{\mathcal{O}}{\otimes} A')_{\beta''} \longrightarrow A_\beta \underset{\mathcal{O}}{\otimes} \mathrm{Res}^G_H(A')$$

(since $\mathsf{C}_\circ(A)^H \otimes 1$ is a unitary $\mathcal{O}$-subalgebra of $\mathsf{C}_\circ(A \otimes_{\mathcal{O}} A')^H$, we can consider a primitive idempotent $i \in \mathsf{C}_\circ(A)^H$ such that $(i \otimes 1)i'' = i'' = i''(i \otimes 1)$ for a suitable $i'' \in B''$) such that we have

$$\text{14.14.4} \qquad \tilde{f}_{\beta''} = (\tilde{f}_\beta \underset{\mathcal{O}}{\otimes} \widetilde{\mathrm{id}}_{\mathrm{Res}^G_H(A')}) \circ \tilde{f}^\beta_{\beta''} \quad ;$$

then, considering the composed $\mathcal{D}G$-interior algebra exoembedding (cf. Proposition 12.9)

14.14.5

$$\begin{array}{ccc} \mathrm{Ind}^G_H((A \underset{\mathcal{O}}{\otimes} A')_{\beta''}) & \xrightarrow{\mathrm{Ind}^G_H(\tilde{f}^\beta_{\beta''})} & \mathrm{Ind}^G_H(A_\beta \underset{\mathcal{O}}{\otimes} \mathrm{Res}^G_H(A')) \\ & & \wr\| \\ & & \mathrm{Ind}^G_H(A_\beta) \underset{\mathcal{O}}{\otimes} A' \xrightarrow{\tilde{h}_\beta \otimes \widetilde{\mathrm{id}}} \mathsf{E}(A) \underset{\mathcal{O}}{\otimes} A' \quad , \end{array}$$

it is easily checked from 3.5.4 and Proposition 12.9 that it fulfills the equality in condition 14.7.1 with respect to $\mathsf{E}(A) \otimes_{\mathcal{O}} A'$ and $\tilde{e}(A) \otimes_{\mathcal{O}} \widetilde{\mathrm{id}}_{A'}$; so, the statement follows from condition 14.7.2.

14.15 Thus, by Proposition 14.14, we have a canonical $\mathcal{D}G$-interior algebra exoembedding

$$\text{14.15.1} \qquad \tilde{e}(\mathcal{O}, A) : \mathsf{E}(A) \longrightarrow \mathsf{E}(\mathcal{O}) \underset{\mathcal{O}}{\otimes} A$$

such that $\tilde{e}(\mathcal{O}, A) \circ \tilde{e}(A) = \tilde{e}(\mathcal{O}) \otimes_{\mathcal{O}} \widetilde{\mathrm{id}}_A$, where $\mathcal{O}$ denotes the trivial $\mathcal{D}G$-interior algebra, and it is not difficult to see that it induces actually a bijection between the sets of pointed groups on $\mathsf{E}(A)$ and on $\mathsf{E}(\mathcal{O}) \otimes_{\mathcal{O}} A$. More generally, if H is a subgroup of G and B a $\mathcal{D}H$-interior algebra, consider the $\mathcal{D}H$-interior algebra exoembedding (cf. 14.8.4)

$$\text{14.15.2} \qquad \tilde{r}_H^G(\mathcal{O}) : \mathsf{E}(\mathcal{O}_H) \longrightarrow \mathrm{Res}_H^G(\mathsf{E}(\mathcal{O})) \quad ,$$

where $\mathcal{O}_H = \mathrm{Res}_H^G(\mathcal{O})$ denotes the trivial $\mathcal{D}H$-interior algebra, and set

$$\text{14.15.3} \qquad \tilde{c}_H^G(B) = \big(\tilde{r}_H^G(\mathcal{O}) \underset{\mathcal{O}}{\otimes} \widetilde{\mathrm{id}}_B\big) \circ \tilde{e}(\mathcal{O}_H, B) \quad ;$$

then, up to suitable identifications (cf. Proposition 12.9), from exoembedding 14.15.1 we get the $\mathcal{D}G$-interior algebra exoembedding

$$\text{14.15.4} \qquad \mathrm{Ind}_H^G\big(\tilde{c}_H^G(B)\big) : \mathrm{Ind}_H^G\big(\mathsf{E}(B)\big) \longrightarrow \mathsf{E}(\mathcal{O}) \underset{\mathcal{O}}{\otimes} \mathrm{Ind}_H^G(B)$$

which allow us to exhibit the following transitivity that we need in Section 16. If H_β and L_ε are pointed groups on A such that $L_\varepsilon \subset H_\beta$, denote by

$$\text{14.15.5} \qquad \tilde{h}_\varepsilon^\beta : \mathrm{Ind}_L^H(A_\varepsilon) \longrightarrow \mathsf{E}(A_\beta)$$

the $\mathcal{D}H$-interior algebra exoembedding provided by Proposition 14.7 (by identifying ε to the corresponding point of L on A_β), which fulfills

$$\text{14.15.6} \qquad \mathrm{Res}_L^H(\tilde{h}_\varepsilon^\beta) \circ \tilde{d}_L^H(A_\varepsilon) = \mathrm{Res}_L^H\big(\tilde{e}(A_\beta)\big) \circ \tilde{f}_\varepsilon^\beta \quad .$$

Proposition 14.16 *With the notation above, for any pair of pointed groups H_β and L_ε on A such that $L_\varepsilon \subset H_\beta$, the following diagram of $\mathcal{D}G$-interior algebra exoembeddings is commutative*

$$\text{14.16.1} \qquad \begin{array}{ccc}
\mathsf{E}(A) & \xrightarrow{\ \tilde{e}(\mathcal{O})\otimes\widetilde{\mathrm{id}}_{\mathsf{E}(A)}\ } & \mathsf{E}(\mathcal{O}) \underset{\mathcal{O}}{\otimes} \mathsf{E}(A) \\
 & & \uparrow {\scriptstyle \widetilde{\mathrm{id}}_{\mathsf{E}(\mathcal{O})}\otimes\tilde{h}_\beta} \\
\tilde{h}_\varepsilon \Big\uparrow & & \mathsf{E}(\mathcal{O}) \underset{\mathcal{O}}{\otimes} \mathrm{Ind}_H^G(A_\beta) \\
 & & \uparrow {\scriptstyle \mathrm{Ind}_H^G(\tilde{c}_H^G(A_\beta))} \\
\mathrm{Ind}_L^G(A_\varepsilon) & \xrightarrow{\ \mathrm{Ind}_H^G(\tilde{h}_\varepsilon^\beta)\ } & \mathrm{Ind}_H^G\big(\mathsf{E}(A_\beta)\big)
\end{array} \quad .$$

Proof: By 3.5.4 and Proposition 12.9, we have

14.16.2

$$\mathrm{Res}^G_H\Big(\mathrm{Ind}^G_H(\tilde{c}^G_H(A_\beta)\circ\tilde{h}^\beta_\varepsilon)\Big)\circ\tilde{d}^G_H(\mathrm{Ind}^H_L(A_\varepsilon)) = (\widetilde{\mathrm{id}}_{\mathsf{E}(\mathcal{O})}\underset{\mathcal{O}}{\otimes}\tilde{d}^G_H(A_\beta))\circ\tilde{c}^G_H(A_\beta)\circ\tilde{h}^\beta_\varepsilon$$

and therefore, since (cf. 2.6.6)

14.16.3 $$\mathrm{Res}^H_L\Big(d^G_H(\mathrm{Ind}^H_L(A_\varepsilon))\Big)\circ d^H_L(A_\varepsilon) = d^G_L(A_\varepsilon) \quad ,$$

from the equalities in condition 14.7.1 and in 14.15.6 we get

14.16.4

$$\begin{aligned}&\mathrm{Res}^G_L\Big((\widetilde{\mathrm{id}}_{\mathsf{E}(\mathcal{O})}\underset{\mathcal{O}}{\otimes}\tilde{h}_\beta)\circ\mathrm{Ind}^G_H(\tilde{c}^G_H(A_\beta))\circ\mathrm{Ind}^G_H(\tilde{h}^\beta_\varepsilon)\Big)\circ\tilde{d}^G_L(A_\varepsilon)\\ &=\Big(\widetilde{\mathrm{id}}_{\mathsf{E}(\mathcal{O})}\underset{\mathcal{O}}{\otimes}\Big(\mathrm{Res}^G_L(\tilde{h}_\beta)\circ\mathrm{Res}^H_L(\tilde{d}^G_H(A_\beta))\Big)\Big)\circ\mathrm{Res}^H_L(\tilde{c}^G_H(A_\beta))\circ\mathrm{Res}^H_L(\tilde{h}^\beta_\varepsilon)\circ\tilde{d}^H_L(A_\varepsilon)\\ &=\Big(\widetilde{\mathrm{id}}_{\mathsf{E}(\mathcal{O})}\underset{\mathcal{O}}{\otimes}\Big(\mathrm{Res}^G_L(\tilde{e}(A))\circ\mathrm{Res}^H_L(\tilde{f}_\beta)\Big)\Big)\circ\mathrm{Res}^H_L(\tilde{c}^G_H(A_\beta))\circ\mathrm{Res}^H_L(\tilde{e}(A_\beta))\circ\tilde{f}^\beta_\varepsilon;\end{aligned}$$

but, it is clear that (cf. 14.14.2)

14.16.5 $$\tilde{c}^G_H(A_\beta)\circ\tilde{e}(A_\beta) = \mathrm{Res}^G_H(\tilde{e}(\mathcal{O}))\underset{\mathcal{O}}{\otimes}\widetilde{\mathrm{id}}_{A_\beta} \quad ;$$

consequently, from the equality in condition 14.7.1 we obtain

$$\begin{aligned}&\mathrm{Res}^G_L\Big((\widetilde{\mathrm{id}}_{\mathsf{E}(\mathcal{O})}\underset{\mathcal{O}}{\otimes}\tilde{h}_\beta)\circ\mathrm{Ind}^G_H(\tilde{c}^G_H(A_\beta))\circ\mathrm{Ind}^G_H(\tilde{h}^\beta_\varepsilon)\Big)\circ\tilde{d}^G_L(A_\varepsilon)\\ &=\mathrm{Res}^G_L(\tilde{e}(\mathcal{O}))\underset{\mathcal{O}}{\otimes}\Big(\mathrm{Res}^G_L(\tilde{e}(A))\circ\tilde{f}_\varepsilon\Big)\\ &=\mathrm{Res}^G_L(\tilde{e}(\mathcal{O}))\underset{\mathcal{O}}{\otimes}(\mathrm{Res}^G_L(\tilde{h}_\varepsilon)\circ\tilde{d}^G_L(A_\varepsilon)) \qquad (14.16.6)\\ &=\mathrm{Res}^G_L\Big((\tilde{e}(\mathcal{O})\underset{\mathcal{O}}{\otimes}\widetilde{\mathrm{id}}_{\mathsf{E}(A)})\circ\tilde{h}_\varepsilon\Big)\circ\tilde{d}^G_L(A_\varepsilon)\end{aligned}$$

and then the commutativity of diagram 14.16.1 follows from Propositions 11.9 and 11.10.

14.17 Let G' be a second finite group and $\varphi : G \to G'$ a group homomorphism; now, we analyze the Higman envelope of the induced $\mathcal{D}G'$-interior algebra $\mathrm{Ind}_\varphi(A)$: as in the restriction case, we will prove that the $\mathcal{D}G'$-interior algebra $\mathrm{Ind}_\varphi(\mathsf{E}(A))$ and the $\mathcal{D}G'$-interior algebra exoembedding $\mathrm{Ind}_\varphi(\tilde{e}(A))$ fulfill condition 14.7.1. Let H_β be a pointed group on A and H' a subgroup of G' such that $\varphi(H)\subset H'$, and denote by $\psi : H \to H'$ the group homomorphism determined by φ; recall that we have a canonical $\mathcal{D}G'$-interior algebra isomorphism (cf. Corollary 12.7)

$$\text{14.17.1} \qquad \mathrm{Ind}_{H'}^{G'}\big(\mathrm{Ind}_\psi(A_\beta)\big) \cong \mathrm{Ind}_\varphi\big(\mathrm{Ind}_H^G(A_\beta)\big)$$

and, up to identify to each other both members of this isomorphism, we consider the $\mathcal{D}H'$-interior algebra exoembedding

$$\text{14.17.2} \qquad \tilde{g}_\beta : \mathrm{Ind}_\psi(A_\beta) \longrightarrow \mathrm{Res}_{H'}^{G'}\Big(\mathrm{Ind}_\varphi\big(\mathsf{E}(A)\big)\Big)$$

given by

$$\text{14.17.3} \qquad \tilde{g}_\beta = \mathrm{Res}_{H'}^{G'}\big(\mathrm{Ind}_\varphi(\tilde{h}_\beta)\big) \circ \tilde{d}_{H'}^{G'}\big(\mathrm{Ind}_\psi(A_\beta)\big) \quad .$$

First of all, we show that $\tilde{g}_\beta$ mainly depends on $\mathrm{Res}_H^G(A)$ and on the group homomorphism $\varphi^{-1}(H') \to H'$ determined by φ, and that in particular, when $H = \varphi^{-1}(H')$, it factorizes throughout $\mathrm{Ind}_\psi(\tilde{f}_\beta)$; in the last case, in order to relate β to the points of H' on $\mathrm{Ind}_\varphi(A)$, we will consider the composed map (cf. 12.3.1)

$$\text{14.17.4} \qquad A^H \xrightarrow{d_\varphi(A)} \mathrm{Ind}_\varphi(A)^{\varphi(H)} \xrightarrow{\mathrm{Tr}_{\varphi(H)}^{H'}} \mathrm{Ind}_\varphi(A)^{H'}$$

which is actually an $\mathcal{O}$-algebra homomorphism. So, let F' be a subgroup of G' containing H', set $F = \varphi^{-1}(F')$ and denote by $\eta : F \to F'$ the group homomorphism determined by φ, and by $\tilde{g}_\beta\big(\mathrm{Res}_F^G(A)\big)$ the corresponding homomorphism 14.17.2 when G, G', φ and A are respectively replaced by F, F', η and $\mathrm{Res}_F^G(A)$.

Proposition 14.18 *With the notation above, we have*

$$\text{14.18.1} \qquad \tilde{g}_\beta = \mathrm{Res}_{H'}^{F'}\Big(\tilde{d}_{F',\varphi}^{G'}\big(\mathsf{E}(A)\big) \circ \mathrm{Ind}_\eta\big(\tilde{r}_F^G(A)\big)\Big) \circ \tilde{g}_\beta\big(\mathrm{Res}_F^G(A)\big) \quad .$$

Moreover, if $H = \varphi^{-1}(H')$ then we get

$$\text{14.18.2} \qquad \tilde{g}_\beta = \mathrm{Res}_{H'}^{G'}\Big(\mathrm{Ind}_\varphi\big(\tilde{e}(A)\big)\Big) \circ \tilde{d}_{H',\varphi}^{G'}(A) \circ \mathrm{Ind}_\psi(\tilde{f}_\beta) \quad .$$

Proof: By the naturality of $d_{F',\varphi}^{G'}$ (cf. 3.22.1 and 12.13), we get

$$\text{14.18.3} \quad \tilde{d}_{F',\varphi}^{G'}\big(\mathsf{E}(A)\big) \circ \mathrm{Ind}_\eta\big(\mathrm{Res}_F^G(\tilde{h}_\beta)\big) = \mathrm{Res}_{F'}^{G'}\big(\mathrm{Ind}_\varphi(\tilde{h}_\beta)\big) \circ \tilde{d}_{F',\varphi}^{G'}\big(\mathrm{Ind}_H^G(A_\beta)\big)$$

and, by the commutativity of diagram 12.13.2 applied to $B = \mathrm{Ind}_H^F(A_\beta)$, we have here the following commutative diagram

14.18.4

$$\begin{array}{ccc}
\mathrm{Res}_{F'}^{G'}\bigg(\mathrm{Ind}_{F'}^{G'}\Big(\mathrm{Ind}_\eta\big(\mathrm{Ind}_H^F(A_\beta)\big)\Big)\bigg) & \cong & \mathrm{Res}_{F'}^{G'}\Big(\mathrm{Ind}_\varphi\big(\mathrm{Ind}_H^G(A_\beta)\big)\Big) \\
{\scriptstyle d_{F'}^{G'}(\mathrm{Ind}_\eta(\mathrm{Ind}_H^F(A_\beta)))}\Big\uparrow & & \Big\uparrow{\scriptstyle d_{F',\varphi}^{G'}(\mathrm{Ind}_H^G(A_\beta))} \\
\mathrm{Ind}_\eta\big(\mathrm{Ind}_H^F(A_\beta)\big) & \xrightarrow{\mathrm{Ind}_\eta(d_F^G(\mathrm{Ind}_H^F(A_\beta)))} & \mathrm{Ind}_\eta\Big(\mathrm{Res}_F^G\big(\mathrm{Ind}_H^G(A_\beta)\big)\Big)
\end{array} \quad .$$

Now, according to definition 14.17.3 applied to $\mathrm{Res}^G_F(A)$, we have

14.18.5
$$\begin{aligned}&\mathrm{Res}^{F'}_{H'}\Big(\tilde d^{G'}_{F',\varphi}(\mathsf{E}(A))\circ \mathrm{Ind}_\eta(\tilde r^G_F(A))\Big)\circ \tilde g_\beta(\mathrm{Res}^G_F(A))\\ &= \mathrm{Res}^{F'}_{H'}\Big(\tilde d^{G'}_{F',\varphi}(\mathsf{E}(A))\circ \mathrm{Ind}_\eta\big(\tilde r^G_F(A)\circ \tilde h_\beta(\mathrm{Res}^G_F(A))\big)\Big)\circ \tilde d^{F'}_{H'}(\mathrm{Ind}_\psi(A_\beta)) \quad ;\end{aligned}$$

but, from 14.8.6, 14.18.3 and 14.18.4, we get

14.18.6
$$\begin{aligned}&\tilde d^{G'}_{F',\varphi}(\mathsf{E}(A))\circ \mathrm{Ind}_\eta\big(\tilde r^G_F(A)\circ \tilde h_\beta(\mathrm{Res}^G_F(A))\big)\\ &= \tilde d^{G'}_{F',\varphi}(\mathsf{E}(A))\circ \mathrm{Ind}_\eta\big(\mathrm{Res}^G_F(\tilde h_\beta)\circ \tilde d^G_F(\mathrm{Ind}^F_H(A_\beta))\big)\\ &= \mathrm{Res}^{G'}_{F'}(\mathrm{Ind}_\varphi(\tilde h_\beta))\circ \tilde d^{G'}_{F',\varphi}(\mathrm{Ind}^G_H(A_\beta))\circ \mathrm{Ind}_\eta\big(\tilde d^G_F(\mathrm{Ind}^F_H(A_\beta))\big)\\ &= \mathrm{Res}^{G'}_{F'}(\mathrm{Ind}_\varphi(\tilde h_\beta))\circ \tilde d^{G'}_{F'}\big(\mathrm{Ind}_\eta(\mathrm{Ind}^F_H(A_\beta))\big)\\ &= \mathrm{Res}^{G'}_{F'}(\mathrm{Ind}_\varphi(\tilde h_\beta))\circ \tilde d^{G'}_{F'}\big(\mathrm{Ind}^{F'}_{H'}(\mathrm{Ind}_\psi(A_\beta))\big) \quad ;\end{aligned}$$

consequently, we obtain

14.18.7
$$\begin{aligned}&\mathrm{Res}^{F'}_{H'}\Big(\tilde d^{G'}_{F',\varphi}(\mathsf{E}(A))\circ \mathrm{Ind}_\eta(r^G_F(A))\Big)\circ \tilde g_\beta(\mathrm{Res}^G_F(A))\\ &= \mathrm{Res}^{G'}_{H'}(\mathrm{Ind}_\varphi(\tilde h_\beta))\circ \mathrm{Res}^{F'}_{H'}\Big(\tilde d^{G'}_{F'}\big(\mathrm{Ind}^{F'}_{H'}(\mathrm{Ind}_\psi(A_\beta))\big)\Big)\circ \tilde d^{F'}_{H'}(\mathrm{Ind}_\psi(A_\beta))\\ &= \tilde g_\beta\end{aligned}$$

Moreover, if $H=\varphi^{-1}(H')$ then, setting $F'=H'$, by the equality in condition 14.7.1, we have

14.18.8
$$\tilde g_\beta(\mathrm{Res}^G_H(A)) = \mathrm{Ind}_\psi\big(\tilde h_\beta(\mathrm{Res}^G_H(A))\big) = \mathrm{Ind}_\psi\big(\tilde e(\mathrm{Res}^G_H(A))\circ \tilde f_\beta\big)$$

and therefore, from 14.8.5 and 14.18.1, we get

14.18.9
$$\begin{aligned}\tilde g_\beta &= \tilde d^{G'}_{H',\varphi}(\mathsf{E}(A))\circ \mathrm{Ind}_\psi\big(\tilde r^G_H(A)\circ \tilde e(\mathrm{Res}^G_H(A))\circ \tilde f_\beta\big)\\ &= \tilde d^{G'}_{H',\varphi}(\mathsf{E}(A))\circ \mathrm{Ind}_\psi\big(\mathrm{Res}^G_H(\tilde e(A))\big)\circ \mathrm{Ind}_\psi(\tilde f_\beta)\\ &= \mathrm{Res}^{G'}_{H'}\big(\mathrm{Ind}_\varphi(\tilde e(A))\big)\circ \tilde d^{G'}_{H',\varphi}(A)\circ \mathrm{Ind}_\psi(\tilde f_\beta) \quad .\end{aligned}$$

Proposition 14.19 *With the notation above, assume that $H=\varphi^{-1}(H')$. For any point β' of H' on $\mathrm{Ind}_\varphi(A)$ such that homomorphism 14.17.4 has multiplicity nonzero at (β,β'), there is a unique $\mathfrak{D}H'$-interior algebra exoembedding*

14.19.1
$$\tilde h^\beta_{\beta'} : \mathrm{Ind}_\varphi(A)_{\beta'} \longrightarrow \mathrm{Ind}_\psi(A_\beta)$$

such that we have

14.19.2
$$\mathrm{Res}^{G'}_{H'}\big(\mathrm{Ind}_\varphi(\tilde e(A))\big)\circ \tilde f_{\beta'} = \tilde g_\beta\circ \tilde h^\beta_{\beta'} \quad .$$

Remark 14.20 Since $\mathrm{Tr}_{\varphi(G)}^{G'}\big(1 \otimes (1 \otimes 1) \otimes 1\big)$ is the unity element in $\mathrm{Ind}_\varphi(A)$, any pointed group on $\mathrm{Ind}_\varphi(A)$ has a G'-conjugate $L'_{\varepsilon'}$ such that

$$s_{\varepsilon'}\big(1 \otimes (1 \otimes 1) \otimes 1\big) \neq 0$$

(cf. 2.8.1) and therefore there is a point ε of $L = \varphi^{-1}(L')$ on A such that homomorphism 14.17.4 (for L') has multiplicity nonzero at $(\varepsilon, \varepsilon')$.

Proof: Choose $i \in \beta$; according to our hypothesis on β', there is $i' \in \beta'$ such that, setting $j' = \mathrm{Tr}_{\varphi(H)}^{H'}\big(1 \otimes (1 \otimes i) \otimes 1\big)$, we have $i'j' = i' = j'i'$ and, in particular, we get $i'\mathrm{Ind}_\varphi(A)i' \subset j'\mathrm{Ind}_\varphi(A)j'$; but, it is easily checked that we have a $\mathfrak{D}H'$-interior algebra isomorphism

14.19.3 $$j'\mathrm{Ind}_\varphi(A)j' \cong \mathrm{Ind}_\psi(iAi)$$

which, for any $a \in A$ such that $K = \mathrm{Ker}(\varphi)$ fixes $1 \otimes a$ in $\mathcal{O} \otimes_{\mathcal{O}K} A$, maps $j'\big(1 \otimes (1 \otimes a) \otimes 1\big)j'$ on $1 \otimes (1 \otimes iai) \otimes 1$; consequently, we get the $\mathfrak{D}H'$-interior algebra embedding

14.19.4 $$h_{\beta'}^{\beta} : i'\mathrm{Ind}_\varphi(A)i' \subset j'\mathrm{Ind}_\varphi(A)j' \cong \mathrm{Ind}_\psi(iAi)$$

and it is clear that the composed $\mathfrak{D}H'$-interior algebra embedding

14.19.5 $$i'\mathrm{Ind}_\varphi(A)i' \xrightarrow{h_{\beta'}^{\beta}} \mathrm{Ind}_\psi(iAi) \subset \mathrm{Ind}_\psi\big(\mathrm{Res}_H^G(A)\big) \xrightarrow{d_{H',\varphi}^{G'}(A)} \mathrm{Res}_{H'}^{G'}\big(\mathrm{Ind}_\varphi(A)\big)$$

coincides with the inclusion $i'\mathrm{Ind}_\varphi(A)i' \subset \mathrm{Ind}_\varphi(A)$; now, equality 14.19.2 follows from equality 14.18.2.

Corollary 14.21 *There is a unique $\mathfrak{D}G'$-interior algebra exoembedding*

14.21.1 $$\tilde{\imath}_\varphi(A) : \mathsf{E}\big(\mathrm{Ind}_\varphi(A)\big) \longrightarrow \mathrm{Ind}_\varphi\big(\mathsf{E}(A)\big)$$

such that we have $\tilde{\imath}_\varphi(A) \circ \tilde{e}\big(\mathrm{Ind}_\varphi(A)\big) = \mathrm{Ind}_\varphi\big(\tilde{e}(A)\big)$. *In particular, if* $A \cong \mathsf{E}(A)$ *then* $\mathrm{Ind}_\varphi(A) \cong \mathsf{E}\big(\mathrm{Ind}_\varphi(A)\big)$.

Proof: Let $H'_{\beta'}$ be a pointed group on $\mathrm{Ind}_\varphi(A)$ and set $H = \varphi^{-1}(H')$; by Remark 14.20, up to replace $H'_{\beta'}$ by a G'-conjugate, we may assume that there is a point β of H on A such that homomorphism 14.17.4 has multiplicity nonzero at (β, β') and we claim that the composed exoembedding (cf. Corollary 12.7)

14.21.2 $$\begin{array}{ccc} \mathrm{Ind}_{H'}^{G'}\big(\mathrm{Ind}_\varphi(A)_{\beta'}\big) \xrightarrow{\mathrm{Ind}_{H'}^{G'}(\tilde{h}_{\beta'}^{\beta})} \mathrm{Ind}_{H'}^{G'}\big(\mathrm{Ind}_\psi(A_\beta)\big) & & \\ \cong & & \\ \mathrm{Ind}_\varphi\big(\mathrm{Ind}_H^G(A_\beta)\big) \xrightarrow{\mathrm{Ind}_\varphi(\tilde{h}_\beta)} \mathrm{Ind}_\varphi\big(\mathsf{E}(A)\big) & & \end{array}$$

fulfills the equality in condition 14.7.1 with respect to $\mathrm{Ind}_\varphi(\mathsf{E}(A))$ and $\mathrm{Ind}_\varphi(\tilde{e}(A))$; indeed, by 3.5.4 and equality 14.19.2, we have

14.21.3
$$\begin{aligned}&\mathrm{Res}^{G'}_{H'}(\mathrm{Ind}_\varphi(\tilde{h}_\beta)) \circ \mathrm{Res}^{G'}_{H'}(\mathrm{Ind}^{G'}_{H'}(\tilde{h}^\beta_{\beta'})) \circ \tilde{d}^{G'}_{H'}(\mathrm{Ind}_\varphi(A)_{\beta'})\\ &= \tilde{g}_\beta \circ \tilde{h}^\beta_{\beta'} = \mathrm{Res}^{G'}_{H'}\Big(\mathrm{Ind}_\varphi\big(\tilde{e}(A)\big)\Big) \circ \tilde{f}_{\beta'} \ ;\end{aligned}$$

that is to say, $\mathrm{Ind}_\varphi(\mathsf{E}(A))$ and $\mathrm{Ind}_\varphi(\tilde{e}(A))$ fulfill condition 14.7.1 and therefore the corollary follows from condition 14.7.2.

Remark 14.22 With the notation of Proposition 14.19, Proposition 14.7 provides us with a $\mathcal{D}G'$-interior algebra exoembedding $\tilde{h}_{\beta'}$ from $\mathrm{Ind}^{G'}_{H'}(\mathrm{Ind}_\varphi(A)_{\beta'})$ to $\mathsf{E}(\mathrm{Ind}_\varphi(A))$ and our proof of Corollary 14.21 proves also that

14.22.1
$$\tilde{\imath}_\varphi(A) \circ \tilde{h}_{\beta'} = \mathrm{Ind}_\varphi(\tilde{h}_\beta) \circ \mathrm{Ind}^{G'}_{H'}(\tilde{h}^\beta_{\beta'}) \quad .$$

14.23 Actually, the arguments above can be pushed further to give a description of the local structure of $\mathrm{Ind}_\varphi(A)$, which prepares the discussion on the local structure of *Hecke $\mathcal{D}G$-interior algebras* in Section 16. According to Remark 14.13, we may assume without loss of generality that $A \cong \mathsf{E}(A)$, so that we have $\mathrm{Ind}_\varphi(A) \cong \mathsf{E}(\mathrm{Ind}_\varphi(A))$ too. In order to carry out this description, it is proving useful to put the following definition: a *local tracing pair on A and* $\mathrm{Ind}_\varphi(A)$ is a pair $(P_\gamma, P'_{\gamma'})$ of local pointed groups P_γ on A and $P'_{\gamma'}$ on $\mathrm{Ind}_\varphi(A)$ such that we have $\varphi(P) = P'$ and, denoting by $\sigma : P \to P'$ the group homomorphism determined by φ, there is a $\mathcal{D}P'$-interior algebra exoembedding (to compare with Proposition 14.19)

14.23.1
$$\tilde{h}^\gamma_{\gamma'} : \mathrm{Ind}_\varphi(A)_{\gamma'} \longrightarrow \mathrm{Ind}_\sigma(A_\gamma)$$

such that we have

14.23.2
$$\tilde{f}_{\gamma'} = \tilde{g}_\gamma \circ \tilde{h}^\gamma_{\gamma'} \quad ;$$

actually, this equality is equivalent to the following one

14.23.3
$$\tilde{h}_{\gamma'} = \mathrm{Ind}_\varphi(\tilde{h}_\gamma) \circ \mathrm{Ind}^{G'}_{P'}(\tilde{h}^\gamma_{\gamma'})$$

since, by 3.5.4 and definition 14.17.3, we have

14.23.4
$$\tilde{g}_\gamma \circ \tilde{h}^\gamma_{\gamma'} = \mathrm{Res}^{G'}_{P'}(\mathrm{Ind}_\varphi(\tilde{h}_\gamma)) \circ \mathrm{Res}^{G'}_{P'}(\mathrm{Ind}^{G'}_{P'}(\tilde{h}^\gamma_{\gamma'})) \circ \tilde{d}^{G'}_{P'}(\mathrm{Ind}_\varphi(A)_{\gamma'})$$

and it suffices to apply the equality in condition 14.7.1 and, in one direction, Propositions 11.9 and 11.10. Then, $\tilde{h}^\gamma_{\gamma'}$ is unique (cf. Proposition 11.10) and, for any $x \in G$, the pair $((P_\gamma)^x, (P'_{\gamma'})^{\varphi(x)})$ is also a local tracing pair on A and $\mathrm{Ind}_\varphi(A)$. Notice that if φ is injective then σ is an isomorphism, which forces $\tilde{h}^\gamma_{\gamma'}$ to be an exoisomorphism as it could be expected.

Theorem 14.24 *The second projection maps surjectively the set of G-conjugacy classes of local tracing pairs on A and* $\mathrm{Ind}_\varphi(A)$ *onto the set of* G'*-conjugacy classes of local pointed groups on* $\mathrm{Ind}_\varphi(A)$.

Proof: Let $P'_{\gamma'}$ be a local pointed group on $\mathrm{Ind}_\varphi(A)$; by Remark 14.20, up to replace it by a G'-conjugate, we may assume that there is a point ε of $L = \varphi^{-1}(P')$ on A such that homomorphism 14.17.4 (for P') has multiplicity nonzero at (ε, γ'); hence, according to Proposition 14.19, there is a $\mathcal{D}P'$-interior algebra exoembedding

14.24.1 $$\tilde{h}^{\varepsilon}_{\gamma'} : \mathrm{Ind}_\varphi(A)_{\gamma'} \longrightarrow \mathrm{Ind}_\rho(A_\varepsilon)$$

such that we have

14.24.2 $$\tilde{f}_{\gamma'} = \tilde{g}_\varepsilon \circ \tilde{h}^{\varepsilon}_{\gamma'} \quad ,$$

where $\rho : L \to P'$ is the group homomorphism determined by φ. Let P_γ be a defect pointed group of L_ε and denote by $\sigma : P \to P'$ the restriction of ρ; by Corollary 14.11, there is a $\mathcal{D}L$-interior algebra exoembedding

14.24.3 $$\tilde{h}^{\gamma}_{\varepsilon} : A_\varepsilon \longrightarrow \mathrm{Ind}^L_P(A_\gamma)$$

such that we have

14.24.4 $$\tilde{h}_\varepsilon = \tilde{h}_\gamma \circ \mathrm{Ind}^G_L(\tilde{h}^{\gamma}_{\varepsilon}) \quad .$$

Now, we claim that the composed $\mathcal{D}P'$-interior algebra exoembedding (cf. Corollary 12.7)

14.24.5 $$\mathrm{Ind}_\varphi(A)_{\gamma'} \xrightarrow{\tilde{h}^{\varepsilon}_{\gamma'}} \mathrm{Ind}_\rho(A_\varepsilon) \xrightarrow{\mathrm{Ind}_\rho(\tilde{h}^{\gamma}_{\varepsilon})} \mathrm{Ind}_\rho\big(\mathrm{Ind}^L_P(A_\gamma)\big) \cong \mathrm{Ind}_\sigma(A_\gamma)$$

fulfills equality 14.23.2; indeed, from 3.5.4, Corollary 12.7 and definition 14.17.3, up to suitable identifications we get

14.24.6 $$\begin{aligned} &\tilde{g}_\gamma \circ \mathrm{Ind}_\rho(\tilde{h}^{\gamma}_{\varepsilon}) \circ \tilde{h}^{\varepsilon}_{\gamma'} \\ &= \mathrm{Res}^{G'}_{P'}(\mathrm{Ind}_\varphi(\tilde{h}_\gamma)) \circ \mathrm{Res}^{G'}_{P'}\Big(\mathrm{Ind}^{G'}_{P'}(\mathrm{Ind}_\rho(\tilde{h}^{\gamma}_{\varepsilon}))\Big) \circ \tilde{d}^{G'}_{P'}(\mathrm{Ind}_\rho(A_\varepsilon)) \circ \tilde{h}^{\varepsilon}_{\gamma'} \\ &= \mathrm{Res}^{G'}_{P'}\Big(\mathrm{Ind}_\varphi(\tilde{h}_\gamma \circ \mathrm{Ind}^G_L(\tilde{h}^{\gamma}_{\varepsilon}))\Big) \circ \tilde{d}^{G'}_{P'}(\mathrm{Ind}_\rho(A_\varepsilon)) \circ \tilde{h}^{\varepsilon}_{\gamma'} \\ &= \tilde{g}_\varepsilon \circ \tilde{h}^{\varepsilon}_{\gamma'} = \tilde{f}_{\gamma'} \quad . \end{aligned}$$

Moreover, since $\mathrm{Tr}^{P'}_{\varphi(P)}\big(1 \otimes (1 \otimes 1) \otimes 1\big)$ is the unity element in $\mathrm{Ind}_\sigma(A_\gamma)$ and γ' is local, we have $\varphi(P) = P'$ (see 2.11.3) and $(P_\gamma, P'_{\gamma'})$ is a local tracing pair on A and $\mathrm{Ind}_\varphi(A)$.

14.25 As we prove below, the inclusion relation between the local pointed groups on $\mathrm{Ind}_\varphi(A)$ can be also lifted to the following order relation between the local tracing pairs on A and $\mathrm{Ind}_\varphi(A)$. As above, we assume that $A \cong \mathsf{E}(A)$. If $(P_\gamma, P'_{\gamma'})$ and

$(Q_\delta, Q'_{\delta'})$ are local tracing pairs on A and $\mathrm{Ind}_\varphi(A)$, we say that $(Q_\delta, Q'_{\delta'})$ *is contained in* $(P_\gamma, P'_{\gamma'})$, and write $(Q_\delta, Q'_{\delta'}) \subset (P_\gamma, P'_{\gamma'})$, if we have

14.25.1 $$Q_\delta \subset P_\gamma \quad \text{and} \quad Q'_{\delta'} \subset P'_{\gamma'} \quad ,$$

and, denoting by $\sigma : P \to P'$ and $\tau : Q \to Q'$ the group homomorphisms determined by φ, there is a $\mathcal{D}Q'$-interior algebra exoembedding

14.25.2 $$\tilde{g}^\gamma_\delta : \mathrm{Ind}_\tau(A_\delta) \longrightarrow \mathrm{Res}^{P'}_{Q'}\big(\mathrm{Ind}_\sigma(A_\gamma)\big)$$

such that we have

14.25.3 $$\tilde{g}_\delta = \mathrm{Res}^{P'}_{Q'}(\tilde{g}_\gamma) \circ \tilde{g}^\gamma_\delta \quad ;$$

then, $\tilde{g}^\gamma_\delta$ is unique and fulfills

14.25.4 $$\tilde{g}^\gamma_\delta \circ \tilde{h}^\delta_{\delta'} = \mathrm{Res}^{P'}_{Q'}(\tilde{h}^\gamma_{\gamma'}) \circ \tilde{f}^{\gamma'}_{\delta'}$$

since, by composing equality 14.25.3 with $\tilde{h}^\delta_{\delta'}$, we get (cf. 14.23.2)

14.25.5 $$\begin{aligned} \mathrm{Res}^{P'}_{Q'}(\tilde{g}_\gamma) \circ \tilde{g}^\gamma_\delta \circ \tilde{h}^\delta_{\delta'} &= \tilde{g}_\delta \circ \tilde{h}^\delta_{\delta'} = \tilde{f}_{\delta'} = \mathrm{Res}^{P'}_{Q'}(\tilde{f}_{\gamma'}) \circ \tilde{f}^{\gamma'}_{\delta'} \\ &= \mathrm{Res}^{P'}_{Q'}(\tilde{g}_\gamma) \circ \mathrm{Res}^{P'}_{Q'}(\tilde{h}^\gamma_{\gamma'}) \circ \tilde{f}^{\gamma'}_{\delta'} \end{aligned}$$

and both uniqueness and equality 14.25.4 follow from Proposition 11.10. Moreover, it is easily checked that:

14.25.6 *The inclusion between local tracing pairs is transitive.*

And, from 14.25.3 and proposition 11.10, for any local tracing pair $(R_\varepsilon, R'_{\varepsilon'})$ contained in $(Q_\delta, Q'_{\delta'})$ we get

14.25.7 $$\mathrm{Res}^{Q'}_{R'}(\tilde{g}^\gamma_\delta) \circ \tilde{g}^\delta_\varepsilon = \tilde{g}^\gamma_\varepsilon \quad .$$

14.26 Actually, assuming that $Q_\delta \subset P_\gamma$ and denoting by $\tilde{g}_\delta(A_\gamma)$ the corresponding exoembedding 14.17.2 when G, G', φ, A, H_β and H' are respectively replaced by P, P', σ, A_γ, Q_δ and Q', equality 14.25.3 is equivalent to the following one

14.26.1 $$\tilde{g}_\delta(A_\gamma) = \mathrm{Res}^{P'}_{Q'}\Big(\mathrm{Ind}_\sigma\big(\tilde{e}(A_\gamma)\big)\Big) \circ \tilde{g}^\gamma_\delta \quad .$$

Indeed, according to Lemma 14.27 below, for suitable $\mathcal{D}P'$-interior algebra exoembeddings $\tilde{e}$ and $\tilde{g}$ we have

14.26.2 $$\mathrm{Res}^{P'}_{Q'}(\tilde{e}) \circ \tilde{g}_\delta = \mathrm{Res}^{P'}_{Q'}(\tilde{g}) \circ \tilde{g}_\delta(A_\gamma) \text{ and } \tilde{e} \circ \tilde{g}_\gamma = \tilde{g} \circ \mathrm{Ind}_\sigma\big(\tilde{e}(A_\gamma)\big) \quad ;$$

hence, respectively composing both members of equalities 14.25.3 and 14.26.1 with $\mathrm{Res}^{P'}_{Q'}(\tilde{e})$ and $\mathrm{Res}^{P'}_{Q'}(\tilde{g})$ on the left, the announced equivalence follows from Proposition 11.10 and equalities 14.26.2. These equalities are in fact independent of localness and exhibit a sort of transitivity which is a consequence of Proposition 14.16;

to state it properly, we borrow notation from 14.15 and, as above, if H_β and L_ε are pointed groups on A such that $L_\varepsilon \subset H_\beta$, we denote by $\tilde{g}_\varepsilon(A_\beta)$ the corresponding exoembedding 14.17.2 when $G, G', \varphi, A, H_\beta$ and H' are respectively replaced by $H, H', \psi, A_\beta, L_\varepsilon$ and a subgroup L' of H' containing $\varphi(L)$; notice that if $L_\varepsilon = H_\beta$ and $L' = H'$ then

14.26.3 $$\tilde{g}_\beta(A_\beta) = \mathrm{Ind}_\psi\big(\tilde{e}(A_\beta)\big) \quad .$$

Lemma 14.27 *With the notation above, for any pair of pointed groups H_β and L_ε on A such that $L_\varepsilon \subset H_\beta$, and any pair of subgroups H' and L' of G' such that $\varphi(H) \subset H'$ and $\varphi(L) \subset L' \subset H'$, we have the following commutative diagram*

14.27.1 $$\begin{array}{ccc} \mathrm{Res}^{G'}_{L'}(\mathrm{Ind}_\varphi(A)) & \xrightarrow{\mathrm{Res}^{G'}_{L'}(\mathrm{Ind}_\varphi(\tilde{e}(\mathcal{O})\otimes\widetilde{\mathrm{id}}_A))} & \mathrm{Res}^{G'}_{L'}\Big(\mathrm{Ind}_\varphi(\mathsf{E}(\mathcal{O}) \underset{\mathcal{O}}{\otimes} A)\Big) \\ \uparrow{\scriptstyle \tilde{g}_\varepsilon} & & \uparrow{\scriptstyle \mathrm{Res}^{H'}_{L'}(\tilde{g})} \\ \mathrm{Ind}_\eta(A_\varepsilon) & \xrightarrow{\tilde{g}_\varepsilon(A_\beta)} & \mathrm{Res}^{H'}_{L'}\Big(\mathrm{Ind}_\psi(\mathsf{E}(A_\beta))\Big) \end{array}$$

where $\psi : H \to H'$ and $\eta : L \to L'$ are the group homomorphisms determined by φ and we set

14.27.2
$$\tilde{g} = \mathrm{Res}^{G'}_{H'}\bigg(\mathrm{Ind}_\varphi\Big((\widetilde{\mathrm{id}}_{\mathsf{E}(\mathcal{O})} \underset{\mathcal{O}}{\otimes} \tilde{h}_\beta) \circ \mathrm{Ind}^G_H(\tilde{c}^G_H(A_\beta))\Big)\bigg) \circ \tilde{d}^{G'}_{H'}\Big(\mathrm{Ind}_\psi(\mathsf{E}(A_\beta))\Big) \quad .$$

Proof: By the naturality of $d^{G'}_{H'}$ (cf. 3.5.4), we have (cf. Corollary 12.7)

14.27.3 $$\begin{aligned} &\tilde{d}^{G'}_{H'}\Big(\mathrm{Ind}_\psi(\mathsf{E}(A_\beta))\Big) \circ \mathrm{Ind}_\psi(\tilde{h}^\beta_\varepsilon) \\ &= \mathrm{Res}^{G'}_{H'}\Big(\mathrm{Ind}^{G'}_{H'}(\mathrm{Ind}_\psi(\tilde{h}^\beta_\varepsilon))\Big) \circ \tilde{d}^{G'}_{H'}\Big(\mathrm{Ind}_\psi(\mathrm{Ind}^H_L(A_\varepsilon))\Big) \\ &= \mathrm{Res}^{G'}_{H'}\Big(\mathrm{Ind}_\varphi(\mathrm{Ind}^G_H(\tilde{h}^\beta_\varepsilon))\Big) \circ \tilde{d}^{G'}_{H'}\Big(\mathrm{Ind}^{H'}_{L'}(\mathrm{Ind}_\eta(A_\varepsilon))\Big) \quad ; \end{aligned}$$

hence, according to Proposition 14.16 (and 14.16.3), we get

14.27.4
$$\begin{aligned} &\mathrm{Res}^{H'}_{L'}(\tilde{g}) \circ \tilde{g}_\varepsilon(A_\beta) \\ &\quad = \mathrm{Res}^{H'}_{L'}(\tilde{g}) \circ \mathrm{Res}^{H'}_{L'}(\mathrm{Ind}_\psi(\tilde{h}^\beta_\varepsilon)) \circ \tilde{d}^{H'}_{L'}(\mathrm{Ind}_\eta(A_\varepsilon)) \\ &\quad = \mathrm{Res}^{G'}_{L'}\bigg(\mathrm{Ind}_\varphi\Big((\widetilde{\mathrm{id}}_{\mathsf{E}(\mathcal{O})} \underset{\mathcal{O}}{\otimes} \tilde{h}_\beta) \circ \mathrm{Ind}^G_H(\tilde{c}^G_H(A_\beta)) \circ \mathrm{Ind}^G_H(\tilde{h}^\beta_\varepsilon)\Big)\bigg) \circ \tilde{d}^{G'}_{L'}(\mathrm{Ind}_\eta(A_\varepsilon)) \\ &\quad = \mathrm{Res}^{G'}_{L'}\Big(\mathrm{Ind}_\varphi((\tilde{e}(\mathcal{O}) \underset{\mathcal{O}}{\otimes} \widetilde{\mathrm{id}}_A) \circ \tilde{h}_\varepsilon)\Big) \circ \tilde{d}^{G'}_{L'}(\mathrm{Ind}_\eta(A_\varepsilon)) \\ &\quad = \mathrm{Res}^{G'}_{L'}\Big(\mathrm{Ind}_\varphi(\tilde{e}(\mathcal{O}) \underset{\mathcal{O}}{\otimes} \widetilde{\mathrm{id}}_A)\Big) \circ \tilde{g}_\varepsilon \quad . \end{aligned}$$

Theorem 14.28 *For any local tracing pair $(P_\gamma, P'_{\gamma'})$ on A and $\mathrm{Ind}_\varphi(A)$, and any local pointed group $Q'_{\delta'}$ on $\mathrm{Ind}_\varphi(A)$ such that $Q'_{\delta'} \subset P'_{\gamma'}$, there is a local pointed group Q_δ on A such that $(Q_\delta, Q'_{\delta'})$ is a local tracing pair on A and $\mathrm{Ind}_\varphi(A)$, contained in $(P_\gamma, P'_{\gamma'})$.*

Proof: Let $\sigma : P \to P'$ be the group homomorphism detemined by φ, set $R = \varphi^{-1}(Q')$ and denote by $\rho : R \to Q'$ the group homomorphism determined by σ; clearly, in 14.17 we can replace respectively G, G', φ, A, H, H' and ψ by $P, P', \sigma, A_\gamma, R, Q'$ and ρ; moreover, δ' can be identified to a local point of Q' on $\mathrm{Ind}_\sigma(A_\gamma)$ throughout the composed $\mathcal{D}Q'$-interior algebra exoembedding (cf. 14.23.1)

14.28.1 $$\mathrm{Ind}_\varphi(A)_{\delta'} \xrightarrow{\tilde{f}^{\gamma'}_{\delta'}} \mathrm{Res}^{P'}_{Q'}\big(\mathrm{Ind}_\varphi(A)_{\gamma'}\big) \xrightarrow{\mathrm{Res}^{P'}_{Q'}(\tilde{h}^{\gamma}_{\gamma'})} \mathrm{Res}^{P'}_{Q'}\big(\mathrm{Ind}_\sigma(A_\gamma)\big) \quad .$$

In this case, since σ is surjective, it follows from Remark 14.20 that there is a point ε of R on A_γ such that homomorphism 14.17.4 has multiplicity nonzero at (ε, δ'); consequently, by Proposition 14.18, we have

14.28.2 $$\tilde{g}_\varepsilon(A_\gamma) = \mathrm{Res}^{P'}_{Q'}\Big(\mathrm{Ind}_\sigma\big(\tilde{e}(A_\gamma)\big)\Big) \circ \tilde{d}^{P'}_{Q',\sigma}(A_\gamma) \circ \mathrm{Ind}_\rho(\tilde{f}^{\gamma}_{\varepsilon})$$

whereas, by Proposition 14.19, there is a $\mathcal{D}Q'$-interior algebra exoembedding

14.28.3 $$\tilde{h}^{\varepsilon}_{\delta'} : \mathrm{Ind}_\varphi(A)_{\delta'} \longrightarrow \mathrm{Ind}_\rho(A_\varepsilon)$$

such that we have (cf. 14.26.1)

14.28.4 $$\mathrm{Res}^{P'}_{Q'}\Big(\mathrm{Ind}_\sigma\big(\tilde{e}(A_\gamma)\big)\Big) \circ \mathrm{Res}^{P'}_{Q'}(\tilde{h}^{\gamma}_{\gamma'}) \circ \tilde{f}^{\gamma'}_{\delta'} = \tilde{g}_\varepsilon(A_\gamma) \circ \tilde{h}^{\varepsilon}_{\delta'} \quad .$$

Let Q_δ be a defect pointed group of R_ε and denote by $\tau : Q \to Q'$ the restriction of ρ; by Corollary 14.11 applied to R_ε, $A_\gamma$and Q_δ, there is a $\mathcal{D}R$-interior algebra exoembedding

14.28.5 $$\tilde{h}^{\delta}_{\varepsilon} : A_\varepsilon \longrightarrow \mathrm{Ind}^{R}_{Q}(A_\delta)$$

such that we have

14.28.6 $$\tilde{h}^{\gamma}_{\varepsilon} = \tilde{h}^{\gamma}_{\delta} \circ \mathrm{Ind}^{P}_{R}(\tilde{h}^{\delta}_{\varepsilon}) \qquad ;$$

moreover, since R is a p-group and we have (cf. 12.4.4)

14.28.7 $$\mathrm{C}_\circ\big(\mathrm{Ind}^{R}_{Q}(A_\delta)\big) \cong \mathrm{Ind}^{R}_{Q}\big(\mathrm{C}_\circ(A_\delta)\big) \qquad ,$$

$\tilde{h}^{\delta}_{\varepsilon}$ is actually an exoisomorphism (cf. 2.12.2). Now, we claim that the composed $\mathcal{D}Q'$-interior algebra exoembeddings

14.28.8
$$\mathrm{Ind}_\varphi(A)_{\delta'} \xrightarrow{\tilde{h}^{\varepsilon}_{\delta'}} \mathrm{Ind}_\rho(A_\varepsilon) \overset{\mathrm{Ind}_\rho(\tilde{h}^{\delta}_{\varepsilon})}{\cong} \mathrm{Ind}_\rho\big(\mathrm{Ind}^{R}_{Q}(A_\delta)\big) \cong \mathrm{Ind}_\tau(A_\delta)$$
$$\mathrm{Ind}_\tau(A_\delta) \cong \mathrm{Ind}_\rho(A_\varepsilon) \xrightarrow{\mathrm{Ind}_\rho(\tilde{f}^{\gamma}_{\varepsilon})} \mathrm{Ind}_\rho\big(\mathrm{Res}^{P}_{R}(A_\delta)\big) \xrightarrow{\tilde{d}^{P'}_{Q',\sigma}(A_\gamma)} \mathrm{Res}^{P'}_{Q'}\big(\mathrm{Ind}_\sigma(A_\gamma)\big)$$

respectively fulfill equality 14.23.2, so that $(Q_\delta, Q'_{\delta'})$ is a local tracing pair on A and $\mathrm{Ind}_\varphi(A)$ (since the first exoembedding together with the localness of δ' force $\tau(Q) = Q'$ by 2.11.3), and equality 14.26.1, so that $(Q_\delta, Q'_{\delta'})$ is contained in $(P_\gamma, P'_{\gamma'})$.

We prove first the second affirmation, namely that the $\mathcal{D}Q'$-interior algebra exoembedding

$$\tilde{g}^\gamma_\delta = \tilde{d}^{P'}_{Q',\sigma}(A_\gamma) \circ \mathrm{Ind}_\rho(\tilde{f}^\gamma_\varepsilon) \circ \mathrm{Ind}_\rho(\tilde{h}^\delta_\varepsilon)^{-1} \tag{14.28.9}$$

fulfills equality 14.26.1 (and therefore, equality 14.25.3); indeed, by 3.5.4, 14.28.2 and 14.28.6, we have

$$\begin{aligned}
\mathrm{Res}^{P'}_{Q'}\Big(\mathrm{Ind}_\sigma\big(\tilde{e}(A_\gamma)\big)\Big) \circ \tilde{g}^\gamma_\delta &= \tilde{g}_\varepsilon(A_\gamma) \circ \mathrm{Ind}_\rho(\tilde{h}^\delta_\varepsilon)^{-1} \\
&= \mathrm{Res}^{P'}_{Q'}(\mathrm{Ind}_\sigma(\tilde{h}^\gamma_\varepsilon)) \circ \tilde{d}^{P'}_{Q'}(\mathrm{Ind}_\rho(A_\varepsilon)) \circ \mathrm{Ind}_\rho(\tilde{h}^\delta_\varepsilon)^{-1} \\
&= \mathrm{Res}^{P'}_{Q'}\Big(\mathrm{Ind}_\sigma(\tilde{h}^\gamma_\varepsilon) \circ \mathrm{Ind}^{P'}_{Q'}(\mathrm{Ind}_\rho(\tilde{h}^\delta_\varepsilon))^{-1}\Big) \circ \tilde{d}^{P'}_{Q'}(\mathrm{Ind}_\tau(A_\delta)) \\
&= \mathrm{Res}^{P'}_{Q'}(\mathrm{Ind}_\sigma(\tilde{h}^\gamma_\delta)) \circ \tilde{d}^{P'}_{Q'}(\mathrm{Ind}_\tau(A_\delta)) = \tilde{g}_\delta(A_\gamma) \quad .
\end{aligned} \tag{14.28.10}$$

We prove now that the $\mathcal{D}Q'$-interior algebra exoembedding

$$\tilde{h}^\delta_{\delta'} = \mathrm{Ind}_\rho(\tilde{h}^\delta_\varepsilon) \circ \tilde{h}^\varepsilon_{\delta'} \tag{14.28.11}$$

fulfills equality 14.23.2; notice that, by 14.28.4, we have

$$\begin{aligned}
\mathrm{Res}^{P'}_{Q'}\Big(\mathrm{Ind}_\sigma\big(\tilde{e}(A_\gamma)\big)\Big) \circ \tilde{g}^\gamma_\delta \circ \tilde{h}^\delta_{\delta'} &= \tilde{g}_\varepsilon(A_\gamma) \circ \tilde{h}^\varepsilon_{\delta'} \\
&= \mathrm{Res}^{P'}_{Q'}\Big(\mathrm{Ind}_\sigma\big(\tilde{e}(A_\gamma)\big)\Big) \circ \mathrm{Res}^{P'}_{Q'}(\tilde{h}^\gamma_{\gamma'}) \circ \tilde{f}^\gamma_{\delta'} \quad ;
\end{aligned} \tag{14.28.12}$$

hence, according to Proposition 11.10, we get

$$\tilde{g}^\gamma_\delta \circ \tilde{h}^\delta_{\delta'} = \mathrm{Res}^{P'}_{Q'}(\tilde{h}^\gamma_{\gamma'}) \circ \tilde{f}^{\gamma'}_{\delta'} \tag{14.28.13}$$

and then, by composing both members with $\mathrm{Res}^{P'}_{Q'}(\tilde{g}_\gamma)$, we obtain (cf. 14.25.3)

$$\begin{aligned}
\tilde{g}_\delta \circ \tilde{h}^\delta_{\delta'} &= \mathrm{Res}^{P'}_{Q'}(\tilde{g}_\gamma) \circ \tilde{g}^\gamma_\delta \circ \tilde{h}^\delta_{\delta'} = \mathrm{Res}^{P'}_{Q'}(\tilde{g}_\gamma \circ \tilde{h}^\gamma_{\gamma'}) \circ \tilde{f}^{\gamma'}_{\delta'} \\
&= \mathrm{Res}^{P'}_{Q'}(\tilde{f}_{\gamma'}) \circ \tilde{f}^{\gamma'}_{\delta'} = \tilde{f}_{\delta'} \quad .
\end{aligned} \tag{14.28.14}$$

14.29 Finally, although we need not it in the sequel, let us explain the action of the bijective *A-fusions* (cf. 14.3) stabilizing $K = \mathrm{Ker}(\varphi)$ on the local tracing pairs on A and $\mathrm{Ind}_\varphi(A)$. If $(P_\gamma, P'_{\gamma'})$ is a local tracing pair on A and $\mathrm{Ind}_\varphi(A)$, Q_δ is a local pointed group on A such that $|Q| = |P|$ and $F_{C_\circ(A)}(Q_\delta, P_\gamma) \neq \emptyset$, and $\tilde{\theta}$ is a $C_\circ(A)$-fusion from Q_δ to P_γ (cf. 2.13) such that $\theta(K \cap Q) = K \cap P$ for some representative θ of $\tilde{\theta}$ then, setting $Q' = \varphi(Q)$ and denoting by $\sigma : P \to P'$ and $\tau : Q \to Q'$ the group homomorphisms determined by φ, θ determines a group isomorphism $\theta' : Q' \cong P'$ and the $\mathcal{D}Q$-interior algebra exoisomorphism $\tilde{f}_\theta : A_\delta \cong \mathrm{Res}_\theta(A_\gamma)$ (cf. 14.3.3) determines a $\mathcal{D}Q'$-interior algebra exoisomorphism (cf. Proposition 12.12)

14.29.1 $$\mathrm{Ind}_\tau(A_\delta) \overset{\mathrm{Ind}_\tau(\tilde{f}_\theta)}{\cong} \mathrm{Ind}_\tau\big(\mathrm{Res}_\theta(A_\gamma)\big) \cong \mathrm{Res}_{\theta'}\big(\mathrm{Ind}_\sigma(A_\gamma)\big) \quad ;$$

now, up to suitable idenifications throughout $\tilde{h}^\gamma_{\gamma'}$ and $\tilde{g}_\delta$, the inverse of this exoisomorphism maps γ' on a local point δ' of Q' on $\mathrm{Ind}_\varphi(A)$ such that $\tilde{\theta}'$ becomes a $\mathrm{C}_\circ\big(\mathrm{Ind}_\varphi(A)\big)$-fusion from $Q'_{\delta'}$ to $P'_{\gamma'}$ and $(Q_\delta, Q'_{\delta'})$ is a local tracing pair on A and $\mathrm{Ind}_\varphi(A)$.

15 Hecke $\mathcal{D}G$-interior algebras and noninjective induction

15.1 As the title suggests, this section is mainly devoted to generalize the results of Section 4 from $\mathcal{O}G$-interior algebras to $\mathcal{D}G$-interior algebras, except that it includes remarks on Higman envelopes which were not needed there. Let G and G' be finite groups, $H^{\cdot\cdot}$ a subgroup of $G \times G'$ and $B^{\cdot\cdot}$ a $\mathcal{D}H^{\cdot\cdot}$-interior algebra; *mutatis mutandis*, we call the $\mathcal{D}G$-interior algebra $\mathrm{Ind}_{H^{\cdot\cdot}}^{G\times G'}(B^{\cdot\cdot})^{1\times G'}$ *Hecke $\mathcal{D}G$-interior algebra* associated with G', $H^{\cdot\cdot}$ and $B^{\cdot\cdot}$; for a suitable choice of $H^{\cdot\cdot}$ and $B^{\cdot\cdot}$, this $\mathcal{D}G$-interior algebra extends to the *Rickard equivalences* between blocks (see Section 18) the role that Hecke $\mathcal{O}G$-interior algebras play in the Morita (stable) equivalences between blocks (see Section 6). As in Section 4, first of all we discuss the particular case where $H^{\cdot\cdot} = G \times G'$; respectively denote by π and π' the first and the second projection maps for $G \times G'$ and consider $\mathcal{O}$ endowed with the extension of the counity map (cf. 9.10) $\mathcal{D} \to \mathcal{O}$ mapping $f \in \mathcal{F}$ on $f(0)$ and d on zero, so that $\mathcal{O}G'$ and $\mathrm{End}_{\mathcal{O}}(\mathcal{O}G') \cong \mathrm{Ind}_1^{G'}(\mathcal{O})$ become $\mathcal{D}G'$-interior algebras whereas $\mathcal{O}(G \times G')$ becomes a $\mathcal{D}(G\times G')$-interior algebra; moreover, for any $\mathcal{D}(G\times G')$-interior algebra $A^{\cdot\cdot}$, we consider $\mathrm{Res}_{G\times 1}^{G\times G'}(A^{\cdot\cdot})$ as a $\mathcal{D}G$-interior algebra.

Proposition 15.2 *Let $A^{\cdot\cdot}$ be a $\mathcal{D}(G \times G')$-interior algebra. There is a unique $\mathcal{D}G$-interior algebra isomorphism*

15.2.1 $$\mathrm{Res}_{G\times 1}^{G\times G'}(A^{\cdot\cdot}) \cong \mathrm{Ind}_\pi\Big(A^{\cdot\cdot} \underset{\mathcal{O}}{\otimes} \mathrm{Res}_{\pi'}(\mathrm{End}_{\mathcal{O}}(\mathcal{O}G'))\Big)$$

mapping $a^{\cdot\cdot} \in A^{\cdot\cdot}$ on $1 \otimes \mathrm{Tr}_{1\times 1}^{1\times G'}\big(a^{\cdot\cdot} \otimes (1\otimes 1\otimes 1)\big)$. In particular, this isomorphism restricts to a $\mathcal{D}G$-algebra isomorphism

15.2.2 $$H_{G,G'}(A^{\cdot\cdot}) : (A^{\cdot\cdot})^{1\times G'} \cong \mathrm{Ind}_\pi\big(A^{\cdot\cdot} \underset{\mathcal{O}}{\otimes} \mathrm{Res}_{\pi'}(\mathcal{O}G')\big)$$

which maps $a^{\cdot\cdot} \in (A^{\cdot\cdot})^{1\times G'}$ on $1 \otimes (a^{\cdot\cdot} \otimes 1)$.

Proof: As in 4.2.3, we have a canonical $\mathcal{D}(G \times G')$-interior algebra isomorphism

15.2.3 $$\mathrm{Res}_{\pi'}\big(\mathrm{End}_{\mathcal{O}}(\mathcal{O}G')\big) \cong \mathrm{Ind}_{G\times 1}^{G\times G'}(\mathcal{O}) \qquad ;$$

hence, by Proposition 12.9, we have a $\mathcal{D}(G \times G')$-interior algebra isomorphism

15.2.4 $$A^{\cdot\cdot} \underset{\mathcal{O}}{\otimes} \mathrm{Res}_{\pi'}\big(\mathrm{End}_{\mathcal{O}}(\mathcal{O}G')\big) \cong \mathrm{Ind}_{G\times 1}^{G\times G'}\big(\mathrm{Res}_{G\times 1}^{G\times G'}(A^{\cdot\cdot})\big)$$

mapping $a^{\cdot\cdot} \otimes (1\otimes 1\otimes 1)$ on $1 \otimes a^{\cdot\cdot} \otimes 1$ for any $a^{\cdot\cdot} \in A^{\cdot\cdot}$; moreover, it follows from Corollary 12.7 that there is a $\mathcal{D}G$-interior algebra isomorphism

15.2.5 $$\mathrm{Ind}_\pi\Big(\mathrm{Ind}_{G\times 1}^{G\times G'}\big(\mathrm{Res}_{G\times 1}^{G\times G'}(A^{\cdot\cdot})\big)\Big) \cong \mathrm{Res}_{G\times 1}^{G\times G'}(A^{\cdot\cdot})$$

mapping $1 \otimes \mathrm{Tr}_{1\times 1}^{1\times G'}(1 \otimes a^{\cdot\cdot} \otimes 1)$ on $a^{\cdot\cdot}$ for any $a^{\cdot\cdot} \in A^{\cdot\cdot}$; now, we get the announced isomorphism 15.2.1 by composing the inverses of isomorphism 15.2.5 and the corresponding induced isomorphism of isomorphism 15.2.4. The existence of a suitable restriction follows from the same statement in Proposition 4.2.

15.3 As in 4.3, in the general situation ρ and ρ' respectively denote the restrictions of π and π' to $H^{\cdot\cdot}$, and we set

15.3.1 $$\mathrm{Ker}(\rho) = 1 \times K' \quad \text{and} \quad \mathrm{Ker}(\rho') = K \times 1 \quad ,$$

where K and K' are respectively suitable subgroups of G and G'; here, $B^{\cdot\cdot} \otimes_{\mathcal{O}} \mathrm{Res}_{\rho'}(\mathcal{O}G')$ is a $\mathcal{D}(1 \times K')$-bimodule by left and right multiplication, and notice that the $\mathcal{O}$-linear map (cf. 4.3.2)

15.3.2 $$q_{G,G'}(B^{\cdot\cdot}) : \mathcal{O} \underset{\mathcal{O}(1\times K')}{\otimes} \big(B^{\cdot\cdot} \underset{\mathcal{O}}{\otimes} \mathrm{Res}_{\rho'}(\mathcal{O}G')\big) \longrightarrow \mathrm{Ind}_{H^{\cdot\cdot}}^{G\times G'}(B^{\cdot\cdot}) \quad ,$$

mapping $1 \otimes (b^{\cdot\cdot} \otimes x')$ on $x'^{-1} \otimes b^{\cdot\cdot} \otimes 1$ for any $b^{\cdot\cdot} \in B^{\cdot\cdot}$ and any $x' \in G'$, is actually $\mathcal{D} \otimes_{\mathcal{O}} \mathcal{D}^{\circ}$-linear.

Theorem 15.4 *With the notation above, there is a unique $\mathcal{D}G$-interior algebra isomorphism*

15.4.1 $$H_{G,G'}(B^{\cdot\cdot}) : \mathrm{Ind}_{H^{\cdot\cdot}}^{G\times G'}(B^{\cdot\cdot})^{1\times G'} \cong \mathrm{Ind}_{\rho}\big(B^{\cdot\cdot} \underset{\mathcal{O}}{\otimes} \mathrm{Res}_{\rho'}(\mathcal{O}G')\big)$$

mapping $\mathrm{Tr}_{1\times K'}^{1\times G'}\big(q_{G,G'}(B^{\cdot\cdot})(1\otimes a)\big)$ *on* $1\otimes(1\otimes a)\otimes 1$ *for any* $a \in B^{\cdot\cdot} \otimes_{\mathcal{O}} \mathrm{Res}_{\rho'}(\mathcal{O}G')$ *such that* $1 \times K'$ *fixes* $1 \otimes a$ *in* $\mathcal{O} \otimes_{\mathcal{O}(1\times K')} \big(B^{\cdot\cdot} \otimes_{\mathcal{O}} \mathrm{Res}_{\rho'}(\mathcal{O}G')\big)$.

Remark 15.5 Since we have a $G \times G'$-interior $\mathcal{F}$-algebra isomorphism (cf. 12.4.4)

15.5.1 $$\mathrm{Ind}_{H^{\cdot\cdot}}^{G\times G'}(\mathrm{C}(B^{\cdot\cdot})) \cong \mathrm{C}\big(\mathrm{Ind}_{H^{\cdot\cdot}}^{G\times G'}(B^{\cdot\cdot})\big) \quad ,$$

isormorphism 15.4.1 applied twice induces the following G-interior $\mathcal{F}$-algebra isomorphisms

15.5.2 $$\begin{aligned} \mathrm{Ind}_{H^{\cdot\cdot}}^{G\times G'}\big(\mathrm{C}(B^{\cdot\cdot})\big)^{1\times G'} &\cong \mathrm{Ind}_{\rho}\big(\mathrm{C}(B^{\cdot\cdot}) \underset{\mathcal{O}}{\otimes} \mathrm{Res}_{\rho'}(\mathcal{O}G')\big) \\ &\cong \mathrm{C}\Big(\mathrm{Ind}_{\rho}\big(B^{\cdot\cdot} \underset{\mathcal{O}}{\otimes} \mathrm{Res}_{\rho'}(\mathcal{O}G')\big)\Big) \quad . \end{aligned}$$

Proof: According to Theorem 4.4, it suffices to prove that isomorphism 4.4.1 is $\mathcal{D} \otimes_{\mathcal{O}} \mathcal{D}^{\circ}$-linear; but for any $x', y' \in G'$, any $a \in B^{\cdot\cdot} \otimes_{\mathcal{O}} \mathrm{Res}_{\rho'}(\mathcal{O}G')$ such that $1 \times K'$ fixes $1 \otimes a$ in $\mathcal{O} \otimes_{\mathcal{O}(1\times K')} \big(B^{\cdot\cdot} \otimes_{\mathcal{O}} \mathrm{Res}_{\rho'}(\mathcal{O}G')\big)$ and any $g, h \in \mathcal{D}$, we have (cf. 12.2.4)

15.4.2 $$g\cdot\big(x' \otimes (1 \otimes a) \otimes y'\big)\cdot h = x' \otimes (1 \otimes g\cdot a\cdot h) \otimes y'$$

and, since $q_{G,G'}(B^{\cdot\cdot})$ is $\mathcal{D} \otimes_{\mathcal{O}} \mathcal{D}^{\circ}$-linear, we have also

15.4.3 $$g\cdot \mathrm{Tr}_{1\times K'}^{1\times G'}\big(q_{G,G'}(B^{\cdot\cdot})(1\otimes a)\big)\cdot h = \mathrm{Tr}_{1\times K'}^{1\times G'}\big(q_{G,G'}(B^{\cdot\cdot})(1\otimes g\cdot a\cdot h)\big) \quad ;$$

consequently, we get

15.4.4 $$\begin{aligned} &H_{G,G'}(B^{\cdot\cdot})\Big(gx'\cdot \mathrm{Tr}_{1\times K'}^{1\times G'}\big(q_{G,G'}(B^{\cdot\cdot})(1\otimes a)\big)\cdot y' h\Big) \\ &= x'\cdot H_{G,G'}(B^{\cdot\cdot})\Big(\mathrm{Tr}_{1\times K'}^{1\times G'}\big(q_{G,G'}(B^{\cdot\cdot})(1\otimes g\cdot a\cdot h)\big)\Big)\cdot y' \\ &= x'\otimes(1\otimes g\cdot a\cdot h)\otimes y' = g\cdot\big(x'\otimes(1\otimes a)\otimes y'\big)\cdot h \\ &= g\cdot H_{G,G'}(B^{\cdot\cdot})\Big(x'\cdot \mathrm{Tr}_{1\times K'}^{1\times G'}\big(q_{G,G'}(B^{\cdot\cdot})(1\otimes a)\big)\cdot y'\Big)\cdot h \quad . \end{aligned}$$

Remark 15.6 Obviously the same kind of proof works for Proposition 15.2, but the direct proof is no longuer.

15.7 If $C^{\cdot\cdot}$ is a second $\mathcal{D}H^{\cdot\cdot}$-interior algebra and $\tilde{f}^{\cdot\cdot} : B^{\cdot\cdot} \to C^{\cdot\cdot}$ a $H^{\cdot\cdot}$-interior $\mathcal{D}$-algebra exomorphism, as in 4.6, we have

15.7.1 $$\tilde{H}_{G,G'}(C^{\cdot\cdot})\circ \mathrm{Ind}_{H^{\cdot\cdot}}^{G\times G'}(\tilde{f}^{\cdot\cdot})^{1\times G'} = \mathrm{Ind}_{\rho}(\tilde{f}^{\cdot\cdot} \underset{\mathcal{O}}{\otimes} \tilde{\mathrm{id}}_{\mathrm{Res}_{\rho'}(\mathcal{O}G')})\circ \tilde{H}_{G,G'}(B^{\cdot\cdot}) \quad .$$

On the other hand, if G'' is a third finite group, $L^{\cdot\cdot\cdot}$ a subgroup of $H^{\cdot\cdot} \times G''$ and $D^{\cdot\cdot\cdot}$ a $\mathcal{D}L^{\cdot\cdot\cdot}$-interior algebra, we have an evident $\mathcal{D}G$-interior algebra isomorphism (cf. 4.6.2)

15.7.2 $$\mathrm{Ind}_{H^{\cdot\cdot}}^{G\times G'}\big(\mathrm{Ind}_{L^{\cdot\cdot\cdot}}^{H^{\cdot\cdot}\times G''}(D^{\cdot\cdot\cdot})^{1\times 1\times G''}\big)^{1\times G'} \cong \mathrm{Ind}_{L^{\cdot\cdot\cdot}}^{G\times G'\times G''}(D^{\cdot\cdot\cdot})^{1\times G'\times G''}$$

which shows the transitive nature of our construction and closes the following commutative diagram of $\mathcal{D}G$-interior algebra isomorphisms (cf. Corollary 12.7, Proposition 12.9 and Theorem 15.4)

15.7.3 $$\begin{array}{ccc} \mathrm{Ind}_{\rho}\big(\mathrm{Ind}_{\sigma^{\cdot\cdot}}(C) \underset{\mathcal{O}}{\otimes} \mathrm{Res}_{\rho'}(\mathcal{O}G')\big) & \cong & \mathrm{Ind}_{\sigma}\Big(D^{\cdot\cdot\cdot} \underset{\mathcal{O}}{\otimes} \mathrm{Res}_{\tau}\big(\mathcal{O}(G'\times G'')\big)\Big) \\ {\scriptstyle H_{G,G'}(\mathrm{Ind}_{\sigma^{\cdot\cdot}}(C))}\ \wr\Vert & & \\ \mathrm{Ind}_{H^{\cdot\cdot}}^{G\times G'}\big(\mathrm{Ind}_{\sigma^{\cdot\cdot}}(C)\big)^{1\times G'} & & \wr\Vert\ {\scriptstyle H_{G,G'\times G''}(D^{\cdot\cdot\cdot})} \\ {\scriptstyle \mathrm{Ind}_{H^{\cdot\cdot}}^{G\times G'}(H_{H^{\cdot\cdot},G''}(D^{\cdot\cdot\cdot}))}\ \wr\Vert & & \\ \mathrm{Ind}_{H^{\cdot\cdot}}^{G\times G'}\big(\mathrm{Ind}_{L^{\cdot\cdot\cdot}}^{H^{\cdot\cdot}\times G''}(D^{\cdot\cdot\cdot})^{1\times 1\times G''}\big)^{1\times G'} & \cong & \mathrm{Ind}_{L^{\cdot\cdot\cdot}}^{G\times G'\times G''}(D^{\cdot\cdot\cdot})^{1\times G'\times G''} \end{array}$$

where $\sigma^{\cdot\cdot} : L^{\cdot\cdot\cdot} \to H^{\cdot\cdot}$ and $\sigma'' : L^{\cdot\cdot\cdot} \to G''$ are respectively the group homomorphisms determined by the first and the second projection maps for $H^{\cdot\cdot} \times G''$, we set $C = D^{\cdot\cdot\cdot} \otimes_{\mathcal{O}} \mathrm{Res}_{\sigma''}(\mathcal{O}G')$ and $\sigma = \rho\circ\sigma^{\cdot\cdot}$, and we denote by $\tau : L^{\cdot\cdot\cdot} \to G'\times G''$ the restriction of $(\rho'\circ\sigma^{\cdot\cdot}, \sigma'')$ to $\Delta(L^{\cdot\cdot\cdot}) \cong L^{\cdot\cdot\cdot}$.

15.8 Concerning the Higman envelope of a Hecke $\mathcal{D}G$-interior algebra, notice that, by Corollary 14.21, there is a canonical $\mathcal{D}(G \times G')$-interior algebra exoembedding

15.8.1 $$\tilde{\imath}_{H^{\cdot\cdot}}^{G\times G'}(B^{\cdot\cdot}) : \mathsf{E}\big(\mathrm{Ind}_{H^{\cdot\cdot}}^{G\times G'}(B^{\cdot\cdot})\big) \longrightarrow \mathrm{Ind}_{H^{\cdot\cdot}}^{G\times G'}\big(\mathsf{E}(B^{\cdot\cdot})\big)$$

such that we have

15.8.2 $$\tilde{\imath}_{H^{\cdot\cdot}}^{G\times G'}(B^{\cdot\cdot}) \circ \tilde{e}\big(\mathrm{Ind}_{H^{\cdot\cdot}}^{G\times G'}(B^{\cdot\cdot})\big) = \mathrm{Ind}_{H^{\cdot\cdot}}^{G\times G'}\big(\tilde{e}(B^{\cdot\cdot})\big) \quad ;$$

similarly, by Proposition 14.14 and Corollary 14.21, there is a canonical $\mathcal{D}G$-interior algebra exoembedding

15.8.3 $$\tilde{e} : \mathsf{E}\Big(\mathrm{Ind}_\rho\big(B^{\cdot\cdot} \underset{\mathcal{O}}{\otimes} \mathrm{Res}_{\rho'}(\mathcal{O}G')\big)\Big) \longrightarrow \mathrm{Ind}_\rho\big(\mathsf{E}(B^{\cdot\cdot}) \underset{\mathcal{O}}{\otimes} \mathrm{Res}_{\rho'}(\mathcal{O}G')\big)$$

given by

15.8.4 $$\tilde{e} = \mathrm{Ind}_\rho\Big(\tilde{e}\big(B^{\cdot\cdot}, \mathrm{Res}_{\rho'}(\mathcal{O}G')\big)\Big) \circ \tilde{\imath}_\rho\big(B^{\cdot\cdot} \underset{\mathcal{O}}{\otimes} \mathrm{Res}_{\rho'}(\mathcal{O}G')\big) \quad ;$$

in particular, we have that:

15.8.5 *If $B^{\cdot\cdot} \cong \mathsf{E}(B^{\cdot\cdot})$ then exoembeddings 15.8.1 and 15.8.3 are exoisomorphisms.*

Moreover, if F is a subgroup of G, setting $L^{\cdot\cdot} = \rho^{-1}(F)$ it is easily checked that the following diagram of $\mathcal{D}F$-interior algebra exoembeddings is commutative (cf. Proposition 12.12)

15.8.6 $$\begin{array}{ccc} \mathrm{Res}_F^G\big(\mathrm{Ind}_{H^{\cdot\cdot}}^{G\times G'}(B^{\cdot\cdot})^{1\times G'}\big) & \overset{h_G}{\cong} & \mathrm{Res}_F^G\Big(\mathrm{Ind}_\rho\big(B^{\cdot\cdot} \underset{\mathcal{O}}{\otimes} \mathrm{Res}_{\rho'}(\mathcal{O}G')\big)\Big) \\ {\scriptstyle d_{F\times G',H^{\cdot\cdot}}^{G\times G'}(B^{\cdot\cdot})}\uparrow & & \uparrow{\scriptstyle d_{F,\rho}^G(B^{\cdot\cdot} \underset{\mathcal{O}}{\otimes} \mathrm{Res}_{\rho'}(\mathcal{O}G'))} \\ \mathrm{Ind}_{L^{\cdot\cdot}}^{F\times G'}\big(\mathrm{Ind}_{L^{\cdot\cdot}}^{H^{\cdot\cdot}}(B^{\cdot\cdot})\big)^{1\times G'} & \overset{h_F}{\cong} & \mathrm{Ind}_\sigma\big(\mathrm{Res}_{L^{\cdot\cdot}}^{H^{\cdot\cdot}}(B^{\cdot\cdot}) \underset{\mathcal{O}}{\otimes} \mathrm{Res}_{\sigma'}(\mathcal{O}G')\big) \end{array}$$

where $\sigma : L^{\cdot\cdot} \to F$ and $\sigma' : L^{\cdot\cdot} \to G'$ are respectively the group homomorphisms determined by ρ and ρ', and we set $h_G = H_{G,G'}(B^{\cdot\cdot})$ and $h_F = H_{F,G'}\big(\mathrm{Res}_{L^{\cdot\cdot}}^{H^{\cdot\cdot}}(B^{\cdot\cdot})\big)$ (writing $d_{F\times G',H^{\cdot\cdot}}^{G\times G'}$ instead of $d_{F\times G',H^{\cdot\cdot}}^{G\times G',G\times G'}$).

15.9 Finally, we claim that the contents of the last part of Section 4, from 4.7 to Proposition 4.10, makes sense and is true when replacing $\mathcal{O}G$- by $\mathcal{D}G$-interior algebras; let us go briefly throughout it. Recall that we consider $\mathcal{O}G'$ as a $\mathcal{D}G'$-interior algebra and that, for any $\mathcal{D}G'$-interior algebra A', the structural map $\mathrm{st}_{A'} : \mathcal{O}G' \to A'$ is a G'-interior $\mathcal{D}$-algebra homomorphism (cf. 11.6); here, we fix an idempotent $\hat{\imath}$ of the following $\mathcal{O}$-algebra (cf. 12.4.4 and 15.5.2)

15.9.1 $$\mathrm{Ind}_\rho\big(\mathsf{C}_\circ(B^{\cdot\cdot}) \underset{\mathcal{O}}{\otimes} \mathrm{Res}_{\rho'}(\mathcal{O}G')\big)^G \cong \mathsf{C}_\circ\big(\mathrm{Ind}_{H^{\cdot\cdot}}^{G\times G'}(B^{\cdot\cdot})\big)^{G\times G'}$$

and then the G-interior $\mathcal{D}$-algebra homomorphism

15.9.2 $$\mathrm{Ind}_\rho(\mathrm{id}_{B^{\cdot\cdot}} \underset{\mathcal{O}}{\otimes} \mathrm{st}_{A'}) : \mathrm{Ind}_\rho\big(B^{\cdot\cdot} \underset{\mathcal{O}}{\otimes} \mathrm{Res}_{\rho'}(\mathcal{O}G')\big) \longrightarrow \mathrm{Ind}_\rho\big(B^{\cdot\cdot} \underset{\mathcal{O}}{\otimes} \mathrm{Res}_{\rho'}(A')\big)$$

maps $\hat{\imath}$ on an idempotent $\hat{\imath}_{A'}$ of $C_\circ\Big(\mathrm{Ind}_\rho\big(B^{\cdot\cdot}\otimes_{\mathcal{O}}\mathrm{Res}_{\rho'}(A')\big)\Big)^G$; moreover, if B' is a second $\mathcal{D}G'$-interior algebra and $f' : A'\to B'$ a G-interior $\mathcal{D}$-algebra homomorphism, equality 4.7.3 remains true and $\mathrm{Ind}_\rho(\mathrm{id}_{B^{\cdot\cdot}}\otimes_{\mathcal{O}} f')$ is a $\mathcal{D}$-bimodule homomorphism whenever f' is so. Consequently, respectively denoting by $\mathrm{Alg}_{\mathcal{D}G}$ and $\mathrm{Alg}_{\mathcal{D}G'}$, the categories of $\mathcal{D}G$- and $\mathcal{D}G'$-interior algebras together with G- and G'-interior $\mathcal{D}$-algebra homomorphisms, we get a functor, still noted

15.9.3 $$\mathcal{F}_{G,G'}(B^{\cdot\cdot})_{\hat{\imath}} : \mathrm{Alg}_{\mathcal{D}G'}\longrightarrow \mathrm{Alg}_{\mathcal{D}G}\quad ,$$

mapping A' on $\hat{\imath}_{A'}\mathrm{Ind}_\rho\big(B^{\cdot\cdot}\otimes_{\mathcal{O}}\mathrm{Res}_{\rho'}(A')\big)\hat{\imath}_{A'}$ and f' on the corresponding restriction of $\mathrm{Ind}_\rho(\mathrm{id}_{B^{\cdot\cdot}}\otimes_{\mathcal{O}} f')$. Notice that $\mathrm{Alg}_{\mathcal{O}G}$ and $\mathrm{Alg}_{\mathcal{O}G'}$ are respectively full subcategories of $\mathrm{Alg}_{\mathcal{D}G}$ and $\mathrm{Alg}_{\mathcal{D}G'}$, and when $B^{\cdot\cdot}$ comes from an $\mathcal{O}H^{\cdot\cdot}$-interior algebra, $\mathcal{F}_{G,G'}(B^{\cdot\cdot})_{\hat{\imath}}$ extends the functor in 4.7.5.

15.10 On the other hand, in the case where $B^{\cdot\cdot} = \mathrm{End}_{\mathcal{O}}(M^{\cdot\cdot})$, for some $\mathcal{D}H^{\cdot\cdot}$-module $M^{\cdot\cdot}$, is *no* longuer true that, for a $\mathcal{D}G'$-module M', the $\mathcal{D}G$-module $\mathrm{Ind}_\rho\big(M^{\cdot\cdot}\otimes_{\mathcal{O}}\mathrm{Res}_{\rho'}(M')\big)$ is the restriction to $\mathcal{D}G$ of the corresponding $\mathrm{Ind}_\rho\big(B^{\cdot\cdot}\otimes_{\mathcal{O}}\mathrm{Res}_{\rho'}(\mathcal{O}G')\big)$-module, since homomorphism 4.8.1 is only credited now as a G-interior $\mathcal{D}$-algebra homomorphism; nevertheless, once again $\hat{\imath}$ determines an idempotent $\hat{\imath}_{M'}$ in $C_\circ\Big(\mathrm{End}_{\mathcal{O}}\Big(\mathrm{Ind}_\rho\big(M^{\cdot\cdot}\otimes_{\mathcal{O}}\mathrm{Res}_{\rho'}(M')\big)\Big)\Big)^G$ and if N' is a second $\mathcal{D}G'$-module and $g' : M'\to N'$ a $\mathcal{D}G'$-module homomorphism then the map $\mathrm{Ind}_\rho(\mathrm{id}_{M^{\cdot\cdot}}\otimes_{\mathcal{O}} g')$ is an $\mathrm{Ind}_\rho\big(B^{\cdot\cdot}\otimes_{\mathcal{O}}\mathrm{Res}_{\rho'}(\mathcal{O}G')\big)$-module homomorphism (and, of course, a $\mathcal{D}G$-module homomorphism), so that inclusion 4.8.3 still holds; hence, respectively denoting by $\mathrm{Mod}_{\mathcal{D}G}$ and $\mathrm{Mod}_{\mathcal{D}G'}$ the categories of $\mathcal{D}G$- and $\mathcal{D}G'$-modules, we have a functor still noted

15.10.1 $$\mathcal{F}_{G,G'}(M^{\cdot\cdot})_{\hat{\imath}} : \mathrm{Mod}_{\mathcal{D}G'}\longrightarrow \mathrm{Mod}_{\mathcal{D}G}$$

which extends the functor in 4.8.4 whenever $M^{\cdot\cdot}$ comes from an $\mathcal{O}H^{\cdot\cdot}$-module and, by Proposition 3.7 and definition 12.3.2, fulfills isomorphism 4.8.5 whenever $M^{\cdot\cdot}$ and M' are $\mathcal{O}$-free. Moreover, this functor is still isomorphic to $\hat{\imath}\big(\mathrm{Ind}_{H^{\cdot\cdot}}^{G\times G'}(M^{\cdot\cdot})\big)\otimes_{\mathcal{O}G'}$ since, as it is easily checked, all the isomorphisms in 4.9 are now $\mathcal{D}$-linear.

15.11 Furthermore, as in 4.10, let G'' be a third finite group, $L^{\cdot\cdot}$ a subgroup of $G'\times G''$ and $C^{\cdot\cdot}$ a $\mathcal{D}L^{\cdot\cdot}$-interior algebra, respectively denote by σ' and σ'' the restrictions to $L^{\cdot\cdot}$ of the first and the second projection maps for $G'\times G''$, set

15.11.1 $$\mathrm{Ker}(\sigma') = 1\times J''\quad\text{and}\quad \mathrm{Ker}(\sigma'') = J'\times 1$$

for suitable subgroups J' of G' and J'' of G'', and consider the pull-back of groups

15.11.2 $$\begin{array}{ccc} H^{\cdot\cdot} & \xrightarrow{\rho'} & G' \\ {\scriptstyle\sigma}\big\uparrow & & \big\uparrow{\scriptstyle\sigma'} \\ H^{\cdot\cdot}\underset{G'}{\times} L^{\cdot\cdot} & \xrightarrow{\rho''} & L^{\cdot\cdot} \end{array}$$

setting $F^{\cdot\cdot} = (\rho \times \sigma'')(H^{\cdot\cdot} \times_{G'} L^{\cdot\cdot})$ and denoting by $\rho \times_{G'} \sigma'' : H^{\cdot\cdot} \times_{G'} L^{\cdot\cdot} \to F^{\cdot\cdot}$ the group homomorphism determined by $\rho \times \sigma''$, and by τ and τ' the respective restrictions to $F^{\cdot\cdot}$ of the first and the second projection maps for $G \times G''$. Now, *mutatis mutandis* Proposition 4.11 remains true with the same proof provided we respectively replace Corollary 3.13 and Propositions 3.16 and 3.21 by Corollary 12.7 and Propositions 12.9 and 12.12.

Proposition 15.12 *With the notation above, assume that $B^{\cdot\cdot}$ and $C^{\cdot\cdot}$ considered respectively as $\mathcal{O}(\mathrm{Ker}(\rho))$- and $\mathcal{O}(\mathrm{Ker}(\sigma'))$-bimodules are both projective. Then the $\mathfrak{D}F^{\cdot\cdot}$-interior algebra $\mathrm{Ind}_{\rho \times_{G'} \sigma''}\big(\mathrm{Res}_\sigma(B^{\cdot\cdot}) \otimes_{\mathcal{O}} \mathrm{Res}_{\rho''}(C^{\cdot\cdot})\big)$ considered as an $\mathcal{O}(\mathrm{Ker}(\tau))$-bimodule is projective and we have a natural embedding*

15.12.1

$$\mathcal{F}_{G,G'}\Big(\mathrm{Ind}_{\rho \underset{G'}{\times} \sigma''}\big(\mathrm{Res}_\sigma(B^{\cdot\cdot}) \underset{\mathcal{O}}{\otimes} \mathrm{Res}_{\rho''}(C^{\cdot\cdot})\big)\Big)_1 \to \mathcal{F}_{G,G'}(B^{\cdot\cdot})_1 \circ \mathcal{F}_{G,G'}(C^{\cdot\cdot})_1 \quad .$$

Moreover, if $B^{\cdot\cdot} = \mathrm{End}_{\mathcal{O}}(M^{\cdot\cdot})$ for some $\mathfrak{D}H^{\cdot\cdot}$-module $M^{\cdot\cdot}$ and $C^{\cdot\cdot} = \mathrm{End}_{\mathcal{O}}(N^{\cdot\cdot})$ for some $\mathfrak{D}L^{\cdot\cdot}$-module $N^{\cdot\cdot}$ then we have a natural direct injection

15.12.2

$$\mathcal{F}_{G,G'}\Big(\mathrm{Ind}_{\rho \times_{G'} \sigma''}\big(\mathrm{Res}_\sigma(M^{\cdot\cdot}) \underset{\mathcal{O}}{\otimes} \mathrm{Res}_{\rho''}(N^{\cdot\cdot})\big)\Big)_1 \longrightarrow \mathcal{F}_{G,G'}(M^{\cdot\cdot})_1 \circ \mathcal{F}_{G,G'}(N^{\cdot\cdot})_1 \quad .$$

16 On the local structure of Hecke $\mathcal{D}G$-interior algebras

16.1 As in Section 15, let G and G' be finite groups, $H^{\cdot\cdot}$ a subgroup of $G \times G'$ and $B^{\cdot\cdot}$ a $\mathcal{D}H^{\cdot\cdot}$-interior algebra, respectively denote by π and π' the first and the second projection maps for $G \times G'$, and set

16.1.1 $$A^{\cdot\cdot} = \mathrm{Ind}_{H^{\cdot\cdot}}^{G\times G'}(B^{\cdot\cdot}), \quad \hat{A} = (A^{\cdot\cdot})^{1\times G'} \quad \text{and} \quad \hat{A}' = (A^{\cdot\cdot})^{G\times 1} \quad ;$$

recall that we have the Hecke $\mathcal{D}G$- and $\mathcal{D}G'$-interior algebra isomorphisms (cf. Theorem 15.4)

16.1.2 $$\hat{A} \cong \mathrm{Ind}_\rho\big(B^{\cdot\cdot} \underset{\mathcal{O}}{\otimes} \mathrm{Res}_{\rho'}(\mathcal{O}G')\big) \quad \text{and} \quad \hat{A}' \cong \mathrm{Ind}_{\rho'}\big(B^{\cdot\cdot} \underset{\mathcal{O}}{\otimes} \mathrm{Res}_{\rho}(\mathcal{O}G)\big)$$

where ρ and ρ' are respectively the restrictions of π and π' to $H^{\cdot\cdot}$. In this section we discuss on the relationship between the local structures of $\hat{A}$ and $B^{\cdot\cdot} \otimes_{\mathcal{O}} \mathrm{Res}_{\rho'}(\mathcal{O}G')$ (and, symmetrically, between those of $\hat{A}'$ and $B^{\cdot\cdot} \otimes_{\mathcal{O}} \mathrm{Res}_\rho(\mathcal{O}G)$) by applying our results on the local structure of induced $\mathcal{D}G$-interior algebras in Section 14; but, as we recall below, the local structure of $B^{\cdot\cdot} \otimes_{\mathcal{O}} \mathrm{Res}_{\rho'}(\mathcal{O}G')$ depends on the local structures of $B^{\cdot\cdot}$ and $\mathrm{Res}_{\rho'}(\mathcal{O}G')$ quite separately, and the first one coincides with the local structure of $A^{\cdot\cdot}$ (cf. [20, Corollary 2.16]), except that the actions of the elements of $G \times G'$ on the local pointed groups on $A^{\cdot\cdot}$ become $B^{\cdot\cdot}$-fusions (cf. 2.13.2) which need not come from elements of $H^{\cdot\cdot}$; hence, we consider $A^{\cdot\cdot}$ and $\mathcal{O}G'$ instead of $B^{\cdot\cdot} \otimes_{\mathcal{O}} \mathrm{Res}_{\rho'}(\mathcal{O}G')$ and consequently, we have to modify our definition in 14.23 in order to take care of our specific context here. As we do there, according to Remark 14.13 we may assume without loss of generality that $B^{\cdot\cdot} \cong \mathsf{E}(B^{\cdot\cdot})$ and then it follows from Corollary 14.21 and 15.7.7 that

16.1.3 $$A^{\cdot\cdot} \cong \mathsf{E}(A^{\cdot\cdot}) \quad , \quad \hat{A} \cong \mathsf{E}(\hat{A}) \quad \text{and} \quad \hat{A}' \cong \mathsf{E}(\hat{A}') \quad .$$

16.2 First of all notice that, since we have

16.2.1 $$\mathsf{C}_\circ(\hat{A})^G = \mathsf{C}_\circ(A^{\cdot\cdot})^{G\times G'} = \mathsf{C}_\circ(\hat{A}')^{G'} \quad ,$$

the points of G on $\hat{A}$ coincide with the points of both $G \times G'$ on $A^{\cdot\cdot}$ and G' on $\hat{A}'$; similarly, since $\mathsf{C}_\circ(\hat{A})^H = \mathsf{C}_\circ(A^{\cdot\cdot})^{H\times G'}$ for any subgroup H of G, $H_{\hat{\beta}}$ is a pointed group on $\hat{A}$ if and only if $(H \times G')_{\hat{\beta}}$ is a pointed group on $A^{\cdot\cdot}$ and then

16.2.2 $$\hat{A}_{\hat{\beta}} = (A^{\cdot\cdot})_{\hat{\beta}}^{1\times G'} \quad \text{and} \quad (\mathsf{C}_\circ(\hat{A}))(H_{\hat{\beta}}) = (\mathsf{C}_\circ(A^{\cdot\cdot}))((H \times G')_{\hat{\beta}}) \quad ;$$

moreover, the last equality induces first a k^*-group isomorphism (cf. 2.8)

16.2.3 $$\hat{\theta} : \hat{\bar{N}}_G(H_{\hat{\beta}}) \cong \hat{\bar{N}}_{G\times G'}\big((H \times G')_{\hat{\beta}}\big)$$

and then a $k_*\hat{\bar{N}}_G(H_{\hat{\beta}})$-module isomorphism between the corresponding multiplicity modules (cf. 2.8)

16.2.4 $$V_{\mathrm{C}_\circ(\hat{A})}(H_{\hat{\beta}}) \cong \mathrm{Res}_{\hat{\theta}}\Big(V_{\mathrm{C}_\circ(A^{\cdot\cdot})}((H\times G')_{\hat{\beta}})\Big) \quad .$$

Furthermore, if $L_{\hat{\varepsilon}}$ is a second pointed group on $\hat{A}$, we have $L_{\hat{\varepsilon}} \subset H_{\hat{\beta}}$ if and only if we have $(L\times G')_{\hat{\varepsilon}} \subset (H\times G')_{\hat{\beta}}$ and then notice that

16.2.5 $$\mathrm{Tr}_L^H\big(\mathrm{C}_\circ(\hat{A})^L\cdot\hat{\varepsilon}\cdot\mathrm{C}_\circ(\hat{A})^L\big) = \mathrm{Tr}_{L\times G'}^{H\times G'}\big(\mathrm{C}_\circ(A^{\cdot\cdot})^{L\times G'}\cdot\hat{\varepsilon}\cdot\mathrm{C}_\circ(A^{\cdot\cdot})^{L\times G'}\big) \quad .$$

16.3 On the other hand recall that, since $\mathcal{O}G'$ obviously has a G'-stable $\mathcal{O}$-basis and we have

16.3.1 $$\mathrm{C}_\circ\big(A^{\cdot\cdot}\underset{\mathcal{O}}{\otimes}\mathrm{Res}_{\pi'}(\mathcal{O}G')\big) = \mathrm{C}_\circ(A^{\cdot\cdot})\underset{\mathcal{O}}{\otimes}\mathrm{Res}_{\pi'}(\mathcal{O}G') \quad ,$$

the set of local pointed groups on $A^{\cdot\cdot}\otimes_{\mathcal{O}}\mathrm{Res}_{\pi'}(\mathcal{O}G')$ (which contains, up to suitable identifications, the set of those on $B^{\cdot\cdot}\otimes_{\mathcal{O}}\mathrm{Res}_{\rho'}(\mathcal{O}G')$) corresponds bijectively with the set of pairs $(P^{\cdot\cdot}_{\gamma^{\cdot\cdot}}, P'_{\gamma'})$ of local pointed groups $P^{\cdot\cdot}_{\gamma^{\cdot\cdot}}$ on $A^{\cdot\cdot}$ and $P'_{\gamma'}$ on $\mathcal{O}G'$ such that $\pi'(P^{\cdot\cdot}) = P'$ (cf. 2.9.1, Lemma 7.10 and 14.5); moreover, denoting by $P^{\cdot\cdot}_{\gamma^{\cdot\cdot}\otimes\gamma'}$ the local pointed group on $A^{\cdot\cdot}\otimes_{\mathcal{O}}\mathrm{Res}_{\pi'}(\mathcal{O}G')$ corresponding to $(P^{\cdot\cdot}_{\gamma^{\cdot\cdot}}, P'_{\gamma'})$ and by $\sigma' : P^{\cdot\cdot}\to P'$ the group homomorphism determined by π', there is a unique $\mathcal{D}P$-interior algebra exoembedding (cf. [21, (5.6.4)])

16.3.2 $$\tilde{f}^{\gamma^{\cdot\cdot},\gamma'}_{\gamma^{\cdot\cdot}\otimes\gamma'} : \big(A^{\cdot\cdot}\underset{\mathcal{O}}{\otimes}\mathrm{Res}_{\pi'}(\mathcal{O}G')\big)_{\gamma^{\cdot\cdot}\otimes\gamma'} \longrightarrow A^{\cdot\cdot}_{\gamma^{\cdot\cdot}}\underset{\mathcal{O}}{\otimes}\mathrm{Res}_{\sigma'}\big((\mathcal{O}G')_{\gamma'}\big)$$

such that we have

16.3.3 $$\big(\tilde{f}_{\gamma^{\cdot\cdot}}\underset{\mathcal{O}}{\otimes}\mathrm{Res}_{\sigma'}(\tilde{f}_{\gamma'})\big)\circ\tilde{f}^{\gamma^{\cdot\cdot},\gamma'}_{\gamma^{\cdot\cdot}\otimes\gamma'} = \tilde{f}_{\gamma^{\cdot\cdot}\otimes\gamma'} \quad .$$

16.4 Furthermore, set $P = \pi(P^{\cdot\cdot})$ and respectively denote by $\sigma : P^{\cdot\cdot}\to P$ and $\psi' : P^{\cdot\cdot}\to G'$ the group homomorphisms determined by π and π'; consider the Hecke $\mathcal{D}G$-interior algebra isomorphism (cf. 15.4.1)

16.4.1 $$H_{P,G'}(A^{\cdot\cdot}_{\gamma^{\cdot\cdot}}) : \mathrm{Ind}_{P^{\cdot\cdot}}^{P\times G'}(A^{\cdot\cdot}_{\gamma^{\cdot\cdot}})^{1\times G'} \cong \mathrm{Ind}_\sigma\big(A^{\cdot\cdot}_{\gamma^{\cdot\cdot}}\underset{\mathcal{O}}{\otimes}\mathrm{Res}_{\psi'}(\mathcal{O}G')\big)$$

and the Higman exoembedding of $\mathcal{D}(P\times G')$-interior algebras (cf. Proposition 14.7, 14.8 and 16.1.3)

16.4.2 $$\tilde{h}_{\gamma^{\cdot\cdot}}\big(\mathrm{Res}_{P\times G'}^{G\times G'}(A^{\cdot\cdot})\big) : \mathrm{Ind}_{P^{\cdot\cdot}}^{P\times G'}(A^{\cdot\cdot}_{\gamma^{\cdot\cdot}}) \longrightarrow \mathrm{Res}_{P\times G'}^{G\times G'}(A^{\cdot\cdot}) \quad ;$$

then, we have the canonical $\mathcal{D}P$-interior algebra exoembedding

16.4.3 $$\tilde g_{\gamma^{\cdot\cdot},\gamma'} : \mathrm{Ind}_\sigma\Big(A^{\cdot\cdot}_{\gamma^{\cdot\cdot}} \underset{\mathcal{O}}{\otimes} \mathrm{Res}_{\sigma'}\big((\mathcal{O}G')_{\gamma'}\big)\Big) \longrightarrow \mathrm{Res}^G_P(\hat A)$$

given by

16.4.4
$$\tilde g_{\gamma^{\cdot\cdot},\gamma'} = \tilde h_{\gamma^{\cdot\cdot}}\big(\mathrm{Res}^{G\times G'}_{P\times G'}(A^{\cdot\cdot})\big)^{1\times G'} \circ \tilde H_{P,G'}(A^{\cdot\cdot}_{\gamma^{\cdot\cdot}})^{-1} \circ \mathrm{Ind}_\sigma\big(\widetilde{\mathrm{id}}_{A^{\cdot\cdot}_{\gamma^{\cdot\cdot}}} \underset{\mathcal{O}}{\otimes} \mathrm{Res}_{\sigma'}(\tilde f_{\gamma'})\big)$$

which will replace exoembedding 14.17.2 in the definition below. Notice that, if $P^{\cdot\cdot}_{\gamma^{\cdot\cdot}}$ comes from $B^{\cdot\cdot}$ (which is true up to $G\times G'$-conjugation), respectively denoting by $\tilde h_{\gamma^{\cdot\cdot}}(B^{\cdot\cdot})$ and by $\tilde g_{\gamma^{\cdot\cdot}}(B^{\cdot\cdot})$ the corresponding exoembeddings in condition 14.7.1 and definition 14.17.2 when G, A, H_β, G', H' and φ are respectively replaced by $H^{\cdot\cdot}$, $B^{\cdot\cdot}$, $P^{\cdot\cdot}_{\gamma^{\cdot\cdot}}$, $G\times G'$, $P\times G'$ and the inclusion $H^{\cdot\cdot}\subset G\times G'$, from 14.8.6 and 14.22.1 we get

16.4.5
$$\begin{aligned}&\tilde h_{\gamma^{\cdot\cdot}}\big(\mathrm{Res}^{G\times G'}_{P\times G'}(A^{\cdot\cdot})\big)\\ &\quad= \mathrm{Res}^{G\times G'}_{P\times G'}\Big(\mathrm{Ind}^{G\times G'}_{H^{\cdot\cdot}}\big(\tilde h_{\gamma^{\cdot\cdot}}(B^{\cdot\cdot})\big)\Big) \circ \tilde d^{G\times G'}_{P\times G'}\big(\mathrm{Ind}^{P\times G'}_{P^{\cdot\cdot}}(A^{\cdot\cdot}_{\gamma^{\cdot\cdot}})\big) = \tilde g_{\gamma^{\cdot\cdot}}(B^{\cdot\cdot})\ ;\end{aligned}$$

in particular, if F is a subgroup of G containing P, setting $L^{\cdot\cdot}=\rho^{-1}(P)$ and denoting by $\tilde g_{\gamma^{\cdot\cdot},\gamma'}\big(\mathrm{Res}^{H^{\cdot\cdot}}_{L^{\cdot\cdot}}(B^{\cdot\cdot})\big)$ the corresponding exoembedding 16.4.3 when G, $H^{\cdot\cdot}$ and $B^{\cdot\cdot}$ are respectively replaced by F, $L^{\cdot\cdot}$ and $\mathrm{Res}^{H^{\cdot\cdot}}_{L^{\cdot\cdot}}(B^{\cdot\cdot})$, the relation between $\tilde g_{\gamma^{\cdot\cdot}}(B^{\cdot\cdot})$ and $\tilde g_{\gamma^{\cdot\cdot}}\big(\mathrm{Res}^{H^{\cdot\cdot}}_{L^{\cdot\cdot}}(B^{\cdot\cdot})\big)$ given by 14.18.1 forces

16.4.6 $$\tilde g_{\gamma^{\cdot\cdot},\gamma'} = \mathrm{Res}^F_P\big(\tilde d^{G\times G'}_{F\times G',H^{\cdot\cdot}}(B^{\cdot\cdot})^{1\times G'}\big) \circ \tilde g_{\gamma^{\cdot\cdot},\gamma'}\big(\mathrm{Res}^{H^{\cdot\cdot}}_{L^{\cdot\cdot}}(B^{\cdot\cdot})\big)\ .$$

16.5 We are ready to adapt the definition of local tracing pairs in 14.23 to our present situation. A *local tracing triple* on $\hat A$, $A^{\cdot\cdot}$ and $\mathcal{O}G'$ is a triple $(P_{\hat\gamma}, P^{\cdot\cdot}_{\gamma^{\cdot\cdot}}, P'_{\gamma'})$ of local pointed groups $P_{\hat\gamma}$ on $\hat A$, $P^{\cdot\cdot}_{\gamma^{\cdot\cdot}}$ on $A^{\cdot\cdot}$ and $P'_{\gamma'}$ on $\mathcal{O}G'$ such that we have $P=\pi(P^{\cdot\cdot})$, $P'=\pi'(P^{\cdot\cdot})$ and there is a $\mathcal{D}P$-interior algebra exoembedding

16.5.1 $$\tilde h^{\gamma^{\cdot\cdot},\gamma'}_{\hat\gamma} : \hat A_{\hat\gamma} \longrightarrow \mathrm{Ind}_\sigma\Big(A^{\cdot\cdot}_{\gamma^{\cdot\cdot}} \underset{\mathcal{O}}{\otimes} \mathrm{Res}_{\sigma'}\big((\mathcal{O}G')_{\gamma'}\big)\Big)$$

fulfilling

16.5.2 $$\tilde f_{\hat\gamma} = \tilde g_{\gamma^{\cdot\cdot},\gamma'} \circ \tilde h^{\gamma^{\cdot\cdot},\gamma'}_{\hat\gamma}\quad ,$$

where $\sigma : P^{\cdot\cdot}\to P$ and $\sigma' : P^{\cdot\cdot}\to P'$ are respectively the group homomorphisms determined by π and π'; then $\tilde h^{\gamma^{\cdot\cdot},\gamma'}_{\hat\gamma}$ is unique by Proposition 11.10 and, for any

$(x, x') \in G \times G'$, the triple $\big((P_{\hat{\gamma}})^x, (P^{\cdot\cdot}_{\gamma^{\cdot\cdot}})^{(x,x')}, (P'_{\gamma'})^{x'}\big)$ is also a local tracing triple on $\hat{A}$, $A^{\cdot\cdot}$ and $\mathcal{O}G'$. Notice that, since Corollary 5.14 guarantees that the set of local points of P on the $\mathcal{O}P$-interior algebra (cf. 15.5.2)

16.5.3
$$\mathrm{Ind}_\sigma\Big(\mathrm{C}_\circ(A^{\cdot\cdot}_{\gamma^{\cdot\cdot}}) \underset{\mathcal{O}}{\otimes} \mathrm{Res}_{\sigma'}\big((\mathcal{O}G')_{\gamma'}\big)\Big) \cong \mathrm{C}_\circ\Big(\mathrm{Ind}_\sigma\Big(A^{\cdot\cdot}_{\gamma^{\cdot\cdot}} \underset{\mathcal{O}}{\otimes} \mathrm{Res}_{\sigma'}\big((\mathcal{O}G')_{\gamma'}\big)\Big)\Big)$$

is never empty, there is always a local point $\hat{\gamma}$ of P on $\hat{A}$ such that $(P_{\hat{\gamma}}, P^{\cdot\cdot}_{\gamma^{\cdot\cdot}}, P'_{\gamma'})$ is a local tracing triple on $\hat{A}$, $A^{\cdot\cdot}$ and $\mathcal{O}G'$. Next two propositions relate the local tracing triples on $\hat{A}$, $A^{\cdot\cdot}$ and $\mathcal{O}G'$ with the local tracing pairs on $B^{\cdot\cdot} \otimes_{\mathcal{O}} \mathrm{Res}_{\rho'}(\mathcal{O}G')$ and $\mathrm{Ind}_\rho\big(B^{\cdot\cdot} \otimes_{\mathcal{O}} \mathrm{Res}_{\rho'}(\mathcal{O}G')\big)$, and in particular, according to Theorem 14.24, they show that we have also a local tracing triple on $\hat{A}$, $A^{\cdot\cdot}$ and $\mathcal{O}G'$ for any choice of a local pointed group on $\hat{A}$; in order to state them, we identify to each other the local pointed groups on $\mathrm{Ind}_\rho\big(B^{\cdot\cdot} \otimes_{\mathcal{O}} \mathrm{Res}_{\rho'}(\mathcal{O}G')\big)$ and on $\hat{A}$ which correspond via $H_{G,G'}(B^{\cdot\cdot})^{-1}$ (cf. 16.1.2), those on $B^{\cdot\cdot}$ and on $A^{\cdot\cdot}$ via $d^{G\times G'}_{H^{\cdot\cdot}}(B^{\cdot\cdot})$, and those on $B^{\cdot\cdot} \otimes_{\mathcal{O}} \mathrm{Res}_{\rho'}(\mathcal{O}G')$ and on $A^{\cdot\cdot} \otimes_{\mathcal{O}} \mathrm{Res}_{\pi'}(\mathcal{O}G')$ via $d^{G\times G'}_{H^{\cdot\cdot}}(B^{\cdot\cdot}) \otimes_{\mathcal{O}} \mathrm{id}_{\mathcal{O}G'}$. In the first one, we give moreover an alternative definition of the exoembedding $\tilde{g}_{\gamma^{\cdot\cdot},\gamma'}$, which will be useful in Theorems 16.15 and 17.9 below; if $P^{\cdot\cdot}_{\gamma^{\cdot\cdot}}$ comes from $B^{\cdot\cdot}$ and $L^{\cdot\cdot}$ is a subgroup of $H^{\cdot\cdot}$ containing $P^{\cdot\cdot}$, we denote by $\tilde{h}_{\gamma^{\cdot\cdot}}\big(\mathrm{Res}^{H^{\cdot\cdot}}_{L^{\cdot\cdot}}(B^{\cdot\cdot})\big)$ the corresponding Higman exoembedding.

Proposition 16.6 *With the notation above, let $P^{\cdot\cdot}_{\gamma^{\cdot\cdot}}$ and $P'_{\gamma'}$ be respectively local pointed groups on $B^{\cdot\cdot}$ and on $\mathcal{O}G'$ such that $\pi'(P^{\cdot\cdot}) = P'$. The following diagram of $\mathcal{D}P$-interior algebra exoembeddings is commutative*

16.6.1
$$\begin{array}{ccc}
\mathrm{Res}^G_P\Big(\mathrm{Ind}_\rho\big(B^{\cdot\cdot} \underset{\mathcal{O}}{\otimes} \mathrm{Res}_{\rho'}(\mathcal{O}G')\big)\Big) & \xleftarrow{\mathrm{Res}^G_P(\tilde{H}_{G,G'}(B^{\cdot\cdot}))} & \mathrm{Res}^G_P(\hat{A}) \\
\uparrow{\scriptstyle \tilde{g}_{\gamma^{\cdot\cdot}\otimes\gamma'}} & & \uparrow{\scriptstyle \tilde{g}_{\gamma^{\cdot\cdot},\gamma'}} \\
\mathrm{Ind}_\sigma\Big(\big(B^{\cdot\cdot} \underset{\mathcal{O}}{\otimes} \mathrm{Res}_{\rho'}(\mathcal{O}G')\big)_{\gamma^{\cdot\cdot}\otimes\gamma'}\Big) & \xrightarrow{\mathrm{Ind}_\sigma(\tilde{f}^{\gamma^{\cdot\cdot},\gamma'}_{\gamma^{\cdot\cdot}\otimes\gamma'})} & \mathrm{Ind}_\sigma\Big(B^{\cdot\cdot}_{\gamma^{\cdot\cdot}} \underset{\mathcal{O}}{\otimes} \mathrm{Res}_{\sigma'}\big((\mathcal{O}G')_{\gamma'}\big)\Big)
\end{array} .$$

Moreover, setting $L^{\cdot\cdot} = \rho^{-1}(P)$ and denoting by $\eta : L^{\cdot\cdot} \to P$ and by $\eta' : L^{\cdot\cdot} \to G'$ the group homomorphisms determined by ρ and by ρ', we have

16.6.2
$$\begin{aligned}
&\mathrm{Res}^G_P\big(\tilde{H}_{G,G'}(B^{\cdot\cdot})\big) \circ \tilde{g}_{\gamma^{\cdot\cdot},\gamma'} \\
&\quad = \tilde{d} \circ \mathrm{Ind}_\eta\Big(\tilde{h}_{\gamma^{\cdot\cdot}}\big(\mathrm{Res}^{H^{\cdot\cdot}}_{L^{\cdot\cdot}}(B^{\cdot\cdot})\big) \underset{\mathcal{O}}{\otimes} \mathrm{Res}_{\eta'}(\widetilde{\mathrm{id}}_{\mathcal{O}G'})\Big) \circ \mathrm{Ind}_\sigma\big(\widetilde{\mathrm{id}}_{A^{\cdot\cdot}_{\gamma^{\cdot\cdot}}} \underset{\mathcal{O}}{\otimes} \mathrm{Res}_{\sigma'}(\tilde{f}_{\gamma'})\big).
\end{aligned}$$

where $d = d^G_{P,\rho}\big(B^{\cdot\cdot} \otimes_{\mathcal{O}} \mathrm{Res}_{\rho'}(\mathcal{O}G')\big)$.

Proof: The naturality of the Hecke isomorphism (cf. 15.7.1) implies that

16.6.3
$$\tilde{H}_{G,G'}(B^{\cdot\cdot}) \circ \mathrm{Ind}_{H^{\cdot\cdot}}^{G\times G'}\big(\tilde{h}_{\gamma^{\cdot\cdot}}(B^{\cdot\cdot})\big)^{1\times G'} = \mathrm{Ind}_\rho\big(\tilde{h}_{\gamma^{\cdot\cdot}}(B^{\cdot\cdot}) \otimes_{\mathcal{O}} \mathrm{Res}_{\rho'}(\mathrm{id}_{\mathcal{O}G'})\big) \circ \tilde{H}_{G,G'}(B^{\cdot\cdot}_{\gamma^{\cdot\cdot}})$$

since, up to suitable identifications (cf. Corollary 12.7 and Proposition 12.9), the commutativity of diagram 15.7.3 applied to G, G' and 1 with the subgroups $H^{\cdot\cdot} \subset G \times G'$ and $P^{\cdot\cdot} \subset H^{\cdot\cdot}$ gives

16.6.4
$$H_{G,G'}\big(\mathrm{Ind}_{P^{\cdot\cdot}}^{H^{\cdot\cdot}}(B^{\cdot\cdot}_{\gamma^{\cdot\cdot}})\big) = H_{G,G'}(B^{\cdot\cdot}_{\gamma^{\cdot\cdot}}) \quad ;$$

similarly, that commutativity applied to G, 1 and G' with the subgroups $P \subset G$ and $P^{\cdot\cdot} \subset P \times G'$ shows, always up to suitable identifications, that

16.6.5
$$H_{G,G'}(B^{\cdot\cdot}_{\gamma^{\cdot\cdot}}) = \mathrm{Ind}_P^G\big(H_{P,G'}(B^{\cdot\cdot}_{\gamma^{\cdot\cdot}})\big) \quad ,$$

so that we get (cf. 3.5.4)

16.6.6
$$\mathrm{Res}_P^G\big(H_{G,G'}(B^{\cdot\cdot})\big) \circ d_P^G\big(\mathrm{Ind}_{P^{\cdot\cdot}}^{P\times G'}(B^{\cdot\cdot}_{\gamma^{\cdot\cdot}})^{1\times G'}\big) = d_P^G\Big(\mathrm{Ind}_\sigma\big(B^{\cdot\cdot}_{\gamma^{\cdot\cdot}} \otimes_{\mathcal{O}} \mathrm{Res}_{\psi'}(\mathcal{O}G')\big)\Big) \circ H_{P,G'}(B^{\cdot\cdot}_{\gamma^{\cdot\cdot}}) \quad .$$

Consequently, by the first equality in 16.4.5, we have

16.6.7
$$\mathrm{Res}_P^G\big(\tilde{H}_{G,G'}(B^{\cdot\cdot})\big) \circ \tilde{g}_{\gamma^{\cdot\cdot},\gamma'} = \mathrm{Res}_P^G\Big(\mathrm{Ind}_\rho\big(\tilde{h}_{\gamma^{\cdot\cdot}}(B^{\cdot\cdot}) \otimes_{\mathcal{O}} \mathrm{Res}_{\rho'}(\tilde{\mathrm{id}}_{\mathcal{O}G'})\big)\Big) \circ \tilde{d}' \circ \mathrm{Ind}_\sigma\big(\tilde{\mathrm{id}}_{A_{\gamma^{\cdot\cdot}}} \otimes_{\mathcal{O}} \mathrm{Res}_{\sigma'}(\tilde{f}_{\gamma'})\big),$$

where $d' = d_P^G\Big(\mathrm{Ind}_\sigma\big(B^{\cdot\cdot}_{\gamma^{\cdot\cdot}} \otimes_{\mathcal{O}} \mathrm{Res}_{\psi'}(\mathcal{O}G')\big)\Big)$; but, since (cf. Corollary 12.7)

16.6.8
$$\mathrm{Ind}_\sigma\big(B^{\cdot\cdot}_{\gamma^{\cdot\cdot}} \otimes_{\mathcal{O}} \mathrm{Res}_{\psi'}(\mathcal{O}G')\big) \cong \mathrm{Ind}_\eta\big(\mathrm{Ind}_{P^{\cdot\cdot}}^{L^{\cdot\cdot}}(B^{\cdot\cdot}_{\gamma^{\cdot\cdot}}) \otimes_{\mathcal{O}} \mathrm{Res}_{\eta'}(\mathcal{O}G')\big) \quad ,$$

it follows from the commutativity of diagram 12.13.2 that

16.6.9
$$d_P^G\Big(\mathrm{Ind}_\sigma\big(B^{\cdot\cdot}_{\gamma^{\cdot\cdot}} \otimes_{\mathcal{O}} \mathrm{Res}_{\psi'}(\mathcal{O}G')\big)\Big) = d_{P,\rho}^G\big(\mathrm{Ind}_{P^{\cdot\cdot}}^{H^{\cdot\cdot}}(B^{\cdot\cdot}_{\gamma^{\cdot\cdot}}) \otimes_{\mathcal{O}} \mathrm{Res}_{\rho'}(\mathcal{O}G')\big) \circ \mathrm{Ind}_\eta\Big(d_{L^{\cdot\cdot}}^{H^{\cdot\cdot}}\big(\mathrm{Ind}_{P^{\cdot\cdot}}^{L^{\cdot\cdot}}(B^{\cdot\cdot}_{\gamma^{\cdot\cdot}}) \otimes_{\mathcal{O}} \mathrm{Res}_{\eta'}(\mathcal{O}G')\big)\Big) \; ;$$

hence, by the naturality of $d_{P,\rho}^G$ (cf. 3.22.1 and 12.13) and equality 14.8.6, we get

16.6.10
$$\begin{aligned}
&\mathrm{Res}_P^G\Big(\mathrm{Ind}_\rho\big(\tilde{h}_{\gamma^{\cdot\cdot}}(B^{\cdot\cdot}) \otimes_{\mathcal{O}} \mathrm{Res}_{\rho'}(\tilde{\mathrm{id}}_{\mathcal{O}G'})\big)\Big) \circ \tilde{d}_P^G\Big(\mathrm{Ind}_\sigma\big(B^{\cdot\cdot}_{\gamma^{\cdot\cdot}} \otimes_{\mathcal{O}} \mathrm{Res}_{\psi'}(\mathcal{O}G')\big)\Big)\\
&= \tilde{d} \circ \mathrm{Ind}_\eta\Big(\Big(\mathrm{Res}_{L^{\cdot\cdot}}^{H^{\cdot\cdot}}\big(\tilde{h}_{\gamma^{\cdot\cdot}}(B^{\cdot\cdot})\big) \circ \tilde{d}_{L^{\cdot\cdot}}^{H^{\cdot\cdot}}\big(\mathrm{Ind}_{P^{\cdot\cdot}}^{L^{\cdot\cdot}}(B^{\cdot\cdot}_{\gamma^{\cdot\cdot}})\big)\Big) \otimes_{\mathcal{O}} \mathrm{Res}_{\eta'}(\tilde{\mathrm{id}}_{\mathcal{O}G'})\Big)\\
&= \tilde{d}_{P,\rho}^G\big(B^{\cdot\cdot} \otimes_{\mathcal{O}} \mathrm{Res}_{\rho'}(\mathcal{O}G')\big) \circ \mathrm{Ind}_\eta\Big(\tilde{h}_{\gamma^{\cdot\cdot}}\big(\mathrm{Res}_{L^{\cdot\cdot}}^{H^{\cdot\cdot}}(B^{\cdot\cdot})\big) \otimes_{\mathcal{O}} \mathrm{Res}_{\eta'}(\tilde{\mathrm{id}}_{\mathcal{O}G'})\Big) \quad .
\end{aligned}$$

Now, equality 16.6.2 follows from equalities 16.6.7 and 16.6.10.

On the other hand, once again from 3.5.4, we have (cf. Corollary 12.7)

16.6.11

$$\tilde{d}_P^G\Big(\mathrm{Ind}_\sigma(B^{\cdot\cdot}_{\gamma^{\cdot\cdot}} \underset{\mathcal{O}}{\otimes} \mathrm{Res}_{\psi'}(\mathcal{O}G'))\Big) \circ \mathrm{Ind}_\sigma(\tilde{f})$$
$$= \mathrm{Res}_P^G\Big(\mathrm{Ind}_P^G(\mathrm{Ind}_\sigma(\tilde{f}))\Big) \circ \tilde{d}_P^G\Big(\mathrm{Ind}_\sigma\big((B^{\cdot\cdot} \underset{\mathcal{O}}{\otimes} \mathrm{Res}_{\rho'}(\mathcal{O}G'))_{\gamma^{\cdot\cdot}\otimes\gamma'}\big)\Big)$$
$$= \mathrm{Res}_P^G\Big(\mathrm{Ind}_\rho(\mathrm{Ind}_{P^{\cdot\cdot}}^{H^{\cdot\cdot}}(\tilde{f}))\Big) \circ \tilde{d}_P^G\Big(\mathrm{Ind}_\sigma\big((B^{\cdot\cdot} \underset{\mathcal{O}}{\otimes} \mathrm{Res}_{\rho'}(\mathcal{O}G'))_{\gamma^{\cdot\cdot}\otimes\gamma'}\big)\Big) \quad ,$$

where $\tilde{f} = (\widetilde{\mathrm{id}}_{B^{\cdot\cdot}_{\gamma^{\cdot\cdot}}} \otimes_{\mathcal{O}} \mathrm{Res}_{\sigma'}(\tilde{f}_{\gamma'})) \circ \tilde{f}^{\gamma^{\cdot\cdot},\gamma'}_{\gamma^{\cdot\cdot}\otimes\gamma'}$, and therefore, setting

16.6.12 $$\tilde{h} = (\tilde{h}_{\gamma^{\cdot\cdot}}(B^{\cdot\cdot}) \underset{\mathcal{O}}{\otimes} \mathrm{Res}_{\rho'}(\widetilde{\mathrm{id}}_{\mathcal{O}G'})) \circ \mathrm{Ind}_{L^{\cdot\cdot}}^{H^{\cdot\cdot}}(\tilde{f}) \quad ,$$

from 16.6.7 and 16.6.11 we get

16.6.13
$$\mathrm{Res}_P^G(\tilde{H}_{G,G'}(B^{\cdot\cdot})) \circ \tilde{g}_{\gamma^{\cdot\cdot},\gamma'} \circ \mathrm{Ind}_\sigma(\tilde{f}^{\gamma^{\cdot\cdot},\gamma'}_{\gamma^{\cdot\cdot}\otimes\gamma'})$$
$$= \mathrm{Res}_P^G(\mathrm{Ind}_\rho(\tilde{h})) \circ \tilde{d}_P^G\Big(\mathrm{Ind}_\sigma\big((B^{\cdot\cdot} \underset{\mathcal{O}}{\otimes} \mathrm{Res}_{\rho'}(\mathcal{O}G'))_{\gamma^{\cdot\cdot}\otimes\gamma'}\big)\Big) \quad .$$

Then, in order to prove the commutativity of diagram 16.6.1, it suffices to prove that $\tilde{h} = \tilde{h}_{\gamma^{\cdot\cdot}\otimes\gamma'}$ (cf. 14.17.3); but from 3.5.4, Proposition 12.9, condition 14.7.1 and equality 16.3.3, we have

16.6.14

$$\mathrm{Res}_{P^{\cdot\cdot}}^{H^{\cdot\cdot}}(\tilde{h}) \circ \tilde{d}_{P^{\cdot\cdot}}^{H^{\cdot\cdot}}\big((B^{\cdot\cdot} \underset{\mathcal{O}}{\otimes} \mathrm{Res}_{\rho'}(\mathcal{O}G'))_{\gamma^{\cdot\cdot}\otimes\gamma'}\big)$$
$$= \Big(\mathrm{Res}_{P^{\cdot\cdot}}^{H^{\cdot\cdot}}(\tilde{h}_{\gamma^{\cdot\cdot}}(B^{\cdot\cdot})) \underset{\mathcal{O}}{\otimes} \mathrm{Res}_{\psi'}(\widetilde{\mathrm{id}}_{\mathcal{O}G'})\Big) \circ \big(\tilde{d}_{P^{\cdot\cdot}}^{H^{\cdot\cdot}}(B^{\cdot\cdot}_{\gamma^{\cdot\cdot}}) \underset{\mathcal{O}}{\otimes} \mathrm{Res}_{\psi'}(\widetilde{\mathrm{id}}_{\mathcal{O}G'})\big) \circ \tilde{f}$$
$$= \Big(\big(\mathrm{Res}_{P^{\cdot\cdot}}^{H^{\cdot\cdot}}(\tilde{h}_{\gamma^{\cdot\cdot}}(B^{\cdot\cdot})) \circ \tilde{d}_{P^{\cdot\cdot}}^{H^{\cdot\cdot}}(B^{\cdot\cdot}_{\gamma^{\cdot\cdot}})\big) \underset{\mathcal{O}}{\otimes} \mathrm{Res}_{\psi'}(\widetilde{\mathrm{id}}_{\mathcal{O}G'})\Big) \circ \tilde{f}$$
$$= (\tilde{f}_{\gamma^{\cdot\cdot}} \underset{\mathcal{O}}{\otimes} \mathrm{Res}_{\psi'}(\widetilde{\mathrm{id}}_{\mathcal{O}G'})) \circ (\widetilde{\mathrm{id}}_{B^{\cdot\cdot}_{\gamma^{\cdot\cdot}}} \underset{\mathcal{O}}{\otimes} \mathrm{Res}_{\psi'}(\tilde{f}_{\gamma'})) \circ \tilde{f}_{\gamma^{\cdot\cdot}\otimes\gamma'}$$
$$= \tilde{f}^{\gamma^{\cdot\cdot},\gamma'}_{\gamma^{\cdot\cdot}\otimes\gamma'} = \mathrm{Res}_{P^{\cdot\cdot}}^{H^{\cdot\cdot}}(\tilde{h}_{\gamma^{\cdot\cdot}\otimes\gamma'}) \circ \tilde{d}_{P^{\cdot\cdot}}^{H^{\cdot\cdot}}\big((B^{\cdot\cdot} \underset{\mathcal{O}}{\otimes} \mathrm{Res}_{\rho'}(\mathcal{O}G'))_{\gamma^{\cdot\cdot}\otimes\gamma'}\big) \quad ;$$

consequently, the equality $\tilde{h} = \tilde{h}_{\gamma^{\cdot\cdot}\otimes\gamma'}$ now follows from Propositions 11.9 and 11.10.

Proposition 16.7 *With the notation above, let $P_{\hat\gamma}$, $P^{\cdot\cdot}_{\gamma^{\cdot\cdot}}$ and $P'_{\gamma'}$ be respectively local pointed groups on $\mathrm{Ind}_\rho\big(B^{\cdot\cdot} \otimes_{\mathcal{O}} \mathrm{Res}_{\rho'}(\mathcal{O}G')\big)$, on $B^{\cdot\cdot}$ and on $\mathcal{O}G'$, and assume that $\pi'(P^{\cdot\cdot}) = P'$. Then $(P^{\cdot\cdot}_{\gamma^{\cdot\cdot}\otimes\gamma'}, P_{\hat\gamma})$ is a local tracing pair on $B^{\cdot\cdot} \otimes_{\mathcal{O}} \mathrm{Res}_{\rho'}(\mathcal{O}G')$ and $\mathrm{Ind}_\rho\big(B^{\cdot\cdot} \otimes_{\mathcal{O}} \mathrm{Res}_{\rho'}(\mathcal{O}G')\big)$ if and only if $(P_{\hat\gamma}, P^{\cdot\cdot}_{\gamma^{\cdot\cdot}}, P'_{\gamma'})$ is a local tracing triple on $\hat A$, $A^{\cdot\cdot}$ and $\mathcal{O}G'$, and in this case, denoting by $\sigma : P^{\cdot\cdot} \to P$ the group homomorphism determined by π, we have*

16.7.1 $$\tilde h_{\hat\gamma}^{\gamma^{\cdot\cdot},\gamma'} = \mathrm{Ind}_\sigma(\tilde f^{\gamma^{\cdot\cdot},\gamma'}_{\gamma^{\cdot\cdot}\otimes\gamma'}) \circ \tilde h_{\hat\gamma}^{\gamma^{\cdot\cdot}\otimes\gamma'} \quad .$$

Proof: Clearly, we may assume that $\pi(P^{\cdot\cdot}) = P$; if $(P^{\cdot\cdot}_{\gamma^{\cdot\cdot}\otimes\gamma'}, P_{\hat\gamma})$ is a local tracing pair on $B^{\cdot\cdot} \otimes_{\mathcal{O}} \mathrm{Res}_{\rho'}(\mathcal{O}G')$ and $\mathrm{Ind}_\rho\big(B^{\cdot\cdot} \otimes_{\mathcal{O}} \mathrm{Res}_{\rho'}(\mathcal{O}G')\big)$, there is a $\mathcal{D}P$-interior algebra exoembedding (cf. 14.23.1)

16.7.2
$$\tilde h_{\hat\gamma}^{\gamma^{\cdot\cdot}\otimes\gamma'} : \mathrm{Ind}_\rho\big(B^{\cdot\cdot} \underset{\mathcal{O}}{\otimes} \mathrm{Res}_{\rho'}(\mathcal{O}G')\big)_{\hat\gamma} \longrightarrow \mathrm{Ind}_\sigma\Big(\big(B^{\cdot\cdot} \underset{\mathcal{O}}{\otimes} \mathrm{Res}_{\rho'}(\mathcal{O}G')\big)_{\gamma^{\cdot\cdot}\otimes\gamma'}\Big)$$

such that we have

16.7.3 $$\tilde f_{\hat\gamma} = \tilde g_{\gamma^{\cdot\cdot}\otimes\gamma'} \circ \tilde h_{\hat\gamma}^{\gamma^{\cdot\cdot}\otimes\gamma'}$$

and therefore, considering the $\mathcal{D}P$-interior algebra exoembedding

16.7.4 $$\tilde h_{\hat\gamma}^{\gamma^{\cdot\cdot},\gamma'} : \hat A_{\hat\gamma} \longrightarrow \mathrm{Ind}_\sigma\Big(B^{\cdot\cdot}_{\gamma^{\cdot\cdot}} \underset{\mathcal{O}}{\otimes} \mathrm{Res}_{\sigma'}\big((\mathcal{O}G')_{\gamma'}\big)\Big)$$

given by (cf. 16.3.2)

16.7.5 $$\tilde h_{\hat\gamma}^{\gamma^{\cdot\cdot},\gamma'} = \mathrm{Ind}_\sigma(\tilde f^{\gamma^{\cdot\cdot},\gamma'}_{\gamma^{\cdot\cdot}\otimes\gamma'}) \circ \tilde h_{\hat\gamma}^{\gamma^{\cdot\cdot}\otimes\gamma'} \quad ,$$

up to suitable identifications, it follows from Proposition 16.6 that

16.7.6 $$\tilde g_{\gamma^{\cdot\cdot},\gamma'} \circ \tilde h_{\hat\gamma}^{\gamma^{\cdot\cdot},\gamma'} = \tilde g_{\gamma^{\cdot\cdot}\otimes\gamma'} \circ \tilde h_{\hat\gamma}^{\gamma^{\cdot\cdot}\otimes\gamma'} = \tilde f_{\hat\gamma} \quad ,$$

so that $(P_{\hat\gamma}, P^{\cdot\cdot}_{\gamma^{\cdot\cdot}}, P'_{\gamma'})$ is a local tracing triple on $\hat A$, $A^{\cdot\cdot}$ and $\mathcal{O}G'$, and equality 16.7.1 holds.

Conversely, assume that $(P_{\hat\gamma}, P^{\cdot\cdot}_{\gamma^{\cdot\cdot}}, P'_{\gamma'})$ is a local tracing triple on $\hat A$, $A^{\cdot\cdot}$ and $\mathcal{O}G'$; then we claim that the corresponding $\mathcal{D}P$-interior algebra exoembedding (cf. 16.5.1)

16.7.7 $$\tilde h_{\hat\gamma}^{\gamma^{\cdot\cdot},\gamma'} : \hat A_{\hat\gamma} \longrightarrow \mathrm{Ind}_\sigma\Big(B^{\cdot\cdot}_{\gamma^{\cdot\cdot}} \underset{\mathcal{O}}{\otimes} \mathrm{Res}_{\sigma'}\big((\mathcal{O}G')_{\gamma'}\big)\Big)$$

factorizes throughout $\mathrm{Ind}_\sigma(\tilde f^{\gamma^{\cdot\cdot},\gamma'}_{\gamma^{\cdot\cdot}\otimes\gamma'})$. Indeed, since by Lemma 7.10 we have

16.7.8
$$\Big(\mathrm{C}_\circ\Big(B^{\cdot\cdot}_{\gamma^{\cdot\cdot}} \underset{\mathcal{O}}{\otimes} \mathrm{Res}_{\sigma'}\big((\mathcal{O}G')_{\gamma'}\big)\Big)\Big)(P^{\cdot\cdot}) \cong \big(\mathrm{C}_\circ(B^{\cdot\cdot}_{\gamma^{\cdot\cdot}})\big)(P^{\cdot\cdot}) \underset{k}{\otimes} \big((\mathcal{O}G')_{\gamma'}\big)(P') \quad ,$$

$\gamma^{\cdot\cdot} \otimes \gamma'$ is the unique local point of $P^{\cdot\cdot}$ on $B^{\cdot\cdot}_{\gamma^{\cdot\cdot}} \otimes_{\mathcal{O}} \mathrm{Res}_{\sigma'}\big((\mathcal{O}G')_{\gamma'}\big)$ and has multiplicity one; hence, if I is a pairwise orthogonal primitive idempotent decomposition of the unity element in $\mathrm{C}_\circ\Big(B^{\cdot\cdot}_{\gamma^{\cdot\cdot}} \otimes_{\mathcal{O}} \mathrm{Res}_{\sigma'}\big((\mathcal{O}G')_{\gamma'}\big)\Big)^{P^{\cdot\cdot}}$, we have $I \cap \gamma^{\cdot\cdot} \otimes \gamma' = \{i\}$ and, for any $j \in I - \{i\}$, there is a proper subgroup $Q^{\cdot\cdot}$ of $P^{\cdot\cdot}$ and an element l of $\mathrm{C}_\circ\Big(B^{\cdot\cdot}_{\gamma^{\cdot\cdot}} \otimes_{\mathcal{O}} \mathrm{Res}_{\sigma'}\big((\mathcal{O}G')_{\gamma'}\big)\Big)^{Q^{\cdot\cdot}}$ such that $j = \mathrm{Tr}_{Q^{\cdot\cdot}}^{P^{\cdot\cdot}}(l)$ (actually, l can be chosen fulfilling $l^2 = l$ and $ll^{(u,u')} = 0$ for any element (u, u') of $P^{\cdot\cdot} - Q^{\cdot\cdot}$), so that in $\mathrm{Ind}_{P^{\cdot\cdot}}^{P\times G'}\big(\mathrm{C}_\circ(B^{\cdot\cdot}_{\gamma^{\cdot\cdot}})\big)^{P^{\cdot\cdot}}$, up to suitable identifications (cf. 2.7 and 16.3.1), we get (cf. Theorem 4.4)

16.7.9
$$\begin{aligned} H_{P,G'}\big(\mathrm{C}_\circ(B^{\cdot\cdot}_{\gamma^{\cdot\cdot}})\big)^{-1}(1 \otimes j) &= \mathrm{Tr}_{P^{\cdot\cdot}}^{P\times G'}\Big(q_{P,G'}\big(\mathrm{C}_\circ(B^{\cdot\cdot}_{\gamma^{\cdot\cdot}})\big)\big(1 \otimes \mathrm{Tr}_{Q^{\cdot\cdot}}^{P^{\cdot\cdot}}(l)\big)\Big) \\ &= \mathrm{Tr}_{Q^{\cdot\cdot}}^{P\times G'}\Big(q_{P,G'}\big(\mathrm{C}_\circ(B^{\cdot\cdot}_{\gamma^{\cdot\cdot}})\big)(1 \otimes l)\Big) \end{aligned}$$

which, since $\gamma^{\cdot\cdot}$ is local, implies that

16.7.10
$$s_{\gamma^{\cdot\cdot}}\Big(H_{P,G'}\big(\mathrm{C}_\circ(B^{\cdot\cdot}_{\gamma^{\cdot\cdot}})\big)^{-1}(1 \otimes j)\Big) = 0 \quad .$$

Consequently, according to Theorem 5.8 applied to $P, G', P^{\cdot\cdot}$ and $\mathrm{C}_\circ(B^{\cdot\cdot}_{\gamma^{\cdot\cdot}})$, $\mathrm{Br}_P(1 \otimes i)$ is the unity element in $\Big(\mathrm{C}_\circ\Big(B^{\cdot\cdot}_{\gamma^{\cdot\cdot}} \otimes_{\mathcal{O}} \mathrm{Res}_{\sigma'}\big((\mathcal{O}G')_{\gamma'}\big)\Big)\Big)(P)$ and, since $\hat{\gamma}$ is local, the image $\hat{\imath}$ of the unity element of $\hat{A}_{\hat{\gamma}}$ by a suitable representative of $\tilde{h}_{\hat{\gamma}}^{\gamma^{\cdot\cdot},\gamma'}$ fulfills $\hat{\imath}(1 \otimes i) = \hat{\imath} = (1 \otimes i)\hat{\imath}$ which proves the claim.

Finally, denoting by

16.7.11
$$\tilde{h}_{\hat{\gamma}}^{\gamma^{\cdot\cdot}\otimes\gamma'} : \hat{A}_{\hat{\gamma}} \longrightarrow \mathrm{Ind}_\sigma\Big(\big(B^{\cdot\cdot} \underset{\mathcal{O}}{\otimes} \mathrm{Res}_{\rho'}(\mathcal{O}G')\big)_{\gamma^{\cdot\cdot}\otimes\gamma'}\Big)$$

the unique $\mathcal{D}P$-interior algebra exoembedding such that

16.7.12
$$\mathrm{Ind}_\sigma(\tilde{f}_{\gamma^{\cdot\cdot}\otimes\gamma'}^{\gamma^{\cdot\cdot},\gamma'}) \circ \tilde{h}_{\hat{\gamma}}^{\gamma^{\cdot\cdot}\otimes\gamma'} = \tilde{h}_{\hat{\gamma}}^{\gamma^{\cdot\cdot},\gamma'} \quad ,$$

up to suitable identifications, it follows from Proposition 16.6 that

16.7.13
$$\tilde{g}_{\gamma^{\cdot\cdot}\otimes\gamma'} \circ \tilde{h}_{\hat{\gamma}}^{\gamma^{\cdot\cdot}\otimes\gamma'} = \tilde{g}_{\gamma^{\cdot\cdot},\gamma'} \circ \tilde{h}_{\hat{\gamma}}^{\gamma^{\cdot\cdot},\gamma'} = \tilde{f}_{\hat{\gamma}} \quad ,$$

so that $(P^{\cdot\cdot}_{\gamma^{\cdot\cdot}\otimes\gamma'}, P_{\hat{\gamma}})$ is a local tracing pair on $B^{\cdot\cdot} \otimes_{\mathcal{O}} \mathrm{Res}_{\rho'}(\mathcal{O}G')$ and $\mathrm{Ind}_\rho\big(B^{\cdot\cdot} \otimes_{\mathcal{O}} \mathrm{Res}_{\rho'}(\mathcal{O}G')\big)$.

16.8 Actually, the first local pointed group in a local tracing triple $(P_{\hat{\gamma}}, P^{\cdot\cdot}_{\gamma^{\cdot\cdot}}, P'_{\gamma'})$ on $\hat{A}$, $A^{\cdot\cdot}$ and $\mathcal{O}G'$ determines the triple up to conjugation; in order to show it, we may assume that $P^{\cdot\cdot}_{\gamma^{\cdot\cdot}}$ comes from a local pointed group on $B^{\cdot\cdot}$ (cf. 2.11.3) and then we are able to apply Theorem 5.8 and Corollary 5.14 to the $\mathcal{O}H^{\cdot\cdot}$-interior algebra $\mathrm{C}_\circ(B^{\cdot\cdot})$.

These results relate indeed $P_{\hat{\gamma}}$ and $P'_{\gamma'}$ to the multiplicity module $V_{\mathrm{C}_\circ(A^{\cdot\cdot})}(P^{\cdot\cdot}_{\gamma^{\cdot\cdot}})$ of $P^{\cdot\cdot}_{\gamma^{\cdot\cdot}}$ on $\mathrm{C}_\circ(A^{\cdot\cdot})$ since, by applying 14.17 to the identity map of $G\times G'$, to $A^{\cdot\cdot}$, to the pointed group $P^{\cdot\cdot}_{\gamma^{\cdot\cdot}}$ and to the subgroup $P \times G'$ of $G \times G'$, we have the $\mathcal{O}(P \times G')$-interior algebra exoembedding (cf. 12.4.4 and 14.17.2)

16.8.1 $$\mathrm{C}_\circ(\tilde{g}_{\gamma^{\cdot\cdot}}) : \mathrm{Ind}_{P^{\cdot\cdot}}^{P\times G'}(\mathrm{C}_\circ(B^{\cdot\cdot})_{\gamma^{\cdot\cdot}}) \longrightarrow \mathrm{Res}^{G\times G'}_{P\times G'}(\mathrm{C}_\circ(A^{\cdot\cdot}))$$

given by (cf. 14.17.)

16.8.2 $$\mathrm{C}_\circ(\tilde{g}_{\gamma^{\cdot\cdot}}) = \mathrm{Res}^{G\times G'}_{P\times G'}(\mathrm{C}_\circ(\tilde{h}_{\gamma^{\cdot\cdot}})) \circ \tilde{d}^{G\times G'}_{P\times G'}\left(\mathrm{Ind}_{P^{\cdot\cdot}}^{P\times G'}(\mathrm{C}_\circ(B^{\cdot\cdot})_{\gamma^{\cdot\cdot}})\right) \quad ,$$

which allows us to identify the multiplicity module of $P^{\cdot\cdot}_{\gamma^{\cdot\cdot}}$ on $\mathrm{Ind}_{P^{\cdot\cdot}}^{P\times G'}(\mathrm{C}_\circ(B^{\cdot\cdot})_{\gamma^{\cdot\cdot}})$ to a direct summand of $V_{\mathrm{C}_\circ(A^{\cdot\cdot})}(P^{\cdot\cdot}_{\gamma^{\cdot\cdot}})$. Notice that, since

16.8.3 $$C_{G\times G'}(P^{\cdot\cdot}) = C_G(P) \times C_{G'}(P') \quad ,$$

we can still consider $V_{\mathrm{C}_\circ(A^{\cdot\cdot})}(P^{\cdot\cdot}_{\gamma^{\cdot\cdot}})$ as a $\Bbbk C_{G'}(P')$-module.

Theorem 16.9 *A triple $(P_{\hat{\gamma}}, P^{\cdot\cdot}_{\gamma^{\cdot\cdot}}, P'_{\gamma'})$ of local pointed groups $P_{\hat{\gamma}}$ on $\hat{A}$, $P^{\cdot\cdot}_{\gamma^{\cdot\cdot}}$ on $A^{\cdot\cdot}$ and $P'_{\gamma'}$ on $\mathcal{O}G'$ is a local tracing triple on $\hat{A}$, $A^{\cdot\cdot}$ and $\mathcal{O}G'$ if and only if $P^{\cdot\cdot}_{\gamma^{\cdot\cdot}}$ is a defect pointed group of $(P \times G')_{\hat{\gamma}}$ on $A^{\cdot\cdot}$, we have $\pi'(P^{\cdot\cdot}) = P'$ and there is a surjective $\Bbbk C_{G'}(P')$-module homomorphism*

16.9.1 $$V_{\mathrm{C}_\circ(A^{\cdot\cdot})_{\hat{\gamma}}}(P^{\cdot\cdot}_{\gamma^{\cdot\cdot}}) \longrightarrow V_{\mathcal{O}G'}(P'_{\gamma'})^* \quad .$$

In particular, the first projection maps the set of $(1 \times G')$-conjugacy classes of local tracing triples on $\hat{A}$, $A^{\cdot\cdot}$ and $\mathcal{O}G'$ bijectively onto the set of local pointed groups on $\hat{A}$.

Proof: Assume first that $(P_{\hat{\gamma}}, P^{\cdot\cdot}_{\gamma^{\cdot\cdot}}, P'_{\gamma'})$ is a local tracing triple on $\hat{A}$, $A^{\cdot\cdot}$ and $\mathcal{O}G'$, and that $P^{\cdot\cdot}_{\gamma^{\cdot\cdot}}$ comes from $B^{\cdot\cdot}$ (cf. 2.11.3); since exoembedding 16.5.1 determines an $\mathcal{O}P$-interior algebra exoembedding

16.9.2 $$\mathrm{C}_\circ(\hat{A})_{\hat{\gamma}} \longrightarrow \mathrm{Ind}_\sigma\big(\mathrm{C}_\circ(B^{\cdot\cdot})_{\gamma^{\cdot\cdot}} \underset{\mathcal{O}}{\otimes} \mathrm{Res}_{\psi'}(\mathcal{O}G')\big)$$

by composing $\mathrm{C}_\circ(\tilde{h}^{\gamma^{\cdot\cdot},\gamma'}_{\hat{\gamma}})$ with $\mathrm{C}_\circ\Big(\mathrm{Ind}_\sigma\big(\widetilde{\mathrm{id}}_{B^{\cdot\cdot}_{\gamma^{\cdot\cdot}}} \otimes_{\mathcal{O}} \mathrm{Res}_{\sigma'}(\tilde{f}_{\gamma'})\big)\Big)$ and isomorphisms 15.5.2, $\hat{\gamma}$ becomes a local point of P on $\mathrm{Ind}_\sigma\big(\mathrm{C}_\circ(B^{\cdot\cdot})_{\gamma^{\cdot\cdot}} \otimes_{\mathcal{O}} \mathrm{Res}_{\psi'}(\mathcal{O}G')\big)$ and, by Theorem 5.8 applied to $\mathrm{C}_\circ(B^{\cdot\cdot})$, we have the inclusion $P^{\cdot\cdot}_{\gamma^{\cdot\cdot}} \subset (P \times G')_{\hat{\gamma}}$ on $\mathrm{Ind}_{P^{\cdot\cdot}}^{P\times G'}(\mathrm{C}_\circ(B^{\cdot\cdot})_{\gamma^{\cdot\cdot}})$ which implies that $P^{\cdot\cdot}_{\gamma^{\cdot\cdot}}$ is a defect pointed group of $(P \times G')_{\hat{\gamma}}$ since $P^{\cdot\cdot}_{\gamma^{\cdot\cdot}}$ is already a maximal local pointed group on this $\mathcal{O}(P \times G')$-interior algebra (cf. 2.11.3); but, throughout exoembedding 16.8.1, $P^{\cdot\cdot}_{\gamma^{\cdot\cdot}}$ and $(P \times G')_{\hat{\gamma}}$ become pointed groups on $\mathrm{C}_\circ(A^{\cdot\cdot})$ and there it is still true that $P^{\cdot\cdot}_{\gamma^{\cdot\cdot}}$ is a defect pointed group of $(P \times G')_{\hat\gamma}$ (cf. 2.11.2). Moreover, since here the equality in condition 14.7.1 implies that

$$16.9.3 \qquad \tilde f_{\gamma^{\cdot\cdot}} = \mathrm{Res}^{G\times G'}_{P^{\cdot\cdot}}(\tilde h_{\gamma^{\cdot\cdot}}) \circ \tilde d^{G\times G'}_{P^{\cdot\cdot}}(A^{\cdot\cdot}_{\gamma^{\cdot\cdot}}) \quad ,$$

it is easily checked that exoembedding 16.8.1 maps the local pointed group $P^{\cdot\cdot}_{\gamma^{\cdot\cdot}}$ on $\mathrm{Ind}^{P\times G'}_{P^{\cdot\cdot}}\big(\mathsf{C}_\circ(B^{\cdot\cdot}_{\gamma^{\cdot\cdot}})\big)$, on the local pointed group $P^{\cdot\cdot}_{\gamma^{\cdot\cdot}}$ on $\mathsf{C}_\circ(A^{\cdot\cdot})$, in other words, that the different identifications of $P^{\cdot\cdot}_{\gamma^{\cdot\cdot}}$ agree. Similarly, since the following diagram is commutative

16.9.4

$$\begin{array}{ccc}
\mathrm{Ind}^{P\times G'}_{P^{\cdot\cdot}}\big(\mathsf{C}_\circ(B^{\cdot\cdot})_{\gamma^{\cdot\cdot}}\big) & \xrightarrow{\mathsf{C}_\circ(\tilde g_{\gamma^{\cdot\cdot}})} & \mathrm{Res}^{G\times G'}_{P\times G'}\big(\mathsf{C}_\circ(A^{\cdot\cdot})\big) \\
\cup & & \\
\mathrm{Ind}^{P\times G'}_{P^{\cdot\cdot}}\big(\mathsf{C}_\circ(B^{\cdot\cdot})_{\gamma^{\cdot\cdot}}\big)^{1\times G'} & & \\
{\scriptstyle H_{G,G'}(\mathsf{C}_\circ(B^{\cdot\cdot})_{\gamma^{\cdot\cdot}})^{-1}}\ \wr\| & & \cup \\
\mathrm{Ind}_\sigma\big(\mathsf{C}_\circ(B^{\cdot\cdot})_{\gamma^{\cdot\cdot}} \otimes_{\mathcal{O}} \mathrm{Res}_{\psi'}(\mathcal{O}G')\big) & & \\
{\scriptstyle \mathrm{Ind}_\sigma(\widetilde{\mathrm{id}}\otimes\mathrm{Res}_{\sigma'}(\tilde f_{\gamma'}))}\ \uparrow & & \\
\mathrm{Ind}_\sigma\Big(\mathsf{C}_\circ(B^{\cdot\cdot})_{\gamma^{\cdot\cdot}} \otimes_{\mathcal{O}} \mathrm{Res}_{\sigma'}\big((\mathcal{O}G')_{\gamma'}\big)\Big) & \xrightarrow{\mathsf{C}_\circ(\tilde g_{\gamma^{\cdot\cdot},\gamma'})} & \mathrm{Res}^{G}_{P}\big(\mathsf{C}_\circ(\hat A)\big)
\end{array}$$

as it is easily checked, it follows from isomorphism 16.5.3 that the different identifications of $P_{\hat\gamma}$ and $(P \times G')_{\hat\gamma}$ agree too.

In particular, $V_{\mathsf{C}_\circ(A^{\cdot\cdot})_{\hat\gamma}}(P^{\cdot\cdot}_{\gamma^{\cdot\cdot}})$ is an indecomposable projective $k_*\hat{\bar N}_{P\times G'}(P^{\cdot\cdot}_{\gamma^{\cdot\cdot}})$-module and, according to Remark 5.9, it is actually an indecomposable direct summand of $k_*\hat{\bar N}_{P\times G'}(P^{\cdot\cdot}_{\gamma^{\cdot\cdot}})\overline{\mathrm{Br}}_{P'}(i')^\circ$, where $i' \in \gamma'$ and, denoting by $\overline{C_{G'}(P')}$ the image of $1 \times C_{G'}(P')$ in $\hat{\bar N}_{P\times G'}(P^{\cdot\cdot}_{\gamma^{\cdot\cdot}})$ and by

$$16.9.5 \qquad \overline{\mathrm{Br}}_{P'} : (\mathcal{O}G')^{P'} \longrightarrow k\overline{C_{G'}(P')}$$

the composition of $\mathrm{Br}_{P'}$ with the canonical map (cf. 2.14.1), $\overline{\mathrm{Br}}_{P'}(i')^\circ$ is the image of $\overline{\mathrm{Br}}_{P'}(i')$ by the antipodal isomorphism $k\overline{C_{G'}(P')} \cong \big(k\overline{C_{G'}(P')}\big)^\circ$; but, it is clear that we have a $k_*\hat{\bar N}_{P\times G'}(P^{\cdot\cdot}_{\gamma^{\cdot\cdot}})$-module isomorphism

$$16.9.6 \qquad k_*\hat{\bar N}_{P\times G'}(P^{\cdot\cdot}_{\gamma^{\cdot\cdot}})\overline{\mathrm{Br}}_{P'}(i')^\circ \cong \mathrm{Ind}^{\hat{\bar N}_{P\times G'}(P^{\cdot\cdot}_{\gamma^{\cdot\cdot}})}_{k^*\times\overline{C_{G'}(P')}}\big(k\overline{C_{G'}(P')}\overline{\mathrm{Br}}_{P'}(i')^\circ\big)$$

and that $kC_{G'}(P')\overline{\mathrm{Br}}_{P'}(i')^\circ$ is a projective cover of the simple $kC_{G'}(P')$-module $V_{\mathcal{O}G'}(P'_{\gamma'})^*$ (cf. [38, (5.9)]), so that, for a suitable n, the maximal semisimple quotient of the restriction of $k_*\hat{\bar N}_{P\times G'}(P^{\cdot\cdot}_{\gamma^{\cdot\cdot}})\overline{\mathrm{Br}}_{P'}(i')^\circ$ to $k\overline{C_{G'}(P')}$ is isomorphic to

$$16.9.7 \qquad \Big(\bigoplus_{x'\in X'} V_{\mathcal{O}G'}(P'_{\gamma'^{x'}})^*\Big)^n \quad ,$$

where, setting $N_{1\times G'}(P^{\cdot\cdot}_{\gamma^{\cdot\cdot}}) = 1 \times N'$ for a suitable subgroup N' of G', X' is a set of representatives for $N'/N_{N'}(P'_{\gamma'})$ in N'; consequently, for a suitable m, the maximal semisimple quotient of the restriction of $V_{\mathrm{C}_\circ(A^{\cdot\cdot})_{\hat\gamma}}(P^{\cdot\cdot}_{\gamma^{\cdot\cdot}})$ to $\overline{kC_{G'}(P')}$ is isomorphic to

16.9.8 $$\Big(\bigoplus_{x'\in X'} V_{\mathcal{O}G'}(P'_{\gamma'^{x'}})^*\Big)^m .$$

Conversely, assume that $P^{\cdot\cdot}_{\gamma^{\cdot\cdot}}$ is a defect pointed group of $(P \times G')_{\hat\gamma}$ on $A^{\cdot\cdot}$, that $\pi'(P^{\cdot\cdot}) = P'$ and that the maximal semisimple quotient of the restriction of $V_{\mathrm{C}_\circ(A^{\cdot\cdot})_{\hat\gamma}}(P^{\cdot\cdot}_{\gamma^{\cdot\cdot}})$ to $\overline{kC_{G'}(P')}$ has a direct summand isomorphic to $V_{\mathcal{O}G'}(P'_{\gamma'})^*$. According to Theorem 14.24 there is a local pointed group $Q^{\cdot\cdot}_{\delta^{\cdot\cdot}\otimes\delta'}$ on $B^{\cdot\cdot} \otimes_{\mathcal{O}} \mathrm{Res}_{\rho'}(\mathcal{O}G')$ such that $(Q^{\cdot\cdot}_{\delta^{\cdot\cdot}\otimes\delta'}, P_{\hat\gamma})$ is a local tracing pair on $B^{\cdot\cdot} \otimes_{\mathcal{O}} \mathrm{Res}_{\rho'}(\mathcal{O}G')$ and $\mathrm{Ind}_\rho\big(B^{\cdot\cdot} \otimes_{\mathcal{O}} \mathrm{Res}_{\rho'}(\mathcal{O}G')\big)$, $\delta^{\cdot\cdot}$ and δ' being respectively local points of $Q^{\cdot\cdot}$ on $B^{\cdot\cdot}$ and of $Q' = \pi'(Q^{\cdot\cdot})$ on $\mathcal{O}G'$; then, we have $\pi(Q^{\cdot\cdot}) = P$ and it follows from Proposition 16.7 that $(P_{\hat\gamma}, Q^{\cdot\cdot}_{\delta^{\cdot\cdot}}, Q'_{\delta'})$ is a local tracing triple on $\hat A$, $A^{\cdot\cdot}$ and $\mathcal{O}G'$; hence, $Q^{\cdot\cdot}_{\delta^{\cdot\cdot}}$ is also a defect pointed group of $(P \times G')_{\hat\gamma}$ and therefore there is $x' \in G'$ such that $P^{\cdot\cdot}_{\gamma^{\cdot\cdot}} = (Q^{\cdot\cdot}_{\delta^{\cdot\cdot}})^{(1,x')}$ (cf. 2.9.5), which in particular implies that $\pi(P^{\cdot\cdot}) = P$; moreover, for a suitable l, the maximal semisimple quotient of the restriction of $V_{\mathrm{C}_\circ(A^{\cdot\cdot})_{\hat\gamma}}(Q^{\cdot\cdot}_{\delta^{\cdot\cdot}})$ to $\overline{kC_{G'}(Q')}$ is isomorphic to

16.9.9 $$\Big(\bigoplus_{y'\in Y'} V_{\mathcal{O}G'}(Q'_{\delta'^{y'}})^*\Big)^l ,$$

where, setting $N_{1\times G'}(Q^{\cdot\cdot}_{\delta^{\cdot\cdot}}) = 1 \times M'$ for a suitable subgroup M' of G', Y' is a set of representatives for $M'/N_{M'}(Q'_{\delta'})$ in M', and therefore there is $y' \in Y'$ such that we have a $kC_{G'}(P')$-module isomorphism

16.9.10 $$V_{\mathcal{O}G'}(P'_{\delta'^{y'x'}}) \cong V_{\mathcal{O}G'}(P'_{\gamma'})$$

which is equivalent to the equality $(\delta')^{y'x'} = \gamma'$ (cf. 2.9.1 and 2.14.1); consequently, we get

16.9.11 $$(P_{\hat\gamma}, P^{\cdot\cdot}_{\gamma^{\cdot\cdot}}, P'_{\gamma'}) = (P_{\hat\gamma}, Q^{\cdot\cdot}_{\delta^{\cdot\cdot}}, Q'_{\delta'})^{(1,y'x')}$$

and, in particular, $(P_{\hat\gamma}, P^{\cdot\cdot}_{\gamma^{\cdot\cdot}}, P'_{\gamma'})$ is a local tracing triple on $\hat A$, $A^{\cdot\cdot}$ and $\mathcal{O}G'$.

Corollary 16.10 *For any point $\alpha^{\cdot\cdot}$ of $G \times G'$ on $A^{\cdot\cdot}$, there is a unique $G \times G'$-conjugacy class of local tracing triples $(P_{\hat\gamma}, P^{\cdot\cdot}_{\gamma^{\cdot\cdot}}, P'_{\gamma'})$ on $\hat A$, $A^{\cdot\cdot}$ and $\mathcal{O}G'$ such that $P_{\hat\gamma}$ and $P^{\cdot\cdot}_{\gamma^{\cdot\cdot}}$ are respectively defect pointed groups of $G_{\alpha^{\cdot\cdot}}$ on $\hat A$ and of $(G \times G')_{\alpha^{\cdot\cdot}}$ on $A^{\cdot\cdot}$.*

Remark 16.11 Actually, Proposition 5.3 is now a consequence of this corollary and equalities 5.2.3.

Proof: If $\alpha^{\cdot\cdot}$ is a point of $G \times G'$ on $A^{\cdot\cdot}$ then it is also a point of G on $\hat{A}$; let $P_{\hat{\gamma}}$ be a defect pointed group of $G_{\alpha^{\cdot\cdot}}$ on $\hat{A}$, so that we have

16.10.1 $$P_{\hat{\gamma}} \subset G_{\alpha^{\cdot\cdot}} \quad \text{and} \quad \alpha^{\cdot\cdot} \subset \mathrm{Tr}_P^G(\hat{A}^P \cdot \hat{\gamma} \cdot \hat{A}^P) \quad ;$$

according to Theorem 16.9, we have a local tracing triple $(P_{\hat{\gamma}}, P^{\cdot\cdot}_{\gamma^{\cdot\cdot}}, P'_{\gamma'})$ on $\hat{A}$, $A^{\cdot\cdot}$ and $\mathcal{O}G'$, and $P^{\cdot\cdot}_{\gamma^{\cdot\cdot}}$ is a defect pointed group of $(P \times G')_{\hat{\gamma}}$ on $A^{\cdot\cdot}$, so that we get (cf. 2.9)

16.10.2 $$P^{\cdot\cdot}_{\gamma^{\cdot\cdot}} \subset (P \times G')_{\hat{\gamma}} \quad \text{and} \quad \hat{\gamma} \subset \mathrm{Tr}_{P^{\cdot\cdot}}^{P \times G'}\big((A^{\cdot\cdot})^{P^{\cdot\cdot}} \cdot \gamma^{\cdot\cdot} \cdot (A^{\cdot\cdot})^{P^{\cdot\cdot}}\big) \quad ;$$

hence, by 16.2, we obtain

16.10.3 $$P^{\cdot\cdot}_{\gamma^{\cdot\cdot}} \subset (G \times G')_{\alpha^{\cdot\cdot}} \quad \text{and} \quad \alpha^{\cdot\cdot} \subset \mathrm{Tr}_{P^{\cdot\cdot}}^{G \times G'}\big((A^{\cdot\cdot})^{P^{\cdot\cdot}} \cdot \gamma^{\cdot\cdot} \cdot (A^{\cdot\cdot})^{P^{\cdot\cdot}}\big) \quad ,$$

so that $P^{\cdot\cdot}_{\gamma^{\cdot\cdot}}$ is a defect pointed group of $(G \times G')_{\alpha^{\cdot\cdot}}$ on $A^{\cdot\cdot}$. The uniqueness follows again from Theorem 16.9.

16.12 As in 14.25, the inclusion relation between the local pointed groups on $\hat{A}$ can be lifted to an order relation between the local tracing triples on $\hat{A}$, $A^{\cdot\cdot}$ and $\mathcal{O}G'$, but we have not found a clear relationship with the order relation between the local tracing pairs on $B^{\cdot\cdot} \otimes_{\mathcal{O}} \mathrm{Res}_{\rho'}(\mathcal{O}G')$ and $\mathrm{Ind}_\rho\big(B^{\cdot\cdot} \otimes_{\mathcal{O}} \mathrm{Res}_{\rho'}(\mathcal{O}G')\big)$; in particular, our proof of Theorem 16.13 below is independent of Theorem 14.26. If $(P_{\hat{\gamma}}, P^{\cdot\cdot}_{\gamma^{\cdot\cdot}}, P'_{\gamma'})$ and $(Q_{\hat{\delta}}, Q^{\cdot\cdot}_{\delta^{\cdot\cdot}}, Q'_{\delta'})$ are local tracing triples on $\hat{A}$, $A^{\cdot\cdot}$ and $\mathcal{O}G'$, we say that $(Q_{\hat{\delta}}, Q^{\cdot\cdot}_{\delta^{\cdot\cdot}}, Q'_{\delta'})$ *is contained in* $(P_{\hat{\gamma}}, P^{\cdot\cdot}_{\gamma^{\cdot\cdot}}, P'_{\gamma'})$, and write $(Q_{\hat{\delta}}, Q^{\cdot\cdot}_{\delta^{\cdot\cdot}}, Q'_{\delta'}) \subset (P_{\hat{\gamma}}, P^{\cdot\cdot}_{\gamma^{\cdot\cdot}}, P'_{\gamma'})$, if we have

16.12.1 $$Q_{\hat{\delta}} \subset P_{\hat{\gamma}} \quad , \quad Q^{\cdot\cdot}_{\delta^{\cdot\cdot}} \subset P^{\cdot\cdot}_{\gamma^{\cdot\cdot}} \quad \text{and} \quad Q'_{\delta'} \subset P'_{\gamma'}$$

and, denoting by $\sigma : P^{\cdot\cdot} \to P$ and $\tau : Q^{\cdot\cdot} \to Q$, and by $\sigma' : P^{\cdot\cdot} \to P'$ and $\tau' : Q^{\cdot\cdot} \to Q'$ the group homomorphisms respectively determined by π and by π', there is a $\mathcal{D}Q$-interior algebra exoembedding

16.12.2
$$\tilde{g}^{\gamma^{\cdot\cdot},\gamma'}_{\delta^{\cdot\cdot},\delta'} : \mathrm{Ind}_\tau\Big(A^{\cdot\cdot}_{\delta^{\cdot\cdot}} \underset{\mathcal{O}}{\otimes} \mathrm{Res}_{\tau'}\big((\mathcal{O}G')_{\delta'}\big)\Big) \to \mathrm{Res}^P_Q\bigg(\mathrm{Ind}_\sigma\Big(A^{\cdot\cdot}_{\gamma^{\cdot\cdot}} \underset{\mathcal{O}}{\otimes} \mathrm{Res}_{\sigma'}\big((\mathcal{O}G')_{\gamma'}\big)\Big)\bigg)$$

such that we have

16.12.3 $$\tilde{g}_{\delta^{\cdot\cdot},\delta'} = \mathrm{Res}^P_Q(\tilde{g}_{\gamma^{\cdot\cdot},\gamma'}) \circ \tilde{g}^{\gamma^{\cdot\cdot},\gamma'}_{\delta^{\cdot\cdot},\delta'} \quad ;$$

then, $\tilde{g}^{\gamma^{\cdot\cdot},\gamma'}_{\delta^{\cdot\cdot},\delta'}$ is unique and fulfills

16.12.4 $$\tilde{g}^{\gamma^{\cdot\cdot},\gamma'}_{\delta^{\cdot\cdot},\delta'} \circ \tilde{h}^{\delta^{\cdot\cdot},\delta'}_{\hat{\delta}} = \mathrm{Res}^P_Q(\tilde{h}^{\gamma^{\cdot\cdot},\gamma'}_{\hat{\gamma}}) \circ \tilde{f}^{\hat{\gamma}}_{\hat{\delta}}$$

since, composing equality 16.12.3 with $\tilde{h}^{\delta^{\cdot\cdot},\delta'}_{\hat{\delta}}$, we get (cf. 16.5.2)

16.12.5 $$\begin{aligned}\mathrm{Res}^P_Q(\tilde{g}_{\gamma^{\cdot\cdot},\gamma'}) \circ \tilde{g}^{\gamma^{\cdot\cdot},\gamma'}_{\delta^{\cdot\cdot},\delta'} \circ \tilde{h}^{\delta^{\cdot\cdot},\delta'}_{\hat{\delta}} &= \tilde{g}_{\delta^{\cdot\cdot},\delta'} \circ \tilde{h}^{\delta^{\cdot\cdot},\delta'}_{\hat{\delta}} = \tilde{f}_{\hat{\delta}} = \mathrm{Res}^P_Q(\tilde{f}_{\hat{\gamma}}) \circ \tilde{f}^{\hat{\gamma}}_{\hat{\delta}} \\ &= \mathrm{Res}^P_Q(\tilde{g}_{\gamma^{\cdot\cdot},\gamma'}) \circ \mathrm{Res}^P_Q(\tilde{h}^{\gamma^{\cdot\cdot},\gamma'}_{\hat{\gamma}}) \circ \tilde{f}^{\hat{\gamma}}_{\hat{\delta}} \quad ,\end{aligned}$$

and both uniqueness and equality 16.12.4 follow from Proposition 11.10. Moreover, it is easily checked that:

16.12.6 *The inclusion between local tracing triples is transitive.*

And, from 16.12.3 and Proposition 11.10, for any local tracing triple $(R_{\hat{\varepsilon}}, R^{\cdot\cdot}_{\varepsilon^{\cdot\cdot}}, R'_{\varepsilon'})$ on $\hat{A}$, $A^{\cdot\cdot}$ and $\mathcal{O}G'$ contained in $(Q_{\hat{\delta}}, Q^{\cdot\cdot}_{\varepsilon^{\cdot\cdot}}, Q'_{\delta'})$, we get

16.12.7
$$\mathrm{Res}^{Q}_{R}(\tilde{g}^{\gamma^{\cdot\cdot},\gamma'}_{\delta^{\cdot\cdot},\delta'}) \circ \tilde{g}^{\delta^{\cdot\cdot},\delta'}_{\varepsilon^{\cdot\cdot},\varepsilon'} = \tilde{g}^{\gamma^{\cdot\cdot},\gamma'}_{\varepsilon^{\cdot\cdot},\varepsilon'} \quad .$$

16.13 Actually, as in 14.26, assuming that $Q^{\cdot\cdot}_{\delta^{\cdot\cdot}} \subset P^{\cdot\cdot}_{\gamma^{\cdot\cdot}}$ and $Q'_{\delta'} \subset P'_{\gamma'}$, and denoting by $\tilde{g}_{\delta^{\cdot\cdot},\delta'}(\mathsf{E}(A^{\cdot\cdot}_{\gamma^{\cdot\cdot}}))$ the corresponding exoembedding 16.4.3 when G, G', $H^{\cdot\cdot}$ and $B^{\cdot\cdot}$ are respectively replaced by P, G', $P^{\cdot\cdot}$ and $\mathsf{E}(A^{\cdot\cdot}_{\gamma^{\cdot\cdot}})$, equality 16.12.3 is equivalent to the following one

16.13.1
$$\begin{aligned}&\tilde{g}_{\delta^{\cdot\cdot},\delta'}(\mathsf{E}(A^{\cdot\cdot}_{\gamma^{\cdot\cdot}}))\\ &= \mathrm{Res}^{P}_{Q}\Big(\tilde{H}_{P,G'}(\mathsf{E}(A^{\cdot\cdot}_{\gamma^{\cdot\cdot}}))^{-1} \circ \mathrm{Ind}_{\sigma}\big(\tilde{e}(A^{\cdot\cdot}_{\gamma^{\cdot\cdot}}) \otimes_{\mathcal{O}} \mathrm{Res}_{\sigma'}(\tilde{f}_{\gamma'})\big)\Big) \circ \tilde{g}^{\gamma^{\cdot\cdot},\gamma'}_{\delta^{\cdot\cdot},\delta'} \, ;\end{aligned}$$

indeed, according to Lemma 16.14 below, for suitable $\mathcal{D}P$-interior algebra exoembeddings $\tilde{e}$ and $\tilde{g}^{1\times G'}$, we have

16.13.2
$$\begin{aligned}\mathrm{Res}^{P}_{Q}(\tilde{e}) \circ \tilde{g}_{\delta^{\cdot\cdot},\delta'} &= \mathrm{Res}^{P}_{Q}(\tilde{g}^{1\times G'}) \circ \tilde{g}_{\delta^{\cdot\cdot},\delta'}(\mathsf{E}(A^{\cdot\cdot}_{\gamma^{\cdot\cdot}}))\\ \tilde{e} \circ \tilde{g}_{\gamma^{\cdot\cdot},\gamma'} &= \tilde{g}^{1\times G'} \circ \tilde{H}_{P,G'}(\mathsf{E}(A^{\cdot\cdot}_{\gamma^{\cdot\cdot}}))^{-1} \circ \mathrm{Ind}_{\sigma}\big(\tilde{e}(A^{\cdot\cdot}_{\gamma^{\cdot\cdot}}) \otimes_{\mathcal{O}} \mathrm{Res}_{\sigma'}(\tilde{f}_{\gamma'})\big) \quad ;\end{aligned}$$

hence, by composing respectively both members of equalities 16.12.3 and 16.13.1 with $\mathrm{Res}^{P}_{Q}(\tilde{e})$ and $\mathrm{Res}^{P}_{Q}(\tilde{g}^{1\times G'})$ on the left, the announced equivalence follows from Proposition 11.10 and equalities 16.13.2. These equalities exhibit a sort of transitivity which is a consequence of Proposition 14.16; in order to state it properly, we borrow notation from 14.15; notice that if $Q^{\cdot\cdot} = P^{\cdot\cdot}$ then

16.13.3
$$\tilde{g}_{\gamma^{\cdot\cdot},\gamma'}(\mathsf{E}(A^{\cdot\cdot}_{\gamma^{\cdot\cdot}})) = \tilde{H}_{P,G'}(\mathsf{E}(A^{\cdot\cdot}_{\gamma^{\cdot\cdot}}))^{-1} \circ \mathrm{Ind}_{\sigma}\big(\tilde{e}(A^{\cdot\cdot}_{\gamma^{\cdot\cdot}}) \otimes_{\mathcal{O}} \mathrm{Res}_{\sigma'}(\tilde{f}_{\gamma'})\big) \quad .$$

Lemma 16.14 *With the notation above, assume that $Q^{\cdot\cdot}_{\delta^{\cdot\cdot}} \subset P^{\cdot\cdot}_{\gamma^{\cdot\cdot}}$ and $Q'_{\delta'} \subset P'_{\gamma'}$. Then, the following diagram is commutative*

16.14.1
$$\begin{array}{ccc}\mathrm{Res}^{G}_{Q}(\hat{A}) & \xrightarrow{\mathrm{Res}^{G}_{Q}(\tilde{e}(\mathcal{O})^{1\times G'} \otimes \tilde{\mathrm{id}}_{\hat{A}})} & \mathrm{Res}^{G}_{Q}(\mathsf{E}(\mathcal{O})^{1\times G'} \otimes_{\mathcal{O}} \hat{A})\\ \uparrow{\scriptstyle \tilde{g}_{\delta^{\cdot\cdot},\delta'}} & & \uparrow{\scriptstyle \mathrm{Res}^{P}_{Q}(\tilde{g}^{1\times G'})}\\ \mathrm{Ind}_{\tau}\big(A^{\cdot\cdot}_{\delta^{\cdot\cdot}} \otimes_{\mathcal{O}} \mathrm{Res}_{\tau'}((\mathcal{O}G')_{\delta'})\big) & \xrightarrow{\tilde{g}_{\delta^{\cdot\cdot},\delta'}(\mathsf{E}(A^{\cdot\cdot}_{\gamma^{\cdot\cdot}}))} & \mathrm{Res}^{P}_{Q}\big(\mathrm{Ind}^{P\times G'}_{P^{\cdot\cdot}}(\mathsf{E}(A^{\cdot\cdot}_{\gamma^{\cdot\cdot}}))^{1\times G'}\big)\end{array}$$

where we set $\tilde{c} = \tilde{c}_{P^{\cdot\cdot}}^{G\times G'}(A^{\cdot\cdot}_{\gamma^{\cdot\cdot}})$ *and*

16.14.2

$$\tilde{g} = \mathrm{Res}^{G\times G'}_{P\times G'}\big((\widetilde{\mathrm{id}}_{\mathsf{E}(\mathcal{O})} \underset{\mathcal{O}}{\otimes} \tilde{h}_{\gamma^{\cdot\cdot}}) \circ \mathrm{Ind}^{G\times G'}_{P^{\cdot\cdot}}(\tilde{c})\big) \circ \tilde{d}^{G\times G'}_{P\times G'}\Big(\mathrm{Ind}^{G\times G'}_{P^{\cdot\cdot}}\big(\mathsf{E}(A^{\cdot\cdot}_{\gamma^{\cdot\cdot}})\big)\Big)\ .$$

Proof: By 3.5.4, we have (cf. definition 14.15.5)

16.14.3

$$\begin{aligned}&\tilde{d}^{G\times G'}_{P\times G'}\Big(\mathrm{Ind}^{P\times G'}_{P^{\cdot\cdot}}\big(\mathsf{E}(A^{\cdot\cdot}_{\gamma^{\cdot\cdot}})\big)\Big) \circ \mathrm{Ind}^{P\times G'}_{P^{\cdot\cdot}}(\tilde{h}^{\gamma^{\cdot\cdot}}_{\delta^{\cdot\cdot}})\\ &\quad = \mathrm{Res}^{G\times G'}_{P\times G'}\big(\mathrm{Ind}^{G\times G'}_{P^{\cdot\cdot}}(\tilde{h}^{\gamma^{\cdot\cdot}}_{\delta^{\cdot\cdot}})\big) \circ \tilde{d}^{G\times G'}_{P\times G'}\big(\mathrm{Ind}^{P\times G'}_{Q^{\cdot\cdot}}(A^{\cdot\cdot}_{\delta^{\cdot\cdot}})\big)\quad ;\end{aligned}$$

hence, according to Proposition 14.16, we get

16.14.4

$$\begin{aligned}&\tilde{g} \circ \mathrm{Ind}^{P\times G'}_{P^{\cdot\cdot}}(\tilde{h}^{\gamma^{\cdot\cdot}}_{\delta^{\cdot\cdot}})\\ &= \mathrm{Res}^{G\times G'}_{P\times G'}\big((\widetilde{\mathrm{id}}_{\mathsf{E}(\mathcal{O})} \underset{\mathcal{O}}{\otimes} \tilde{h}_{\gamma^{\cdot\cdot}}) \circ \mathrm{Ind}^{G\times G'}_{P^{\cdot\cdot}}(\tilde{c}) \circ \mathrm{Ind}^{G\times G'}_{P^{\cdot\cdot}}(\tilde{h}^{\gamma^{\cdot\cdot}}_{\delta^{\cdot\cdot}})\big) \circ \tilde{d}\\ &= \mathrm{Res}^{G\times G'}_{P\times G'}\Big(\big(\tilde{e}(\mathcal{O}) \underset{\mathcal{O}}{\otimes} \widetilde{\mathrm{id}}_{A^{\cdot\cdot}}\big) \circ \tilde{h}_{\delta^{\cdot\cdot}}\Big) \circ \tilde{d}\quad ,\end{aligned}$$

where $d = d^{G\times G'}_{P\times G'}\big(\mathrm{Ind}^{P\times G'}_{Q^{\cdot\cdot}}(A^{\cdot\cdot}_{\delta^{\cdot\cdot}})\big)$. Consequently, we obtain (cf. 14.8.6 and 16.4.4)

16.14.5

$$\begin{aligned}&\mathrm{Res}^{P}_{Q}(\tilde{g}^{1\times G'}) \circ \tilde{g}_{\delta^{\cdot\cdot},\delta'}\big(\mathsf{E}(A^{\cdot\cdot}_{\gamma^{\cdot\cdot}})\big)\\ &= \mathrm{Res}^{P}_{Q}\Big(\big(\tilde{g} \circ \mathrm{Ind}^{P\times G'}_{P^{\cdot\cdot}}(\tilde{h}^{\gamma^{\cdot\cdot}}_{\delta^{\cdot\cdot}})\big)^{1\times G'}\Big) \circ \tilde{d}^{P}_{Q}\big(\mathrm{Ind}^{Q\times G'}_{Q^{\cdot\cdot}}(A^{\cdot\cdot}_{\delta^{\cdot\cdot}})^{1\times G'}\big) \circ \tilde{f}\\ &= \mathrm{Res}^{G}_{Q}\Big(\big(\tilde{e}(\mathcal{O})^{1\times G'} \underset{\mathcal{O}}{\otimes} \widetilde{\mathrm{id}}_{\hat{A}}\big) \circ (\tilde{h}_{\delta^{\cdot\cdot}})^{1\times G'}\Big) \circ \tilde{d}^{G}_{Q}\big(\mathrm{Ind}^{Q\times G'}_{Q^{\cdot\cdot}}(A^{\cdot\cdot}_{\delta^{\cdot\cdot}})^{1\times G'}\big) \circ \tilde{f}\\ &= \mathrm{Res}^{G}_{Q}\big(\tilde{e}(\mathcal{O})^{1\times G'} \underset{\mathcal{O}}{\otimes} \widetilde{\mathrm{id}}_{\hat{A}}\big) \circ \tilde{g}_{\delta^{\cdot\cdot},\delta'}\end{aligned}$$

where $\tilde{f} = \tilde{H}_{Q,G'}(A^{\cdot\cdot}_{\delta^{\cdot\cdot}})^{-1} \circ \mathrm{Ind}_{\tau}\big(\widetilde{\mathrm{id}}_{A^{\cdot\cdot}_{\delta^{\cdot\cdot}}} \otimes_{\mathcal{O}} \mathrm{Res}_{\tau'}(\tilde{f}_{\delta'})\big)$.

Theorem 16.15 *For any local tracing triple $(P_{\hat{\gamma}}, P^{\cdot\cdot}_{\gamma^{\cdot\cdot}}, P'_{\gamma'})$ on $\hat{A}$, $A^{\cdot\cdot}$ and $\mathcal{O}G'$, and any local pointed group $Q_{\hat{\delta}}$ on $\hat{A}$ such that $Q_{\hat{\delta}} \subset P_{\hat{\gamma}}$, there are local pointed groups $Q^{\cdot\cdot}_{\delta^{\cdot\cdot}}$ on $A^{\cdot\cdot}$ and $Q'_{\delta'}$ on $\mathcal{O}G'$ such that $(Q_{\hat{\delta}}, Q^{\cdot\cdot}_{\delta^{\cdot\cdot}}, Q'_{\delta'})$ is a local tracing triple on $\hat{A}$, $A^{\cdot\cdot}$ and $\mathcal{O}G'$, contained in $(P_{\hat{\gamma}}, P^{\cdot\cdot}_{\gamma^{\cdot\cdot}}, P'_{\gamma'})$.*

Proof: Let $(P_{\hat{\gamma}}, P^{\cdot\cdot}_{\gamma^{\cdot\cdot}}, P'_{\gamma'})$ be a local tracing triple on $\hat{A}$, $A^{\cdot\cdot}$ and $\mathcal{O}G'$, and $Q_{\hat{\delta}}$ a local pointed group on $\hat{A}$ such that $Q_{\hat{\delta}} \subset P_{\hat{\gamma}}$, set $R^{\cdot\cdot} = P^{\cdot\cdot} \cap (Q \times G')$ and $R' = \pi'(R^{\cdot\cdot})$, and respectively denote by $\sigma : P^{\cdot\cdot} \to P$ and $\eta : R^{\cdot\cdot} \to Q$, and by $\sigma' : P^{\cdot\cdot} \to P'$ and $\eta' : R^{\cdot\cdot} \to R'$ the group homomorphisms determined by π and by π'; since $Q_{\hat{\delta}} \subset P_{\hat{\gamma}}$, $\hat{\delta}$ comes from a local point of Q on $\hat{A}_{\hat{\gamma}}$ (cf. 2.7.2) and then exoembedding 16.5.1 allow us to identify $\hat{\delta}$ to a local point of Q on $\mathrm{Ind}_\sigma\Big(A^{\cdot\cdot}_{\gamma^{\cdot\cdot}} \otimes_{\mathcal{O}} \mathrm{Res}_{\sigma'}\big((\mathcal{O}G')_{\gamma'}\big)\Big)$ (cf. 2.11. 2). On the other hand, Proposition 12.12 provides a canonical $\mathcal{D}Q$-interior algebra isomorphism

16.15.1
$$\begin{aligned}&\mathrm{Ind}_\eta\Big(\mathrm{Res}^{P^{\cdot\cdot}}_{R^{\cdot\cdot}}\Big(A^{\cdot\cdot}_{\gamma^{\cdot\cdot}} \underset{\mathcal{O}}{\otimes} \mathrm{Res}_{\sigma'}\big((\mathcal{O}G')_{\gamma'}\big)\Big)\Big)\\ &\cong \mathrm{Res}^P_Q\Big(\mathrm{Ind}_\sigma\Big(A^{\cdot\cdot}_{\gamma^{\cdot\cdot}} \underset{\mathcal{O}}{\otimes} \mathrm{Res}_{\sigma'}\big((\mathcal{O}G')_{\gamma'}\big)\Big)\Big)\end{aligned}$$

and, respectively considering pairwise orthogonal primitive idempotent decompositions $I^{\cdot\cdot}$ and I' of the unity elements in $\mathrm{C}_{\mathcal{O}}(A^{\cdot\cdot}_{\gamma^{\cdot\cdot}})^{R^{\cdot\cdot}}$ and in $\big((\mathcal{O}G')_{\gamma'}\big)^{R'}$, the set $\{1 \otimes (i^{\cdot\cdot} \otimes i')\}_{i^{\cdot\cdot}\in I^{\cdot\cdot}, i'\in I'}$ is clearly a Q-fixed pairwise orthogonal idempotent decomposition of the unity element in the left member of isomorphism 16.15.1. Consequently, there are respectively points $\varepsilon^{\cdot\cdot}$ and ε' of $R^{\cdot\cdot}$ and R' on $A^{\cdot\cdot}_{\gamma^{\cdot\cdot}}$ and $(\mathcal{O}G')_{\gamma'}$ such that $\hat{\delta}$, considered as a local point of Q on that member, comes from $\mathrm{Ind}_\eta\Big(A^{\cdot\cdot}_{\varepsilon^{\cdot\cdot}} \otimes_{\mathcal{O}} \mathrm{Res}_{\eta'}\big((\mathcal{O}G')_{\varepsilon'}\big)\Big)$ via the canonical exoembedding $\mathrm{Ind}_\eta(f^{\gamma^{\cdot\cdot}}_{\varepsilon^{\cdot\cdot}} \otimes_{\mathcal{O}} \mathrm{Res}_{\eta'}(f^{\gamma'}_{\varepsilon'}))$; that is to say, we have the following commutative diagram of $\mathcal{D}Q$-interior algebra exoembeddings (cf. [20, 1.6])

16.15.2
$$\begin{array}{ccc}
\mathrm{Res}^P_Q(\hat{A}_{\hat{\gamma}}) & \xrightarrow{\ \mathrm{Res}^P_Q(\tilde{h}^{\gamma^{\cdot\cdot},\gamma'}_{\hat{\gamma}})\ } & \mathrm{Res}^P_Q\Big(\mathrm{Ind}_\sigma\Big(A^{\cdot\cdot}_{\gamma^{\cdot\cdot}} \underset{\mathcal{O}}{\otimes} \mathrm{Res}_{\sigma'}\big((\mathcal{O}G')_{\gamma'}\big)\Big)\Big)\\
\uparrow & & \wr\| \quad \tilde{d}^P_{Q,\sigma}(A^{\cdot\cdot}_{\gamma^{\cdot\cdot}} \underset{\mathcal{O}}{\otimes} \mathrm{Res}_{\sigma'}((\mathcal{O}G')_{\gamma'}))\\
\tilde{f}_{\hat{\delta}} & & \mathrm{Ind}_\eta\Big(\mathrm{Res}^{P^{\cdot\cdot}}_{R^{\cdot\cdot}}\Big(A^{\cdot\cdot}_{\gamma^{\cdot\cdot}} \underset{\mathcal{O}}{\otimes} \mathrm{Res}_{\sigma'}\big((\mathcal{O}G')_{\gamma'}\big)\Big)\Big)\\
| & & \uparrow \quad \mathrm{Ind}_\eta(\tilde{f}^{\gamma^{\cdot\cdot}}_{\varepsilon^{\cdot\cdot}} \underset{\mathcal{O}}{\otimes} \mathrm{Res}_{\eta'}(\tilde{f}^{\gamma'}_{\varepsilon'}))\\
\hat{A}_{\hat{\delta}} & \xrightarrow{\ \tilde{h}^{\varepsilon^{\cdot\cdot},\varepsilon'}_{\hat{\delta}}\ } & \mathrm{Ind}_\eta\Big(A^{\cdot\cdot}_{\varepsilon^{\cdot\cdot}} \underset{\mathcal{O}}{\otimes} \mathrm{Res}_{\eta'}\big((\mathcal{O}G')_{\varepsilon'}\big)\Big)
\end{array}$$

where we set (cf. 2.7)

16.15.3
$$\tilde{h}^{\varepsilon^{\cdot\cdot},\varepsilon'}_{\hat{\delta}} = \tilde{f}_{\hat{\delta}}\Big(\mathrm{Ind}_\eta\Big(A^{\cdot\cdot}_{\varepsilon^{\cdot\cdot}} \underset{\mathcal{O}}{\otimes} \mathrm{Res}_{\eta'}\big((\mathcal{O}G')_{\varepsilon'}\big)\Big)\Big) \quad .$$

Let $Q^{\cdot\cdot}_{\delta^{\cdot\cdot}}$ be a defect pointed group of $R^{\cdot\cdot}_{\varepsilon^{\cdot\cdot}}$, set $Q' = \pi'(Q^{\cdot\cdot})$ and respectively denote by $\tau : Q^{\cdot\cdot} \to Q$ and $\tau' : Q^{\cdot\cdot} \to Q'$ the group homomorphisms determined by π and π'; by Corollary 14.11, there is a $\mathcal{D}R^{\cdot\cdot}$-interior algebra exoembedding

16.15.4
$$\tilde{h}^{\delta^{\cdot\cdot}}_{\varepsilon^{\cdot\cdot}} : A^{\cdot\cdot}_{\varepsilon^{\cdot\cdot}} \longrightarrow \mathrm{Ind}^{R^{\cdot\cdot}}_{Q^{\cdot\cdot}}(A^{\cdot\cdot}_{\delta^{\cdot\cdot}})$$

such that we have

16.15.5
$$\tilde{h}_{\varepsilon^{\cdot\cdot}} = \tilde{h}_{\delta^{\cdot\cdot}} \circ \mathrm{Ind}^{G\times G'}_{R^{\cdot\cdot}}(\hat{h}^{\delta^{\cdot\cdot}}_{\varepsilon^{\cdot\cdot}}) \quad ;$$

moreover, since $R^{\cdot\cdot}$ is a p-group and we have (cf. 12.4.4)

$$\text{16.15.6} \qquad \mathsf{C}_\circ\big(\mathrm{Ind}_{Q^{\cdot\cdot}}^{R^{\cdot\cdot}}(A^{\cdot\cdot}_{\delta^{\cdot\cdot}})\big) = \mathrm{Ind}_{Q^{\cdot\cdot}}^{R^{\cdot\cdot}}\big(\mathsf{C}_\circ(A^{\cdot\cdot}_{\delta^{\cdot\cdot}})\big) \quad ,$$

$\tilde{h}^{\delta^{\cdot\cdot}}_{\varepsilon^{\cdot\cdot}}$ is actually an exoisomorphism (cf. 2.12.2); hence, we have a canonical $\mathcal{D}Q$-interior algebra isomorphism (cf. Corollary 12.7 and Proposition 12.9)

$$\text{16.15.7} \quad \mathrm{Ind}_\eta\Big(A^{\cdot\cdot}_{\varepsilon^{\cdot\cdot}} \mathop{\otimes}_{\mathcal{O}} \mathrm{Res}_{\eta'}\big((\mathcal{O}G')_{\varepsilon'}\big)\Big) \cong \mathrm{Ind}_\tau\bigg(A^{\cdot\cdot}_{\delta^{\cdot\cdot}} \mathop{\otimes}_{\mathcal{O}} \mathrm{Res}_{\tau'}\Big(\mathrm{Res}_{Q'}^{R'}\big((\mathcal{O}G')_{\varepsilon'}\big)\Big)\bigg)$$

and therefore, once again, there is a point δ' of Q' on $(\mathcal{O}G')_{\varepsilon'}$ such that $\hat{\delta}$, considered as a local point of Q on the right member of isomorphism 16.15.7, comes from $\mathrm{Ind}_\tau\big(A^{\cdot\cdot}_{\delta^{\cdot\cdot}} \otimes_{\mathcal{O}} \mathrm{Res}_{\tau'}((\mathcal{O}G')_{\delta'})\big)$ via the canonical exoembedding $\mathrm{Ind}_\tau\big(\widetilde{\mathrm{id}}_{A^{\cdot\cdot}_{\delta^{\cdot\cdot}}} \otimes_{\mathcal{O}} \mathrm{Res}_{\tau'}(\tilde{f}^{\varepsilon'}_{\delta'})\big)$, and we set (cf. 2.7)

$$\text{16.15.8} \qquad \tilde{h}^{\delta^{\cdot\cdot},\delta'}_{\hat{\delta}} = \tilde{f}_{\hat{\delta}}\bigg(\mathrm{Ind}_\tau\Big(A^{\cdot\cdot}_{\delta^{\cdot\cdot}} \mathop{\otimes}_{\mathcal{O}} \mathrm{Res}_{\tau'}\big((\mathcal{O}G')_{\delta'}\big)\Big)\bigg) \quad .$$

Notice that, since $\hat{\delta}$ is local, the existence of this exoembedding forces τ to be surjective (cf. 2.11.3) and δ' to be local (cf. Theorem 5.8 applied to the $\mathcal{O}Q$-interior algebra $\mathsf{C}_\circ(A^{\cdot\cdot}_{\delta^{\cdot\cdot}}) = \mathsf{C}_\circ(A^{\cdot\cdot})_{\delta^{\cdot\cdot}}$, and Remark 5.9); moreover, we have

$$\text{16.15.9} \qquad Q^{\cdot\cdot}_{\delta^{\cdot\cdot}} \subset R^{\cdot\cdot}_{\varepsilon^{\cdot\cdot}} \subset P^{\cdot\cdot}_{\gamma^{\cdot\cdot}} \quad \text{and} \quad Q'_{\delta'} \subset R'_{\varepsilon'} \subset P'_{\gamma'} \quad .$$

Now, we claim that $\tilde{h}^{\delta^{\cdot\cdot},\delta'}_{\hat{\delta}}$ fulfills equality 16.5.2, so that $(Q_{\hat{\delta}}, Q^{\cdot\cdot}_{\delta^{\cdot\cdot}}, Q'_{\delta'})$ is a local tracing triple on $\hat{A}$, $A^{\cdot\cdot}$ and $\mathcal{O}G'$, and that, setting $d = d^P_{Q,\sigma}\big(A^{\cdot\cdot}_{\gamma^{\cdot\cdot}} \otimes_{\mathcal{O}} \mathrm{Res}_{\sigma'}((\mathcal{O}G')_{\gamma'})\big)$ and $\tilde{f} = \widetilde{\mathrm{id}}_{A^{\cdot\cdot}_{\delta^{\cdot\cdot}}} \otimes_{\mathcal{O}} \mathrm{Res}_{\tau'}(\tilde{f}^{\varepsilon'}_{\delta'})$, the $\mathcal{D}Q$-interior algebra exoembedding

$$\text{16.15.10} \quad \tilde{g}^{\gamma^{\cdot\cdot},\gamma'}_{\delta^{\cdot\cdot},\delta'} = \tilde{d} \circ \mathrm{Ind}_\eta\Big(\big(\tilde{f}^{\gamma^{\cdot\cdot}}_{\varepsilon^{\cdot\cdot}} \circ (\tilde{h}^{\delta^{\cdot\cdot}}_{\varepsilon^{\cdot\cdot}})^{-1}\big) \mathop{\otimes}_{\mathcal{O}} \mathrm{Res}_{\eta'}(\tilde{f}^{\gamma'}_{\varepsilon'})\Big) \circ \mathrm{Ind}_\tau(\tilde{f})$$

fulfills equality 16.13.1, so that $(Q_{\hat{\delta}}, Q^{\cdot\cdot}_{\delta^{\cdot\cdot}}, Q'_{\delta'})$ is contained in $(P_{\hat{\gamma}}, P^{\cdot\cdot}_{\gamma^{\cdot\cdot}}, P'_{\gamma'})$. We prove first that $\tilde{g}^{\gamma^{\cdot\cdot},\gamma'}_{\delta^{\cdot\cdot},\delta'}$ fulfills equality 16.13.1; actually, by equality 16.6.2, it suffices to prove that (cf. 14.3.2)

16.15.11

$$\tilde{d}^P_{Q,\sigma}\big(\mathsf{E}(A^{\cdot\cdot}_{\gamma^{\cdot\cdot}}) \mathop{\otimes}_{\mathcal{O}} \mathrm{Res}_{\psi'}(\mathcal{O}G')\big) \circ \mathrm{Ind}_\eta\bigg(\tilde{h}_{\delta^{\cdot\cdot}}\Big(\mathrm{Res}^{P^{\cdot\cdot}}_{R^{\cdot\cdot}}\big(\mathsf{E}(A^{\cdot\cdot}_{\gamma^{\cdot\cdot}})\big)\Big) \mathop{\otimes}_{\mathcal{O}} \mathrm{Res}_{\eta'}(\tilde{f}_{\varepsilon'})\bigg)$$
$$= \mathrm{Res}^P_Q\Big(\mathrm{Ind}_\sigma\big(\tilde{e}(A^{\cdot\cdot}_{\gamma^{\cdot\cdot}}) \mathop{\otimes}_{\mathcal{O}} \mathrm{Res}_{\sigma'}(\tilde{f}_{\gamma'})\big)\Big) \circ \tilde{d} \circ \mathrm{Ind}_\eta\Big(\big(\tilde{f}^{\gamma^{\cdot\cdot}}_{\varepsilon^{\cdot\cdot}} \circ (\tilde{h}^{\delta^{\cdot\cdot}}_{\varepsilon^{\cdot\cdot}})^{-1}\big) \mathop{\otimes}_{\mathcal{O}} \mathrm{Res}_{\eta'}(\tilde{f}^{\gamma'}_{\varepsilon'})\Big) ;$$

but, on one hand, by the naturality of $d^P_{Q,\sigma}$ (cf. 3.22 and 12.13), we have

16.15.12

$$\mathrm{Res}^P_Q\Big(\mathrm{Ind}_\sigma\big(\tilde{e}(A^{\cdot\cdot}_{\gamma^{\cdot\cdot}}) \mathop{\otimes}_{\mathcal{O}} \mathrm{Res}_{\sigma'}(\tilde{f}_{\gamma'})\big)\Big) \circ \tilde{d}^P_{Q,\sigma}\Big(A^{\cdot\cdot}_{\gamma^{\cdot\cdot}} \mathop{\otimes}_{\mathcal{O}} \mathrm{Res}_{\sigma'}\big((\mathcal{O}G')_{\gamma'}\big)\Big)$$
$$= \tilde{d}^P_{Q,\sigma}\big(\mathsf{E}(A^{\cdot\cdot}_{\gamma^{\cdot\cdot}}) \mathop{\otimes}_{\mathcal{O}} \mathrm{Res}_{\psi'}(\mathcal{O}G')\big) \circ \mathrm{Ind}_\eta\Big(\mathrm{Res}^{P^{\cdot\cdot}}_{R^{\cdot\cdot}}\big(\tilde{e}(A^{\cdot\cdot}_{\gamma^{\cdot\cdot}}) \mathop{\otimes}_{\mathcal{O}} \mathrm{Res}_{\sigma'}(\tilde{f}_{\gamma'})\big)\Big) \quad ;$$

on the other hand, it follows easily from equalities in condition 14.7.1, 14.8.6 and 16.15.5 that

16.15.13 $$\begin{aligned}\tilde h_{\delta^{\cdot\cdot}}\Big(\mathrm{Res}^{P^{\cdot\cdot}}_{R^{\cdot\cdot}}(\mathsf{E}(A^{\cdot\cdot}_{\gamma^{\cdot\cdot}}))\Big)\circ \tilde h^{\delta^{\cdot\cdot}}_{\varepsilon^{\cdot\cdot}} &= \tilde h_{\varepsilon^{\cdot\cdot}}\Big(\mathrm{Res}^{P^{\cdot\cdot}}_{R^{\cdot\cdot}}(\mathsf{E}(A^{\cdot\cdot}_{\gamma^{\cdot\cdot}}))\Big)\\ &= \mathrm{Res}^{P^{\cdot\cdot}}_{R^{\cdot\cdot}}(\tilde e(A^{\cdot\cdot}_{\gamma^{\cdot\cdot}}))\circ \tilde f^{\gamma^{\cdot\cdot}}_{\varepsilon^{\cdot\cdot}} \quad ;\end{aligned}$$

now, it is easy to check equality 16.15.11.

Finally, we prove that $\tilde h^{\delta^{\cdot\cdot},\delta'}_{\hat\delta}$ fulfills equality 16.5.2; according to our identification of $\hat\delta$ to a local point of Q on the right member of isomorphism 16.15.7, we have (cf. 16.15.5 and 16.15.8)

16.15.14 $$\mathrm{Ind}_\eta\big(\tilde h^{\delta^{\cdot\cdot}}_{\varepsilon^{\cdot\cdot}} \underset{\mathcal{O}}{\otimes} \mathrm{Res}_{\eta'}(\widetilde{\mathrm{id}}_{\mathcal{O}G'})\big)^{-1}\circ \mathrm{Ind}_\tau(\tilde f)\circ \tilde h^{\delta^{\cdot\cdot},\delta'}_{\hat\delta} = \tilde h^{\varepsilon^{\cdot\cdot},\varepsilon'}_{\hat\delta} \quad ;$$

hence, from equality 16.12.3 and the commutativity of diagram 16.15.2, we get now

16.15.15 $$\begin{aligned}\tilde g_{\delta^{\cdot\cdot},\delta'}\circ \tilde h^{\delta^{\cdot\cdot},\delta'}_{\hat\delta} &= \mathrm{Res}^P_Q(\tilde g_{\gamma^{\cdot\cdot},\gamma'})\circ \tilde g^{\gamma^{\cdot\cdot},\gamma'}_{\delta^{\cdot\cdot},\delta'}\circ \tilde h^{\delta^{\cdot\cdot},\delta'}_{\hat\delta}\\ &= \mathrm{Res}^P_Q(\tilde g_{\gamma^{\cdot\cdot},\gamma'})\circ \tilde d\circ \mathrm{Ind}_\eta\big(\tilde f^{\gamma^{\cdot\cdot}}_{\varepsilon^{\cdot\cdot}} \underset{\mathcal{O}}{\otimes} \mathrm{Res}_{\eta'}(\tilde f^{\gamma'}_{\varepsilon'})\big)\circ \tilde h^{\varepsilon^{\cdot\cdot},\varepsilon'}_{\hat\delta}\\ &= \mathrm{Res}^P_Q(\tilde g_{\gamma^{\cdot\cdot},\gamma'})\circ \mathrm{Res}^P_Q(\tilde h^{\gamma^{\cdot\cdot},\gamma'}_{\hat\gamma})\circ \tilde f^{\hat\gamma}_{\hat\delta}\\ &= \mathrm{Res}^P_Q(\tilde f_{\hat\gamma})\circ \tilde f^{\hat\gamma}_{\hat\delta} = \tilde f_{\hat\delta} \quad .\end{aligned}$$

Corollary 16.16 *Let $(P_{\hat\gamma}, P^{\cdot\cdot}_{\gamma^{\cdot\cdot}}, P'_{\gamma'})$ and $(Q_{\hat\delta}, Q^{\cdot\cdot}_{\delta^{\cdot\cdot}}, Q'_{\delta'})$ be local tracing triples on $\hat A$, $A^{\cdot\cdot}$ and $\mathcal{O}G'$. For any $x\in G$ such that $(Q_{\hat\delta})^x \subset P_{\hat\gamma}$ there is $x'\in G'$ such that*

16.16.1 $$(Q_{\hat\delta}, Q^{\cdot\cdot}_{\delta^{\cdot\cdot}}, Q'_{\delta'})^{(x,x')} \subset (P_{\hat\gamma}, P^{\cdot\cdot}_{\gamma^{\cdot\cdot}}, P'_{\gamma'}) \quad .$$

Proof: If $(Q_{\hat\delta})^x \subset P_{\hat\gamma}$ for some $x\in G$, according to Theorem 16.15 there are local pointed groups $R^{\cdot\cdot}_{\varepsilon^{\cdot\cdot}}$ on $A^{\cdot\cdot}$ and $R'_{\varepsilon'}$ on $\mathcal{O}G'$ such that $\big((Q_{\hat\delta})^x, R^{\cdot\cdot}_{\varepsilon^{\cdot\cdot}}, R'_{\varepsilon'}\big)$ is a local tracing triple on $\hat A$, $A^{\cdot\cdot}$ and $\mathcal{O}G'$, contained in $(P_{\hat\gamma}, P^{\cdot\cdot}_{\gamma^{\cdot\cdot}}, P'_{\gamma'})$; but, since $\big((Q_{\hat\delta})^x, (Q^{\cdot\cdot}_{\delta^{\cdot\cdot}})^{(x,1)}, (Q'_{\delta'})\big)$ is also a local tracing triple on $\hat A$, $A^{\cdot\cdot}$ and $\mathcal{O}G'$, it follows from the last statement of Theorem 16.9 that there is $x'\in G'$ such that

16.16.2 $$\big((Q_{\hat\delta})^x, R^{\cdot\cdot}_{\varepsilon^{\cdot\cdot}}, R'_{\varepsilon'}\big) = \big((Q_{\hat\delta})^x, (Q^{\cdot\cdot}_{\delta^{\cdot\cdot}})^{(x,x')}, (Q'_{\delta'})^{x'}\big) \quad .$$

17 Brauer sections in basic Hecke $\mathcal{D}G$-interior algebras

17.1 We keep all the notation of Section 16 and we assume again that $B^{\cdot\cdot} \cong \mathsf{E}(B^{\cdot\cdot})$, so that $A^{\cdot\cdot}$, $\hat{A}$ and $\hat{A}'$ still coincide with their Higman envelopes (cf. Corollary 14.21 and 15.7.1). In this section we analyze a particular kind of local tracing triples on $\hat{A}$, $A^{\cdot\cdot}$ and $\mathcal{O}G'$ (cf. 16.4) which occurs without exception in the Hecke $\mathcal{D}G$-interior algebras associated with the *basic Rickard equivalences* between blocks introduced in Section 19 below. Precisely, we say that a local tracing triple $(P_{\hat{\gamma}}, P^{\cdot\cdot}_{\gamma^{\cdot\cdot}}, P'_{\gamma'})$ on $\hat{A}$, $A^{\cdot\cdot}$ and $\mathcal{O}G'$ is *basic* if $\hat{A}_{\hat{\gamma}}$ is a *split* $\mathcal{D}P$-module (cf. 10.12 and 11.2) and $A^{\cdot\cdot}_{\gamma^{\cdot\cdot}}$ is a $\mathrm{Ker}(\sigma)$-*basic* $\mathcal{D}P^{\cdot\cdot}$-interior algebra (cf. 13.2), where $\sigma : P^{\cdot\cdot} \longrightarrow P$ is the group homomorphism determined by π; in this case, it is clear that any local tracing triple on $\hat{A}$, $A^{\cdot\cdot}$ and $\mathcal{O}G'$ contained in $(P_{\hat{\gamma}}, P^{\cdot\cdot}_{\gamma}, P'_{\gamma'})$ is basic too (cf. 14.3.1 and 16.12.1). Moreover, since $(\mathcal{O}G')_{\gamma'}$ has a $P' \times P'$-stable $\mathcal{O}$-basis by left and right multiplication, the basic condition on $A^{\cdot\cdot}_{\gamma^{\cdot\cdot}}$ is inherited by the tensor product $A^{\cdot\cdot}_{\gamma^{\cdot\cdot}} \otimes_{\mathcal{O}} \mathrm{Res}_{\sigma'}\big((\mathcal{O}G')_{\gamma'}\big)$, where $\sigma' : P^{\cdot\cdot} \longrightarrow P'$ is the group homomorphism determined by π', so that, by Lemma 13.4, the induced $\mathcal{D}P$-interior algebra $\mathrm{Ind}_{\sigma}\Big(A^{\cdot\cdot}_{\gamma^{\cdot\cdot}} \otimes_{\mathcal{O}} \mathrm{Res}_{\sigma'}\big((\mathcal{O}G')_{\gamma'}\big)\Big)$ is basic and then, by the existence of the structural $\mathcal{D}P$-interior algebra exoembedding $\tilde{h}_{\hat{\gamma}}^{\gamma^{\cdot\cdot},\gamma'}$ (cf. 16.4.1), $\hat{A}_{\hat{\gamma}}$ is a basic $\mathcal{D}P$-interior algebra too: before going further, let us collect some elementary facts on basic $\mathcal{D}P$-interior algebras which are split $\mathcal{D}P$-modules.

Lemma 17.2 *Let H be a finite group, Q a p-subgroup of H and B a basic $\mathcal{D}$-interior H-algebra which is also a split $\mathcal{D}H$-module. Then the inclusion $\mathsf{C}_\circ(B)^Q \subset B^Q$ induces a $N_H(Q)$-algebra isomorphism*

17.2.1 $$\big(\mathsf{C}_\circ(B)\big)(Q) \cong \mathsf{C}_\circ\big(B(Q)\big)$$

and for any subgroup L of $N_H(Q)$ containing Q we have

17.2.2 $$\mathrm{Br}_Q^B\big(\mathsf{C}_\circ(B^L)\big) = \mathsf{C}_\circ\big(B(Q)^L\big) \ \text{and}\ \mathrm{Br}_Q^B\big(\mathsf{B}_\circ(B^L)\big) = \mathsf{B}_\circ\big(B(Q)^L\big) \quad .$$

In particular, Br_Q^B induces a bijection between the set of pointed groups L_ε on B such that $Q \subset L \subset N_H(Q)$ and $\mathrm{Br}_Q^B(\varepsilon) \neq \{0\}$, and the set of all the pointed groups on the $\mathcal{D}$-interior $\bar{N}_H(Q)$-algebra $B(Q)$, which maps L_ε on $(L/Q)_{\mathrm{Br}_Q^B(\varepsilon)}$ preserving inclusion, localness and contractility.

Proof: Isomorphism 17.2.1 follows from 11.5.1. As there, we have a $\mathcal{D}H$-module isomorphism $B \cong B' \oplus B''$ where B' is a $\mathcal{D}H$-module which coincides with $\mathsf{C}(B')$ and B'' is a contractile $\mathcal{D}H$-module; hence, if L is a subgroup of $N_H(Q)$ containing Q, we get a $\mathcal{D}$-module isomorphism

17.2.3 $$B(Q)^L \cong B'(Q)^L \oplus B''(Q)^L$$

where we have $\mathsf{C}\big(B'(Q)^L\big) = B'(Q)^L$ and $B''(Q)^L$ is contractile, so that

17.2.4

$$\mathsf{C}\big(B(Q)^L\big) \cong B'(Q)^L \oplus \mathsf{B}\big(B''(Q)^L\big) \text{ and } \mathsf{B}\big(B(Q)^L\big) \cong \mathsf{B}\big(B''(Q)^L\big) \quad .$$

On the other hand, according to our hypothesis, if R is a Sylow p-subgroup of L then B is a permutation $\mathcal{O}R$-module (cf. 2.2) and therefore so are B' and B'' (cf. [38, (27.2)]), so that we have (cf. 2.2.5)

17.2.5 $$\mathrm{Br}_Q^B(B'^R) = B'(Q)^R \quad \text{and} \quad \mathrm{Br}_Q^B(B''^R) = B''(Q)^R \quad ;$$

now, applying Tr_R^L to both members of both equalities, we get (cf. 2.2.3)

17.2.6 $$\mathrm{Br}_Q^B(B'^L) = B'(Q)^L \quad \text{and} \quad \mathrm{Br}_Q^B(B''^L) = B''(Q)^L$$

and the second one implies clearly

17.2.7 $$\mathrm{Br}_Q^B\big(\mathsf{B}(B''^L)\big) = \mathsf{B}\big(B''(Q)^L\big) \quad .$$

Consequently, equalities 17.2.2 follow from equalities 17.2.4, 17.2.6 and 17.2.7.

17.3 From now on, let $(P_{\hat\gamma}, P^{\cdot\cdot}_{\gamma^{\cdot\cdot}}, P'_{\gamma'})$ be a basic local tracing triple on $\hat A$, $A^{\cdot\cdot}$ and $\mathcal{O}G'$, and denote by

17.3.1 $$\tilde h_{\hat\gamma}^{\gamma^{\cdot\cdot},\gamma'} : \hat A_{\hat\gamma} \longrightarrow \mathrm{Ind}_\sigma\Big(A^{\cdot\cdot}_{\gamma^{\cdot\cdot}} \underset{\mathcal{O}}{\otimes} \mathrm{Res}_{\sigma'}\big((\mathcal{O}G')_{\gamma'}\big)\Big)$$

the structural exoembedding (cf. 16.4.1) with σ and σ' as above. According to Lemma 17.2 we have

17.3.2 $$\{0\} \neq \big(\mathsf{C}_\circ(\hat A_{\hat\gamma})\big)(P) \cong \mathsf{C}_\circ\big(\hat A_{\hat\gamma}(P)\big)$$

and then, since the unity element is primitive in $\big(\mathsf{C}_\circ(\hat A_{\hat\gamma})\big)(P)$, P has a unique point on $\hat A_{\hat\gamma}(P)$ (with multiplicity one), which determines a point of P on $\mathrm{Ind}_\sigma\Big(A^{\cdot\cdot}_{\gamma^{\cdot\cdot}} \otimes_{\mathcal{O}} \mathrm{Res}_{\sigma'}\big((\mathcal{O}G')_{\gamma'}\big)(P)\Big)$ via the $\mathcal{D}Z(P)$-interior P-algebra exoembedding

17.3.3 $$\tilde h_{\hat\gamma}^{\gamma^{\cdot\cdot},\gamma'}(P) : \hat A_{\hat\gamma}(P) \longrightarrow \mathrm{Ind}_\sigma\Big(A^{\cdot\cdot}_{\gamma^{\cdot\cdot}} \underset{\mathcal{O}}{\otimes} \mathrm{Res}_{\sigma'}\big((\mathcal{O}G')_{\gamma'}\big)\Big)(P) \quad .$$

But, another consequence of the fact that the tensor product $A^{\cdot\cdot}_{\gamma^{\cdot\cdot}} \otimes_{\mathcal{O}} \mathrm{Res}_{\sigma'}\big((\mathcal{O}G')_{\gamma'}\big)$ is still a $\mathrm{Ker}(\sigma)$-basic $\mathcal{D}P^{\cdot\cdot}$-interior algebra is that Theorem 13.9 applies providing a pairwise orthogonal idempotent decomposition of the unity element in the k-algebra $\mathsf{C}_\circ\Big(\mathrm{Ind}_\sigma\Big(A^{\cdot\cdot}_{\gamma^{\cdot\cdot}} \otimes_{\mathcal{O}} \mathrm{Res}_{\sigma'}\big((\mathcal{O}G')_{\gamma'}\big)\Big)(P)\Big)$. Hence, that point of P comes already from at least one term of this decomposition.

17.4 More generally, for any subgroup Q of P, let us call any group homomorphism $\mu : Q \longrightarrow \sigma^{-1}(Q)$ such that $\sigma\big(\mu(u)\big) = u$, for any $u \in Q$,

Q-section of σ and then set $\mu' = \sigma' \circ \mu$, $Q^\mu = \mu(Q)$ and $Q^{\mu'} = \mu'(Q)$; thus, for any Q-section μ of σ, Theorem 13.9 guarantees the existence of a canonical $\mathcal{D}\Big(\sigma\big(C_{P^{\cdot\cdot}}(Q^\mu)\big)\Big)$-interior $\sigma\big(N_{P^{\cdot\cdot}}(Q^\mu)\big)$-algebra embedding

17.4.1 $$e_\mu^{\gamma^{\cdot\cdot},\gamma'} : \mathrm{Ind}_{\sigma_\mu}\Big(A^{\cdot\cdot}_{\gamma^{\cdot\cdot}}(Q^\mu) \underset{k}{\otimes} \mathrm{Res}_{\sigma'_\mu}\big((\mathcal{O}G')_{\gamma'}(Q^{\mu'})\big)\Big) \longrightarrow \hat{A}_{\gamma^{\cdot\cdot},\gamma'}(Q)$$

where $\hat{A}_{\gamma^{\cdot\cdot},\gamma'} = \mathrm{Ind}_\sigma\Big(A^{\cdot\cdot}_{\gamma^{\cdot\cdot}} \otimes_{\mathcal{O}} \mathrm{Res}_{\sigma'}\big((\mathcal{O}G')_{\gamma'}\big)\Big)$ and

$$\sigma_\mu : C_{P^{\cdot\cdot}}(Q^\mu) \longrightarrow \sigma\big(C_{P^{\cdot\cdot}}(Q^\mu)\big) \quad \text{and} \quad \sigma'_\mu : C_{P^{\cdot\cdot}}(Q^\mu) \longrightarrow C_{P'}(Q^{\mu'})$$

are respectively the group homomorphisms determined by σ and σ'; moreover, denoting by $\mathfrak{M}_{\sigma,Q}$ a set of representatives for the set of orbits of $\mathrm{Ker}(\sigma)$ on the set of Q-sections of σ, the family $\Big\{e_\mu^{\gamma^{\cdot\cdot},\gamma'}\big(1 \otimes (1 \otimes 1)\big)\Big\}_{\mu \in \mathfrak{M}_{\sigma,Q}}$ is a $N_P(Q)$-stable pairwise orthogonal idempotent decomposition of the unity element in the $N_P(Q)$-algebra

17.4.2 $$\mathrm{C}_\circ\Big(\mathrm{Ind}_\sigma\Big(A^{\cdot\cdot}_{\gamma^{\cdot\cdot}} \underset{\mathcal{O}}{\otimes} \mathrm{Res}_{\sigma'}\big((\mathcal{O}G')_{\gamma'}\big)\Big)(Q)\Big)$$

which does not depend on the choice of $\mathfrak{M}_{\sigma,Q}$ (cf. Remark 13.10). Notice that if $N_P(Q)$ stabilizes the $\mathrm{Ker}(\sigma)$-orbit of a Q-section μ of σ or, equivalently whenever $e_\mu^{\gamma^{\cdot\cdot},\gamma'} \neq 0$, fixes the idempotent $e_\mu^{\gamma^{\cdot\cdot},\gamma'}\big(1 \otimes (1 \otimes 1)\big)$ then we have

17.4.3 $$N_{P^{\cdot\cdot}}\big(\sigma^{-1}(Q)\big) = \mathrm{Ker}(\sigma)\cdot N_{P^{\cdot\cdot}}(Q^\mu)$$

and therefore, since σ induces an isomorphism $Q^\mu \cong Q$, we get

17.4.4 $$\sigma\big(N_{P^{\cdot\cdot}}(Q^\mu)\big) = N_P(Q) \quad \text{and} \quad \sigma\big(C_{P^{\cdot\cdot}}(Q^\mu)\big) = C_P(Q) \quad .$$

If Q is a normal subgroup of P, we say that a Q-section μ of σ is a *Q-section of* $(P_{\hat{\gamma}}, P^{\cdot\cdot}_{\gamma^{\cdot\cdot}}, P'_{\gamma'})$ if P stabilizes the $\mathrm{Ker}(\sigma)$-orbit of μ and there is a $\mathcal{D}C_P(Q)$-interior P-algebra exoembedding

17.4.5 $$\tilde{h}_{\hat{\gamma}}^\mu : \hat{A}_{\hat{\gamma}}(Q) \longrightarrow \mathrm{Ind}_{\sigma_\mu}\Big(A^{\cdot\cdot}_{\gamma^{\cdot\cdot}}(Q^\mu) \underset{k}{\otimes} \mathrm{Res}_{\sigma'_\mu}\big((\mathcal{O}G')_{\gamma'}(Q^{\mu'})\big)\Big)$$

such that we have

17.4.6 $$\tilde{e}_\mu^{\gamma^{\cdot\cdot},\gamma'} \circ \tilde{h}_{\hat{\gamma}}^\mu = \tilde{h}_{\hat{\gamma}}^{\gamma^{\cdot\cdot},\gamma'}(Q) \qquad ;$$

in that case $\tilde{h}_{\hat{\gamma}}^\mu$ is unique (cf. Proposition 11.10). When $Q = P$ we say simply that μ is a *section of* $(P_{\hat{\gamma}}, P^{\cdot\cdot}_{\gamma^{\cdot\cdot}}, P'_{\gamma'})$. Notice that the unique 1-section of σ is a 1-section of $(P_{\hat{\gamma}}, P^{\cdot\cdot}_{\gamma^{\cdot\cdot}}, P'_{\gamma'})$ too and, since $\hat{A}_{\hat{\gamma}}(P) \neq \{0\}$ (cf. 17.3.2), we have that (cf. 17.4.5):

17.4.7 *If μ is a section of $(P_{\hat{\gamma}}, P^{\cdot\cdot}_{\gamma^{\cdot\cdot}}, P'_{\gamma'})$ then $A^{\cdot\cdot}_{\gamma^{\cdot\cdot}}(P^\mu) \neq \{0\}$.*

Proposition 17.5 *Let Q and R be normal subgroups of P such that $R \subset Q$. Any R-section of $(P_{\hat\gamma}, P''_{\gamma''}, P'_{\gamma'})$ can be extended to a Q-section of $(P_{\hat\gamma}, P''_{\gamma''}, P'_{\gamma'})$.*

Remark 17.6 Notice that, if μ and ν are respectively a Q-section and a R-section of $(P_{\hat\gamma}, P''_{\gamma''}, P'_{\gamma'})$ such that μ extends ν, it follows from Proposition 13.16 that there is a $\mathcal{D}C_P(Q)$-interior P-algebra embedding

17.6.1
$$\begin{array}{c}\mathrm{Ind}_{\sigma_\nu}\Big(A''_{\gamma''}(R^\nu)\underset{k}{\otimes}\mathrm{Res}_{\sigma'_\nu}\big((\mathcal{O}G')_{\gamma'}(R^{\nu'})\big)\Big)(Q)\\ e\Big\uparrow\\ \mathrm{Ind}_{\sigma_\mu}\Big(A''_{\gamma''}(Q^\mu)\underset{k}{\otimes}\mathrm{Res}_{\sigma'_\mu}\big((\mathcal{O}G')_{\gamma'}(Q^{\mu'})\big)\Big)\end{array}$$

such that we have

17.6.2
$$e_\mu^{\gamma'',\gamma'} = e_\nu^{\gamma'',\gamma'}(Q)\circ e \quad\text{and}\quad \tilde h^\nu_{\hat\gamma}(Q) = \tilde e\circ \tilde h^\mu_{\hat\gamma} \qquad ,$$

the second equality being a consequence of the first one together with Proposition 11.10 and equality 17.4.6 (we may identify $\big(\hat A_{\hat\gamma}(R)\big)(Q)$ to $\hat A_{\hat\gamma}(Q)$ by Lemma 7.10).

Proof: Let $\nu : R \longrightarrow \sigma^{-1}(R)$ be a R-section of $(P_{\hat\gamma}, P''_{\gamma''}, P'_{\gamma'})$, so that we have a $\mathcal{D}C_P(R)$-interior P-algebra exoembedding (cf. 17.4.5)

17.5.1
$$\tilde h^\nu_{\hat\gamma} : \hat A_{\hat\gamma}(R) \longrightarrow \mathrm{Ind}_{\sigma_\nu}\Big(A''_{\gamma''}(R^\nu)\underset{k}{\otimes}\mathrm{Res}_{\sigma'_\nu}\big((\mathcal{O}G')_{\gamma'}(R^{\nu'})\big)\Big)$$

fulfilling $\tilde e_\nu^{\gamma'',\gamma'}\circ \tilde h^\nu_{\hat\gamma} = \tilde h^{\gamma'',\gamma'}_{\hat\gamma}(R)$. It is quite clear that $\hat A_{\hat\gamma}(R)$ remains a split $\mathcal{D}P$-module (cf. 10.12), whereas it is easily checked that $A''_{\gamma''}(R^\nu)$ is a Ker(σ_ν)-basic $k\otimes_{\mathcal{O}}\mathcal{D}C_{P''}(R^\nu)$-interior $N_{P''}(R^\nu)$-algebra (cf. 13.7); consequently, the tensor product

17.5.2
$$B_\nu = A''_{\gamma''}(R^\nu)\underset{k}{\otimes}\mathrm{Res}_{\sigma'_\nu}\big((\mathcal{O}G')_{\gamma'}(R^{\nu'})\big)$$

is also Ker(σ_ν)-basic (over k!) and therefore both Lemma 13.4 and Theorem 13.9 apply again (notice that, since $\mathrm{Ker}(\sigma)\cap R^\nu = 1$, we have $N_{\mathrm{Ker}(\sigma)}(R^\nu) = \mathrm{Ker}(\sigma_\nu)$). So, on one hand, it follows from Lemmas 7.10 and 13.4 that we have $\big(\hat A_{\hat\gamma}(R)\big)(Q)\cong \hat A_{\hat\gamma}(Q)$ and from Lemma 17.2 applied to P, Q and $\hat A_{\hat\gamma}$ (cf. 17.1) that $\{1\}$ is a local point of P (or P/Q) on $\hat A_{\hat\gamma}(Q)$; hence, from 17.5.1, we get a $\mathcal{D}C_P(Q)$-interior P-algebra exoembedding

17.5.3
$$\tilde h^\nu_{\hat\gamma}(Q) : \hat A_{\hat\gamma}(Q) \longrightarrow \big(\mathrm{Ind}_{\sigma_\nu}(B_\nu)\big)(Q)$$

which determines a local point of P on the right end.

On the other hand, if follows from Lemma 7.10 and Theorem 13.9 that, for any Q-section μ of the restriction of σ to $N_{P^{\cdot\cdot}}(R^\nu)$, there is a canonical $\mathcal{D}\Big(\sigma\big(C_{P^{\cdot\cdot}}(R^\nu{\cdot}Q^\mu)\big)\Big)$-interior $\sigma\big(N_{P^{\cdot\cdot}}(R^\nu)\cap N_{P^{\cdot\cdot}}(Q^\mu)\big)$-algebra embedding

17.5.4
$$\mathrm{Ind}_{\sigma_{\nu,\mu}}\Big(A^{\cdot\cdot}_{\gamma^{\cdot\cdot}}(R^\nu{\cdot}Q^\mu)\underset{k}{\otimes}\mathrm{Res}_{\sigma'_{\nu,\mu}}\big((\mathcal{O}G')_{\gamma'}(R^{\nu'}{\cdot}Q^{\mu'})\big)\Big)\longrightarrow\big(\mathrm{Ind}_{\sigma_\nu}(B_\nu)\big)(Q)$$

noted e_μ, where

$$\sigma_{\nu,\mu}:C_{P^{\cdot\cdot}}(R^\nu{\cdot}Q^\mu)\longrightarrow\sigma\big(C_{P^{\cdot\cdot}}(R^\nu{\cdot}Q^\mu)\big)\quad\text{and}$$
$$\sigma'_{\nu,\mu}:N_{P^{\cdot\cdot}}(R^\nu)\cap N_{P^{\cdot\cdot}}(Q^\mu)\longrightarrow N_{P'}(R^{\nu'})\cap N_{P'}(Q^{\mu'})$$

are respectively the group homomorphisms determined by σ and σ', in such a way that, choosing a set of representatives $\mathfrak{M}_\nu$ for the set of orbits of $\mathrm{Ker}(\sigma_\nu)$ on the set of Q-sections of the restriction of σ to $N_{P^{\cdot\cdot}}(R^\nu)$, the family $E=\Big\{e_\mu\big(1\otimes(1\otimes1)\big)\Big\}_{\mu\in\mathfrak{M}_\nu}$ is a P-stable pairwise orthogonal idempotent decomposition of the unity element in $\mathrm{C}_\circ\Big(\big(\mathrm{Ind}_{\sigma_\nu}(B_\nu)\big)(Q)\Big)$. But, since $A^{\cdot\cdot}_{\gamma^{\cdot\cdot}}$ is a projective $\mathcal{O}\big(\mathrm{Ker}(\sigma)\times\mathrm{Ker}(\sigma)\big)$-module by left and right multiplication (cf. 13.2), for any nontrivial subgroup $T^{\cdot\cdot}$ of $\mathrm{Ker}(\sigma)$, we have $A^{\cdot\cdot}_{\gamma^{\cdot\cdot}}(T^{\cdot\cdot})=\{0\}$ (cf. 2.2.6); hence, according to Lemma 7.10, in the left end of e_μ we have $A^{\cdot\cdot}_{\gamma^{\cdot\cdot}}(R^\nu{\cdot}Q^\mu)\neq\{0\}$ only if $R^\nu\subset Q^\mu$ or, equivalently, if μ extends ν, and in that case it is clear that

17.5.5 $\quad R^\nu{\cdot}Q^\mu=Q^\mu\quad,\quad R^{\nu'}{\cdot}Q^{\mu'}=Q^{\mu'}\quad\text{and}\quad N_{P^{\cdot\cdot}}(Q^\mu)\subset N_{P^{\cdot\cdot}}(R^\nu)\quad.$

Now, since P stabilizes the family E, a suitable idempotent i of the point of P on $\big(\mathrm{Ind}_{\sigma_\nu}(B_\nu)\big)(Q)$ determined by exoembedding 17.5.3 centralizes E (cf. [38, (24.1)]) and then, since we have $i=\Sigma_{e\in E}\,ie$ and i is primitive in $\mathrm{C}_\circ\Big(\big(\mathrm{Ind}_{\sigma_\nu}(B_\nu)\big)(Q)^P\Big)$, P acts transitively on the set F of $e\in E$ such that $ie\neq0$; moreover, since that point is local, we have $F=\Big\{e_\mu\big(1\otimes(1\otimes1)\big)\Big\}$ and $ie_\mu\big(1\otimes(1\otimes1)\big)=i$ for a suitable Q-section μ of σ which extends ν (by the argument above) and has a P-stable $\mathrm{Ker}(\sigma_\mu)$-orbit; hence, the image of $\hat{A}_{\hat{\gamma}}(Q)$ by a suitable representative h^ν of $\tilde{h}^\nu_{\hat{\gamma}}(Q)$ is contained in $\mathrm{Im}(e_\mu)$, which allow us to define by restriction a $\mathcal{D}C_P(Q)$-interior P-algebra embedding

17.5.6 $$h^\mu:\hat{A}_{\hat{\gamma}}(Q)\longrightarrow\mathrm{Ind}_{\sigma_\mu}\Big(A^{\cdot\cdot}_{\gamma^{\cdot\cdot}}(Q^\mu)\underset{k}{\otimes}\mathrm{Res}_{\sigma'_\mu}\big((\mathcal{O}G')_{\gamma'}(Q^{\mu'})\big)\Big)$$

such that $e_\mu\circ h^\mu=h^\nu$ and therefore, since by Lemma 7.10 we have

17.5.7 $$\tilde{e}^{\gamma^{\cdot\cdot},\gamma'}_\nu(Q)\circ\tilde{h}^\nu_{\hat{\gamma}}(Q)=\tilde{h}^{\gamma^{\cdot\cdot},\gamma'}_{\hat{\gamma}}(Q)\quad,$$

by Proposition 13.16 we get

17.5.8 $$\tilde h_{\hat\gamma}^{\gamma^{\cdot\cdot},\gamma'}(Q) = \tilde e_{\nu}^{\gamma^{\cdot\cdot},\gamma'}(Q)\circ\tilde e_\mu\circ\tilde h^\mu = \tilde e_{\mu}^{\gamma^{\cdot\cdot},\gamma'}\circ\tilde h^\mu \quad .$$

Consequently, μ is actually a Q-section of $(P_{\hat\gamma}, P^{\cdot\cdot}_{\gamma^{\cdot\cdot}}, P'_{\gamma'})$.

17.7 For a normal subgroup Q of P, the fact that Theorem 13.9 applies to the $\mathcal{D}$-interior algebra $\Big(\mathrm{Ind}_\sigma\big(A^{\cdot\cdot}_{\gamma^{\cdot\cdot}}\otimes_{\mathcal{O}}\mathrm{Res}_{\sigma'}((\mathcal{O}G')_{\gamma'})\big)\Big)(Q)$, whereas we have only a classification of the points of the k-algebra $\Big(\mathrm{C}_\circ\Big(\mathrm{Ind}_\sigma\big(A^{\cdot\cdot}_{\gamma^{\cdot\cdot}}\otimes_{\mathcal{O}}\mathrm{Res}_{\sigma'}((\mathcal{O}G')_{\gamma'})\big)\Big)\Big)(Q)$ (cf. Corollary 5.14) has been a serious handicap to analyze the eventual uniqueness up to conjugacy of the Q-sections of $(P_{\hat\gamma}, P^{\cdot\cdot}_{\gamma^{\cdot\cdot}}, P'_{\gamma'})$, and we do not know whether our next result is the best answer to this question. Actually, this result depends only on the two last terms of the triple (so, ignoring the "tracing" condition) and on the "basic" part of our hypothesis; in order to state it independently of our setting, let $R^{\cdot\cdot}$ be a p-subgroup of $G\times G'$, set $R=\pi(R^{\cdot\cdot})$ and $R'=\pi'(R^{\cdot\cdot})$, and respectively denote by $\rho: R^{\cdot\cdot}\to R$ and $\rho': R^{\cdot\cdot}\to R'$ the group homomorphisms determined by π and π'; now, let Q be a normal subgroup of R, $D^{\cdot\cdot}$ a $\mathrm{Ker}(\rho)$-basic $\mathcal{D}R^{\cdot\cdot}$-interior algebra and ε' a local point of R' on $\mathcal{O}G'$, set $Q^{\cdot\cdot}=\rho^{-1}(Q)$ and $Q'=\rho'(Q^{\cdot\cdot})$, and denote by f' the block of $C_{G'}(Q')$ such that (cf. 2.14.2)

17.7.1 $$(Q', f')\subset (R', b(\varepsilon')) \quad ,$$

where $b(\varepsilon')$ is the block of $C_{G'}(R')$ determined by ε', and by N' the set of $x'\in N_{G'}(Q',f')$ such that $(1,x')$ normalizes $Q^{\cdot\cdot}$. Since $D^{\cdot\cdot}\otimes_{\mathcal{O}}\mathrm{Res}_{\rho'}((\mathcal{O}G')_{\varepsilon'})$ is also a $\mathrm{Ker}(\rho)$-basic $\mathcal{D}R^{\cdot\cdot}$-interior algebra, Theorem 13.9 still applies to the $\mathcal{D}C_R(Q)$-interior $N_R(Q)$-algebra

17.7.2 $$\Big(\mathrm{Ind}_\rho\big(D^{\cdot\cdot}\underset{\mathcal{O}}{\otimes}\mathrm{Res}_{\rho'}((\mathcal{O}G')_{\varepsilon'})\big)\Big)(Q)$$

and, for any Q-section μ of ρ, it guarantees the existence of a canonical $\mathcal{D}\big(\rho(C_{R^{\cdot\cdot}}(Q^\mu))\big)$-interior algebra embedding

17.7.3 $$e_\mu^{D^{\cdot\cdot},\varepsilon'}:\mathrm{Ind}_{\rho_\mu}\big(D^{\cdot\cdot}(Q^\mu)\underset{\mathcal{O}}{\otimes}\mathrm{Res}_{\rho'_\mu}((\mathcal{O}G')_{\varepsilon'}(Q^{\mu'}))\big)\longrightarrow D(Q)$$

where $D=\mathrm{Ind}_\rho\big(D^{\cdot\cdot}\otimes_{\mathcal{O}}\mathrm{Res}_{\rho'}((\mathcal{O}G')_{\varepsilon'})\big)$ and $\rho_\mu: C_{R^{\cdot\cdot}}(Q^\mu)\to\rho\big(C_{R^{\cdot\cdot}}(Q^\mu)\big)$ and $\rho'_\mu: C_{R^{\cdot\cdot}}(Q^\mu)\to C_{R'}(Q^{\mu'})$ are respectively the group homomorphisms determined by ρ and ρ'. Moreover, the evident action of N' on $Q^{\cdot\cdot}$ induces an action of N' on the set of Q-sections μ of ρ and we denote by $\mu^{x'}$ the image of μ by $x'\in N'$.

Theorem 17.8 *With the notation above, if $\mathfrak{M}$ is a set of representatives for the orbits of* $\mathrm{Ker}(\rho)$ *in a N'-stable set of Q-sections of ρ then the idempotent* $\Sigma_{\mu\in\mathfrak{M}} e_\mu^{D^{\cdot\cdot},\varepsilon'}(1\otimes(1\otimes 1))$ *is central in* $\mathrm{Ind}_\rho\Big(D^{\cdot\cdot}\otimes_{\mathcal{O}}\mathrm{Res}_{\rho'}\big((\mathcal{O}G')_{\varepsilon'}\big)\Big)(Q)$*. Moreover, for any Q-section μ of ρ and any $x'\in N_{G'}(R'_{\varepsilon'})$ such that $(1,x')$ normalizes $R^{\cdot\cdot}$ and stabilizes the isomorphism class of $D^{\cdot\cdot}$, the idempotents $e_\mu^{D^{\cdot\cdot},\varepsilon'}(1\otimes(1\otimes 1))$ and $e_{\mu^{x'}}^{D^{\cdot\cdot},\varepsilon'}(1\otimes(1\otimes 1))$ are conjugate in* $\mathrm{C}_\circ\Bigg(\Big(\mathrm{Ind}_\rho\Big(D^{\cdot\cdot}\otimes_{\mathcal{O}}\mathrm{Res}_{\rho'}\big((\mathcal{O}G')_{\varepsilon'}\big)\Big)\Big)(Q)\Bigg)^R$.

Proof: It is quite clear that if $W^{\cdot\cdot}$ is a $\big(\mathrm{Ker}(\rho)\times\mathrm{Ker}(\rho)\big)\cdot\Delta(R^{\cdot\cdot})$-stable $\mathcal{O}$-basis of $D^{\cdot\cdot}$ then, since $R\times G' = R^{\cdot\cdot}\cdot(1\times G')$, the set $(1\times G')\otimes W^{\cdot\cdot}\otimes(1\times G')$ is a $\big((1\times G')\times(1\times G')\big)\cdot\Delta(R\times G')$-stable $\mathcal{O}$-basis of $\mathrm{Ind}_{R^{\cdot\cdot}}^{R\times G'}(D^{\cdot\cdot})$, so that this $\mathcal{D}(R\times G')$-interior algebra is $(1\times G')$-basic; in particular, the $\mathcal{D}R^{\cdot\cdot}$-interior algebra

17.8.1 $$C^{\cdot\cdot} = \mathrm{Res}_{R^{\cdot\cdot}}^{R\times G'}\big(\mathrm{Ind}_{R^{\cdot\cdot}}^{R\times G'}(D^{\cdot\cdot})\big)\underset{\mathcal{O}}{\otimes}\mathrm{Res}_{\rho'}\big(\mathrm{Res}_{R'}^{G'}(\mathcal{O}G')\big)$$

is $\mathrm{Ker}(\rho)$-basic too and therefore, according to Theorem 13.9 applied to $C^{\cdot\cdot}$ and Q, for any Q-section μ of ρ we have a canonical $\mathcal{D}\Big(\rho\big(C_{R^{\cdot\cdot}}(Q^\mu)\big)\Big)$-interior $\rho\big(N_{R^{\cdot\cdot}}(Q^\mu)\big)$-algebra embedding

17.8.2 $$e_{\rho,\mu}(C^{\cdot\cdot}) : \mathrm{Ind}_{\rho_\mu}\big(C^{\cdot\cdot}(Q^\mu)\big) \longrightarrow \big(\mathrm{Ind}_\rho(C^{\cdot\cdot})\big)(Q)$$

and we denote by i_μ the image of the unity element. Moreover, denoting by L' the set of $x'\in G'$ such that $(1,x')$ normalizes $Q^{\cdot\cdot}$, L' is a subgroup of G' and it follows easily from Lemma 5.6 that

17.8.3 $$\big(\mathrm{Ind}_\rho(C^{\cdot\cdot})\big)(Q) = \sum_{x'\in L'}\mathrm{Br}_Q\Big(1\otimes\big((1,x')\cdot\mathrm{Ind}_{R^{\cdot\cdot}}^{R\times G'}(D^{\cdot\cdot})^{Q^{\cdot\cdot}}\otimes x'\big)\Big) \quad .$$

Now, we claim that, for any $x'\in L'$, any $a^{\cdot\cdot}\in\mathrm{Ind}_{R^{\cdot\cdot}}^{R\times G'}(D^{\cdot\cdot})^{Q^{\cdot\cdot}}$, any $a'\in(\mathcal{O}G')^{Q'}$ and any Q-section μ of ρ, we have

17.8.4
$$i_\mu\mathrm{Br}_Q\Big(1\otimes\big((1,x')\cdot a^{\cdot\cdot}\otimes x'a'\big)\Big) = \mathrm{Br}_Q\Big(1\otimes\big((1,x')\cdot a^{\cdot\cdot}\otimes x'a'\big)\Big)i_{\mu^{x'}} \quad .$$

Indeed, by linearity we may assume that $a^{\cdot\cdot}$ and a' are invertible elements (cf. 2.1.1); in that case, denoting by $\varphi_{x'} : Q^{\cdot\cdot}\cong Q^{\cdot\cdot}$ the group automorphism mapping $(u,u')\in Q^{\cdot\cdot}$ on $(u,u'^{x'})$, we have a $\mathcal{D}Q^{\cdot\cdot}$-interior algebra isomorphism

17.8.5 $$f_{\varphi_{x'}} : \mathrm{Res}_{Q^{\cdot\cdot}}^{R^{\cdot\cdot}}(C^{\cdot\cdot}) = \mathrm{Res}_{\varphi_{x'}}\big(\mathrm{Res}_{Q^{\cdot\cdot}}^{R^{\cdot\cdot}}(C^{\cdot\cdot})\big)$$

mapping $a\in C^{\cdot\cdot}$ on $a^{(1,x')\cdot a^{\cdot\cdot}\otimes x'a'}$ and then, since we have a canonical

$\mathcal{D}\Big(\rho\big(C_{R^{\cdot\cdot}}(Q^{\mu})\big)\Big)$-interior algebra isomorphism (cf. Proposition 12.12)

17.8.6 $$\Big(\mathrm{Res}^{R}_{Q}\big(\mathrm{Ind}_{\rho}(C^{\cdot\cdot})\big)\Big)(Q) \cong \Big(\mathrm{Ind}_{\tau}\big(\mathrm{Res}^{R^{\cdot\cdot}}_{Q^{\cdot\cdot}}(C^{\cdot\cdot})\big)\Big)(Q)$$

where $\tau : Q^{\cdot\cdot} \to Q$ is the group homomorphism determined by ρ, up to suitable identifications equality 17.8.4 becomes

17.8.7 $$\Big(\big(\mathrm{Ind}_{\tau}(f_{\varphi_{x'}})\big)(Q)\Big)(i_{\mu}) = i_{\varphi_{x'}\circ\mu}$$

which follows from Proposition 13.17 applied to $f_{\varphi_{x'}}$.

On the other hand, choosing $i' \in \varepsilon'$ such that we have $i'f' = i' = f'i'$ in $(\mathcal{O}G')^{Q'}$ (cf. 2.15.5) and denoting by $f_{\varepsilon'}$ the representative of $\tilde{f}_{\varepsilon'}$ which maps the unity element on i', we consider the $\mathcal{D}R^{\cdot\cdot}$-interior algebra embedding

17.8.8 $$g : D^{\cdot\cdot} \underset{\mathcal{O}}{\otimes} \mathrm{Res}_{\rho'}\big((\mathcal{O}G')_{\varepsilon'}\big) \longrightarrow C^{\cdot\cdot}$$

given by $g = d^{R\times G'}_{R^{\cdot\cdot}}(D^{\cdot\cdot}) \otimes f_{\varepsilon'}$; from Proposition 13.17, for any Q-section μ of ρ we get then the following commutative diagram

17.8.9 $$\begin{array}{ccc} \Big(\mathrm{Ind}_{\rho}\Big(D^{\cdot\cdot} \underset{\mathcal{O}}{\otimes} \mathrm{Res}_{\rho'}\big((\mathcal{O}G')_{\varepsilon'}\big)\Big)\Big)(Q) & \xrightarrow{(\mathrm{Ind}_{\rho}(g))(Q)} & \big(\mathrm{Ind}_{\rho}(C^{\cdot\cdot})\big)(Q) \\ \uparrow{\scriptstyle e^{D^{\cdot\cdot},\varepsilon'}_{\mu}} & & \uparrow{\scriptstyle e_{\rho,\mu}(C^{\cdot\cdot})} \\ \mathrm{Ind}_{\rho_{\mu}}\Big(D^{\cdot\cdot}(Q^{\mu}) \underset{\mathcal{O}}{\otimes} \mathrm{Res}_{\rho'_{\mu}}\big((\mathcal{O}G')_{\varepsilon'}(Q^{\mu'})\big)\Big) & \xrightarrow{(\mathrm{Ind}_{\rho_{\mu}}(g(Q^{\mu}))} & \mathrm{Ind}_{\rho_{\mu}}\big(C^{\cdot\cdot}(Q^{\mu})\big) \end{array}$$

and therefore, setting $j_{\mu} = \Big(\big(\mathrm{Ind}_{\rho}(g)\big)(Q)\Big)\Big(e^{D^{\cdot\cdot},\varepsilon'}_{\mu}\big(1\otimes(1\otimes 1)\big)\Big)$, we obtain

17.8.10 $$i_{\mu}j_{\mu} = j_{\mu} = j_{\mu}i_{\mu} \qquad ,$$

so that, since from the last statement in Theorem 13.9 we have

17.8.11 $$\sum_{\mu\in\mathfrak{M}_{\rho,Q}} j_{\mu} = \mathrm{Br}_{Q}\Big(1\otimes\Big(\big((1,1)\otimes 1\otimes(1,1)\big)\otimes 1\Big)\Big) = j$$

where $\mathfrak{M}_{\rho,Q}$ is a set of representatives for the orbits of $\mathrm{Ker}(\rho)$ in the set of Q-sections of ρ, we have also

17.8.12 $$ji_{\mu} = j_{\mu} = i_{\mu}j \qquad .$$

In conclusion, since for any $x' \in L'$ we have (cf. 2.3)

17.8.13 $$\begin{aligned} &\big((1,1)\otimes 1\otimes(1,x')\big)\cdot\mathrm{Ind}^{R\times G'}_{R^{\cdot\cdot}}(D^{\cdot\cdot})^{Q^{\cdot\cdot}}\cdot\big((1,1)\otimes 1\otimes(1,1)\big) \\ &\qquad = (1,1)\otimes N^{\varphi_{x'}}_{D^{\cdot\cdot}}(Q^{\cdot\cdot})\otimes(1,1) \qquad , \end{aligned}$$

from equality 17.8.3 and via the embedding $\big(\mathrm{Ind}_\rho(g)\big)(Q)$ we get

17.8.14
$$\Big(\mathrm{Ind}_\rho\Big(D^{\cdot\cdot} \underset{\mathcal{O}}{\otimes} \mathrm{Res}_{\rho'}\big((\mathcal{O}G')_{\varepsilon'}\big)\Big)\Big)(Q) = \sum_{x'\in N'} \mathrm{Br}_Q\Big(1\otimes \big(N_{D^{\cdot\cdot}}^{\varphi_{x'}}(Q^{\cdot\cdot})\otimes i'x'i'\big)\Big)$$

since, for any $x' \in L' - N'$, we have

17.8.15
$$i'x'i' = i'f'x'f'i' = i'x'(f')^{x'}f'x' = 0 \quad ;$$

on the other hand, from equalities 17.8.4 and 17.8.12, for any $x' \in N'$, any $a^{\cdot\cdot} \in N_{D^{\cdot\cdot}}^{\varphi_{x'}}(Q^{\cdot\cdot})$ and any Q-section μ of ρ, we obtain

17.8.16
$$\begin{aligned} &e_\mu^{D^{\cdot\cdot},\varepsilon'}\big(1\otimes(1\otimes 1)\big)\mathrm{Br}_Q\big(1\otimes(a^{\cdot\cdot}\otimes i'x'i')\big)\\ &= \mathrm{Br}_Q\big(1\otimes(a^{\cdot\cdot}\otimes i'x'i')\big)e_{\mu^{x'}}^{D^{\cdot\cdot},\varepsilon'}\big(1\otimes(1\otimes 1)\big) \quad ; \end{aligned}$$

now, equalities 17.8.14 and 17.8.16 prove the first statement of the theorem.

Finally, for any $x' \in G'$ such that $(1, x')$ normalizes $R^{\cdot\cdot}$ and stabilizes the isomorphism class of $D^{\cdot\cdot}$, there is $c^{\cdot\cdot} \in C_{\mathcal{O}}(D^{\cdot\cdot})^*$ such that $(u, u')\cdot c^{\cdot\cdot} = c^{\cdot\cdot}\cdot(u, u'^{x'})$ for any $(u, u') \in R^{\cdot\cdot}$; on the other hand, for any $x' \in N_{G'}(R'_{\varepsilon'})$, there is $c' \in \big((\mathcal{O}G')^{R'}\big)^*$ such that $i'^{x'c'} = i'$; hence, for any $x' \in N_{G'}(R'_{\varepsilon'})$ such that $(1, x')$ normalizes $R^{\cdot\cdot}$ (so that it normalizes $Q^{\cdot\cdot}$ too) and stabilizes the isomorphism class of $D^{\cdot\cdot}$, choosing in 17.8.4

17.8.17
$$a^{\cdot\cdot} = (1, x')^{-1}\otimes c^{\cdot\cdot}\otimes(1,1) \quad \text{and} \quad a' = c'i'$$

(notice that $(u, u')\cdot a^{\cdot\cdot} = a^{\cdot\cdot}\cdot(u, u')$ for any $(u, u') \in R^{\cdot\cdot}$), it is clear that $\mathrm{Br}_Q\Big(1\otimes\big((1, x')\cdot a^{\cdot\cdot}\otimes x'a'\big)\Big)$ is an invertible element in the image of embedding $\big(\mathrm{Ind}_\rho(g)\big)(Q)$ and the last statement follows from equalities 17.8.4 and 17.8.12, via this embedding.

Corollary 17.9 *Let Q be a normal subgroup of P, set $Q^{\cdot\cdot} = \sigma^{-1}(Q)$ and $Q' = \sigma'(Q^{\cdot\cdot})$, and denote by f' the block of $C_{G'}(Q')$ such that $(Q', f') \subset \big(P', b(\gamma')\big)$. If λ and μ are Q-sections of $(P_{\hat\gamma}, P^{\cdot\cdot}_{\gamma^{\cdot\cdot}}, P'_{\gamma'})$ then there is $x' \in N_{G'}(Q', f')$ such that $(1, x')$ normalizes $Q^{\cdot\cdot}$ and we have $\mu = \lambda^{x'}$.*

Proof: With the notation above, fixing $R^{\cdot\cdot} = P^{\cdot\cdot}$, $D^{\cdot\cdot} = A^{\cdot\cdot}_{\gamma^{\cdot\cdot}}$ and $R'_{\varepsilon'} = P'_{\gamma'}$, setting $i_\nu = e_\nu^{\gamma^{\cdot\cdot},\gamma'}\big(1\otimes(1\otimes 1)\big)$ (cf. 17.4.1) for any Q-section ν of σ and choosing $X' \subset N'$ such that $1\times X'$ is a set of representatives for $(1\times N')/\mathrm{Ker}(\sigma)$, it follows from Theorem 17.8 that $l_\lambda = \Sigma_{x'\in X'} i_{\lambda^{x'}}$, and $l_\mu = \Sigma_{x'\in X'} i_{\mu^{x'}}$ are central idempotents in $\Big(\mathrm{Ind}_\sigma\Big(A^{\cdot\cdot}_{\gamma^{\cdot\cdot}}\otimes_{\mathcal{O}}\mathrm{Res}_{\sigma'}\big((\mathcal{O}G')_{\gamma'}\big)\Big)\Big)(Q)$; but from 17.4.6 we get

17.9.1
$$e_\lambda^{\gamma^{\cdot\cdot},\gamma'}\circ h_{\hat\gamma}^{\lambda} = h_{\hat\gamma}^{\gamma^{\cdot\cdot},\gamma'}(Q) = e_\mu^{\gamma^{\cdot\cdot},\gamma'}\circ h_{\hat\gamma}^{\mu} \quad ;$$

hence, as it is easily checked, l_λ and l_μ are not orthogonal, so that $l_\lambda = l_\mu$.

Proposition 17.10 *Let $(Q_{\hat\delta}, Q^{\cdot\cdot}_{\delta^{\cdot\cdot}}, Q'_{\delta'})$ be a local tracing triple on $\hat A$, $A^{\cdot\cdot}$ and $\mathcal{O}G'$, contained in $(P_{\hat\gamma}, P^{\cdot\cdot}_{\gamma^{\cdot\cdot}}, P'_{\gamma'})$ and respectively denote by $\tau : Q^{\cdot\cdot} \to Q$ and $\tau' : Q^{\cdot\cdot} \longrightarrow Q'$ the group homomorphisms determined by π and π'. There is a representative h of*

17.10.1 $$\mathrm{Res}^P_Q\Big(\mathrm{Ind}_\sigma\big(\tilde e(A^{\cdot\cdot}_{\gamma^{\cdot\cdot}}) \underset{\mathcal{O}}{\otimes} \mathrm{Res}_{Q'}(\widetilde{\mathrm{id}}_{(\mathcal{O}G')_{\gamma'}})\big)\Big) \circ \tilde g^{\gamma^{\cdot\cdot},\gamma'}_{\delta^{\cdot\cdot},\delta'}$$

such that, for any subgroup R of Q and any R-section ν of τ, we have

17.10.2 $$\mathrm{Im}\big(h(R) \circ e_\nu^{\delta^{\cdot\cdot},\delta'}\big) \subset \mathrm{Im}\big(e_\nu^{\mathsf{E}(A^{\cdot\cdot}_{\gamma^{\cdot\cdot}}),\gamma'}\big)$$

and, in particular, there is a unique $\mathcal{D}\big(\sigma(C_{Q^{\cdot\cdot}}(R^\nu))\big)$-interior $\sigma(N_{Q^{\cdot\cdot}}(R^\nu))$-algebra embedding

17.10.3 $$\begin{array}{c}\mathrm{Res}^{C_P(R)}_{C_Q(R)}\Big(\mathrm{Ind}_{\sigma_\nu}\Big(\mathsf{E}(A^{\cdot\cdot}_{\gamma^{\cdot\cdot}})(R^\nu) \underset{k}{\otimes} \mathrm{Res}_{\sigma'_\nu}\big((\mathcal{O}G')_{\gamma'}(R^{\nu'})\big)\Big)\Big)\\ \Big\uparrow {\scriptstyle g_\nu}\\ \mathrm{Ind}_{\tau_\nu}\Big(A^{\cdot\cdot}_{\delta^{\cdot\cdot}}(R^\nu) \underset{k}{\otimes} \mathrm{Res}_{\tau'_\nu}\big((\mathcal{O}G')_{\delta'}(R^{\nu'})\big)\Big)\end{array}$$

fulfilling

17.10.4 $$\mathrm{Res}^{C_P(R)}_{C_Q(R)}\big(e_\nu^{\mathsf{E}(A^{\cdot\cdot}_{\gamma^{\cdot\cdot}}),\gamma'}\big) \circ g_\nu = h(R) \circ e_\nu^{\delta^{\cdot\cdot},\delta'} \quad .$$

Remark 17.11 The representative h is certainly the composition of representatives of both terms in 17.10.1 but we do not know whether the representative of the first term could be chosen coming from one of $\tilde e(A^{\cdot\cdot}_{\gamma^{\cdot\cdot}}) \otimes_{\mathcal{O}} \mathrm{Res}_{\sigma'}(\widetilde{\mathrm{id}}_{(\mathcal{O}G')_{\gamma'}})$; this lack of knowledge forces us to consider $\mathsf{E}(A^{\cdot\cdot}_{\gamma^{\cdot\cdot}})$ instead of $A^{\cdot\cdot}_{\gamma^{\cdot\cdot}}$. Notice that, by Proposition 14.7, the $\mathcal{D}P^{\cdot\cdot}$-interior algebra $\mathsf{E}(A^{\cdot\cdot}_{\gamma^{\cdot\cdot}})$ is $\mathrm{Ker}(\sigma)$-basic too and, by 17.1, the local tracing triple $(Q_{\hat\delta}, Q^{\cdot\cdot}_{\delta^{\cdot\cdot}}, Q'_{\delta'})$ is also basic.

Proof: Set $D^{\cdot\cdot} = \mathsf{E}(A^{\cdot\cdot}_{\gamma^{\cdot\cdot}})$ and $T^{\cdot\cdot} = \sigma^{-1}(Q)$, and respectively denote by $\eta : T^{\cdot\cdot} \to Q$ and $\eta' : T^{\cdot\cdot} \to P'$ the group homomorphisms determined by σ and σ'; first of all we claim that

17.10.5
$$\mathrm{Res}^P_Q\Big(\mathrm{Ind}_\sigma\big(\tilde e(A^{\cdot\cdot}_{\gamma^{\cdot\cdot}}) \underset{\mathcal{O}}{\otimes} \mathrm{Res}_{\sigma'}(\widetilde{\mathrm{id}}_{(\mathcal{O}G')_{\gamma'}})\big)\Big) \circ \tilde g^{\gamma^{\cdot\cdot},\gamma'}_{\delta^{\cdot\cdot},\delta'}$$
$$= \tilde d_{\gamma'} \circ \mathrm{Ind}_\eta\Big(\tilde h_{\delta^{\cdot\cdot}}(\mathrm{Res}^{P^{\cdot\cdot}}_{T^{\cdot\cdot}}(D^{\cdot\cdot})) \underset{\mathcal{O}}{\otimes} \mathrm{Res}_{\eta'}(\widetilde{\mathrm{id}}_{(\mathcal{O}G')_{\gamma'}})\Big) \circ \mathrm{Ind}_\tau\big(\widetilde{\mathrm{id}}_{A^{\cdot\cdot}_{\delta^{\cdot\cdot}}} \underset{\mathcal{O}}{\otimes} \mathrm{Res}_{\tau'}(\tilde f^{\gamma'}_{\delta'})\big)$$

where $d_{\gamma'} = d^P_{Q,\sigma}\big(D^{\cdot\cdot} \otimes_{\mathcal{O}} \mathrm{Res}_{\sigma'}((\mathcal{O}G')_{\gamma'})\big)$; indeed, from equality 16.13.1 we get

17.10.6
$$\mathrm{Res}^P_Q\Big(\mathrm{Ind}_\sigma\Big(\big(\widetilde{\mathrm{id}}_{D^{\cdot\cdot}} \underset{\mathcal{O}}{\otimes} \mathrm{Res}_{\sigma'}(\tilde f_{\gamma'})\big) \circ \big(\tilde e(A^{\cdot\cdot}_{\gamma^{\cdot\cdot}}) \underset{\mathcal{O}}{\otimes} \mathrm{Res}_{\sigma'}(\widetilde{\mathrm{id}}_{(\mathcal{O}G')_{\gamma'}})\big)\Big)\Big) \circ \tilde g^{\gamma^{\cdot\cdot},\gamma'}_{\delta^{\cdot\cdot},\delta'}$$
$$= \mathrm{Res}^P_Q\big(\tilde H_{P,G'}(D^{\cdot\cdot})\big) \circ \tilde g_{\delta^{\cdot\cdot},\delta'}(D^{\cdot\cdot})$$

and from equality 16.6.2 that

17.10.7

$$\begin{aligned}&\mathrm{Res}^P_Q\big(\tilde{H}_{P,G'}(D^{\cdot\cdot})\big)\circ\tilde{g}_{\delta^{\cdot\cdot},\delta'}(D^{\cdot\cdot})\\&\quad=\tilde{d}\circ\mathrm{Ind}_\eta\Big(\tilde{h}_{\delta^{\cdot\cdot}}\big(\mathrm{Res}^{P^{\cdot\cdot}}_{T^{\cdot\cdot}}(D^{\cdot\cdot})\big)\underset{\mathcal{O}}{\otimes}\mathrm{Res}_{\theta'\circ\eta'}(\tilde{\mathrm{id}}_{\mathcal{O}G'})\Big)\circ\mathrm{Ind}_\tau\big(\tilde{\mathrm{id}}_{A^{\cdot\cdot}_{\delta^{\cdot\cdot}}}\underset{\mathcal{O}}{\otimes}\mathrm{Res}_{\tau'}(\tilde{f}_{\delta'})\big)\end{aligned}$$

where $\theta' : P' \to G'$ is the inclusion map and $d = d^P_{Q,\sigma}(D^{\cdot\cdot} \otimes_{\mathcal{O}} \mathrm{Res}_{\theta'\circ\sigma'}(\mathcal{O}G'))$; on the other hand, by the naturality of $d^P_{Q,\sigma}$ (cf. 3.22.1 and 12.13), we have

17.10.8
$$\begin{aligned}&\mathrm{Res}^P_Q\Big(\mathrm{Ind}_\sigma\big(\tilde{\mathrm{id}}_{D^{\cdot\cdot}}\underset{\mathcal{O}}{\otimes}\mathrm{Res}_{\sigma'}(\tilde{f}_{\gamma'})\big)\Big)\circ\tilde{d}^P_{Q,\sigma}\Big(D^{\cdot\cdot}\underset{\mathcal{O}}{\otimes}\mathrm{Res}_{\sigma'}\big((\mathcal{O}G')_{\gamma'}\big)\Big)\\&\quad=\tilde{d}^P_{Q,\sigma}\big(D^{\cdot\cdot}\underset{\mathcal{O}}{\otimes}\mathrm{Res}_{\theta'\circ\sigma'}(\mathcal{O}G')\big)\circ\mathrm{Ind}_\eta\Big(\mathrm{Res}^{P^{\cdot\cdot}}_{T^{\cdot\cdot}}\big(\tilde{\mathrm{id}}_{D^{\cdot\cdot}}\underset{\mathcal{O}}{\otimes}\mathrm{Res}_{\sigma'}(f_{\gamma'})\big)\Big)\end{aligned}$$

and it is clear that (cf. Proposition 12.9)

17.10.9

$$\begin{aligned}&\big(\tilde{\mathrm{id}}_{\mathrm{Res}^{P^{\cdot\cdot}}_{T^{\cdot\cdot}}(D^{\cdot\cdot})}\underset{\mathcal{O}}{\otimes}\mathrm{Res}_{\eta'}(\tilde{f}_{\gamma'})\big)\circ\Big(\tilde{h}_{\delta^{\cdot\cdot}}\big(\mathrm{Res}^{P^{\cdot\cdot}}_{T^{\cdot\cdot}}(D^{\cdot\cdot})\big)\underset{\mathcal{O}}{\otimes}\mathrm{Res}_{\eta'}\big(\tilde{\mathrm{id}}_{(\mathcal{O}G')_{\gamma'}}\big)\Big)\\&\quad=\Big(\tilde{h}_{\delta^{\cdot\cdot}}\big(\mathrm{Res}^{P^{\cdot\cdot}}_{T^{\cdot\cdot}}(D^{\cdot\cdot})\big)\underset{\mathcal{O}}{\otimes}\mathrm{Res}_{\theta'\circ\eta'}(\tilde{\mathrm{id}}_{\mathcal{O}G'})\Big)\circ\mathrm{Ind}^{T^{\cdot\cdot}}_{Q^{\cdot\cdot}}\Big(\tilde{\mathrm{id}}_{A^{\cdot\cdot}_{\delta^{\cdot\cdot}}}\underset{\mathcal{O}}{\otimes}\mathrm{Res}_{\tau'}\big(\mathrm{Res}^{P'}_{Q'}(\tilde{f}_{\gamma'})\big)\Big);\end{aligned}$$

now, equality 17.10.5 follows from Proposition 11.10 and equalities 17.10.6, 17.10.7, 17.10.8 and 17.10.9.

Let R be a subgroup of Q and ν an R-section of τ; it is clear that ν is also an R-section of σ. Choosing a representative $f^{\gamma'}_{\delta'}$ of $\tilde{f}^{\gamma'}_{\delta'}$, it follows from Proposition 13.17 applied to the tensor product of $\mathrm{id}_{A^{\cdot\cdot}_{\delta^{\cdot\cdot}}}$ and $\mathrm{Res}_{\tau'}(f^{\gamma'}_{\delta'})$ that

17.10.10

$$\begin{aligned}&\Big(\mathrm{Ind}_\tau\big(\mathrm{id}_{A^{\cdot\cdot}_{\delta^{\cdot\cdot}}}\underset{\mathcal{O}}{\otimes}\mathrm{Res}_{\tau'}(f^{\gamma'}_{\delta'})\big)\Big)(R)\circ e^{\delta^{\cdot\cdot},\delta'}_\nu\\&\quad=e_{\tau,\nu}\Big(A^{\cdot\cdot}_{\delta^{\cdot\cdot}}\underset{\mathcal{O}}{\otimes}\mathrm{Res}_{\zeta'}\big((\mathcal{O}G')_{\gamma'}\big)\Big)\circ\mathrm{Ind}_{\tau_\nu}\Big(\tilde{\mathrm{id}}_{A^{\cdot\cdot}_{\delta^{\cdot\cdot}}(R^\nu)}\underset{\mathcal{O}}{\otimes}\mathrm{Res}_{\tau'_\nu}\big(\tilde{f}^{\gamma'}_{\delta'}(R^{\nu'})\big)\Big)\end{aligned}$$

where $\zeta' : Q^{\cdot\cdot} \to P'$ is the restriction of σ' and $e_{\tau,\nu}\Big(A^{\cdot\cdot}_{\delta^{\cdot\cdot}} \otimes_{\mathcal{O}} \mathrm{Res}_{\tau'_\nu}\big(f^{\gamma'}_{\delta'}(R^{\nu'})\big)\Big)$ is the corresponding embedding provided by Theorem 13.9; secondly, up to suitable identifications (cf. Corollary 12.7 and Proposition 12.9), it follows from Corollary 13.20 applied to the $\mathcal{D}Q^{\cdot\cdot}$-interior algebra $C^{\cdot\cdot} = A^{\cdot\cdot}_{\delta^{\cdot\cdot}} \otimes_{\mathcal{O}} \mathrm{Res}_{\zeta'}\big((\mathcal{O}G')_{\gamma'}\big)$ and to the inclusion $Q^{\cdot\cdot} \subset T^{\cdot\cdot}$ that

17.10.11
$$e_{\tau,\nu}(C^{\cdot\cdot}) = e_{\eta,\nu}\big(\mathrm{Ind}^{T^{\cdot\cdot}}_{Q^{\cdot\cdot}}(C^{\cdot\cdot})\big)\circ\mathrm{Ind}_{\eta_\nu}(d_{R^\nu})$$

where $e_{\eta,\nu}\big(\mathrm{Ind}^{T^{\cdot\cdot}}_{Q^{\cdot\cdot}}(C^{\cdot\cdot})\big)$ is provided again by Theorem 13.9 and d_{R^ν} is the canonical $\mathcal{D}C_{T^{\cdot\cdot}}(R^\nu)$-interior $N_{T^{\cdot\cdot}}(R^\nu)$-algebra embedding (cf. 13.18.4)

17.10.12
$$\mathrm{Ind}^{C_{T^{\cdot\cdot}}(R^\nu)}_{C_{Q^{\cdot\cdot}}(R^\nu)}\big(C^{\cdot\cdot}(R^\nu)\big)\longrightarrow\big(\mathrm{Ind}^{T^{\cdot\cdot}}_{Q^{\cdot\cdot}}(C^{\cdot\cdot})\big)(R^\nu)\qquad;$$

thirdly, choosing a representative $h_{\delta^{\cdot\cdot}}$ of $\tilde{h}_{\delta^{\cdot\cdot}}\big(\mathrm{Res}^{P^{\cdot\cdot}}_{T^{\cdot\cdot}}(D^{\cdot\cdot})\big)$, it follows from Proposition 13.17 applied to the tensor product of $h_{\delta^{\cdot\cdot}}$ and $\mathrm{Res}_{\eta'}(\mathrm{id}_{(\mathcal{O}G')_{\gamma'}})$ that, setting $E^{\cdot\cdot} = D^{\cdot\cdot} \otimes_{\mathcal{O}} \mathrm{Res}_{\sigma'}\big((\mathcal{O}G')_{\gamma'}\big)$, we have

17.10.13
$$\begin{aligned}&\Big(\mathrm{Ind}_\eta\big(h_{\delta^{\cdot\cdot}} \underset{\mathcal{O}}{\otimes} \mathrm{Res}_{\eta'}(\mathrm{id}_{(\mathcal{O}G')_{\gamma'}})\big)\Big)(R) \circ e_{\eta,\nu}\big(\mathrm{Ind}^{T^{\cdot\cdot}}_{Q^{\cdot\cdot}}(C^{\cdot\cdot})\big)\\ &= e_{\eta,\nu}\big(\mathrm{Res}^{P^{\cdot\cdot}}_{T^{\cdot\cdot}}(E^{\cdot\cdot})\big) \circ \mathrm{Ind}_{\eta_\nu}\big(h_{\delta^{\cdot\cdot}}(R^\nu) \underset{k}{\otimes} \mathrm{Res}_{\eta'_\nu}(\mathrm{id}_{(\mathcal{O}G')_{\gamma'}(R^{\nu'})})\big) \quad ;\end{aligned}$$

fourthly, it follows from Corollary 13.13 applied to the $\mathcal{D}P^{\cdot\cdot}$-interior algebra $E^{\cdot\cdot} = D^{\cdot\cdot} \otimes_{\mathcal{O}} \mathrm{Res}_{\sigma'}\big((\mathcal{O}G')_{\gamma'}\big)$ that

17.10.14
$$\begin{aligned}&\big(d^P_{Q,\sigma}(E^{\cdot\cdot})\big)(R) \circ e_{\eta,\nu}\big(\mathrm{Res}^{P^{\cdot\cdot}}_{T^{\cdot\cdot}}(E^{\cdot\cdot})\big)\\ &= e^{\gamma^{\cdot\cdot},\gamma'}_\nu \circ d^{\sigma(C_{P^{\cdot\cdot}}(R^\nu))}_{\sigma(C_{T^{\cdot\cdot}}(R^\nu)),\sigma_\nu}\Big(D^{\cdot\cdot}(R^\nu) \underset{k}{\otimes} \mathrm{Res}_{\sigma'_\nu}\big((\mathcal{O}G')_{\gamma'}(R^{\nu'})\big)\Big) \quad .\end{aligned}$$

Finally, it suffices to choose

17.10.15
$$h = d^P_{Q,\sigma}(E^{\cdot\cdot}) \circ \mathrm{Ind}_\eta\big(h_{\delta^{\cdot\cdot}} \underset{\mathcal{O}}{\otimes} \mathrm{Res}_{\eta'}(\mathrm{id}_{(\mathcal{O}G')_{\gamma'}})\big) \circ \mathrm{Ind}_\tau\big(\widetilde{\mathrm{id}}_{A^{\cdot\cdot}_{\delta^{\cdot\cdot}}} \underset{\mathcal{O}}{\otimes} \mathrm{Res}_{\tau'}(\tilde{f}^{\gamma'}_{\delta'})\big)$$

to get inclusion 17.10.2 and, in particular, g_ν equal to the composition of the second terms in the right members of equalities 17.10.10, 17.10.11, 17.10.13 and 17.10.14; now, by equality 17.10.5, h is indeed a representative of exoembedding 17.10.1.

Theorem 17.12 *Let λ be a section of $(P_{\hat\gamma}, P^{\cdot\cdot}_{\gamma^{\cdot\cdot}}, P'_{\gamma'})$, $(Q_{\hat\delta}, Q^{\cdot\cdot}_{\delta^{\cdot\cdot}}, Q'_{\delta'})$ a local tracing triple on $\hat A$, $A^{\cdot\cdot}$ and $\mathcal{O}G'$, and x an element of G such that $(Q_{\hat\delta})^x \subset P_{\hat\gamma}$. Then $(Q_{\hat\delta}, Q^{\cdot\cdot}_{\delta^{\cdot\cdot}}, Q'_{\delta'})$ is basic and, for any section μ of $(Q_{\hat\delta}, Q^{\cdot\cdot}_{\delta^{\cdot\cdot}}, Q'_{\delta'})$, there is $x' \in G'$ such that*

17.12.1
$$(Q^{\cdot\cdot})^{(x,x')} \subset P^{\cdot\cdot} \quad , \quad \big(Q', b(\delta')\big)^{x'} \subset \big(P', b(\gamma')\big) \quad \textit{and}$$
$$\lambda(u^x) = \mu(u)^{(x,x')} \textit{ for any } u \in Q \quad .$$

Proof: By Corollary 16.16, there is $y' \in G'$ such that

17.12.2
$$(Q_{\hat\delta}, Q^{\cdot\cdot}_{\delta^{\cdot\cdot}}, Q'_{\delta'})^{(x,y')} \subset (P_{\hat\gamma}, P^{\cdot\cdot}_{\gamma^{\cdot\cdot}}, P'_{\gamma'}) \quad ,$$

so that, up to replace $(Q_{\hat\delta}, Q^{\cdot\cdot}_{\delta^{\cdot\cdot}}, Q'_{\delta'})$ by $(Q_{\hat\delta}, Q^{\cdot\cdot}_{\delta^{\cdot\cdot}}, Q'_{\delta'})^{(x,y')}$ and any section μ of $(Q_{\hat\delta}, Q^{\cdot\cdot}_{\delta^{\cdot\cdot}}, Q'_{\delta'})$ by $\mu^{(x,y')}$, we may assume that $x = 1 = y'$, and it is clear that $(Q_{\hat\delta}, Q^{\cdot\cdot}_{\delta^{\cdot\cdot}}, Q'_{\delta'})$ is basic too (cf. 17.1). If $Q = P$ then, by Theorem 16.9, inclusion 17.12.2 is an equality and the existence of $x' \in G'$ fulfilling condition 17.12.1 follows from Corollary 17.9.

Arguing by induction on $|P : Q|$, we may already assume that $Q \neq P$ and assume first that there is a local pointed group $R_{\hat{\varepsilon}}$ on $\hat{A}$ such that

17.12.3 $$Q_{\hat{\delta}} \subsetneq R_{\hat{\varepsilon}} \subsetneq P_{\hat{\gamma}} \quad ;$$

then, it follows from Theorem 16.15 that there are local pointed groups $R^{\cdot\cdot}_{\varepsilon^{\cdot\cdot}}$ on $A^{\cdot\cdot}$ and $R'_{\varepsilon'}$, on $\mathcal{O}G'$ such that $(R_{\hat{\varepsilon}}, R^{\cdot\cdot}_{\varepsilon^{\cdot\cdot}}, R'_{\varepsilon'})$ is a local tracing triple on $\hat{A}$, $A^{\cdot\cdot}$ and $\mathcal{O}G'$, contained in $(P_{\hat{\gamma}}, P^{\cdot\cdot}_{\gamma^{\cdot\cdot}}, P'_{\gamma'})$. By the inductive hypothesis, if ν is a section of $(R_{\hat{\varepsilon}}, R^{\cdot\cdot}_{\varepsilon^{\cdot\cdot}}, R'_{\varepsilon'})$ then there is $y' \in G'$ such that

17.12.4 $$(R^{\cdot\cdot})^{(1,y')} \subset P^{\cdot\cdot} \quad , \quad \big(R', b(\varepsilon')\big)^{y'} \subset \big(P', b(\gamma')\big) \quad \text{and}$$
$$\lambda(u) = \nu(u)^{(1,y')} \text{ for any } u \in R \quad ;$$

moreover, once again by the induction hypothesis, if μ is a section of $(Q_{\hat{\delta}}, Q^{\cdot\cdot}_{\delta^{\cdot\cdot}}, Q'_{\delta'})$ then there is $z' \in G'$ such that

17.12.5 $$(Q^{\cdot\cdot})^{(1,z')} \subset R^{\cdot\cdot} \quad , \quad \big(Q', b(\delta')\big)^{z'} \subset \big(R', b(\varepsilon')\big) \quad \text{and}$$
$$\nu(u) = \mu(u)^{(1,z')} \text{ for any } u \in Q \quad .$$

Consequently, setting $x' = z'y'$ we get

17.12.6 $$(Q^{\cdot\cdot})^{(1,x')} \subset (R^{\cdot\cdot})^{(1,y')} \subset P^{\cdot\cdot} \quad , \quad \big(Q', b(\delta')\big)^{x'} \subset \big(R', b(\varepsilon')\big)^{y'} \subset \big(P', b(\gamma')\big)$$

and, for any $u \in Q$, we obtain

17.12.7 $$\lambda(u) = \nu(u)^{(1,y')} = \mu(u)^{(1,x')} \quad .$$

Assume now that $Q_{\hat{\delta}}$ is maximal in $P_{\hat{\gamma}}$, which implies that P normalizes $Q_{\hat{\delta}}$ (cf. [19, Corollary 1.5]); in particular, according to Proposition 17.5 successively applied to the trivial subgroup and Q, and to Q and P, there is a Q-section $\mu_\circ$ of $(P_{\hat{\gamma}}, P^{\cdot\cdot}_{\gamma^{\cdot\cdot}}, P'_{\gamma'})$ and $\mu_\circ$ can be extended to a section $\lambda_\circ$ of $(P_{\hat{\gamma}}, P^{\cdot\cdot}_{\gamma^{\cdot\cdot}}, P'_{\gamma'})$; then, it follows from Corollary 17.9 that there is $y' \in N_{G'}\big(P', b(\gamma')\big)$ such that $(1, y')$ normalizes $P^{\cdot\cdot}$ and we have $\lambda = (\lambda_\circ)^{y'}$. On the other hand, set $D^{\cdot\cdot} = \mathsf{E}(A^{\cdot\cdot}_{\gamma^{\cdot\cdot}})$, $T^{\cdot\cdot} = \sigma^{-1}(Q)$ and $T' = \sigma'(T^{\cdot\cdot})$, denote by f' the block of $C_{G'}(T')$ such that $(T', f') \subset \big(P', b(\gamma')\big)$ (cf. 2.14.2) and by N' the set of $x' \in N_{G'}(T', f')$ such that $(1, x')$ normalizes $T^{\cdot\cdot}$, and choose $X' \subset N'$ such that $1 \times X'$ is a set of representatives for $(1 \times N')/\mathrm{Ker}(\sigma)$; it follows from Theorem 17.8 that, for any Q-section μ of σ, the idempotent

17.12.8 $$l_\mu = \sum_{x' \in X'} e^{D^{\cdot\cdot},y'}_{\mu^{x'}} \big(1 \otimes (1 \otimes 1)\big)$$

is central in the k-algebra $D = \Big(\mathrm{Ind}_\sigma\Big(D^{\cdot\cdot} \otimes_{\mathcal{O}} \mathrm{Res}_{\sigma'}\big((\mathcal{O}G')_{\gamma'}\big)\Big)\Big)(Q)$ and we claim that if μ is a section of $(Q_{\hat{\delta}}, Q^{\cdot\cdot}_{\delta^{\cdot\cdot}}, Q'_{\delta'})$ then we have $l_{\mu_\circ} = l_\mu$.

So let μ be a section of $(Q_{\hat\delta}, Q^{\cdot\cdot}_{\delta^{\cdot\cdot}}, Q'_{\delta'})$; it suffices to prove that $l_{\mu_\circ} l_\mu \neq 0$ or, equivalently, that $i_\circ l_{\mu_\circ} = i_\circ$ and $i l_\mu = i$ for suitable nonzero conjugate idempotents $i_\circ$ and i in D. Since $\mu_\circ$ is a Q-section of $(P_{\hat\gamma}, P^{\cdot\cdot}_{\gamma^{\cdot\cdot}}, P'_{\gamma'})$, it follows from Proposition 13.17 applied to a representative of $\tilde e = \tilde e(A^{\cdot\cdot}_{\gamma^{\cdot\cdot}}) \otimes_{\mathcal{O}} \mathrm{Res}_{\sigma'}(\widetilde{\mathrm{id}}_{(\mathcal{O}G')_{\gamma'}})$ that

17.12.9
$$\begin{aligned}\tilde e^{D^{\cdot\cdot},\gamma'}_{\mu_\circ} \circ \mathrm{Ind}_{\sigma_{\mu_\circ}}\big(\tilde e(Q^{\mu_\circ})\big) \circ \tilde h^{\mu_\circ}_{\hat\gamma} &= \big(\mathrm{Ind}_\sigma(\tilde e)\big)(Q) \circ \tilde e^{\gamma^{\cdot\cdot},\gamma'}_{\mu_\circ} \circ \tilde h^{\mu_\circ}_{\hat\gamma} \\ &= \big(\mathrm{Ind}_\sigma(\tilde e) \circ \tilde h^{\gamma^{\cdot\cdot},\gamma'}_{\hat\gamma}\big)(Q) \quad ;\end{aligned}$$

moreover, since μ is a section of $(Q_{\hat\delta}, Q^{\cdot\cdot}_{\delta^{\cdot\cdot}}, Q'_{\delta'})$, according to equality 16.12.4, we have (cf. 17.4.6)

17.12.10
$$\begin{aligned}\big(\mathrm{Res}^P_Q(\tilde h^{\gamma^{\cdot\cdot},\gamma'}_{\hat\gamma}) \circ \tilde f^{\hat\gamma}_{\hat\delta}\big)(Q) &= (\tilde g^{\gamma^{\cdot\cdot},\gamma'}_{\delta^{\cdot\cdot},\delta'} \circ \tilde h^{\delta^{\cdot\cdot},\delta'}_{\hat\delta})(Q) \\ &= (\tilde g^{\gamma^{\cdot\cdot},\gamma'}_{\delta^{\cdot\cdot},\delta'})(Q) \circ \tilde e^{\delta^{\cdot\cdot},\delta'}_{\mu} \circ \tilde h^{\mu}_{\hat\delta} \quad ;\end{aligned}$$

hence, from Proposition 17.10 and equalities 17.12.9 and 17.12.10, we get

17.12.11
$$\begin{aligned}&\mathrm{Res}^{C_P(Q)}_{Z(Q)}\Big(\tilde e^{D^{\cdot\cdot},\gamma'}_{\mu_\circ} \circ \mathrm{Ind}_{\sigma_{\mu_\circ}}\big(\tilde e(Q^{\mu_\circ})\big) \circ \tilde h^{\mu_\circ}_{\hat\gamma}\Big) \circ \tilde f^{\hat\gamma}_{\hat\delta}(Q) \\ &\quad = \Big(\mathrm{Res}^P_Q\big(\mathrm{Ind}_\sigma(\tilde e)\big) \circ \tilde g^{\gamma^{\cdot\cdot},\gamma'}_{\delta^{\cdot\cdot},\delta'}\Big)(Q) \circ \tilde e^{\delta^{\cdot\cdot},\delta'}_{\mu} \circ \tilde h^{\mu}_{\hat\delta} \\ &\quad = \mathrm{Res}^{C_P(Q)}_{Z(Q)}(\tilde e^{D^{\cdot\cdot},\gamma'}_{\mu}) \circ \tilde g \circ \tilde h^{\mu}_{\hat\delta} \quad ,\end{aligned}$$

for a suitable $\mathcal{D}Z(Q)$-interior Q-algebra exoembedding $\tilde g$, and therefore there are representatives $h_\circ$ and h of any member of equalities 17.12.1 such that

17.12.12
$$\begin{aligned}e^{D^{\cdot\cdot},\gamma'}_{\mu_\circ}\big(1 \otimes (1 \otimes 1)\big)h_\circ(1) &= h_\circ(1) = h_\circ(1)e^{D^{\cdot\cdot},\gamma'}_{\mu_\circ}\big(1 \otimes (1 \otimes 1)\big) \\ e^{D^{\cdot\cdot},\gamma'}_{\mu}\big(1 \otimes (1 \otimes 1)\big)h(1) &= h(1) = h(1)e^{D^{\cdot\cdot},\gamma'}_{\mu}\big(1 \otimes (1 \otimes 1)\big) \quad ;\end{aligned}$$

now, it suffices to set $i_\circ = h_\circ(1)$ and $i = h(1)$.

In conclusion, for a suitable $z' \in X'$, we have $\mu_\circ = \mu^{z'}$ and, setting $x' = z'y'$, on one hand, for any $u \in Q$, we get

17.12.13
$$\mu(u)^{(1,x')} = \mu_0(u)^{(1,y')} = \lambda_\circ(u)^{(1,y')} = \lambda(u) \quad ;$$

on the other hand, we have

17.12.14
$$(Q^{\cdot\cdot})^{(1,x')} \subset (T^{\cdot\cdot})^{(1,y')} \subset P^{\cdot\cdot}$$

since $(1, z')$ and $(1, y')$ respectively normalize $T^{\cdot\cdot}$ and $P^{\cdot\cdot}$; similarly, since $Q'_{\delta'} \subset P'_{\gamma'}$ we have (cf. 2.14.2)

17.12.15
$$\big(Q', b(\delta')\big) \subset (T', f') \subset \big(P', b(\gamma')\big)$$

and, since z' and y' respectively normalize (T', f') and $\big(P', b(\gamma')\big)$, we get

17.12.16
$$\big(Q', b(\delta')\big)^{x'} \subset (T', f')^{y'} \subset \big(P', b(\gamma')\big) \quad .$$

Corollary 17.13 *Let λ be a section of $(P_{\hat\gamma}, P''_{\gamma''}, P'_{\gamma'})$ and denote by e' the block of $C_{G'}(P^{\lambda'})$ such that $(P^{\lambda'}, e') \subset (P', b(\gamma'))$. For any basic local tracing triple $(Q_{\hat\delta}, Q''_{\delta''}, Q'_{\delta'})$ on $\hat{A}$, A'' and $\mathcal{O}G'$, and any section μ of it, denoting by f' the block of $C_{G'}(Q^{\mu'})$ such that $(Q^{\mu'}, f') \subset (Q', b(\delta'))$ we have*

$$\tilde\lambda' \circ E_G(Q_{\hat\delta}, P_{\hat\gamma}) \subset E_{G'}\big((Q^{\mu'}, f'), (P^{\lambda'}, e')\big) \circ \tilde\mu' \quad . \tag{17.13.1}$$

Proof: For any representative φ of an element of $E_G(Q_{\hat\delta}, P_{\hat\gamma})$ (cf. 2.13) and any $x \in G$ such that $(Q_{\hat\delta})^x \subset P_{\hat\gamma}$ and $\varphi(u) = u^x$ for any $u \in Q$, according to Theorem 17.12 there is $x' \in G'$ fulfilling condition 17.12.1; in particular for any $u \in Q$, we have

$$\lambda'(\varphi(u)) = \lambda'(u^x) = \pi'\big(\mu(u)^{(x,x')}\big) = \mu'(u)^{x'} \tag{17.13.2}$$

and $\big(Q', b(\delta')\big)^{x'} \subset \big(P', b(\gamma')\big)$ which forces that $(Q^{\mu'}, f')^{x'} \subset (P^{\lambda'}, e')$ since $(Q^{\mu'})^{x'} \subset P^{\lambda'}$ and $(P^{\lambda'}, e') \subset \big(P', b(\gamma')\big)$ (see 2.14.2).

18 Rickard equivalences between Brauer blocks

18.1 This section is similar to Section 6 in many aspects and we make no effort to avoid redundancy in order to emphasize it. Let G and G' be finite groups, b and b' blocks of G and G' respectively, and $M^{\cdot\cdot}$ an indecomposable $\mathcal{D}(G \times G')$-module which has projective restrictions to $\mathcal{O}(G \times 1)$ and to $\mathcal{O}(1 \times G')$, and is associated with $b \otimes (b')^{\circ}$ (recall that $(b')^{\circ}$ is the image of b' by the antipodal isomorphism $\mathcal{O}G' \cong (\mathcal{O}G')^{\circ}$). Respectively denote by $\mathrm{Mod}_{\mathcal{D}Gb}$ and $\mathrm{Mod}_{\mathcal{D}G'b'}$ the categories of ($\mathcal{O}$-finite as usual) $\mathcal{O}$-free $\mathcal{D}Gb$- and $\mathcal{D}G'b'$-modules, and as in 10.7, by $\overline{\mathrm{Mod}}_{\mathcal{D}Gb}$ and $\overline{\mathrm{Mod}}_{\mathcal{D}G'b'}$ the *homotopy categories* of $A = \mathcal{O}Gb$ and $A' = \mathcal{O}G'b'$; it is clear that $M^{\cdot\cdot}$ determines a functor from $\mathrm{Mod}_{\mathcal{D}G'b'}$ to $\mathrm{Mod}_{\mathcal{D}Gb}$, namely the functor

$$\mathcal{F}_{M^{\cdot\cdot}} : \mathrm{Mod}_{\mathcal{D}G'b'} \longrightarrow \mathrm{Mod}_{\mathcal{D}Gb} \tag{18.1.1}$$

mapping any $\mathcal{O}$-free $\mathcal{D}G'b'$-module M' on $M^{\cdot\cdot} \otimes_{\mathcal{O}G'} M'$ (cf. 10.2 provided we consider M' as a $\mathcal{D}(G \times G')$-module with the trivial action of G and identify G' to the normal subgroup $1 \times G'$ of $G \times G'$), which is $\mathcal{O}$-free too since $M^{\cdot\cdot}$ is a projective $\mathcal{O}(1 \times G')$-module, and any $\mathcal{D}G'b'$-module homomorphism $f' : M' \to N'$ where N' is a second $\mathcal{O}$-free $\mathcal{D}G'b'$-module, on the $\mathcal{D}Gb$-module homomorphism

$$\mathrm{id}_{M^{\cdot\cdot}} \underset{\mathcal{O}G'}{\otimes} f' : M^{\cdot\cdot} \underset{\mathcal{O}G'}{\otimes} M' \longrightarrow M^{\cdot\cdot} \underset{\mathcal{O}G'}{\otimes} N' \quad . \tag{18.1.2}$$

It is easily checked that $\mathcal{F}_{M^{\cdot\cdot}}$ preserves contractility (cf. 10.7) and therefore it induces a functor

$$\bar{\mathcal{F}}_{M^{\cdot\cdot}} : \overline{\mathrm{Mod}}_{\mathcal{D}G'b'} \longrightarrow \overline{\mathrm{Mod}}_{\mathcal{D}Gb} \quad . \tag{18.1.3}$$

18.2 It is clear that the $\mathcal{O}$-dual $(M^{\cdot\cdot})^*$ is an indecomposable $\mathcal{D}(G \times G')$-module associated with $b^{\circ} \otimes b'$ and, since $\mathcal{O}G$ and $\mathcal{O}G'$ are symmetric $\mathcal{O}$-algebras, its restrictions to $\mathcal{O}(G \times 1)$ and to $\mathcal{O}(1 \times G')$ are both projective too; hence, it defines also a functor

$$\bar{F}_{(M^{\cdot\cdot})^*} : \overline{\mathrm{Mod}}_{\mathcal{D}Gb} \longrightarrow \overline{\mathrm{Mod}}_{\mathcal{D}G'b'} \quad ; \tag{18.2.1}$$

moreover, the projectivity of the restrictions of both $M^{\cdot\cdot}$ and $(M^{\cdot\cdot})^*$ to both $\mathcal{O}(G \times 1)$ and $\mathcal{O}(1 \times G')$ implies easily that $\bar{F}_{M^{\cdot\cdot}}$ and $\bar{F}_{(M^{\cdot\cdot})^*}$ are both left and right adjoints to each other. Consequently, $\bar{F}_{M^{\cdot\cdot}}$ is an equivalence of categories if and only if the compositions $\bar{F}_{M^{\cdot\cdot}} \circ \bar{F}_{(M^{\cdot\cdot})^*}$ and $\bar{F}_{(M^{\cdot\cdot})^*} \circ \bar{F}_{M^{\cdot\cdot}}$ are respectively isomorphic to the identity functors on $\overline{\mathrm{Mod}}_{\mathcal{D}Gb}$ and $\overline{\mathrm{Mod}}_{\mathcal{D}G'b'}$. A *sufficient condition* for it is that, in the homotopy categories of $\mathcal{O}(G \times G)$ and $\mathcal{O}(G' \times G')$, the $\mathcal{D}(G \times G)$-module A and the $\mathcal{D}(G' \times G')$-module A' (where $\mathcal{D}$ acts via the canonical map $\mathcal{D} \to \mathcal{O}$) are respectively isomorphic to $M^{\cdot\cdot} \otimes_{\mathcal{O}G'} (M^{\cdot\cdot})^*$ and $(M^{\cdot\cdot})^* \otimes_{\mathcal{O}G} M^{\cdot\cdot}$: in that case, we say that $M^{\cdot\cdot}$ *defines a Rickard equivalence between* b and b'; that is to say, according to Corollary 10.11 and to the fact that the $\mathcal{O}(G \times G)$-module A and the $\mathcal{O}(G' \times G')$-module A' are both indecomposable (cf. 2.14), we have that:

18.2.2 *$M^{\cdot\cdot}$ defines a Rickard equivalence between b and b′ if and only if, for suitable contractile $\mathcal{D}(G \times G)$- and $\mathcal{D}(G' \times G')$-modules C and C′, we respectively have $\mathcal{D}(G \times G)$- and $\mathcal{D}(G' \times G')$-module isomorphisms*

$$M^{\cdot\cdot} \underset{\mathcal{O}G'}{\otimes} (M^{\cdot\cdot})^* \cong A \oplus C \quad \text{and} \quad (M^{\cdot\cdot})^* \underset{\mathcal{O}G}{\otimes} M^{\cdot\cdot} \cong A' \oplus C' \quad .$$

If so, notice that $M^{\cdot\cdot}$ is *not* contractile. Rickard's main result in [30] and [31] states that the so called *derived categories* of the A- and A'-module categories are equivalent as *triangulated categories* if and only if there is such an $M^{\cdot\cdot}$ defining a Rickard equivalence between b and b'.

18.3 Our first proposition translates the condition in 18.2.2 in terms of $\mathcal{D}G$-interior algebras and *homotopy isomorphisms* between G-interior $\mathcal{D}$-algebras (cf. 11.7), and provides the link with Hecke $\mathcal{D}G$-interior algebras. Recall that $\mathrm{End}_{\mathcal{O}}(M^{\cdot\cdot})$ has a canonical $\mathcal{D}(G \times G')$-interior algebra structure; hence, $\mathrm{End}_{\mathcal{O}}(M^{\cdot\cdot})^{1\times G'}$ is a $\mathcal{D}G$-interior algebra and notice that the structural map st : $A \to \mathrm{End}_{\mathcal{O}}(M^{\cdot\cdot})^{1\times G'}$ is a unitary homomorphism of G-interior $\mathcal{D}$-algebras (cf. 11.6).

Proposition 18.4 *The $\mathcal{D}(G \times G')$-module $M^{\cdot\cdot}$ defines a Rickard equivalence between b and b′ if and only if the structural map* st : $A \to \mathrm{End}_{\mathcal{O}}(M^{\cdot\cdot})^{1\times G'}$ *is a G-interior $\mathcal{D}$-algebra homotopy isomorphism. In this case, we have an $\mathcal{O}G$-interior algebra isomorphism*

18.4.1 $$A \cong \mathrm{H}_{\circ}\big(\mathrm{End}_{\mathcal{O}}(M^{\cdot\cdot})^{1\times G'}\big) \quad ,$$

$\mathrm{End}_{\mathcal{O}}(M^{\cdot\cdot})^{1\times G'}$ *is a 0-split $\mathcal{D}(G \times G)$-module and* st *induces a bijection between the set of pointed groups on A and the set of noncontractile pointed groups on* $\mathrm{End}_{\mathcal{O}}(M^{\cdot\cdot})^{1\times G'}$*, which preserves inclusion and localness.*

Proof: Since the restriction of $M^{\cdot\cdot}$ to $\mathcal{O}(1 \times G')$ is projective, $M^{\cdot\cdot}$ is $\mathcal{O}$-free and the trace map $\mathrm{Tr}_{1\times 1}^{1\times G'}$ on $M^{\cdot\cdot} \otimes_{\mathcal{O}} (M^{\cdot\cdot})^* \cong \mathrm{End}_{\mathcal{O}}(M^{\cdot\cdot})$ induces a $\mathcal{D}(G \times G)$-module isomorphism

18.4.2 $$M^{\cdot\cdot} \underset{\mathcal{O}G'}{\otimes} (M^{\cdot\cdot})^* \cong \big(M^{\cdot\cdot} \underset{\mathcal{O}}{\otimes} (M^{\cdot\cdot})^*\big)^{1\times G'} \quad .$$

Then, if $M^{\cdot\cdot}$ defines a Rickard equivalence between b and b', we have a $\mathcal{D}(G \times G)$-module isomorphism (cf. 18.2.2)

18.4.3 $$A \oplus C \cong \mathrm{End}_{\mathcal{O}}(M^{\cdot\cdot})^{1\times G'}$$

for a suitable contractile $\mathcal{D}(G \times G)$-module C, and we denote by

18.4.4 $$f : A \longrightarrow \mathrm{End}_{\mathcal{O}}(M^{\cdot\cdot})^{1\times G'} \text{ and } g : \mathrm{End}_{\mathcal{O}}(M^{\cdot\cdot})^{1\times G'} \longrightarrow A$$

the corresponding split injection and split surjection; since for any $x \in G$ we have

18.4.5 $$x{\cdot}f(b) = f(xb) = f(bx) = f(b){\cdot}x$$

and for any $\mathscr{f}, \mathscr{g} \in \mathcal{F}$ we get

18.4.6 $$(\mathscr{f} + \mathscr{g}d)\big(f(b)\big) = f\big((\mathscr{f} + \mathscr{g}d)(b)\big) = \mathscr{f}(0)f(b) \quad ,$$

the multiplication by $f(b)$ on the right determines a $\mathcal{D}(G\times G)$-module endomorphism $m_{f(b)}$ of $\mathrm{End}_{\mathcal{O}}(M^{\cdot\cdot})^{1\times G'}$ and it is easily checked that f is the composition of the structural map $\mathrm{st}_A : A \to \mathrm{End}_{\mathcal{O}}(M^{\cdot\cdot})^{1\times G'}$ and $m_{f(b)}$; consequently, we obtain

18.4.7 $$\mathrm{id}_A = g \circ f = (g \circ m_{f(b)}) \circ \mathrm{st}_A \quad ,$$

so that st_A is a split injection; in particular, $\mathrm{Im}(\mathrm{st}_A)$ is a direct summand of $\mathrm{End}_{\mathcal{O}}\,(M^{\cdot\cdot})^{1\times G'}$ as $\mathcal{D}(G \times G)$-modules and any complement of it is isomorphic to C.

Conversely, assume that $\mathrm{st}_A : A \to \mathrm{End}_{\mathcal{O}}(M^{\cdot\cdot})^{1\times G'}$ is a G-interior $\mathcal{D}$-algebra homotopy isomorphism; since A is an indecomposable $\mathcal{O}G$-bimodule and st_A is unitary, st_A is a split injection (cf. Corollary 10.11) and therefore, by 18.4.2, we have a $\mathcal{D}(G \times G)$-module isomorphism

18.4.8 $$M^{\cdot\cdot} \underset{\mathcal{O}G'}{\otimes} (M^{\cdot\cdot})^* \cong A \oplus C \quad ,$$

where $C = \big(\mathrm{End}_{\mathcal{O}}(M^{\cdot\cdot})^{1\times G'}\big)/\mathrm{Im}(\mathrm{st}_A)$ is contractile. In order to prove the second isomorphism in 18.2.2, we follow an argument of Rickard (see the proof of Theorem 2.1 in [33]); precisely, we will prove that the structural homomorphism $\mathrm{st}_{A'} : A' \to \mathrm{End}_{\mathcal{O}}(M^{\cdot\cdot})^{G\times 1}$ is also a split injection and that its cokernel is a contractile $\mathcal{D}(G' \times G')$-module. By symmetry, the trace map $\mathrm{Tr}_{1\times 1}^{G\times 1}$ on $\mathrm{End}_{\mathcal{O}}(M^{\cdot\cdot})$ induces also a $\mathcal{D}(G' \times G')$-module isomorphism

18.4.9 $$(M^{\cdot\cdot})^* \underset{\mathcal{O}G}{\otimes} M^{\cdot\cdot} \cong \big((M^{\cdot\cdot})^* \underset{\mathcal{O}}{\otimes} M^{\cdot\cdot}\big)^{G\times 1} \quad ;$$

in particular, since each member is canonically the dual of the other one, we get a $G' \times G'$-stable nondegenerate symmetric scalar product in $\mathrm{End}_{\mathcal{O}}(M^{\cdot\cdot})^{G\times 1}$; since A' admits also a $G' \times G'$-stable nondegenerate symmetric scalar product, it makes sense to consider the adjoint map $(\mathrm{st})^{\mathrm{ad}}$ and the composition $(\mathrm{st}_{A'})^{\mathrm{ad}} \circ \mathrm{st}_{A'}$ is a $\mathcal{D}(G' \times G')$-module endomorphism of A', so that it is the multiplication by a suitable $z' \in Z(A')$.

We claim that z' is invertible in $Z(A')$; indeed, $\mathrm{st}_{A'}$ and $(\mathrm{st}_{A'})^{\mathrm{ad}}$ respectively correspond to the unit and counit of the adjunctions $(\mathcal{F}_{M^{\cdot\cdot}}, \mathcal{F}_{(M^{\cdot\cdot})^*})$ and $(\mathcal{F}_{(M^{\cdot\cdot})^*}, \mathcal{F}_{M^{\cdot\cdot}})$, and therefore, up to suitable identifications, the $\mathcal{D}(G \times G')$-module homomorphisms

18.4.10 $$M^{\cdot\cdot} \xrightarrow{\mathrm{id}_{M^{\cdot\cdot}} \otimes\, \mathrm{st}_{A'}} M^{\cdot\cdot} \underset{\mathcal{O}G'}{\otimes} (M^{\cdot\cdot})^* \underset{\mathcal{O}G}{\otimes} M^{\cdot\cdot} \xrightarrow{\mathrm{id}_{M^{\cdot\cdot}} \otimes (\mathrm{st}_{A'})^{\mathrm{ad}}} M^{\cdot\cdot}$$

are respectively a split injection and a split surjection by the general properties of adjoint functors (see the proof of Theorem 1.2 in [33]); moreover, by 18.4.8, we have a $\mathcal{D}(G \times G')$-module isomorphism

18.4.11 $$M^{\cdot\cdot} \underset{\mathcal{O}G'}{\otimes} (M^{\cdot\cdot})^* \underset{\mathcal{O}G}{\otimes} M^{\cdot\cdot} \cong M^{\cdot\cdot} \oplus (C \underset{\mathcal{O}G}{\otimes} M^{\cdot\cdot}) \quad ;$$

consequently, since $C \otimes_{\mathcal{O}G} M^{\cdot\cdot}$ is a contractile $\mathcal{D}(G \times G')$-module (cf. 10.7.3) and $M^{\cdot\cdot}$ is noncontractile, both $\mathrm{id}_{M^{\cdot\cdot}} \otimes_{\mathcal{O}} \mathrm{st}_{A'}$ and $\mathrm{id}_{M^{\cdot\cdot}} \otimes_{\mathcal{O}} (\mathrm{st}_{A'})^{\mathrm{ad}}$ determine nonzero isomorphisms in the homotopy category $\overline{\mathrm{Mod}}_{\mathcal{D}(G\times G')}$ and therefore their composition $z'\mathrm{id}_{A'}$ is invertible in $\mathrm{End}_{\mathcal{O}}(M^{\cdot\cdot})^{G\times G'}$, so that z' does not belong to $J\big(Z(A')\big)$.

In particular, $\mathrm{st}_{A'}$ is a split injection, so that we have a $\mathcal{D}(G' \times G')$-module isomorphism

18.4.12 $$(M^{\cdot\cdot})^* \underset{\mathcal{O}G}{\otimes} M^{\cdot\cdot} \cong A' \oplus C'$$

for a suitable $\mathcal{D}(G' \times G')$-module C'; hence, by 18.4.8 and 18.4.12, we get the following $\mathcal{D}(G' \times G')$-module isomorphisms

18.4.13 $$\begin{aligned} A' \oplus C' \oplus C' \oplus (C' \underset{\mathcal{O}G'}{\otimes} C') &\cong (M^{\cdot\cdot})^* \underset{\mathcal{O}G}{\otimes} M^{\cdot\cdot} \underset{\mathcal{O}G'}{\otimes} (M^{\cdot\cdot})^* \underset{\mathcal{O}G}{\otimes} M^{\cdot\cdot} \\ &\cong \big((M^{\cdot\cdot})^* \underset{\mathcal{O}G}{\otimes} M^{\cdot\cdot}\big) \oplus \big((M^{\cdot\cdot})^* \underset{\mathcal{O}G}{\otimes} C \underset{\mathcal{O}G}{\otimes} M^{\cdot\cdot}\big) \\ &\cong A' \oplus C' \oplus \big((M^{\cdot\cdot})^* \underset{\mathcal{O}G}{\otimes} C \underset{\mathcal{O}G}{\otimes} M^{\cdot\cdot}\big) \end{aligned}$$

and therefore the $\mathcal{D}(G' \times G')$-module C' is a direct summand of $(M^{\cdot\cdot})^* \otimes_{\mathcal{O}G} C \otimes_{\mathcal{O}G} M^{\cdot\cdot}$ which is contractile (cf. 10.7.3).

Finally, since $\mathrm{H}_\circ(A) = A$, isomorphism 18.4.1 follows from the very definition of the homotopy isomorphisms (cf. 11.7) and isomorphism 18.4.3 proves the 0-splitness of $\mathrm{End}_{\mathcal{O}}(M^{\cdot\cdot})^{1\times G'}$; moreover, since $\mathrm{Im}(\mathrm{st}_A)$ is contained in $\mathrm{C}_\circ\big(\mathrm{End}_{\mathcal{O}}(M^{\cdot\cdot})^{1\times G'}\big)$, for any subgroup H of G, isomorphism 18.4.1 induces a surjective $\mathcal{O}$-algebra homomorphism

18.4.14 $$\mathrm{C}_\circ(\mathrm{End}_{\mathcal{O}}(M^{\cdot\cdot})^{H\times G'}) \longrightarrow A^H$$

and, in particular, it induces a bijective map between the set of points of H on A and the set of noncontractile points of H on $\mathrm{End}_{\mathcal{O}}(M^{\cdot\cdot})^{1\times G'}$.

18.5 Since $M^{\cdot\cdot}$ is an indecomposable $\mathcal{D}(G \times G')$-module, $\alpha^{\cdot\cdot} = \{\mathrm{id}_{M^{\cdot\cdot}}\}$ is the unique point of $G \times G'$ on $\mathrm{End}_{\mathcal{O}}(M^{\cdot\cdot})$ and we denote by $P^{\cdot\cdot}_{\gamma^{\cdot\cdot}}$ a defect pointed group of $(G \times G')_{\alpha^{\cdot\cdot}}$, which is unique up to $G \times G'$-conjugation (cf. 2.9.5 and 14.5); as in the ordinary module case, we call $P^{\cdot\cdot}$ *vertex* of $M^{\cdot\cdot}$ and any indecomposable direct summand $N^{\cdot\cdot}$ determined by $\gamma^{\cdot\cdot}$ in the $\mathcal{D}P^{\cdot\cdot}$-module $\mathrm{Res}^{G\times G'}_{P^{\cdot\cdot}}(M^{\cdot\cdot})$ *$\mathcal{D}P^{\cdot\cdot}$-source* of $M^{\cdot\cdot}$. Denote by $S^{\cdot\cdot}$ the Higman envelope of $\mathrm{End}_{\mathcal{O}}(N^{\cdot\cdot})$ (cf. Proposition 14.7) and set

18.5.1 $$A^{\cdot\cdot} = \mathrm{Ind}^{G\times G'}_{P^{\cdot\cdot}}(S^{\cdot\cdot}) \quad , \quad \hat{A} = (A^{\cdot\cdot})^{1\times G'} \quad \text{and} \quad \hat{A}' = \big((A^{\cdot\cdot})^\circ\big)^{G\times 1} \quad ;$$

according to Corollary 14.11, we have a Higman exoembedding of $\mathcal{D}(G\times G')$-interior algebras

18.5.2 $$\tilde{h}^{\gamma^{\cdot\cdot}}_{\alpha^{\cdot\cdot}} : \mathrm{End}_{\mathcal{O}}(M^{\cdot\cdot}) \longrightarrow \mathrm{Ind}^{G\times G'}_{P^{\cdot\cdot}}\big(\mathrm{End}_{\mathcal{O}}(N^{\cdot\cdot})\big)$$

such that $\tilde{h}_{\gamma^{\cdot\cdot}} \circ \tilde{h}^{\gamma^{\cdot\cdot}}_{\alpha^{\cdot\cdot}} = \tilde{h}_{\alpha^{\cdot\cdot}} = \tilde{e}\big(\mathrm{End}_{\mathcal{O}}(M^{\cdot\cdot})\big)$; in particular, denoting by $\tilde{e}_{\gamma^{\cdot\cdot}} : \mathrm{End}_{\mathcal{O}}(N^{\cdot\cdot}) \to S^{\cdot\cdot}$ the canonical exoembedding, the composed $\mathcal{D}(G \times G')$-interior algebra exoembedding

$$\text{18.5.3} \qquad \mathrm{End}_{\mathcal{O}}(M^{\cdot\cdot}) \xrightarrow{\tilde{h}^{\gamma^{\cdot\cdot}}_{\alpha^{\cdot\cdot}}} \mathrm{Ind}_{P^{\cdot\cdot}}^{G\times G'}\big(\mathrm{End}_{\mathcal{O}}(N^{\cdot\cdot})\big) \xrightarrow{\mathrm{Ind}_{P^{\cdot\cdot}}^{G\times G'}(\tilde{e}_{\gamma^{\cdot\cdot}})} A^{\cdot\cdot}$$

allows us to identify $\alpha^{\cdot\cdot}$ to a point of $G \times G'$ on $A^{\cdot\cdot}$ and then, since $A^{\cdot\cdot} \cong \mathsf{E}(A^{\cdot\cdot})$ (cf. Corollary 14.21), we get a $\mathcal{D}(G\times G')$-interior algebra exoembedding (cf. Proposition 14.7)

$$\text{18.5.4} \qquad \mathsf{E}\big(\mathrm{End}_{\mathcal{O}}(M^{\cdot\cdot})\big) \longrightarrow A^{\cdot\cdot} \quad ,$$

simply denoting by $\tilde{h}_{\gamma^{\cdot\cdot}}$ and $\tilde{h}_{\alpha^{\cdot\cdot}}$ the corresponding Higman exoembeddings into $A^{\cdot\cdot}$. Now, $\alpha^{\cdot\cdot}$ can also be considered as a point of G on $\hat{A}$ and we have $b{\cdot}\alpha^{\cdot\cdot} = \alpha^{\cdot\cdot}$; then, according to Proposition 18.4, we have that:

18.5.5 *$M^{\cdot\cdot}$ defines a Rickard equivalence between b and b' if and only if the structural map $A \to \hat{A}_{\alpha^{\cdot\cdot}}$ is a G-interior $\mathcal{D}$-algebra homotopy isomorphism.*

18.6 At this point, we can apply the results of Sections 15 and 16 in order to analyze this condition. Respectively denote by $\rho : P^{\cdot\cdot} \to G$ and $\rho' : P^{\cdot\cdot} \to G'$ the restrictions of the first and the second projection maps for $G \times G'$, by P and P' the images of ρ and ρ', and by $\sigma : P^{\cdot\cdot} \to P$ and $\sigma' : P^{\cdot\cdot} \to P'$ the group homomorphisms determined by ρ and ρ'. Recall that, according to Theorem 15.4, we have the Hecke $\mathcal{D}G$-interior algebra isomorphism

$$\text{18.6.1} \qquad H_{G,G'}(S^{\cdot\cdot}) : \hat{A} \cong \mathrm{Ind}_{\rho}\big(S^{\cdot\cdot} \underset{\mathcal{O}}{\otimes} \mathrm{Res}_{\rho'}(\mathcal{O}G')\big)$$

which, for any $a \in S^{\cdot\cdot} \otimes_{\mathcal{O}} \mathrm{Res}_{\rho'}(\mathcal{O}G')$ such that $\mathrm{Ker}(\rho)$ fixes $1 \otimes a$ in $\mathcal{O} \otimes_{\mathcal{O}\mathrm{Ker}(\rho)} \big(S^{\cdot\cdot} \otimes_{\mathcal{O}} \mathrm{Res}_{\rho'}(\mathcal{O}G')\big)$, maps $\mathrm{Tr}^{1\times G'}_{\mathrm{Ker}(\rho)}\Big(\big(q_{G,G'}(S^{\cdot\cdot})\big)(1 \otimes a)\Big)$ on $(1,1) \otimes (1 \otimes a) \otimes (1,1)$, where $q_{G,G'}(S^{\cdot\cdot})$ maps an element $1 \otimes (a^{\cdot\cdot} \otimes x')$ of $\mathcal{O} \otimes_{\mathcal{O}\mathrm{Ker}(\rho)} \big(S^{\cdot\cdot} \otimes_{\mathcal{O}} \mathrm{Res}_{\rho'}(\mathcal{O}G')\big)$ on $(1,x')^{-1} \otimes a^{\cdot\cdot} \otimes (1,1) \in A^{\cdot\cdot}$ for any $a^{\cdot\cdot} \in S^{\cdot\cdot}$ and any $x' \in G'$; notice that, according to 15.10, the functor $\mathcal{F}_{M^{\cdot\cdot}}$ above coincides, up to isomorphism, with the restriction to $\mathrm{Mod}_{\mathcal{D}G'b'}$ of the functor $\mathcal{F}_{G,G'}(N^{\cdot\cdot})_{i^{\cdot\cdot}}$ where $i^{\cdot\cdot} \in \alpha^{\cdot\cdot}$. Moreover, it follows from Corollary 16.10 that there are local points $\hat{\gamma}$ of P on $\hat{A}$ and γ' of P' on $\mathcal{O}G'$ such that $(P_{\hat{\gamma}}, P^{\cdot\cdot}_{\gamma^{\cdot\cdot}}, P'_{\gamma'})$ is a local tracing triple on $\hat{A}$, $A^{\cdot\cdot}$ and $\mathcal{O}G'$, and that $P_{\hat{\gamma}}$ is a defect pointed group of $G_{\alpha^{\cdot\cdot}}$ on $\hat{A}$; in particular, we have a canonical $\mathcal{D}P$-interior algebra exoembedding (cf. 16.5.1)

$$\text{18.6.2} \qquad \tilde{h}^{\gamma^{\cdot\cdot}}_{\hat{\gamma}}, \gamma' : \hat{A}_{\hat{\gamma}} \longrightarrow \mathrm{Ind}_{\sigma}\Big(S^{\cdot\cdot}_{\gamma^{\cdot\cdot}} \underset{\mathcal{O}}{\otimes} \mathrm{Res}_{\sigma'}\big((\mathcal{O}G')_{\gamma'}\big)\Big)$$

which allows us to identify $\hat{\gamma}$ to a local point of P on $\mathrm{Ind}_{\sigma}\Big(S^{\cdot\cdot}_{\gamma^{\cdot\cdot}} \otimes_{\mathcal{O}} \mathrm{Res}_{\sigma'}\big((\mathcal{O}G')_{\gamma'}\big)\Big)$. Furthermore, denoting by $e_{\gamma^{\cdot\cdot}}$ a representative of $\tilde{e}_{\gamma^{\cdot\cdot}} = \tilde{e}\big(\mathrm{End}_{\mathcal{O}}(N^{\cdot\cdot})\big)$, since

$H_{G,G'}(S^{\cdot\cdot})$ maps the element

$$\left(1 \otimes (b')^{\circ}\right)\cdot \mathrm{Tr}_{P^{\cdot\cdot}}^{G\times G'}\left((1,1)\otimes e_{\gamma^{\cdot\cdot}}(\mathrm{id}_{N^{\cdot\cdot}})\otimes(1,1)\right)$$

of $\hat{A}$ on the element

$$\mathrm{Tr}_P^G\Big((1,1)\otimes\Big(1\otimes\big(e_{\gamma^{\cdot\cdot}}(\mathrm{id}_{N^{\cdot\cdot}})\otimes b'\big)\Big)\otimes(1,1)\Big)$$

of $\mathrm{Ind}_\rho\big(S^{\cdot\cdot}_{\gamma^{\cdot\cdot}}\otimes_{\mathcal{O}}\mathrm{Res}_{\rho'}(\mathcal{O}G')\big)$ and $M^{\cdot\cdot}$ is associated with $b\otimes(b')^{\circ}$, which implies that $b\otimes(b')^{\circ}\cdot\alpha^{\cdot\cdot}=\alpha^{\cdot\cdot}$, the image of $\hat{A}_{\hat{\gamma}}$ by any representative of $\tilde{h}_{\hat{\gamma}}^{\gamma^{\cdot\cdot},\gamma'}$ is contained in $\mathrm{Ind}_\sigma\Big(S^{\cdot\cdot}_{\gamma^{\cdot\cdot}}\otimes_{\mathcal{O}}\mathrm{Res}_{\sigma'}\big((\mathcal{O}G')_{\gamma'}b'\big)\Big)$, so that γ' is already a local point of P' on A'; notice that, since $\hat{\gamma}$ can also be considered as a point of $P\times G'$ on $A^{\cdot\cdot}$, the inclusion $\mathsf{C}_\circ(\hat{A})\subset\mathsf{C}_\circ(A^{\cdot\cdot})$ induces a k^*-group isomorphism (cf. 2.8)

18.6.3 $$\hat{\nu}:\hat{\bar{N}}_G(P_{\hat{\gamma}})\cong\hat{\bar{N}}_{G\times G'}\big((P\times G')_{\hat{\gamma}}\big)$$

together with a $k_*\hat{\bar{N}}_G(P_{\hat{\gamma}})$-module isomorphism (cf. 2.8)

18.6.4 $$V_{\mathsf{C}_\circ(\hat{A})}(P_{\hat{\gamma}})\cong V_{\mathsf{C}_\circ(A^{\cdot\cdot})}\big((P\times G')_{\hat{\gamma}}\big)\quad .$$

18.7 If $M^{\cdot\cdot}$ defines a Rickard equivalence between b and b', although any pointed group H_β on A determines via the structural map $\mathrm{st}: A\to A^{\cdot\cdot}_{\alpha^{\cdot\cdot}}$ a noncontractile pointed group $H_{\hat{\beta}}$ on $\hat{A}_{\alpha^{\cdot\cdot}}$ (cf. Proposition 18.14), $\hat{\beta}$ need not contain $\mathrm{st}(\beta)$; in particular, if γ is the local point of P on A corresponding to $\hat{\gamma}$ (notice that, since $\alpha^{\cdot\cdot}$ is noncontractile and $P_{\hat{\gamma}}$ is a defect pointed group of $G_{\alpha^{\cdot\cdot}}$ on $\hat{A}$, by 14.5.4 the point $\hat{\gamma}$ is noncontractile too) and we choose $i\in\gamma$ then we have $\mathrm{st}(i)=\hat{\imath}+\hat{l}$, where $\hat{\imath}\in\hat{\gamma}$, $\hat{l}\in\mathsf{B}_\circ(\hat{A}^P_{\alpha^{\cdot\cdot}})$ and $\hat{\imath}\hat{l}=0=\hat{l}\hat{\imath}$. In order to take care of this fact, for any $n\geq 0$, we introduce the family $\mathfrak{Q}_n$ formed by the pair (P,id_P) and, for any local pointed group $Q_{\hat{\delta}}$ on $\hat{A}$ such that $Q_{\hat{\delta}}\subsetneqq P_{\hat{\gamma}}$ and any group homomorphism $\varphi: Q\to P$ such that $\tilde{\varphi}\in E_G(Q_{\hat{\delta}},P_{\hat{\gamma}})$, by the pair (Q,φ) repeated n times, and then we consider the corresponding $\mathfrak{Q}_n$-induced $\mathcal{D}P$-interior algebra (cf. 12.15)

18.7.1 $$\mathrm{Ind}_{\mathfrak{Q}_n}^P\big(\mathrm{Res}_{\mathfrak{Q}_n}^P(\hat{A}_{\hat{\gamma}})\big)\cong\mathrm{Ind}_{\mathfrak{Q}_n}^P\Big(\mathrm{Res}_{\mathfrak{Q}_n}^P\Big(\mathrm{Ind}_\sigma\big(S^{\cdot\cdot}_{\gamma^{\cdot\cdot}}\underset{\mathcal{O}}{\otimes}\mathrm{Res}_{\sigma'}(A'_{\gamma'})\big)_{\hat{\gamma}}\Big)\Big)\quad;$$

recall that we have a canonical $\mathcal{D}P$-interior algebra embedding (cf. 12.16.2)

18.7.2 $$\hat{A}_{\hat{\gamma}}\longrightarrow\mathrm{Ind}_{\mathfrak{Q}_n}^P\big(\mathrm{Res}_{\mathfrak{Q}_n}^P(\hat{A}_{\hat{\gamma}})\big)$$

and notice that this embedding is associated with the unique local point of P on that $\mathcal{D}P$-interior algebra (cf. 2.11.3 and 12.4.4); similarly, by 12.15.4, we get a canonical $\mathcal{O}P$-interior algebra embedding

18.7.3 $$\mathsf{H}_\circ(\hat{A}_{\hat{\gamma}})\longrightarrow\mathrm{Ind}_{\mathfrak{Q}_n}^P\Big(\mathrm{Res}_{\mathfrak{Q}_n}^P\big(\mathsf{H}_\circ(\hat{A}_{\hat{\gamma}})\big)\Big)$$

and all the local points of P on the right end come from the left end. We are ready to state our main result; in its statement we prefer to emphasize the right member of exoisomorphism 18.7.1 since it is more explicitly related to the given date G', b' and $M^{\cdot\cdot}$.

Theorem 18.8 *With the notation above, $M^{\cdot\cdot}$ defines a Rickard equivalence between b and b' if and only if P and P' are respectively defect groups of b and b', for a suitable local point γ of P on A one of the following two conditions holds:*

18.8.1 *For some $n \geq 0$ there is a P-interior $\mathfrak{D}$-algebra homotopy embedding*

$$e : A_\gamma \longrightarrow \mathrm{Ind}_{\Omega_n}^P\Big(\mathrm{Res}_{\Omega_n}^P\Big(\mathrm{Ind}_\sigma\big(S_{\gamma^{\cdot\cdot}}^{\cdot\cdot} \underset{\mathcal{O}}{\otimes} \mathrm{Res}_{\sigma'}(A'_{\gamma'})\big)_{\hat{\gamma}}\Big)\Big) \quad .$$

18.8.2 $\mathrm{Ind}_\sigma\big(S_{\gamma^{\cdot\cdot}}^{\cdot\cdot} \otimes_{\mathcal{O}} \mathrm{Res}_{\sigma'}(A'_{\gamma'})\big)_{\hat{\gamma}}$ *is a 0-split $\mathfrak{D}(P \times P)$-module and there is an $\mathcal{O}P$-interior algebra isomorphism*

$$A_\gamma \cong \mathsf{H}_0\Big(\mathrm{Ind}_\sigma\big(S_{\gamma^{\cdot\cdot}}^{\cdot\cdot} \underset{\mathcal{O}}{\otimes} \mathrm{Res}_{\sigma'}(A'_{\gamma'})\big)_{\hat{\gamma}}\Big) \quad .$$

and, denoting by $\hat{\theta} : \hat{\bar{N}}_G(P_{\hat{\gamma}}) \cong \hat{\bar{N}}_G(P_\gamma)$ the k^*-group isomorphism induced by that condition, we have a $k_*\hat{\bar{N}}_G(P_\gamma)$-module isomorphism

18.8.3 $$V_A(P_\gamma) \cong \mathrm{Res}_{\hat{\upsilon}\circ\hat{\theta}^{-1}}\Big(V_{\mathsf{C}_\circ(\mathrm{End}_{\mathcal{O}}(M^{\cdot\cdot}))}\big((P \times G')_{\hat{\gamma}}\big)\Big) \quad .$$

Moreover, in that case both conditions hold and, denoting by $\hat{\gamma}'$ the noncontractile point of P' on $\hat{A}'_{\alpha^{\cdot\cdot}}$ corresponding to γ', $(P'_{\hat{\gamma}'}, P^{\cdot\cdot}_{\gamma^{\cdot\cdot}}, P_\gamma)$ is a local tracing triple on $\hat{A}'$, $(A^{\cdot\cdot})^\circ$ and $\mathcal{O}G$.

Remark 18.9 Since $\mathsf{H}_\circ(A_\gamma) \cong A_\gamma$, the existence of the homotopy embedding e above implies, by Corollary 10.11, that the $\mathfrak{D}P$-interior algebra (cf. 11.7)

18.9.1 $$\hat{B} = e(1)\mathrm{Ind}_{\Omega_n}^P\Big(\mathrm{Res}_{\Omega_n}^P\Big(\mathrm{Ind}_\sigma\big(S_{\gamma^{\cdot\cdot}}^{\cdot\cdot} \underset{\mathcal{O}}{\otimes} \mathrm{Res}_{\sigma'}(A'_{\gamma'})\big)_{\hat{\gamma}}\Big)\Big)e(1)$$

is a 0-split $\mathfrak{D}(P \times P)$-module (cf. 10.12) and that we have an $\mathcal{O}P$-interior algebra isomorphism (cf. 11.7)

18.9.2 $$A_\gamma \cong \mathsf{H}_\circ(\hat{B}) \quad ;$$

in particular, since γ is local, P has a local point on $\hat{B}$, which forces the factorization of embedding 18.7.2 throughout some $\mathfrak{D}P$-interior algebra embedding (cf. 2.11.3, 12.4.4 and 12.16.2)

18.9.3 $$\mathrm{Ind}_\sigma\big(S_{\gamma^{\cdot\cdot}}^{\cdot\cdot} \underset{\mathcal{O}}{\otimes} \mathrm{Res}_{\sigma'}(A'_{\gamma'})\big)_{\hat{\gamma}} \longrightarrow \hat{B}$$

and then this embedding induces an $\mathcal{O}P^{\cdot\cdot}$-interior algebra isomorphism

18.9.4 $$\mathrm{H}_\circ\Big(\mathrm{Ind}_\sigma\big(S^{\cdot\cdot}_{\gamma^{\cdot\cdot}} \otimes_{\mathcal{O}} \mathrm{Res}_{\sigma'}(A'_{\gamma'})\big)_{\hat\gamma}\Big) \cong A_\gamma \quad .$$

Hence, condition 18.8.1 implies condition 18.8.2, and isomorphism 18.9.4 determines a surjective $\mathcal{O}P$-interior algebra homomorphism

18.9.5 $$\mathrm{C}_\circ\Big(\mathrm{Ind}_\sigma\big(S^{\cdot\cdot}_{\gamma^{\cdot\cdot}} \otimes_{\mathcal{O}} \mathrm{Res}_{\sigma'}(A'_{\gamma'})\big)_{\hat\gamma}\Big) \longrightarrow A_\gamma \quad ;$$

then, condition 18.8.2 implies moreover that this homomorphism has an $\mathcal{O}(P \times P)$-module section (cf. Proposition 10.13) and therefore the induced homomorphism

18.9.6 $$\mathrm{Ind}_P^G\Big(\mathrm{C}_\circ\Big(\mathrm{Ind}_\sigma\big(S^{\cdot\cdot}_{\gamma^{\cdot\cdot}} \otimes_{\mathcal{O}} \mathrm{Res}_{\sigma'}(A'_{\gamma'})\big)\Big)_{\hat\gamma}\Big) \longrightarrow \mathrm{Ind}_P^G(A_\gamma)$$

admits an $\mathcal{O}(G \times G)$-module section, so that we have $N_G(P_{\hat\gamma}) = N_G(P_\gamma)$ (cf. 2.11.5) and homomorphism 18.9.6 induces a k^*-group isomorphism (cf. 2.11.5)

18.9.7 $$\hat\theta : \hat{\bar N}_G(P_{\hat\gamma}) \cong \hat{\bar N}_G(P_\gamma) \quad .$$

Proof: Assume first that $M^{\cdot\cdot}$ defines a Rickard equivalence between b and b'. Since $P_{\hat\gamma}$ is a defect pointed group of $G_{\alpha^{\cdot\cdot}}$ which is noncontractile (cf. 18.2), $P_{\hat\gamma}$ is noncontractile too (cf. 14.5.4) and then it follows from Proposition 18.4 that there is a point γ of P on A corresponding to $\hat\gamma$ and that P_γ is a maximal local pointed group on A, so that P is a defect group of b. Since this situation and our hypotheses are symmetric on b and b', it is also true that P' is defect group of b'.

In particular, setting $\bar C_G(P) = C_G(P)/Z(P)$, $\bar C_{G'}(P') = C_{G'}(P')/Z(P')$ and $\bar C_{G\times G'}(P^{\cdot\cdot}) = C_{G\times G'}(P^{\cdot\cdot})/Z(P^{\cdot\cdot})$, and denoting by $\overline{\mathrm{Br}}_P : (\mathcal{O}G)^P \to k\bar C_G(P)$, $\overline{\mathrm{Br}}_{P'} : (\mathcal{O}G')^{P'} \to k\bar C_{G'}(P')$ and

18.8.4 $$\omega : k\bar C_{G\times G'}(P^{\cdot\cdot}) \longrightarrow k\bar C_G(P)\overline{\mathrm{Br}}_P(b) \otimes_k k\bar C_{G'}(P')\overline{\mathrm{Br}}_{P'}(b')^\circ$$

the corresponding canonical $\mathcal{O}$-algebra homomorphisms, $k\bar C_G(P)\overline{\mathrm{Br}}_P(b)$ and $k\bar C_{G'}(P')\overline{\mathrm{Br}}_{P'}(b')^\circ$ are semisimple k-algebras and

18.8.5 $$\bar V = \mathrm{Ind}_\omega\Big(\mathrm{Res}^{\hat{\bar N}_{G\times G'}(P^{\cdot\cdot}_{\gamma^{\cdot\cdot}})}_{\bar C_{G\times G'}(P^{\cdot\cdot})}\big(V_{\mathrm{C}_\circ(A^{\cdot\cdot})_{\alpha^{\cdot\cdot}}}(P^{\cdot\cdot}_{\gamma^{\cdot\cdot}})\big)\Big)$$

is the maximal semisimple quotient of the restriction to $k\bar C_{G\times G'}(P^{\cdot\cdot})$ of the multiplicity $k_*\hat{\bar N}_{G\times G'}(P^{\cdot\cdot}_{\gamma^{\cdot\cdot}})$-module $V_{\mathrm{C}_\circ(A^{\cdot\cdot})_{\alpha^{\cdot\cdot}}}(P^{\cdot\cdot}_{\gamma^{\cdot\cdot}})$ (cf. 2.9). Hence, since

$$Z(P) \times Z(P') = Z(P^{\cdot\cdot})\big(1 \times Z(P')\big),$$

choosing $i \in \gamma$ and $\hat{\imath} \in \hat{\gamma}$ such that $\hat{\imath}{\cdot}i = \hat{\imath} = i{\cdot}\hat{\imath}$, and still denoting by ω the corresponding group homomorphism from $\bar{C}_{G\times G'}(P^{\cdot\cdot})$ onto $\bar{C}_G(P)\mathrm{Br}_P(b) \otimes \bar{C}_{G'}(P')\overline{\mathrm{Br}}_{P'}(b')$ and by

$$\bar{s}_{\gamma^{\cdot\cdot}} : \mathrm{C}_\circ(A^{\cdot\cdot}_{\alpha^{\cdot\cdot}})^{P^{\cdot\cdot}} \longrightarrow \mathrm{Ind}_\omega\Big(\big(\mathrm{C}_\circ(A^{\cdot\cdot}_{\alpha^{\cdot\cdot}})\big)(P^{\cdot\cdot}_{\gamma^{\cdot\cdot}})\Big)$$

the canonical $\mathcal{O}$-algebra homomorphism, $\big(\bar{s}_{\gamma^{\cdot\cdot}}(\hat{\imath})\big)(\bar{V})$ (cf. Remark 3.9) is the maximal semisimple quotient of the restriction to $kC_{G'}(P')$ of the multiplicity $k_*\hat{\bar{N}}_{P\times G'}(P^{\cdot\cdot}_{\gamma^{\cdot\cdot}})$-module $V_{\mathrm{C}_\circ(A^{\cdot\cdot})_{\hat{\gamma}}}(P^{\cdot\cdot}_{\gamma^{\cdot\cdot}})$ and therefore, according to Theorem 16.9, for any $i' \in \gamma'$, we have

$$\big(\overline{\mathrm{Br}}_P(b) \otimes \overline{\mathrm{Br}}_{P'}(i')^\circ\big){\cdot}\big(\bar{s}_{\gamma^{\cdot\cdot}}(\hat{\imath})\big)(\bar{V}) \neq \{0\} \quad , \tag{18.8.6}$$

where $\overline{\mathrm{Br}}_{P'}(i')^\circ$ is the image of $\overline{\mathrm{Br}}_{P'}(i')$ by the antipodal isomorphism $k\bar{C}_{G'}(P') \cong \big(k\bar{C}_{G'}(P')\big)^\circ$ (notice that the canonical image of $(A')^\circ$ in $A^{\cdot\cdot}$ centralizes $(A^{\cdot\cdot})^{P\times G'}$).

A fortiori we get

$$\big(\overline{\mathrm{Br}}_P(i) \otimes \overline{\mathrm{Br}}_{P'}(i')^\circ\big){\cdot}\bar{V} \neq \{0\} \tag{18.8.7}$$

and then, considering a pairwise orthogonal primitive idempotent decomposition $\hat{I}'$ of the canonical image of i' in $\mathrm{C}_\circ(\hat{A}'_{\alpha^{\cdot\cdot}})^{P'} = \mathrm{C}_\circ\big((A^{\cdot\cdot}_{\alpha^{\cdot\cdot}})^\circ\big)^{G\times P'}$, there is $\hat{\imath}' \in \hat{I}'$ such that we still have

$$\big(\overline{\mathrm{Br}}_P(i)^\circ \otimes \overline{\mathrm{Br}}_{P'}(b')\big){\cdot}\big(\bar{s}_{\gamma^{\cdot\cdot}}(\hat{\imath}')\big)(\bar{V}^*) \neq \{0\} \quad ; \tag{18.8.8}$$

in particular, denoting by $\hat{\gamma}'$ the point of P' on $\hat{A}'$ which contains $\hat{\imath}'$, we have $P^{\cdot\cdot}_{\gamma^{\cdot\cdot}} \subset (G\times P')_{\hat{\gamma}'} \subset (G\times G')_{\alpha^{\cdot\cdot}}$ since $\bar{s}_{\gamma^{\cdot\cdot}}(\hat{\imath}') \neq 0$, and therefore $\hat{\gamma}'$ is noncontractile (cf. 14.3.5), so that $P'_{\hat{\gamma}'}$ is the pointed group on $\hat{A}'$ corresponding to $P'_{\gamma'}$ by Proposition 18.4, which forces $P'_{\hat{\gamma}'}$ to be local. Moreover, since $P^{\cdot\cdot}_{\gamma^{\cdot\cdot}}$ is a defect pointed group of $(G\times G')_{\alpha^{\cdot\cdot}}$ and is contained in $(G\times P')_{\hat{\gamma}'}$, $P^{\cdot\cdot}_{\gamma^{\cdot\cdot}}$ is a defect pointed group of $(G\times P')_{\hat{\gamma}'}$ too and, *mutatis mutandis*, $\big(\bar{s}_{\gamma^{\cdot\cdot}}(\hat{\imath}')\big)(\bar{V}^*)$ is the maximal semisimple quotient of the restriction to $kC_G(P)$ of the multiplicity $k_*\hat{\bar{N}}_{G\times P'}(P^{\cdot\cdot}_{\gamma^{\cdot\cdot}})$-module $V_{\mathrm{C}_\circ(A^{\cdot\cdot})^\circ_{\hat{\gamma}'}}(P^{\cdot\cdot}_{\gamma^{\cdot\cdot}})$, so that inequality 18.8.8 guarantees the existence of a surjective $kC_G(P)$-module homomorphism

$$V_{\mathrm{C}_\circ(A^{\cdot\cdot})^\circ_{\hat{\gamma}'}}(P^{\cdot\cdot}_{\gamma^{\cdot\cdot}}) \longrightarrow V_{\mathcal{O}G}(P_\gamma)^* \quad . \tag{18.8.9}$$

In conclusion, by Theorem 16.9 again, $(P'_{\hat{\gamma}'}, P^{\cdot\cdot}_{\gamma^{\cdot\cdot}}, P_\gamma)$ is a local tracing triple on $\hat{A}'$, $(A^{\cdot\cdot})^\circ$ and $\mathcal{O}G$.

On the other hand, denoting by $\mathrm{st}(i)$ the canonical image of i in $\mathrm{C}_\circ(\hat{A}_{\alpha^{\cdot\cdot}})^P$ and setting $\hat{A}_\gamma = \mathrm{st}(i)\hat{A}\mathrm{st}(i)$ considered as a $\mathcal{D}P$-interior algebra, by Proposition 18.4 the structural map $A \longrightarrow \hat{A}_{\alpha^{\cdot\cdot}}$ induces a P-interior $\mathcal{D}$-algebra homotopy isomorphism

18.8.10 $$\mathrm{st}_\gamma : A_\gamma \longrightarrow \hat{A}_\gamma$$

and we have a canonical $\mathcal{D}P$-interior algebra exoembedding

18.8.11 $$\tilde{f}_{\hat{\gamma}}^{\gamma} : \hat{A}_{\hat{\gamma}} \longrightarrow \hat{A}_\gamma \quad .$$

Set $\hat{l} = \mathrm{st}(i) - \hat{\imath}$ (see 18.7) and denote by n the maximal natural number (zero included) in the set of multiplicities $m_{\hat{\varepsilon}}(\mathrm{C}_\circ(\hat{l}\hat{A}\hat{l}))$ (cf. 2.8) when $\hat{\varepsilon}$ runs over the set of points of P on $\hat{l}\hat{A}\hat{l}$; then, we claim that there is a $\mathcal{D}P$-interior algebra exoembedding

18.8.12 $$\tilde{e} : \hat{A}_\gamma \longrightarrow \mathrm{Ind}_{\Omega_n}^P(\mathrm{Res}_{\Omega_n}^P(\hat{A}_{\hat{\gamma}}))$$

such that the following diagram of $\mathcal{D}P$-interior algebra exoembeddings, obtained from 18.8.11 and 12.16.8, is commutative

18.8.13 $$\begin{array}{ccc} \hat{A}_\gamma & \longrightarrow & \mathrm{Ind}_{\Omega_n}^P(\mathrm{Res}_{\Omega_n}^P(\hat{A}_\gamma)) \\ {\scriptstyle \tilde{f}_{\hat{\gamma}}^{\gamma}}\uparrow & \searrow^{\tilde{e}} & \uparrow{\scriptstyle \mathrm{Ind}_{\Omega_n}^P(\mathrm{Res}_{\Omega_n}^P(\tilde{f}_{\hat{\gamma}}^{\gamma}))} \\ \hat{A}_{\hat{\gamma}} & \longrightarrow & \mathrm{Ind}_{\Omega_n}^P(\mathrm{Res}_{\Omega_n}^P(\hat{A}_{\hat{\gamma}})) \end{array} \quad .$$

Indeed, we know already that any point $\hat{\varepsilon}$ of P on $\hat{A}_\gamma$ different from $\hat{\gamma}$ is contractile (cf. Proposition 18.4) and in particular, since $P_{\hat{\varepsilon}} \subset G_{\alpha^{\cdot\cdot}}$, if $Q_{\hat{\delta}}$ is a defect pointed group of $P_{\hat{\varepsilon}}$ then $Q \neq P$ (cf. 2.9.5); once again since $P_{\hat{\varepsilon}}$ is contained in $G_{\alpha^{\cdot\cdot}}$, there is $x \in G$ such that $(Q_{\hat{\delta}})^x \subset P_{\hat{\gamma}}$, so that we have a canonical $\mathcal{D}P$-interior algebra exoembedding (cf. 2.13.2, 14.3.3 and Corollary 14.11)

18.8.14 $$\hat{A}_{\hat{\varepsilon}} \longrightarrow \mathrm{Ind}_Q^P(\mathrm{Res}_\varphi(\hat{A}_{\hat{\gamma}})) \quad ,$$

where $\varphi : Q \to P$ maps $u \in Q$ on u^x, and we claim that, identifying $\hat{\varepsilon}$ to the corresponding point of P on $\mathrm{Ind}_{\Omega_n}^P(\mathrm{Res}_{\Omega_n}^P(\hat{A}_\gamma))$ throughout the canonical $\mathcal{O}P$-interior algebra exoembedding (cf. 12.16.2)

18.8.15 $$\tilde{f}_{(P,\mathrm{id})}^{\gamma} : \hat{A}_\gamma \longrightarrow \mathrm{Ind}_{\Omega_n}^P(\mathrm{Res}_{\Omega_n}^P(\hat{A}_\gamma)) \quad ,$$

this point comes from the point of P on $\mathrm{Ind}_{\Omega_n}^P(\mathrm{Res}_{\Omega_n}^P(\hat{A}_{\hat{\gamma}}))$ determined by exoembedding 18.8.14 and the canonical $\mathcal{O}P$-interior algebra exoembedding (cf. 12.16)

18.8.16 $$\tilde{f}_{(Q,\varphi)}^{\hat{\gamma}} : \mathrm{Ind}_Q^P(\mathrm{Res}_\varphi(\hat{A}_{\hat{\gamma}})) \longrightarrow \mathrm{Ind}_{\Omega_n}^P(\mathrm{Res}_{\Omega_n}^P(\hat{A}_{\hat{\gamma}})) \quad .$$

It suffices to prove that

18.8.17 $$\mathrm{Ind}_{\Omega_n}^P(\mathrm{Res}_{\Omega_n}^P(\tilde{f}_{\hat{\gamma}}^{\gamma})) \circ \tilde{f}_{(Q,\varphi)}^{\hat{\gamma}} \circ \mathrm{Ind}_Q^P(\tilde{f}_\varphi) \circ \tilde{h}_{\hat{\varepsilon}}^{\hat{\delta}} = \tilde{f}_{(P,\mathrm{id})}^{\gamma} \circ \tilde{f}_{\hat{\varepsilon}}^{\gamma} \quad ,$$

where $\tilde{h}_{\hat{\varepsilon}}^{\hat{\delta}}$ is the corresponding Higman exoembedding (cf. Corollary 14.11), $\tilde{f}_\varphi$ comes from the A-fusion $\tilde{\varphi}$ (cf. 14.3.3) and $\tilde{f}_{\hat{\varepsilon}}^{\gamma}$ factorizes $\tilde{f}_{\hat{\varepsilon}}$ (cf. 14.2.1); but, by

3.5.4, 12.16.3, 12.16.8, 14.3.4 and Corollary 14.11, we get

18.8.18
$$\begin{aligned}&\mathrm{Res}_1^P\Big(\mathrm{Ind}_{\Omega_n}^P(\mathrm{Res}_{\Omega_n}^P(\tilde f_{\hat\gamma}^{\gamma}))\circ \tilde f_{(Q,\varphi)}^{\hat\gamma}\circ \mathrm{Ind}_Q^P(\tilde f_\varphi)\circ \tilde h_{\hat\varepsilon}^{\hat\delta}\Big)\circ \mathrm{Res}_1^Q(\tilde f_{\hat\delta}^{\hat\varepsilon})\\ &= \mathrm{Res}_1^P\Big(\tilde f_{(Q,\varphi)}^{\gamma}\circ \mathrm{Ind}_Q^P(\mathrm{Res}_\varphi(\tilde f_{\hat\gamma}^{\gamma})\circ \tilde f_\varphi)\Big)\circ \mathrm{Res}_1^Q(\tilde d_Q^P(\hat A_{\hat\delta}))\\ &= \mathrm{Res}_1^P(\tilde f_{(Q,\varphi)}^{\gamma})\circ \mathrm{Res}_1^Q\Big(\tilde d_Q^P(\mathrm{Res}_\varphi(\hat A_\gamma))\Big)\circ \mathrm{Res}_1^P(\tilde f_{\hat\gamma}^{\gamma})\circ \mathrm{Res}_1^Q(\tilde f_\varphi)\\ &= \mathrm{Res}_1^P(\tilde f_{(P,\mathrm{id})}^{\gamma})\circ \mathrm{Res}_1^Q(\tilde f_{\hat\delta}^{\gamma}) = \mathrm{Res}_1^P(\tilde f_{(P,\mathrm{id})}^{\gamma}\circ \tilde f_{\hat\varepsilon}^{\gamma})\circ \mathrm{Res}_1^Q(\tilde f_{\hat\delta}^{\hat\varepsilon})\end{aligned}$$

and now our claim follows from Propositions 11.9 and 11.10. Consequently, according to our choice of n, we can find an idempotent $\hat k$ in $\mathsf{C}_\circ\Big(\mathrm{Ind}_{\Omega_n}^P(\mathrm{Res}_{\Omega_n}^P(\hat A_{\hat\gamma}))\Big)^P$ having the same multiplicity than $\hat l$ at each point of P on $\mathrm{Ind}_{\Omega_n}^P(\mathrm{Res}_{\Omega_n}^P(\hat A_\gamma))$ and being orthogonal to the image of $\hat\imath$ (cf. [19, Corollary 2.4]); then, there is $\hat c\in\bigg(\mathsf{C}_\circ\Big(\mathrm{Ind}_{\Omega_n}^P(\mathrm{Res}_{\Omega_n}^P(\hat A_\gamma))\Big)^P\bigg)^*$ such that $\hat\imath^{\hat c}=\hat\imath$ and $\hat l^{\hat c}=\hat k$ (cf. 2.8.4); now, identifying $\hat A_\gamma$ to its image in $\mathrm{Ind}_{\Omega_n}^P(\mathrm{Res}_{\Omega_n}^P(\hat A_\gamma))$ by $f_{(P,\mathrm{id})}^{\gamma}$ (cf. 12.16.2) and $\hat A_{\hat\gamma}$ to $\hat\imath\hat A\hat\imath=\hat\imath(\hat A_\gamma)\hat\imath$, it suffices to set $e(\hat a)=\hat a^{\hat c}$ for any $\hat a\in\hat A_\gamma$.

Now, since $\tilde h_{\hat\gamma}^{\gamma^{\cdot\cdot},\gamma'}$ determines a $\mathcal{D}P$-interior algebra exoembedding

18.8.19
$$\hat A_{\hat\gamma}\cong \mathrm{Ind}_\sigma\big(S_{\gamma^{\cdot\cdot}}^{\cdot\cdot}\underset{\mathcal{O}}{\otimes}\mathrm{Res}_{\sigma'}(A'_{\gamma'})\big)_{\hat\gamma}\qquad,$$

the composition of the P-interior $\mathcal{D}$-algebra homotopic isomorphism $\mathrm{st}_\gamma : A_\gamma\to\hat A_\gamma$ and the embedding e above is a P-interior $\mathcal{D}$-algebra homotopy embedding as announced in condition 18.8.1; then, by Remark 18.9, condition 18.8.2 is also fulfilled, both conditions inducing the same k^*-group isomorphism

18.8.20
$$\hat\theta : \hat{\bar N}_G(P_{\hat\gamma})\cong \hat{\bar N}_G(P_\gamma)\qquad;$$

moreover, isomorphism 18.4.1 determines a surjective $\mathcal{O}G$-interior algebra homomorphism

18.8.21
$$\mathsf{C}_\circ(\hat A_{\alpha^{\cdot\cdot}})\longrightarrow A$$

which induces a $k_*\hat{\bar N}_G(P_{\hat\gamma})$-interior algebra isomorphism (cf. 2.11.5 and 10.13.2)

18.8.22
$$\big(\mathsf{C}_\circ(\hat A_{\alpha^{\cdot\cdot}})\big)(P_{\hat\gamma})\cong \mathrm{Res}_{\hat\theta}\big(A(P_\gamma)\big)$$

and this isomorphism together with isomorphism 18.6.4 give isomorphism 18.8.3.

Conversely, by Remark 18.9, we may assume that condition 18.8.2 holds; then, we have an $\mathcal{O}G$-interior algebra isomorphism (cf. 12.4.4)

18.8.23
$$\mathrm{Ind}_P^G(A_\gamma)\cong \mathsf{H}_\circ\Big(\mathrm{Ind}_P^G\Big(\mathrm{Ind}_\sigma\big(S_{\gamma^{\cdot\cdot}}^{\cdot\cdot}\underset{\mathcal{O}}{\otimes}\mathrm{Res}_{\sigma'}(A'_{\gamma'})\big)_{\hat\gamma}\Big)\Big)\quad,$$

so that we have a surjective $\mathcal{O}G$-interior algebra homomorphism

18.8.24 $$\mathrm{C}_\circ\Big(\mathrm{Ind}_P^G\big(\mathrm{Ind}_\sigma(S_{\gamma^{\cdot\cdot}}^{\cdot\cdot}\otimes_{\mathcal{O}}\mathrm{Res}_{\sigma'}(A'_{\gamma'}))_{\hat{\gamma}}\big)\Big)\longrightarrow \mathrm{Ind}_P^G(A_\gamma)$$

having the kernel equal to $\mathrm{B}_\circ\Big(\mathrm{Ind}_P^G\big(\mathrm{Ind}_\sigma(S_{\gamma^{\cdot\cdot}}^{\cdot\cdot}\otimes_{\mathcal{O}}\mathrm{Res}_{\sigma'}(A'_{\gamma'}))_{\hat{\gamma}}\big)\Big)$ and inducing the same k^*-group isomorphism

18.8.25 $$\hat{\theta}:\hat{\bar{N}}_G(P_{\hat{\gamma}})\cong\hat{\bar{N}}_G(P_\gamma)\quad ,$$

and it is quite clear that $\mathrm{Ind}_P^G\big(\mathrm{Ind}_\sigma(S_{\gamma^{\cdot\cdot}}^{\cdot\cdot}\otimes_{\mathcal{O}}\mathrm{Res}_{\sigma'}(A'_{\gamma'}))_{\hat{\gamma}}\big)$ is an 0-split $\mathcal{D}(G\times G)$-module, so that homomorphism 18.8.24 has an $\mathcal{O}(G\times G)$-module section (cf. Proposition 10.13). But, since P is a defect group of b, A_γ is a source $\mathcal{O}P$-interior algebra of the $\mathcal{O}G$-interior algebra A and therefore we have the Higman exoembedding of $\mathcal{O}G$-interior algebras (cf. Corollary 14.11)

18.8.26 $$\tilde{h}_\alpha^\gamma : A\longrightarrow \mathrm{Ind}_P^G(A_\gamma)$$

which determines a point α of G on $\mathrm{Ind}_P^G(A_\gamma)$; similarly, since $P_{\hat{\gamma}}$ is a defect pointed group of $G_{\alpha^{\cdot\cdot}}$ on $\hat{A}$, we have the Higman exoembedding of $\mathcal{D}G$-interior algebras (cf. Corollary 14.11)

18.8.27 $$\tilde{h}_{\alpha^{\cdot\cdot}}^{\hat{\gamma}} : \hat{A}_{\alpha^{\cdot\cdot}}\longrightarrow \mathrm{Ind}_P^G(\hat{A}_{\hat{\gamma}})$$

which determines a point $\alpha^{\cdot\cdot}$ of G on $\mathrm{Ind}_P^G\big(\mathrm{Ind}_\sigma(S_{\gamma^{\cdot\cdot}}^{\cdot\cdot}\otimes_{\mathcal{O}}\mathrm{Res}_{\sigma'}(A'_{\gamma'}))_{\hat{\gamma}}\big)$ and forces $\hat{A}_{\alpha^{\cdot\cdot}}$ to be an 0-split $\mathcal{D}(G\times G)$-module.

The point is that, since isomorphism 18.8.23 induces the k^*-group isomorphism $\hat{\theta}:\hat{\bar{N}}_G(P_{\hat{\gamma}})\cong\hat{\bar{N}}_G(P_\gamma)$, isomorphisms 18.6.4 and 18.8.3 induce a $k_*\hat{\bar{N}}_G(P_\gamma)$-module isomorphism

18.8.28 $$V_A(P_\gamma)\cong\mathrm{Res}_{\hat{\theta}^{-1}}\big(V_{\mathrm{C}_\circ(\hat{A}_{\alpha^{\cdot\cdot}})}(P_{\hat{\gamma}})\big)$$

and therefore, since homomorphism 18.8.24 admits an $\mathcal{O}(G\times G)$-module section, this homomorphism maps the point $\alpha^{\cdot\cdot}$ of G on $\mathrm{C}_\circ\Big(\mathrm{Ind}_P^G\big(\mathrm{Ind}_\sigma(S_{\gamma^{\cdot\cdot}}^{\cdot\cdot}\otimes_{\mathcal{O}}\mathrm{Res}_{\sigma'}(A'_{\gamma'}))_{\hat{\gamma}}\big)\Big)$ onto the point α of G on $\mathrm{Ind}_P^G(A_\gamma)$ (cf. 2.9.3 and 2.11.5). Consequently, isomorphism 18.8.23 induces an $\mathcal{O}G$-interior algebra isomorphism

18.8.29 $$A\cong\mathrm{H}_\circ(\hat{A}_{\alpha^{\cdot\cdot}})$$

and therefore, since $\hat{A}_{\alpha^{\cdot\cdot}}$ is an 0-split $\mathcal{D}(G\times G)$-module, the structural map $A\to\hat{A}_{\alpha^{\cdot\cdot}}$ is a G-interior $\mathcal{D}$-algebra homotopy isomorphism; since $A_{\alpha^{\cdot\cdot}}^{\cdot\cdot}\cong\mathrm{End}_{\mathcal{O}}(M^{\cdot\cdot})$, it suffices to apply now Proposition 18.4 to conclude.

18.10 Now, we discuss explicitly how Theorem 18.8 provides a description of all the choices of $M^{\cdot\cdot}$ defining a Rickard equivalence between b and b'. We fix defect pointed groups P_γ of b and $P'_{\gamma'}$ of b', and we consider all the subgroups $Q^{\cdot\cdot}$ of $P \times P'$ surjectively mapped onto P and P' by the projection maps, noted $\tau : Q^{\cdot\cdot} \to P$ and $\tau' : Q^{\cdot\cdot} \to P'$; for such a subgroup $Q^{\cdot\cdot}$, we look for all the indecomposable $\mathcal{O}Q^{\cdot\cdot}$-modules $L^{\cdot\cdot}$ of vertex $Q^{\cdot\cdot}$ such that their restrictions to $\mathcal{O}\mathrm{Ker}(\tau)$ and to $\mathcal{O}\mathrm{Ker}(\tau')$ are both projective and, for some $n \geq 0$, there is a P-interior $\mathcal{O}$-algebra homotopy embedding

18.10.1 $$f : A_\gamma \to \mathrm{Ind}^P_{\mathfrak{Q}_n}\Big(\mathrm{Res}^P_{\mathfrak{Q}_n}\big(\mathrm{Ind}_\tau(\mathrm{End}_\mathcal{O}(L^{\cdot\cdot}) \underset{\mathcal{O}}{\otimes} \mathrm{Res}_{\tau'}(A'_{\gamma'}))\big)\Big) \quad ,$$

where $\mathfrak{Q}_n$ is the family formed by the pair (P, id_P) and, for any local pointed group R_ε on A such that $R_\varepsilon \subsetneqq P_\gamma$ and any group homomorphism $\varphi : R \to P$ such that $\tilde{\varphi} \in E_G(R_\varepsilon, P_\gamma)$, by the pair (R, φ) repeated n times. Then, we claim that:

18.10.2 *The triple $(Q^{\cdot\cdot}, L^{\cdot\cdot}, f)$ determines one choice for $M^{\cdot\cdot}$ defining a Rickard equivalence between b and b'.*

18.11 Indeed, arguing as in Remark 18.9, it is not difficult to see that f determines a unique noncontractile local point $\hat{\gamma}$ of P on $\mathrm{Ind}_\tau\big(\mathrm{End}_\mathcal{O}(L^{\cdot\cdot}) \otimes_\mathcal{O} \mathrm{Res}_{\tau'}(A'_{\gamma'})\big)$ such that f induces an $\mathcal{O}P$-interior algebra isomorphism

18.11.1 $$A_\gamma \cong \mathrm{H}_\circ\Big(\mathrm{Ind}_\tau\big(\mathrm{End}_\mathcal{O}(L^{\cdot\cdot}) \underset{\mathcal{O}}{\otimes} \mathrm{Res}_{\tau'}(A'_{\gamma'})\big)_{\hat{\gamma}}\Big)$$

and $\mathrm{Ind}_\tau\big(\mathrm{End}_\mathcal{O}(L^{\cdot\cdot}) \otimes_\mathcal{O} \mathrm{Res}_{\tau'}(A'_{\gamma'})\big)_{\hat{\gamma}}$ is an 0-split $\mathcal{O}(P \times P)$-module; in particular, the surjective $\mathcal{O}P$-interior algebra homomorphism

18.11.2 $$\mathrm{C}_\circ\Big(\mathrm{Ind}_\tau\big(\mathrm{End}_\mathcal{O}(L^{\cdot\cdot}) \underset{\mathcal{O}}{\otimes} \mathrm{Res}_{\tau'}(A'_{\gamma'})\big)_{\hat{\gamma}}\Big) \longrightarrow A_\gamma$$

determined by isomorphism 18.11.1 admits an $\mathcal{O}(P \times P)$-module section (cf. Proposition 10.13) and therefore induces a bijection between the set of noncontractile local pointed groups $R_{\hat{\varepsilon}}$ on $\mathrm{C}_\circ\Big(\mathrm{Ind}_\tau\big(\mathrm{End}_\mathcal{O}(L^{\cdot\cdot}) \otimes_\mathcal{O} \mathrm{Res}_{\tau'}(A'_{\gamma'})\big)_{\hat{\gamma}}\Big)$ and the set of local pointed groups R_ε on A_γ, which preserves k^*-fusions (cf. 2.11.5 and 2.13.7), so that f induces a k^*-group isomorphism

18.11.3 $$\hat{F}_A(R_\varepsilon) \cong \hat{F}_{\mathrm{C}_\circ(\mathrm{Ind}_\tau(\mathrm{End}_\mathcal{O}(L^{\cdot\cdot})\otimes_\mathcal{O}\mathrm{Res}_{\tau'}(A'_{\gamma'})))}(R_{\hat{\varepsilon}}) \quad ;$$

18.12 On the other hand, the canonical exoembedding $\tilde{f}_{\gamma'} : A'_{\gamma'} \to \mathrm{Res}^{G'}_{P'}(\mathcal{O}G')$ together with the Hecke $\mathcal{O}P$-interior algebra isomorphism (cf. Theorem 15.4)

18.12.1 $$\mathrm{Ind}^{P\times G'}_{Q^{\cdot\cdot}}(\mathrm{End}_\mathcal{O}(L^{\cdot\cdot}))^{1\times G'} \cong \mathrm{Ind}_\tau\Big(\mathrm{End}_\mathcal{O}(L^{\cdot\cdot}) \underset{\mathcal{O}}{\otimes} \mathrm{Res}_{\tau'}\big(\mathrm{Res}^{G'}_{P'}(\mathcal{O}G')\big)\Big)$$

and with the canonical embedding (cf. 2.6.6)

18.12.2 $$\mathrm{Ind}_{Q^{\cdot\cdot}}^{P\times G'}(\mathrm{End}_{\mathcal{O}}(L^{\cdot\cdot})) \to \mathrm{Res}_{P\times G'}^{G\times G'}\Big(\mathrm{Ind}_{Q^{\cdot\cdot}}^{G\times G'}(\mathrm{End}_{\mathcal{O}}(L^{\cdot\cdot}))\Big) \quad ,$$

determine a canonical exoembedding (cf. 2.6.5)

$$\mathrm{Ind}_{\tau}(\mathrm{End}_{\mathcal{O}}(L^{\cdot}) \underset{\mathcal{O}}{\otimes} \mathrm{Res}_{\tau'}(A'_{\gamma'})) \longrightarrow \mathrm{Res}_P^G\Big(\mathrm{End}_{\mathcal{O}G'}(\mathrm{Ind}_{Q^{\cdot\cdot}}^{G\times G'}(L^{\cdot\cdot}))\Big)$$

which allows us to identify $\hat{\gamma}$ to a noncontractile local point of P on $\mathrm{End}_{\mathcal{O}G'}(\mathrm{Ind}_{Q^{\cdot\cdot}}^{G\times G'}(L^{\cdot\cdot}))$ and, equivalently, to a noncontractile point of $P \times G'$ on $\mathrm{End}_{\mathcal{O}}(\mathrm{Ind}_{Q^{\cdot\cdot}}^{G\times G'}(L^{\cdot\cdot}))$; then, as in 18.6 and Remark 18.9, we have

18.12.3 $$\bar{N}_G(P_\gamma) = \bar{N}_G(P_{\hat{\gamma}}) \cong \bar{N}_{G\times G'}\big((P\times G')_{\hat{\gamma}}\big)$$

and, by 18.11.3, we get k^*-group isomorphisms

18.12.4 $$\hat{\theta} : \hat{\bar{N}}_G(P_{\hat{\gamma}}) \cong \hat{\bar{N}}_G(P_\gamma) \quad \text{and} \quad \hat{\upsilon} : \hat{\bar{N}}_G(P_{\hat{\gamma}}) \cong \hat{\bar{N}}_{G\times G'}\big((P\times G')_{\hat{\gamma}}\big) \quad .$$

Moreover notice that, denoting by $\delta^{\cdot\cdot}$ the noncontractile local point of $Q^{\cdot\cdot}$ on $\mathrm{End}_{\mathcal{O}}(\mathrm{Ind}_{Q^{\cdot\cdot}}^{G\times G'}(L^{\cdot\cdot}))$ determined by $L^{\cdot\cdot}$, it follows from Theorem 5.8 that

18.12.5 $$Q^{\cdot\cdot}_{\delta^{\cdot\cdot}} \subset (P\times G')_{\hat{\gamma}} \quad .$$

18.13 Now, since $N_{G\times G'}(Q^{\cdot\cdot}_{\delta^{\cdot\cdot}})$ normalizes $P \times G'$, and $Q^{\cdot\cdot}_{\delta^{\cdot\cdot}}$ is a maximal local pointed group on $\mathrm{End}_{\mathcal{O}}(\mathrm{Ind}_{Q^{\cdot\cdot}}^{G\times G'}(L^{\cdot\cdot}))$ or, equivalently, on the $\mathcal{O}(G\times G')$-interior algebra $\mathrm{C}_\circ\Big(\mathrm{End}_{\mathcal{O}}(\mathrm{Ind}_{Q^{\cdot\cdot}}^{G\times G'}(L^{\cdot\cdot}))\Big)$ (cf. 2.11.3, Proposition 3.7, 12.3 and 12.4.4), it follows from Theorem 5.1 in [2] that, for any point $\beta^{\cdot\cdot}$ of $G\times G'$ on this $\mathcal{O}(G\times G')$-interior algebra such that $(P\times G')_{\hat{\gamma}} \subset (G\times G')_{\beta^{\cdot\cdot}}$, the multiplicity module of $(P\times G')_{\hat{\gamma}}$ on $\mathrm{C}_\circ\Big(\mathrm{End}_{\mathcal{O}}(\mathrm{Ind}_{Q^{\cdot\cdot}}^{G\times G'}(L^{\cdot\cdot}))\Big)$ is an indecomposable projective $k_*\hat{\bar{N}}_{G\times G'}\big((P\times G')_{\hat{\gamma}}\big)$-module and that this correspondence defines a bijection between the set of such points and the set of the isomorphism classes of such modules. Finally, notice that the restriction of $\mathrm{Ind}_{Q^{\cdot\cdot}}^{G\times G'}(L^{\cdot\cdot})$ to $\mathcal{O}(G\times 1)$ and to $\mathcal{O}(1\times G')$ are both projective. Consequently, it follows from 2.9.3 and Theorem 18.8 that:

18.13.1 *The indecomposable projective* $k_*\hat{\bar{N}}_{G\times G'}\big((P\times G')_{\hat{\gamma}}\big)$*-module* $\mathrm{Res}_{\hat{\theta}\circ\hat{\upsilon}^{-1}}\big(V_A(P_\gamma)\big)$ *determines an indecomposable direct summand* $M^{\cdot\cdot}$ *of the* $\mathcal{D}(G\times G')$*-module* $\mathrm{Ind}_{Q^{\cdot\cdot}}^{G\times G'}(L^{\cdot\cdot})$ *which defines a Rickard equivalence between* b *and* b'.

19 Basic Rickard equivalences between Brauer blocks

19.1 As announced in the introduction, in this last section we consider a particular kind of Rickard equivalences between blocks which, at least conjecturally, seems to be very frequent and which includes the case of the *basic Morita equivalences* discussed in Section 7. We keep all the notation of Section 18; in particular, $M^{\cdot\cdot}$ is an indecomposable $\mathcal{D}(G \times G')$-module associated with $b \otimes (b')^{\circ}$, which has projective restrictions to $\mathcal{O}(G \times 1)$ and to $\mathcal{O}(1 \times G')$, $P^{\cdot\cdot}$ is a vertex of $M^{\cdot\cdot}$ and $N^{\cdot\cdot}$ is a $\mathcal{D}P^{\cdot\cdot}$-source of it, we set $P = \pi(P^{\cdot\cdot})$ and $P' = \pi'(P^{\cdot\cdot})$, and we respectively denote by $\sigma : P^{\cdot\cdot} \to P$ and $\sigma' : P^{\cdot\cdot} \to P'$ the group homomorphisms determined by π and π'. Assume that $M^{\cdot\cdot}$ defines a Rickard equivalence between b and b': we say that this equivalence is *basic* if the $\mathcal{D}P^{\cdot\cdot}$-interior algebra $\mathrm{End}_{\mathcal{O}}(N^{\cdot\cdot})$ is $\mathrm{Ker}(\sigma)$- and $\mathrm{Ker}(\sigma')$-basic (cf. 13.2); that is to say, since the restrictions of $N^{\cdot\cdot}$ to $\mathrm{Ker}(\sigma)$ and to $\mathrm{Ker}(\sigma')$ are both projective, this equivalence is basic if and only if, denoting by $\Delta(P^{\cdot\cdot})$ the diagonal subgroup of $P^{\cdot\cdot} \times P^{\cdot\cdot}$ and considering the action of $P^{\cdot\cdot} \times P^{\cdot\cdot}$ on the $\mathcal{O}$-module $\mathrm{End}_{\mathcal{O}}(N^{\cdot\cdot})$ by left and right multiplication, $\mathrm{End}_{\mathcal{O}}(N^{\cdot\cdot})$ has $\big(\mathrm{Ker}(\sigma) \times \mathrm{Ker}(\sigma)\big)\cdot\Delta(P^{\cdot\cdot})$- and $\big(\mathrm{Ker}(\sigma') \times \mathrm{Ker}(\sigma')\big)\cdot\Delta(P^{\cdot\cdot})$-stable $\mathcal{O}$-bases.

19.2 It is quite clear that this equivalence is basic whenever $N^{\cdot\cdot}$ is a permutation $\mathcal{O}P^{\cdot\cdot}$-module; but notice that if $M^{\cdot\cdot}$ has a trivial $\mathcal{D}$-structure and defines a basic Morita equivalence between b and b' (cf. 7.1), it defines actually a basic Rickard equivalence between them and nevertheless $N^{\cdot\cdot}$ is an endopermutation $\mathcal{O}P^{\cdot\cdot}$-module (cf. 2.2) which need not have a $P^{\cdot\cdot}$-stable $\mathcal{O}$-basis; actually, the general case is a mixed situation (see Proposition 19.4 below). If $\mathcal{O}$ has characteristic zero, since the restriction of $N^{\cdot\cdot}$ to $\mathcal{O}\mathrm{Ker}(\sigma)$ is projective, it follows from Weiss' Criterion (cf. Theorem A1.2 below) that $\mathrm{End}_{\mathcal{O}}(N^{\cdot\cdot})$ is a permutation $\mathcal{O}\Big(\big(\mathrm{Ker}(\sigma) \times \mathrm{Ker}(\sigma)\big)\cdot\Delta(P^{\cdot\cdot})\Big)$-module if and only if $\mathrm{End}_{\mathcal{O}}(N^{\cdot\cdot})^{\mathrm{Ker}(\sigma)\times\mathrm{Ker}(\sigma)}$ is a permutation $\mathcal{O}P$-module; but because $N^{\cdot\cdot}$ has a projective restriction to $\mathcal{O}\mathrm{Ker}(\sigma)$, we have an $\mathcal{O}P$-module isomorphism

19.2.1 $$\mathrm{End}_{\mathcal{O}}(N^{\cdot\cdot})^{\mathrm{Ker}(\sigma)\times\mathrm{Ker}(\sigma)} \cong \mathrm{Ind}_{\sigma}\big(\mathrm{End}_{\mathcal{O}}(N^{\cdot\cdot})\big) \quad ;$$

consequently, by symmetry, we get (cf. Proposition 3.7)

19.2.2 *If $\mathcal{O}$ has characteristic zero, the Rickard equivalence defined by $M^{\cdot\cdot}$ between b and b' is basic if and only if* $\mathrm{Ind}_{\sigma}(N^{\cdot\cdot})$ *and* $\mathrm{Ind}_{\sigma'}(N^{\cdot\cdot})$ *are respectively endopermutation $\mathcal{O}P$- and $\mathcal{O}P'$-modules.*

Notice that, by Proposition 14.7, if $M^{\cdot\cdot}$ defines a basic Rickard equivalence between b and b', the $\mathcal{D}P^{\cdot\cdot}$-interior algebra $S^{\cdot\cdot} = \mathsf{E}\big(\mathrm{End}_{\mathcal{O}}(N^{\cdot\cdot})\big)$ and therefore the $\mathcal{D}P^{\cdot\cdot}$-interior algebra $S^{\cdot\cdot} \otimes_{\mathcal{O}} \mathrm{Res}_{\sigma'}\big(\mathrm{Res}^{G'}_{P'}(\mathcal{O}G')\big)$ are $\mathrm{Ker}(\sigma)$- and $\mathrm{Ker}(\sigma')$-basic too, so that by Lemma 13.4 the $\mathcal{D}P$-interior algebra (cf. 15.4.1)

19.2.3 $$\mathrm{Ind}_{P^{\cdot\cdot}}^{G\times G'}(S^{\cdot\cdot})^{1\times G'} \cong \mathrm{Ind}_{\sigma}\Big(S^{\cdot\cdot} \underset{\mathcal{O}}{\otimes} \mathrm{Res}_{\sigma'}\big(\mathrm{Res}^{G'}_{P'}(\mathcal{O}G')\big)\Big)$$

is basic; actually, it is easily proved that $\mathrm{Ind}_{P^{\cdot\cdot}}^{P\times G'}(S^{\cdot\cdot})$ is a $(1 \times G')$-basic $\mathcal{D}(P \times G')$-interior algebra.

19.3 From now on, we assume that $M^{\cdot\cdot}$ *defines a basic Rickard equivalence between* b *and* b'. Then, since any local pointed group $Q^{\cdot\cdot}_{\delta^{\cdot\cdot}}$ on $A^{\cdot\cdot} = \mathrm{Ind}_{P^{\cdot\cdot}}^{G\times G'}(S^{\cdot\cdot})$ has a $G \times G'$-conjugate coming from $S^{\cdot\cdot}$ (cf. 2.11.3 and 12.4.4), $A^{\cdot\cdot}_{\delta^{\cdot\cdot}}$ is a $Q^{\cdot\cdot} \cap \mathrm{Ker}(\sigma)$- and $Q^{\cdot\cdot} \cap \mathrm{Ker}(\sigma')$-basic $\mathcal{D}Q^{\cdot\cdot}$-interior algebra; on the other hand, according to Proposition 18.4, the $\mathcal{D}G$-interior algebra $\hat{A}_{\alpha^{\cdot\cdot}} \cong \mathrm{End}_{\mathcal{O}(1\times G')}(M^{\cdot\cdot})$ is a 0-split $\mathcal{D}(G \times G')$-module (cf. 10.12) and therefore, for any local pointed group $Q_{\hat{\delta}}$ on $\hat{A}$ such that $Q_{\hat{\delta}} \subset G_{\alpha^{\cdot\cdot}}$, $\hat{A}_{\hat{\delta}}$ is a 0-split $\mathcal{D}(Q \times Q)$-module; consequently, any local tracing triple $(Q_{\hat{\delta}}, Q^{\cdot\cdot}_{\delta^{\cdot\cdot}}, Q'_{\delta'})$ on $\hat{A}$, $A^{\cdot\cdot}$ and $\mathcal{O}G'$ such that $Q_{\hat{\delta}} \subset G_{\alpha^{\cdot\cdot}}$ is basic (cf. 17.1), so that all the results of Section 17 apply to it, namely $(Q_{\hat{\delta}}, Q^{\cdot\cdot}_{\delta^{\cdot\cdot}}, Q'_{\delta'})$ admits a section (cf. Proposition 17.5). Moreover, recall that the structural map $A \to \hat{A}_{\alpha^{\cdot\cdot}}$ induces a bijection between the set of pointed groups on A and the set of noncontractile pointed groups on $\hat{A}_{\alpha^{\cdot\cdot}}$ (cf. Proposition 18.4), and we denote by $H_{\hat{\beta}}$ the pointed group on $\hat{A}$ corresponding to a pointed group H_β on A.

Proposition 19.4 *There are an indecomposable endopermutation $\mathcal{O}P$-module N of vertex P and an indecomposable $\mathcal{D}P^{\cdot\cdot}$-module $L^{\cdot\cdot}$ of vertex $P^{\cdot\cdot}$ having a $P^{\cdot\cdot}$-stable $\mathcal{O}$-basis where* $\mathrm{Ker}(\sigma)$ *and* $\mathrm{Ker}(\sigma')$ *act freely, both unique up to isomorphisms, such that $N^{\cdot\cdot}$ is a direct summand of* $\mathrm{Res}_\sigma(N) \otimes_{\mathcal{O}} L^{\cdot\cdot}$.

Remark 19.5 Notice that the $\mathcal{D}P^{\cdot\cdot}$-interior algebra $\mathrm{End}_{\mathcal{O}}\big(\mathrm{Res}_\sigma(N) \otimes_{\mathcal{O}} L^{\cdot\cdot}\big)$ is Ker (σ)-basic.

Proof: Since $P^{\cdot\cdot}$ stabilizes an $\mathcal{O}$-basis of $\mathrm{End}_{\mathcal{O}}(N^{\cdot\cdot})$, the restriction $N^{\cdot\cdot}_{\mathcal{O}}$ of $N^{\cdot\cdot}$ to $\mathcal{O}P^{\cdot\cdot}$ is already an endopermutation $\mathcal{O}P^{\cdot\cdot}$-module (cf. 2.2) and therefore there is a family $\{P^{\cdot\cdot}_i\}_{i\in I}$ of subgroups of $P^{\cdot\cdot}$ and, for any $i \in I$, an indecomposable endopermutation $\mathcal{O}P^{\cdot\cdot}_i$-module $N^{\cdot\cdot}_i$ of vertex $P^{\cdot\cdot}_i$ such that we have an $\mathcal{O}P^{\cdot\cdot}$-module isomorphism

19.4.1
$$N^{\cdot\cdot}_{\mathcal{O}} \cong \bigoplus_{i\in I} \mathrm{Ind}_{P^{\cdot\cdot}_i}^{P^{\cdot\cdot}}(N^{\cdot\cdot}_i) \quad .$$

But, on one hand, since we have an $\mathcal{O}P$-module isomorphism

19.4.2
$$\mathrm{End}_{\mathcal{O}}(N^{\cdot\cdot}_{\mathcal{O}})^{\mathrm{Ker}(\sigma)\times\mathrm{Ker}(\sigma')} \cong \mathrm{Ind}_\sigma\big(\mathrm{End}_{\mathcal{O}}(N^{\cdot\cdot}_{\mathcal{O}})\big)$$

and, by Proposition 3.7, an $\mathcal{O}P$-interior algebra isomorphism

19.4.3
$$\mathrm{Ind}_\sigma\big(\mathrm{End}_{\mathcal{O}}(N^{\cdot\cdot}_{\mathcal{O}})\big) \cong \mathrm{End}_{\mathcal{O}}\big(\mathrm{Ind}_\sigma(N^{\cdot\cdot}_{\mathcal{O}})\big) \quad ,$$

$\mathrm{Ind}_\sigma(N^{\cdot\cdot}_{\mathcal{O}})$ is an endopermutation $\mathcal{O}P$-module. On the other hand, since $\mathrm{Res}_{Z^{\cdot\cdot}}^{P^{\cdot\cdot}}(N^{\cdot\cdot}_{\mathcal{O}})$ is projective for any subgroup $Z^{\cdot\cdot}$ of $\mathrm{Ker}(\sigma)$ and $\big(\mathrm{End}_{\mathcal{O}}(N^{\cdot\cdot}_i)\big)(P^{\cdot\cdot}_i) \neq \{0\}$ for any $i \in I$, it follows from Lemma 7.10 that, for any $i \in I$, we have

19.4.4
$$P^{\cdot\cdot}_i \cap \mathrm{Ker}(\sigma) = 1 \quad ,$$

so that the restriction σ_i of σ to $P_i^{\cdot\cdot}$ is injective and, denoting by $\lambda_i : \sigma(P_i^{\cdot\cdot}) \cong P_i^{\cdot\cdot}$ the corresponding inverse isomorphism, we have an $\mathcal{O}P$-module isomorphism (cf. Corollary 3.13)

19.4.5
$$\mathrm{Ind}_\sigma(N_{\mathcal{O}}^{\cdot\cdot}) = \bigoplus_{i\in I} \mathrm{Ind}_{\sigma(P^{\cdot\cdot}_i)}^P\big(\mathrm{Res}_{\lambda_i}(N_i^{\cdot\cdot})\big) \quad ;$$

moreover, according to Corollary 16.10 and 19.3 above, there is a basic local tracing triple $(P_{\hat{\gamma}}, P_{\gamma^{\cdot\cdot}}^{\cdot\cdot}, P'_{\gamma'})$ on $\hat{A}$, $A^{\cdot\cdot}$ and $\mathcal{O}G'$, where $\gamma^{\cdot\cdot}$ corresponds to $N^{\cdot\cdot}$ (cf. 2.10), which admits a section λ, so that we have $\big(\mathrm{End}_{\mathcal{O}}(N^{\cdot\cdot})\big)(P^\lambda) \neq \{0\}$, where $P^\lambda = \lambda(P)$ (cf. 17.4.7), forcing the existence of $j \in I$ such that $P_j^{\cdot\cdot}$ is $P^{\cdot\cdot}$-conjugate to P^λ.

Consequently, since $\mathrm{Ind}_\sigma(N_{\mathcal{O}}^{\cdot\cdot})$ is an endopermutation $\mathcal{O}P$-module and there is $j \in I$ such that $\sigma(P_j^{\cdot\cdot}) = P$, setting $N = \mathrm{Res}_{\lambda_j}(N_j^{\cdot\cdot})$, $N_i^{\cdot\cdot}$ is a direct summand of $\mathrm{Res}_{\sigma_i}(N)$ (cf. [38, (28.9)]), so that $\mathrm{Res}_{\sigma_i}(N)^* \otimes_{\mathcal{O}} N_i^{\cdot\cdot}$ is a permutation $\mathcal{O}P_i^{\cdot\cdot}$-module for any $i \in I$ (cf. [23, 1.5]); hence, $\mathrm{Res}_\sigma(N)^* \otimes_{\mathcal{O}} N_{\mathcal{O}}^{\cdot\cdot}$ is a permutation $\mathcal{O}P^{\cdot\cdot}$-module (cf. 19.4.1) having projective restrictions to $\mathrm{Ker}(\sigma)$ and to $\mathrm{Ker}(\sigma')$. On the other hand, setting $S = \mathrm{End}_{\mathcal{O}}(N)$, since we have a canonical $\mathcal{O}$-algebra isomorphism

19.4.6
$$\mathrm{C}_\circ\big(\mathrm{Res}_\sigma(S)^\circ \otimes_{\mathcal{O}} S_{\gamma^{\cdot\cdot}}^{\cdot\cdot}\big) \cong \mathrm{Res}_\sigma(S)^\circ \otimes_{\mathcal{O}} \mathrm{C}_\circ(S_{\gamma^{\cdot\cdot}}^{\cdot\cdot}) \quad ,$$

it follows from Lemma 7.10 that we have a k-algebra isomorphism

19.4.7
$$\Big(\mathrm{C}_\circ\big(\mathrm{Res}_\sigma(S)^\circ \otimes_{\mathcal{O}} S_{\gamma^{\cdot\cdot}}^{\cdot\cdot}\big)\Big)(P^{\cdot\cdot}) \cong \big(\mathrm{C}_\circ(S_{\gamma^{\cdot\cdot}}^{\cdot\cdot})\big)(P^{\cdot\cdot})$$

and therefore the $\mathcal{D}P^{\cdot\cdot}$-module $\mathrm{Res}_\sigma(N)^* \otimes_{\mathcal{O}} N^{\cdot\cdot}$ has a unique indecomposable direct summand $L^{\cdot\cdot}$ of vertex $P^{\cdot\cdot}$, up to isomorphisms. Finally, since $\mathcal{O}$ is a direct summand of $\mathrm{End}_{\mathcal{O}}(N) = S$ as $\mathcal{O}P$-modules (cf. [23, 1.5]), $N^{\cdot\cdot}$ is a direct summand of the $\mathcal{D}P^{\cdot\cdot}$-module

19.4.8
$$\mathrm{Res}_\sigma(S) \otimes_{\mathcal{O}} N^{\cdot\cdot} \cong \mathrm{Res}_\sigma(N) \otimes_{\mathcal{O}} \big(\mathrm{Res}_\sigma(N)^* \otimes_{\mathcal{O}} N^{\cdot\cdot}\big)$$

and therefore, it is a direct summand of $\mathrm{Res}_\sigma(N) \otimes_{\mathcal{O}} L^{\cdot\cdot}$.

The uniqueness of N follows from the fact that $\mathrm{Ind}_\sigma(N_{\mathcal{O}}^{\cdot\cdot})$ is a direct summand of

19.4.9
$$\mathrm{Ind}_\sigma\big(\mathrm{Res}_\sigma(N) \otimes_{\mathcal{O}} L_{\mathcal{O}}^{\cdot\cdot}\big) \cong N \otimes_{\mathcal{O}} \mathrm{Ind}_\sigma(L_{\mathcal{O}}^{\cdot\cdot}) \quad ,$$

where $L_{\mathcal{O}}^{\cdot\cdot}$ is the restriction of $L^{\cdot\cdot}$ to $\mathcal{O}P^{\cdot\cdot}$, which is an endopermutation $\mathcal{O}P$-module having N as a direct summand (cf. [38, (28.4)]); then, the uniqueness of $L^{\cdot\cdot}$ follows from the fact that $L^{\cdot\cdot}$ is a direct summand of the $\mathcal{D}P^{\cdot\cdot}$-module $\mathrm{Res}_\sigma(N)^* \otimes_{\mathcal{O}} N^{\cdot\cdot}$ (cf. 19.4.7).

19.6 Fix a maximal local pointed group P_γ on A; by Corollary 16.10, up to modify our choices of $P^{\cdot\cdot}$ and $N^{\cdot\cdot}$, we have a basic local tracing triple $(P_{\hat\gamma}, P^{\cdot\cdot}_{\gamma^{\cdot\cdot}}, P'_{\gamma'})$ on $\hat A$, $A^{\cdot\cdot}$ and $\mathcal{O}G'$ where $\gamma^{\cdot\cdot}$ corresponds to $N^{\cdot\cdot}$, and we denote by $\lambda : P \to P^{\cdot\cdot}$ a section of it (cf. Proposition 17.5). Now, it follows from Theorem 17.12 that, for any local pointed group Q_δ on A such that $Q_\delta \subset P_\gamma$, there is a basic local tracing triple $(Q_{\hat\delta}, Q^{\cdot\cdot}_{\delta^{\cdot\cdot}}, Q'_{\delta'})$ on $\hat A$, $A^{\cdot\cdot}$ and $\mathcal{O}G'$ such that:

19.6.1 *We have $Q^{\cdot\cdot} \subset P^{\cdot\cdot}$, $\big(Q', b(\delta')\big) \subset \big(P', b(\gamma')\big)$ and $\lambda(Q) \subset Q^{\cdot\cdot}$, and the group homomorphism $\lambda_Q : Q \to Q^{\cdot\cdot}$ determined by λ is a section of $(Q_{\hat\delta}, Q^{\cdot\cdot}_{\delta^{\cdot\cdot}}, Q'_{\delta'})$.*

Set $e = b(\gamma)$, $f = b(\delta)$, $\lambda' = \sigma' \circ \lambda$ and $Q^{\lambda'} = \lambda'(Q) \subset Q'$, and denote by $\lambda'_Q : Q \to Q^{\lambda'}$ the group homomorphism determined by λ', by $f^{\lambda'}$ the block of $C_{G'}(Q^{\lambda'})$ such that $(Q^{\lambda'}, f^{\lambda'}) \subset \big(Q', b(\delta')\big)$ (cf. 2.14.2) and by $(Q, f)^{\lambda'}$ the Brauer pair $(Q^{\lambda'}, f^{\lambda'})$, which does not depend on the choices of δ and $(Q_{\hat\delta}, Q^{\cdot\cdot}_{\delta^{\cdot\cdot}}, Q'_{\delta'})$ since we have (cf. 2.14.2)

19.6.2 $$(Q, f)^{\lambda'} \subset \big(Q', b(\delta')\big) \subset \big(P', b(\gamma')\big) \quad .$$

Notice that, if R_ε is a local pointed group on A such that $R_\varepsilon \subset Q_\delta$ and $(R_{\hat\varepsilon}, R^{\cdot\cdot}_{\varepsilon^{\cdot\cdot}}, R'_{\varepsilon'})$ a basic local tracing triple on $\hat A$, $A^{\cdot\cdot}$ and $\mathcal{O}G'$ fulfilling condition 19.6.1, it follows once again from Theorem 17.12 that

19.6.3 *There is $x' \in C_{G'}(Q^{\lambda'})$ such that $(R^{\cdot\cdot})^{(1,x')} \subset Q^{\cdot\cdot}$ and*
$$\big(R', b(\varepsilon')\big)^{x'} \subset \big(Q', b(\delta')\big) \quad .$$

Theorem 19.7 *With the notation above, the group homomorphism $\lambda' : P \to P'$ is bijective and, for any G, b-Brauer pairs (R, g) and (Q, f) such that $(R, g) \subset (P, e)$ and $(Q, f) \subset (P, e)$, we have*

19.7.1 $$\tilde\lambda'_Q \circ E_G\big((R, g), (Q, f)\big) = E_{G'}\big((R, g)^{\lambda'}, (Q, f)^{\lambda'}\big) \circ \tilde\lambda'_R \quad .$$

In particular, λ' induces an equivalence between the Brauer categories of b and b'.

Proof: Since $S^{\cdot\cdot}_{\gamma^{\cdot\cdot}}(P^\lambda) \neq \{0\}$ (cf. 17.4.7) and the restriction of $N^{\cdot\cdot}$ to $\mathcal{O}\,\mathrm{Ker}(\sigma')$ is projective, by Lemma 7.10 we have

19.7.2 $$P^\lambda \cap \mathrm{Ker}(\sigma') = 1 \quad ;$$

consequently, λ' is injective and in particular we have $|P| \le |P'|$; then, by symmetry, we get the equality $|P| = |P'|$ and therefore λ' is bijective.

Let (Q, f) and (R, g) be (G, b)-Brauer pairs such that $(Q, f) \subset (P, e)$ and $(R, g) \subset (P, e)$, and $\varphi : R \to Q$ a group homomorphism such that $\tilde\varphi \in E_G\big((R, g), (Q, f)\big)$; then, we claim that

19.7.3 $$\tilde\lambda'_Q \circ \tilde\varphi \circ (\tilde\lambda'_R)^{-1} \in E_{G'}\big((R, g)^{\lambda'}, (Q, f)^{\lambda'}\big) \quad .$$

Consider $x \in G$ such that $(R, g)^x \subset (Q, f)$ and $\varphi(u) = u^x$ for any $u \in R$; since $(R^x, g^x) \subset (P, e)$, we have also $(R^x, g^x)^{\lambda'} \subset (Q, f)^{\lambda'}$ (cf. 2.14.2) and therefore we may assume that $|R| = |Q|$.

Recall that f is also a block of $Q{\cdot}C_G(Q)$ and consider a maximal $\big(Q{\cdot}C_G(Q), f\big)$-Brauer pair (T, h); since (T, h) is also a (G, b)-Brauer pair, there is $y \in G$ such that $(T, h)^y \subset (P, e)$ (cf. [38, (40.13)]) and therefore, since we have $(Q, f) \subset (T, h)$ both as $\big(Q{\cdot}C_G(Q), f\big)$- and as (G, b)-Brauer pairs, we have also $(Q, f)^y \subset (P, e)$ (cf. 2.14.2); now, denoting by $\psi : Q \cong Q^y$ the group isomorphism mapping $u \in Q$ on u^y, it suffices to prove that $\tilde{\lambda}'_{Q^y} \circ (\tilde{\psi} \circ \tilde{\varphi}) \circ (\tilde{\lambda}'_R)^{-1}$ and $\tilde{\lambda}'_{Q^y} \circ \tilde{\psi} \circ (\tilde{\lambda}'_Q)^{-1}$ respectively belong to $E_{G'}\big((R, g)^{\lambda'}, (Q^y, f^y)^{\lambda'}\big)$ and to $E_{G'}\big((Q, f)^{\lambda'}, (Q^y, f^y)^{\lambda'}\big)$. Hence, we may assume that $Q{\cdot}C_P(Q)$ is a defect group of f in $Q{\cdot}C_G(Q)$.

In this case, according to 2.15.3, for any local point δ of Q on $\mathcal{O}G$ associated with f, we have $Q_\delta \subset P_\gamma$; let ε be a local point of R on $\mathcal{O}G$ associated with g such that $R_\varepsilon \subset P_\gamma$ (cf. 2.9.1 and 2.14.2) and set $\delta = \varepsilon^x$; then, δ is a local point of Q on $\mathcal{O}G$ associated with f, so that $Q_\delta \subset P_\gamma$ and, according to Proposition 18.4, $\tilde{\varphi}$ belongs to $E_G(R_{\hat{\varepsilon}}, Q_{\hat{\delta}})$. Moreover, there are basic local tracing triples $(Q_{\hat{\delta}}, Q_{\ddot{\delta}}, Q'_{\delta'})$ and $(R_{\hat{\varepsilon}}, R_{\ddot{\varepsilon}}, R'_{\varepsilon'})$ on $\hat{A}$, $A^{\cdot\cdot}$ and $\mathcal{O}G'$ fulfilling condition 19.6.1; in particular, we have (cf. 2.14.2)

19.7.4 $$(Q, f)^{\lambda'} \subset \big(Q', b(\delta')\big) \quad \text{and} \quad (R, g)^{\lambda'} \subset \big(R', b(\varepsilon')\big)$$

and the group homomorphisms λ_Q and λ_R are respectively sections of $(Q_{\hat{\delta}}, Q_{\ddot{\delta}}, Q'_{\delta'})$ and $(R_{\hat{\varepsilon}}, R_{\ddot{\varepsilon}}, R'_{\varepsilon'})$. Consequently, it follows from Corollary 17.13 that $\tilde{\lambda}'_Q \circ \tilde{\varphi} \circ (\tilde{\lambda}'_R)^{-1}$ belongs indeed to $E_{G'}\big((R, g)^{\lambda'}, (Q, f)^{\lambda'}\big)$.

In conclusion, considering the sum

19.7.5 $$s(G, b) = \sum_{Q,R} |E_G\big((R, e_R), (Q, e_Q)\big)| \quad ,$$

where Q and R run on the set of subgroups of P and, for any subgroup Q of P, we denote by e_Q the block of $C_G(Q)$ such that $(Q, e_Q) \subset (P, e)$, it is clear that $s(G, b)$ does not depend on the choice of P_γ and, since for any subgroups Q and R of P we have proved that

19.7.6 $$\tilde{\lambda}'_Q \circ E_G\big((R, e_R), (Q, e_Q)\big) \circ (\tilde{\lambda}'_R)^{-1} \subset E_{G'}\big((R, e_R)^{\lambda'}, (Q, e_Q)^{\lambda'}\big) \quad ,$$

we get $s(G, b) \leq s(G', b')$. By symmetry, the equality holds, which forces all the inclusions 19.7.6 to be equalities.

19.8 Let Q_δ be a local pointed group on A such that $Q_\delta \subset P_\gamma$, and $(Q_{\hat{\delta}}, Q_{\ddot{\delta}}, Q'_{\delta'})$ a basic local tracing triple on $\hat{A}$, $A^{\cdot\cdot}$ and $\mathcal{O}G'$ fulfilling condition 19.6.1; although we have no clear idea on the nature of the inclusion $Q^{\lambda'} \subset Q'$, our next result gives a sufficient condition to get equality. Recall that Q_δ is called *selfcentralizing* if, for any $x \in G$ such that $(Q_\delta)^x \subset P_\gamma$, we have $C_P(Q^x) \subset Q^x$ (cf. 2.16.4) and in that case δ is the unique local point of Q on A associated with $b(\delta)$ (cf. 2.16.1).

Corollary 19.9 *With the notation above, let Q_δ be a selfcentralizing local pointed group on A such that $Q_\delta \subset P_\gamma$ and $(Q_{\hat{\delta}}, Q_{\ddot{\delta}}, Q'_{\delta'})$ a local tracing triple on $\hat{A}$, $A^{\cdot\cdot}$*

and $\mathcal{O}G'$ fulfilling condition 19.6.1, and set $f = b(\delta)$ and $f' = b(\delta')$. Then $(Q, f)^{\lambda'}$ and (Q', f') are selfcentralizing (G, b)-Brauer pairs and we have

19.9.1 $$N_{G'}\big((Q, f)^{\lambda'}\big) = C_{G'}(Q^{\lambda'})\Big(N_{G'}(Q', f') \cap N_{G'}\big((Q, f)^{\lambda'}\big)\Big) \quad .$$

In particular, $Q^{\lambda'}$ has a unique local point $\delta^{\lambda'}$ on A' such that $(Q^{\lambda'})_{\delta^{\lambda'}} \subset Q'_{\delta'}$, and if $\mathrm{O}_p\big(E_G(Q_\delta)\big) = 1$ then $Q^{\lambda'} = Q'$.

Proof: By Theorem 19.7, for any $x' \in G'$ such that $\big((Q, f)^{\lambda'}\big)^{x'} \subset \big(P', b(\gamma')\big)$ there is $x \in G$ such that $(Q, f)^x \subset (P, e)$ and $\lambda'(u^x) = \lambda'(u)^{x'}$ for any $u \in Q$, and therefore we have $C_P(Q^x) \subset Q^x$ (cf. 2.16.4), so that we get

19.9.2 $$C_{P'}\big((Q^{\lambda'})^{x'}\big) = C_{P'}\big(\lambda'(Q^x)\big) \subset \lambda'(Q^x) = (Q^{\lambda'})^{x'}$$

which proves that $(Q, f)^{\lambda'}$ is a selfcentralizing (G', b')-Brauer pair (cf. 2.16.4); moreover, since $\big(P', b(\gamma')\big)$ contains both $(Q, f)^{\lambda'}$ and (Q', f'), we have (cf. 2.14.2)

19.9.3 $$(Q, f)^{\lambda'} \subset (Q', f')$$

and therefore (Q', f') is a selfcentralizing (G', b')-Brauer pair too (cf. 2.16.4).

On the other hand, if x' normalizes $(Q, f)^{\lambda'}$ then, always by Theorem 19.7, we can choose x normalizing (Q, f) or, equivalently, normalizing Q_δ; then, by Proposition 18.4, x normalizes $Q_{\hat\delta}$ too and therefore $\big(Q_{\hat\delta}, (Q_{\ddot\delta})^{(x,x')}, (Q'_{\delta'})^{x'}\big)$ is also a (basic) local tracing triple on $\hat A$, $A^{\cdot\cdot}$ and $\mathcal{O}G'$, and λ_Q is still a section of this triple; hence, it follows from Theorem 17.12 that there is $z' \in G'$ such that $(Q', f')^{z'} = (Q', f')^{x'}$ and $\lambda'(u) = \lambda'(u)^{z'}$ for any $u \in Q$; that is to say, we have $x' = y'z'$ where $y' \in N_{G'}(Q', f')$ and $z' \in C_{G'}(Q^{\lambda'})$.

In particular, since $(Q, f)^{\lambda'}$ is selfcentralizing, there is a unique local point $\delta^{\lambda'}$ of $Q^{\lambda'}$ associated with $f^{\lambda'}$ (cf. 2.16.1) or, equivalently, fulfilling $(Q^{\lambda'})_{\delta^{\lambda'}} \subset Q'_{\delta'}$ (cf. 2.16.1 and 19.9.3), since δ' is also the unique local point of Q' associated with f'; moreover, we have $C_{Q'}(Q^{\lambda'}) = Z(Q^{\lambda'})$ (cf. 2.16.4) and therefore, by equality 19.9.1, the quotient $\bar N_{Q'}(Q^{\lambda'}) = N_{Q'}(Q^{\lambda'})/Q^{\lambda'}$ is isomorphic to a normal subgroup of the group (cf. Proposition 18.4)

19.9.4 $$E_{G'}\big((Q, f)^{\lambda'}\big) \cong E_G(Q, f) = E_G(Q_\delta) \quad .$$

19.10 Recall that, for any (G, b)-Brauer pair (Q, f), any maximal $\big(Q{\cdot}C_G(Q), f\big)$-Brauer pair (R, g) contains (Q, f) (cf. [38, (40.13)]) and is also a selfcentralizing (G, b)-Brauer pair (cf. 2.16.2); in particular, up to G-conjugation, we may assume that

19.10.1 $$(Q, f) \subset (R, g) \subset (P, e) \quad ;$$

in that case, we have $R = Q{\cdot}C_P(Q)$ (cf. 2.14.2) and $(R, g)^{\lambda'}$ is also a maximal $\big(Q^{\lambda'}{\cdot}C_{G'}(Q^{\lambda'}), f^{\lambda'}\big)$-Brauer pair since, as it is easily checked using $(\lambda')^{-1}$

and Theorem 19.7, for any $x' \in G'$ such that $\big((Q,f)^{\lambda'}\big)^{x'} \subset \big(P', b(\gamma')\big)$, we have (cf. 2.16.2)

19.10.2 $$|C_{P'}\big((Q^{\lambda'})^{x'}\big)| \le |C_{P'}(Q^{\lambda'})| \quad .$$

On the other hand, since $P^{\cdot\cdot}$ stabilizes an $\mathcal{O}$-basis of $A^{\cdot\cdot}_{\gamma^{\cdot\cdot}} \cong \mathrm{End}_{\mathcal{O}}(N^{\cdot\cdot})$, for any local pointed group $Q^{\cdot\cdot}_{\delta^{\cdot\cdot}}$ on $A^{\cdot\cdot}_{\alpha^{\cdot\cdot}} \cong \mathrm{End}_{\mathcal{O}}(M^{\cdot\cdot})$ (cf. 18.5) and any subgroup $R^{\cdot\cdot}$ of $Q^{\cdot\cdot}$, the $\mathcal{D}C_{Q^{\cdot\cdot}}(R^{\cdot\cdot})$-interior algebra $A^{\cdot\cdot}_{\delta^{\cdot\cdot}}(R^{\cdot\cdot})$ is a simple k-algebra (cf. 2.2.7 and 2.9.5) and therefore, for a suitable $k \otimes_{\mathcal{O}} \mathcal{D}C_{Q^{\cdot\cdot}}(R^{\cdot\cdot})$-module $M^{\cdot\cdot}_{\delta^{\cdot\cdot},R^{\cdot\cdot}}$, we have

19.10.3 $$A^{\cdot\cdot}_{\delta^{\cdot\cdot}}(R^{\cdot\cdot}) \cong \mathrm{End}_k(M^{\cdot\cdot}_{\delta^{\cdot\cdot},R^{\cdot\cdot}}) \quad .$$

Moreover, it is quite clear that the $\mathcal{D}Q^{\cdot\cdot}$-interior algebra $A^{\cdot\cdot}_{\delta^{\cdot\cdot}}$ is $Q^{\cdot\cdot} \cap \mathrm{Ker}(\sigma)$- and $Q^{\cdot\cdot} \cap \mathrm{Ker}(\sigma')$-basic, so that $A^{\cdot\cdot}_{\delta^{\cdot\cdot}}(R^{\cdot\cdot})$ is a $C_{Q^{\cdot\cdot}}(R^{\cdot\cdot}) \cap \mathrm{Ker}(\sigma)$- and $C_{Q^{\cdot\cdot}}(R^{\cdot\cdot}) \cap \mathrm{Ker}(\sigma')$-basic $k \otimes_{\mathcal{O}} \mathcal{D}C_{Q^{\cdot\cdot}}(R^{\cdot\cdot})$-interior algebra.

Theorem 19.11 *With the notation above, any conjugacy class of (G,b)-Brauer pairs has a representative (Q,f) such that $(Q,f) \subset (P,e)$, $C_P(Q)$ is a defect group of f in $C_G(Q)$ and, setting $R = Q{\cdot}C_P(Q)$ and respectively denoting by ε and $\varepsilon^{\lambda'}$ the local points of R and $R^{\lambda'}$ on A and A' fulfilling $R_\varepsilon \subset P_\gamma$ and $(R^{\lambda'})_{\varepsilon^{\lambda'}} \subset P'_{\gamma'}$, for a suitable subgroup $R^{\cdot\cdot}$ of $Q^\lambda{\cdot}C_{P^{\cdot\cdot}}(Q^\lambda)$ containing R^λ, a suitable local point $\varepsilon^{\cdot\cdot}$ of $R^{\cdot\cdot}$ on $A^{\cdot\cdot}_{\alpha^{\cdot\cdot}}$ and a suitable indecomposable direct summand $N^{\cdot\cdot}_Q$ of $M^{\cdot\cdot}_{\varepsilon^{\cdot\cdot},Q^\lambda}$ with vertex, $C_{R^{\cdot\cdot}}(Q^\lambda)$, we have a $\mathcal{D}C_P(Q)$-interior algebra embedding*

19.11.1 $$\hat{A}_{\hat{\varepsilon}}(Q) \longrightarrow \mathrm{Ind}_\rho\Big(\mathrm{End}_k(N^{\cdot\cdot}_Q) \underset{k}{\otimes} \mathrm{Res}_{\rho'}\big(A'_{\varepsilon^{\lambda'}}(Q^{\lambda'})\big)\Big) \quad ,$$

where $\rho : C_{R^{\cdot\cdot}}(Q^\lambda) \to C_P(Q)$ and $\rho' : C_{R^{\cdot\cdot}}(Q^\lambda) \to C_{P'}(Q^{\lambda'})$ are respectively the group homomorphisms determined by π and π'. In particular, a suitable indecomposable direct summand $M^{\cdot\cdot}_Q$ of the $k \otimes_{\mathcal{O}} \mathcal{D}\big(C_G(Q) \times C_{G'}(Q^{\lambda'})\big)$-module $\mathrm{Ind}_{C_{R^{\cdot\cdot}}(Q^\lambda)}^{C_G(Q)\times C_{G'}(Q^\lambda)}(N^{\cdot\cdot}_Q)$ defines a basic Rickard equivalence between the blocks of $C_G(Q)$ and $C_{G'}(Q^{\lambda'})$ over k respectively determined by f and $f^{\lambda'}$.

Remark 19.12 Setting

$$C_G(Q) = G_Q,\ C_P(Q) = P_Q,\ C_{R^{\cdot\cdot}}(Q^\lambda) = R^{\cdot\cdot}_Q,\ C_{G'}(Q^{\lambda'}) = G'_Q,\ C_{P'}(Q^{\lambda'}) = P'_Q,$$
$$\mathrm{Br}_Q(\hat{\varepsilon}) = \hat{\varepsilon}_Q,\ \mathrm{Br}_{Q^{\lambda'}}(\varepsilon^{\lambda'}) = \varepsilon'_Q,\ S^{\cdot\cdot}_Q = \mathrm{End}_k(N^{\cdot\cdot}_Q),\ A^{\cdot\cdot}_Q = \mathrm{Ind}_{R^{\cdot\cdot}_Q}^{G_Q\times G'_Q}(S^{\cdot\cdot}_Q)$$

and $\hat{A}_Q = (A^{\cdot\cdot}_Q)^{1\times G'_Q}$, and denoting by $\varepsilon^{\cdot\cdot}_Q$ the unique point of $R^{\cdot\cdot}_Q$ on $S^{\cdot\cdot}_Q$, embedding 19.11.1 allows us to identify $\hat{\varepsilon}_Q$ to a local point of P_Q (localness follows from Lemma 17.2) on the Hecke $k \otimes_{\mathcal{O}} \mathcal{D}G_Q$-interior algebra $\hat{A}_Q$ since we have a canonical $\mathcal{D}P_Q$-interior algebra exoembedding (cf. 16.4.3)

19.12.1 $$\tilde{g} : \mathrm{Ind}_\rho\Big(S^{\cdot\cdot}_Q \underset{k}{\otimes} \mathrm{Res}_{\rho'}\big(A'_{\varepsilon^{\lambda'}}(Q^{\lambda'})\big)\Big) \longrightarrow \mathrm{Res}_{P_Q}^{G_Q}(\hat{A}_Q)$$

given by (cf. 16.4.4)

19.12.2 $$\tilde{g} = \tilde{d}^{G_Q\times G'_Q}_{P_Q\times G'_Q}(S^{\cdot\cdot}_Q)^{1\times G'_Q} \circ \tilde{H}_{P_Q,G_Q}(S^{\cdot\cdot}_Q)^{-1} \circ \mathrm{Ind}_\rho\big(\widetilde{\mathrm{id}}_{S^{\cdot\cdot}_Q} \underset{k}{\otimes} \mathrm{Res}_{\rho'}(\tilde{f}_{\varepsilon'_Q})\big)$$

and then it is quite clear that $\big((P_Q)_{\hat{\varepsilon}_Q}, (R^{\cdot\cdot}_Q)_{\varepsilon^{\cdot\cdot}_Q}, (P'_Q)_{\varepsilon'_Q}\big)$ is a maximal basic local tracing triple on $\hat{A}_Q$, $A^{\cdot\cdot}_Q$ and kG'_Q; moreover, the inclusion $R^\lambda \subset R^{\cdot\cdot}$ guarantees that the group homomorphism $P_Q \to R^{\cdot\cdot}_Q$ determined by λ is a section of this triple.

Proof: Let (Q, f) be a (G, b)-Brauer pair and (R, g) a maximal $\big(Q{\cdot}C_G(Q), f\big)$-Brauer pair, denote by ε the local point of R on A associated with g (cf. 19.10) and consider a local tracing triple $(R_{\hat{\varepsilon}}, R^{\cdot\cdot}_{\varepsilon^{\cdot\cdot}}, R'_{\varepsilon'})$ on $\hat{A}$, $A^{\cdot\cdot}$ and $\mathcal{O}G'$ (which is basic), and a Q-section μ of it; it follows from Proposition 17.5 that μ can be extended to a section of $(R_{\hat{\varepsilon}}, R^{\cdot\cdot}_{\varepsilon^{\cdot\cdot}}, R'_{\varepsilon'})$ and then, according to Theorem 17.12, up to replace (Q, f) by a suitable G-conjugate, we may assume that we have $R_\varepsilon \subset P_\gamma$ and $\lambda_Q = \mu$, and that $(R_{\hat{\varepsilon}}, R^{\cdot\cdot}_{\varepsilon^{\cdot\cdot}}, R'_{\varepsilon'})$ fulfills condition 19.6.1; in particular, by 19.10 we have

19.11.2 $$R = Q{\cdot}C_P(Q) \quad\text{and}\quad C_R(Q)^\lambda = C_{R^\lambda}(Q^\lambda) = C_{R'}(Q^\lambda) = C_{P'}(Q^\lambda)$$

and, denoting by $\rho : R^{\cdot\cdot} \to R$ and $\rho' : R^{\cdot\cdot} \to R'$ the group homomorphisms determined by π and π', we have a canonical $\mathcal{D}C_P(Q)$-interior R-algebra exoembedding (cf. 17.4.5)

19.11.3 $$\tilde{h}^{\lambda Q}_{\hat{\varepsilon}} : \hat{A}_{\hat{\varepsilon}}(Q) \longrightarrow \mathrm{Ind}_{\rho_{\lambda_Q}}\Big(A^{\cdot\cdot}_{\varepsilon^{\cdot\cdot}}(Q^\lambda) \underset{k}{\otimes} \mathrm{Res}_{\rho'_{\lambda_Q}}\big(A'_{\varepsilon'}(Q^{\lambda'})\big)\Big) \quad ;$$

moreover, since λ_R is a section of $(R_{\hat{\varepsilon}}, R_{\varepsilon^{\cdot\cdot}}, R'_{\varepsilon'})$, it follows from Remark 17.6 that we have the following commutative diagram of $\mathcal{D}Z(R)$-interior algebra exoembeddings

19.11.4 $$\begin{array}{ccc} \big(\hat{A}_{\hat{\varepsilon}}(Q)\big)(R) & \xrightarrow{\tilde{h}^{\lambda Q}_{\hat{\varepsilon}}(R)} & \Big(\mathrm{Ind}_{\rho_{\lambda_Q}}\Big(A^{\cdot\cdot}_{\varepsilon^{\cdot\cdot}}(Q^\lambda) \underset{k}{\otimes} \mathrm{Res}_{\rho'_{\lambda_Q}}\big(A'_{\varepsilon'}(Q^{\lambda'})\big)\Big)\Big)(R) \\ \wr\| & & \uparrow \tilde{e} \\ \hat{A}_{\hat{\varepsilon}}(R) & \xrightarrow{\tilde{h}^{\lambda R}_{\hat{\varepsilon}}} & \mathrm{Ind}_{\rho_{\lambda_R}}\Big(A^{\cdot\cdot}_{\varepsilon^{\cdot\cdot}}(R^\lambda) \underset{k}{\otimes} \mathrm{Res}_{\rho'_{\lambda_R}}\big(A'_{\varepsilon'}(R^{\lambda'})\big)\Big) \end{array} .$$

Since $R = Q{\cdot}C_P(Q)$, Lemma 17.2 and exoembedding 19.11.3 allow us to identify $\mathrm{Br}_Q(\hat{\varepsilon})$ to a local point $\hat{\varepsilon}_Q$ of $C_P(Q)$ on the right end of this exoembedding and therefore, setting $D^{\cdot\cdot} = A^{\cdot\cdot}_{\varepsilon^{\cdot\cdot}}(Q^\lambda)$ and $D' = A'_{\varepsilon'}(Q^{\lambda'})$, it is quite clear, according to equalities 19.11.2, that there is a point β' of $R^{\lambda'}$ on $A'_{\varepsilon'}$ or, equivalently, a point $\beta'_Q = \mathrm{Br}_{Q^{\lambda'}}(\beta')$ of $C_P(Q)^{\lambda'}$ on D' (cf. 2.15.4) such that $\tilde{h}^{\lambda Q}_{\hat{\varepsilon}}$ factorizes throughout the $\mathcal{D}C_P(Q)$-interior R-algebra exoembedding

19.11.5

$$\mathrm{Ind}_{\rho_{\lambda_Q}}(\tilde{g}_{\beta'_Q}) : \mathrm{Ind}_{\rho_{\lambda_Q}}\big(D^{\cdot\cdot} \underset{k}{\otimes} \mathrm{Res}_{\rho'_{\lambda_Q}}(D'_{\beta'_Q})\big) \to \mathrm{Ind}_{\rho_{\lambda_Q}}\big(D^{\cdot\cdot} \underset{k}{\otimes} \mathrm{Res}_{\rho'_{\lambda_Q}}(D')\big) \quad ,$$

where we set

19.11.6 $$s\tilde{g}_{\beta'_Q} = \widetilde{\mathrm{id}}_{D^{\cdot\cdot}} \underset{k}{\otimes} \mathrm{Res}_{\rho'_{\lambda_Q}}\big(\tilde{f}^{\tilde{\varepsilon}'}_{\beta'}(Q^{\lambda'})\big) \quad ,$$

and a new $\mathcal{D}C_P(Q)$-interior R-algebra exoembedding

19.11.7 $$\tilde{h}^{\lambda_Q,\beta'_Q}_{\hat{\varepsilon}} : \hat{A}_{\hat{\varepsilon}}(Q) \longrightarrow \mathrm{Ind}_{\rho_{\lambda_Q}}\big(D^{\cdot\cdot} \underset{k}{\otimes} \mathrm{Res}_{\rho'_{\lambda_Q}}(D'_{\beta'_Q})\big) \quad .$$

Similarly, there is a point $\beta^{\cdot\cdot}_Q$ of $C_{R^{\cdot\cdot}}(Q^\lambda)$ on $D^{\cdot\cdot}$ such that $\tilde{h}^{\lambda_Q,\beta'_Q}_{\hat{\varepsilon}}$ factorizes throughout the $\mathcal{D}C_P(Q)$-interior R-algebra exoembedding

19.11.8

$$\mathrm{Ind}_{\rho_{\lambda_Q}}(\tilde{g}_{\beta^{\cdot\cdot}_Q}) : \mathrm{Ind}_{\rho_{\lambda_Q}}\big(D^{\cdot\cdot}_{\beta^{\cdot\cdot}_Q} \underset{k}{\otimes} \mathrm{Res}_{\rho'_{\lambda_Q}}(D'_{\beta'_Q})\big) \to \mathrm{Ind}_{\rho_{\lambda_Q}}\big(D^{\cdot\cdot} \underset{k}{\otimes} \mathrm{Res}_{\rho'_{\lambda_Q}}(D'_{\beta'_Q})\big) \quad ,$$

where we set

19.11.9 $$\tilde{g}_{\beta^{\cdot\cdot}_Q} = \tilde{f}_{\beta^{\cdot\cdot}_Q}(D^{\cdot\cdot}) \underset{k}{\otimes} \mathrm{Res}_{\rho'_{\lambda_Q}}(\widetilde{\mathrm{id}}_{D'_{\beta'_Q}}) \quad ,$$

and a new $\mathcal{D}C_P(Q)$-interior R-algebra exoembedding

19.11.10 $$\tilde{h}^{\beta^{\cdot\cdot}_Q,\beta'_Q}_{\hat{\varepsilon}} : \hat{A}_{\hat{\varepsilon}}(Q) \longrightarrow \mathrm{Ind}_{\rho_{\lambda_Q}}\big(D^{\cdot\cdot}_{\beta^{\cdot\cdot}_Q} \underset{k}{\otimes} \mathrm{Res}_{\rho'_{\lambda_Q}}(D'_{\beta'_Q})\big)$$

(notice that $R^\lambda \subset Q^\lambda{\cdot}C_{R^{\cdot\cdot}}(Q^\lambda)$).

Then, considering a defect pointed group $T^{\cdot\cdot}_{\delta^{\cdot\cdot}_Q}$ of $C_{R^{\cdot\cdot}}(Q^\lambda)_{\beta^{\cdot\cdot}_Q}$, it follows from Corollary 14.11 that we have a canonical $\mathcal{D}C_{R^{\cdot\cdot}}(Q^\lambda)$-interior algebra exoembedding

19.11.11 $$\tilde{h}^{\delta^{\cdot\cdot}_Q}_{\beta^{\cdot\cdot}_Q} : D^{\cdot\cdot}_{\beta^{\cdot\cdot}_Q} \longrightarrow \mathrm{Ind}^{C_{R^{\cdot\cdot}}(Q^\lambda)}_{T^{\cdot\cdot}}(D^{\cdot\cdot}_{\delta^{\cdot\cdot}_Q})$$

and, since $C_{R^{\cdot\cdot}}(Q^\lambda)$ is a p-group, this exoembedding is actually an exoisomorphism (cf. 2.12.2 and 12.4.4). Consequently, up to suitable identifications (cf. Corollary 12.7 and Proposition 12.9), exoembedding 19.11.10 becomes (cf. 19.11.3)

19.11.12 $$\tilde{h}^{\delta^{\cdot\cdot}_Q,\beta'_Q}_{\hat{\varepsilon}} : \hat{A}_{\hat{\varepsilon}}(Q) \longrightarrow \mathrm{Ind}_\tau\Big(D^{\cdot\cdot}_{\delta^{\cdot\cdot}_Q} \underset{k}{\otimes} \mathrm{Res}_{\tau'}\big(\mathrm{Res}^{T^{\lambda'}}_{T'}(D'_{\beta'_Q})\big)\Big) \quad ,$$

where $T = C_P(Q)$, $T' = \rho'(T^{\cdot\cdot})$ and we respectively denote by $\tau : T^{\cdot\cdot} \to T$ and $\tau' : T^{\cdot\cdot} \to T'$ the group homomorphisms determined by π and π'.

Now, we claim that

19.11.13 $$\tau(T^{\cdot\cdot}) = T \quad , \quad T' = T^{\lambda'} \quad \text{and} \quad \beta' = \varepsilon^{\lambda'} \quad ;$$

indeed, since $\hat{\varepsilon}_Q$ is still a local point of $T = C_P(Q)$ on the right end of exoembedding 19.11.12, we have necessarily $\tau(T^{\cdot\cdot}) = T$ (cf. 2.11.3 and 12.4.4) and it follows from Theorem 13.9 applied to the $\mathrm{Ker}(\tau)$-basic $k \otimes_{\mathcal{O}} \mathcal{D}C_{R^{\cdot\cdot}}(Q^\lambda)$-interior algebra $D^{\cdot\cdot}_{\delta^{\cdot\cdot}_Q} \otimes_k \mathrm{Res}_{\tau'}\big(\mathrm{Res}^{T^{\lambda'}}_{T'}(D'_{\beta'_Q})\big)$ that τ admits a section $\nu_Q : T \to T^{\cdot\cdot}$ such that $\tilde{h}^{\delta^{\cdot\cdot}_Q,\beta'_Q}_{\hat{\varepsilon}}(T)$ factorizes according to the following commutative diagram of $\mathcal{D}Z(R)$-interior algebra exoembeddings

19.11.14

$$\begin{array}{ccc} \big(\hat{A}_{\hat{\varepsilon}}(Q)\big)(T) & \xrightarrow{\tilde{h}^{\delta^{\cdot\cdot}_Q,\beta'_Q}_{\hat{\varepsilon}}(T)} & \Big(\mathrm{Ind}_\tau\Big(D^{\cdot\cdot}_{\delta^{\cdot\cdot}_Q} \underset{k}{\otimes} \mathrm{Res}_{\tau'}\big(\mathrm{Res}^{T^{\lambda'}}_{T'}(D'_{\beta'_Q})\big)\Big)\Big)(T) \\ & \searrow{\scriptstyle \tilde{h}^{\nu_Q}_{\hat{\varepsilon}}} & \uparrow{\scriptstyle \tilde{e}^{\delta^{\cdot\cdot}_Q}_{\nu_Q}} \\ & & \mathrm{Ind}_{\tau_{\nu_Q}}\Big(D^{\cdot\cdot}_{\delta^{\cdot\cdot}_Q}(T^{\nu_Q}) \underset{k}{\otimes} \mathrm{Res}_{\tau'_{\nu_Q}}\big(D'_{\beta'_Q}(T^{\nu'_Q})\big)\Big) \end{array} \quad ;$$

in particular, we have

19.11.15
$$D^{\cdot\cdot}_{\delta^{\cdot\cdot}_Q}(T^{\nu_Q}) \neq \{0\}$$

whereas, since the restriction of $N^{\cdot\cdot}$ to $\mathcal{O}\mathrm{Ker}(\sigma')$ is projective, for any nontrivial subgroup $Z^{\cdot\cdot}$ of $\mathrm{Ker}(\tau')$ we have (cf. 2.2.6)

19.11.16
$$D^{\cdot\cdot}_{\delta^{\cdot\cdot}_Q}(Z^{\cdot\cdot}) = \{0\} \quad ;$$

consequently, by Lemma 7.10, τ' maps T^{ν_Q} injectively into T', which forces

19.11.17
$$T' = T^{\lambda'} = T^{\nu'_Q}$$

(a similar argument applied to $\mathrm{Ker}(\sigma)$ shows that $\nu_Q(z) = \lambda(z)$ for any $z \in Z(Q)$). Moreover, it follows from Theorem 5.8 and Remark 5.9 applied to the $kT^{\cdot\cdot}$-interior algebra $\mathrm{C}_\circ(D^{\cdot\cdot}_{\delta^{\cdot\cdot}_Q}) = \mathrm{C}_\circ(D^{\cdot\cdot})_{\delta^{\cdot\cdot}_Q}$ (cf. 15.5.2) that β'_Q is a local point of $T^{\lambda'}$ on $kC_{G'}(Q^{\lambda'})$ (recall that we have a canonical exoembedding from D' to $kC_{G'}(Q^{\lambda'})$), which forces β' to be a local point of $R^{\lambda'}$ on $A'_{\varepsilon'}$ (cf. Lemma 7.10), so that we have $\beta' = \varepsilon^{\lambda'}$ (cf. Corollary 19.9).

Our purpose now is to relate ν_Q with λ. First of all, according to Proposition 13.17 applied to exoembedding 19.11.6, we have the following commutative diagram of $\mathcal{D}Z(R)$-interior algebra exoembeddings (notice that, by Lemma 7.10, we have canonical isomorphisms $A^{\cdot\cdot}_{\varepsilon^{\cdot\cdot}}(R^\lambda) \cong D^{\cdot\cdot}(T^\lambda)$ and $A'_{\varepsilon'}(R^{\lambda'}) \cong D'(T^{\lambda'})$ as $\mathcal{D}C_{R^{\cdot\cdot}}(R^\lambda)$- and $\mathcal{D}C_{R'}(R^{\lambda'})$-interior algebras respectively)

19.11.18

$$\begin{array}{ccc} \mathrm{Ind}_{\rho_{\lambda_R}}\Big(D^{\cdot\cdot}(T^\lambda) \underset{k}{\otimes} \mathrm{Res}_{\rho'_{\lambda_R}}\big(D'(T^{\lambda'})\big)\Big) & \xrightarrow{\tilde{e}} & \Big(\mathrm{Ind}_{\rho_{\lambda_Q}}\big(D^{\cdot\cdot} \underset{k}{\otimes} \mathrm{Res}_{\rho'_{\lambda_Q}}(D')\big)\Big)(T) \\ \uparrow{\scriptstyle \tilde{f}_l} & & \uparrow{\scriptstyle \tilde{f}_r} \\ \mathrm{Ind}_{\rho_{\lambda_R}}\Big(D^{\cdot\cdot}(T^\lambda) \underset{k}{\otimes} \mathrm{Res}_{\rho'_{\lambda_R}}\big(D'_{\varepsilon'_Q}(T^{\lambda'})\big)\Big) & \xrightarrow{\tilde{e}^{D^{\cdot\cdot}}_{\nu}} & \Big(\mathrm{Ind}_{\rho_{\lambda_Q}}\big(D^{\cdot\cdot} \underset{k}{\otimes} \mathrm{Res}_{\rho'_{\lambda_Q}}(D'_{\varepsilon'_Q})\big)\Big)(T) \end{array}$$

where we set $\tilde{f}_l = \mathrm{Ind}_{\rho_{\lambda_R}}(\tilde{g}_{\beta'_Q}(T^\lambda))$, $\tilde{f}_r = (\mathrm{Ind}_{\rho_{\lambda_Q}}(\tilde{g}_{\beta'_Q}))(T)$ and $\varepsilon'_Q = \mathrm{Br}_{Q^{\lambda'}}(\varepsilon^{\lambda'})$ (so that $\varepsilon'_Q = \beta'_Q$), we denote by $\nu : T \to C_{R^{\cdot\cdot}}(Q^\lambda)$ the group homomorphism determined by λ, and $\tilde{e}_\nu^{D^{\cdot\cdot}}$ comes from Theorem 13.9. Moreover, by Lemma 17.2, the unity element in $\mathrm{C}_\circ(\hat{A}_{\hat{\varepsilon}}(R))$ is primitive and therefore, since $\varepsilon^{\lambda'}$ is the unique local point of $R^{\lambda'}$ on $A'_{\varepsilon'}$ (cf. Corollary 19.9) or, equivalently, ε'_Q is the unique local point of $T^{\lambda'}$ on D' (cf. 2.9.1 and Lemma 7.10), $\tilde{h}_{\hat{\varepsilon}}^{\lambda_R}$ factorizes throughout the exoembedding $\tilde{f}_l$ and a new $\mathcal{D}Z(R)$-interior algebra exoembedding

19.11.19 $$\tilde{h}_{\hat{\varepsilon}}^{\lambda_R,\varepsilon'_Q} : \hat{A}_{\hat{\varepsilon}}(R) \longrightarrow \mathrm{Ind}_{\rho_{\lambda_R}}\Big(D^{\cdot\cdot}(T^\lambda) \underset{k}{\otimes} \mathrm{Res}_{\rho'_{\lambda_R}}\big(D'_{\varepsilon'_Q}(T^{\lambda'})\big)\Big) \quad ;$$

then, we claim that the following diagram of $\mathcal{D}Z(R)$-interior algebra exoembeddings is commutative

19.11.20 $$\begin{array}{ccc} (\hat{A}_{\hat{\varepsilon}}(Q))(T) & \xrightarrow{\tilde{h}_{\hat{\varepsilon}}^{\lambda_Q,\varepsilon'_Q(T)}} & \Big(\mathrm{Ind}_{\rho_{\lambda_Q}}\big(D^{\cdot\cdot} \underset{k}{\otimes} \mathrm{Res}_{\rho'_{\lambda_Q}}(D'_{\varepsilon'_Q})\big)\Big)(T) \\ \wr\Vert & & \uparrow{\scriptstyle \tilde{e}_\nu^{D^{\cdot\cdot}}} \\ \hat{A}_{\hat{\varepsilon}}(R) & \xrightarrow{\tilde{h}_{\hat{\varepsilon}}^{\lambda_R,\varepsilon'_Q}} & \mathrm{Ind}_{\rho_{\lambda_R}}\Big(D^{\cdot\cdot}(T^\lambda) \underset{k}{\otimes} \mathrm{Res}_{\rho'_{\lambda_R}}\big(D'_{\varepsilon'_Q}(T^{\lambda'})\big)\Big) \end{array} \quad ;$$

indeed, by Proposition 11.10, it suffices to prove that the composition of $\tilde{f}_r$ with both diagonal exoembeddings in diagram 19.11.20 coincide, which is an easy consequence of the factorization of $\tilde{h}_{\hat{\varepsilon}}^{\lambda_Q}$ and $\tilde{h}_{\hat{\varepsilon}}^{\lambda_R}$, and the commutativity of diagrams 19.11.4 and 19.11.18 (notice that, up to suitable identifications, the exoembeddings $\tilde{h}_{\hat{\varepsilon}}^{\lambda_Q}(R)$ and $\tilde{h}_{\hat{\varepsilon}}^{\lambda_Q}(T)$ coincide).

Similarly, according again to Proposition 13.17 applied this time to exoembedding 19.11.9, since ν_Q is still a section of ρ_{λ_Q} we have the following commutative diagram of $\mathcal{D}Z(R)$-interior algebra exoembeddings

19.11.21

$$\begin{array}{ccc} \mathrm{Ind}_\zeta\Big(D^{\cdot\cdot}(T^{\nu_Q}) \underset{k}{\otimes} \mathrm{Res}_{\zeta'}\big(D'_{\varepsilon'_Q}(T^{\nu'_Q})\big)\Big) & \xrightarrow{\tilde{e}_{\nu_Q}^{D^{\cdot\cdot}}} & \Big(\mathrm{Ind}_{\rho_{\lambda_Q}}\big(D^{\cdot\cdot} \underset{k}{\otimes} \mathrm{Res}_{\rho'_{\lambda_Q}}(D'_{\varepsilon'_Q})\big)\Big)(T) \\ \uparrow{\scriptstyle \tilde{g}_l} & & \uparrow{\scriptstyle \tilde{g}_r} \\ \mathrm{Ind}_\zeta\Big(D^{\cdot\cdot}_{\beta^{\cdot\cdot}_Q}(T^{\nu_Q}) \underset{k}{\otimes} \mathrm{Res}_{\zeta'}\big(D'_{\varepsilon'_Q}(T^{\nu'_Q})\big)\Big) & \xrightarrow{\tilde{e}_{\nu_Q}^{\beta^{\cdot\cdot}_Q}} & \Big(\mathrm{Ind}_{\rho_{\lambda_Q}}\big(D^{\cdot\cdot}_{\beta^{\cdot\cdot}_Q} \underset{k}{\otimes} \mathrm{Res}_{\rho'_{\lambda_Q}}(D'_{\varepsilon'_Q})\big)\Big)(T) \end{array}$$

where we set $\tilde{g}_l = \mathrm{Ind}_\zeta(\tilde{g}_{\beta^{\cdot\cdot}_Q}(T^{\nu_Q}))$ and $\tilde{g}_r = (\mathrm{Ind}_{\rho_{\lambda_Q}}(\tilde{g}_{\beta^{\cdot\cdot}_Q}))(T)$, we respectively denote by $\zeta : C_{R^{\cdot\cdot}}(Q^\lambda{\cdot}T^{\nu_Q}) \to Z(R)$ and $\zeta' : C_{R^{\cdot\cdot}}(Q^\lambda{\cdot}T^{\nu_Q}) \to Z(R)^{\lambda'}$ the group homomorphisms determined by π and π', and the exoembeddings $\tilde{e}_{\nu_Q}^{D^{\cdot\cdot}}$ and

$\tilde{e}_{\nu_Q}^{\beta^{\cdot\cdot}_Q}$ come from Theorem 13.9. Moreover, according to Corollary 13.20 applied to exoisomorphism 19.11.11, we have also the following commutative diagram of $\mathcal{D}Z(R)$-interior algebra exoembeddings (notice that $Z(T) = Z(R)$)

19.11.22

$$\begin{array}{ccc}
\mathrm{Ind}_\zeta\Big(D^{\cdot\cdot}_{\beta^{\cdot\cdot}_Q}(T^{\nu_Q}) \underset{k}{\otimes} \mathrm{Res}_{\zeta'}\big(D'_{\varepsilon'_Q}(T^{\nu'_Q})\big)\Big) & \overset{\tilde{e}_{\nu_Q}^{\beta^{\cdot\cdot}_Q}}{\longrightarrow} & \Big(\mathrm{Ind}_{\rho_{\lambda_Q}}\big(D^{\cdot\cdot}_{\beta^{\cdot\cdot}_Q} \underset{k}{\otimes} \mathrm{Res}_{\rho'_{\lambda_Q}}(D'_{\varepsilon'_Q})\big)\Big)(T) \\
\tilde{h}_l \big\uparrow & & \wr\| \; \tilde{h}_r \\
\mathrm{Ind}_{\tau_{\nu_Q}}\Big(D^{\cdot\cdot}_{\delta^{\cdot\cdot}_Q}(T^{\nu_Q}) \underset{\mathcal{O}}{\otimes} \mathrm{Res}_{\tau'_{\nu_Q}}\big(D'_{\varepsilon'_Q}(T^{\nu'_Q})\big)\Big) & \overset{\tilde{e}_{\nu_Q}^{\delta^{\cdot\cdot}_Q}}{\longrightarrow} & \Big(\mathrm{Ind}_\tau\big(D^{\cdot\cdot}_{\delta^{\cdot\cdot}_Q} \underset{k}{\otimes} \mathrm{Res}_{\tau'}(D'_{\varepsilon'_Q})\big)\Big)(T)
\end{array}$$

where we set $\tilde{h}_l = \mathrm{Ind}_\zeta\Big(\tilde{h}^{\delta^{\cdot\cdot}_Q}_{\beta^{\cdot\cdot}_Q}(T^{\nu_Q}) \otimes_k \mathrm{Res}_{\zeta'}\big(\widetilde{\mathrm{id}}_{D'_{\varepsilon'_Q}}(T^{\nu'_Q})\big)\Big) \circ \tilde{d}$ together with $d = \Big(d^{C_{R^{\cdot\cdot}}(Q^\lambda)}_{C_{R^{\cdot\cdot}}(Q_\lambda\cdot T^{\nu_Q}),T^{\cdot\cdot}}\big(D^{\cdot\cdot}_{\delta^{\cdot\cdot}_Q} \otimes_k \mathrm{Res}_{\tau'}(D'_{\varepsilon'_Q})\big)\Big)(T)$ (up to suitable identifications via Corollary 12.7 and Proposition 12.9), and *mutatis mutandis* $\tilde{h}_r = \Big(\mathrm{Ind}_{\rho_{\lambda_Q}}\big(\tilde{h}^{\delta^{\cdot\cdot}_Q}_{\beta^{\cdot\cdot}_Q} \underset{k}{\otimes} \mathrm{Res}_{\rho'_{\lambda_Q}}(\widetilde{\mathrm{id}}_{D'_{\varepsilon'_Q}})\big)\Big)(T)$, and once again $\tilde{e}^{\delta^{\cdot\cdot}_Q}_{\nu_Q}$ comes from Theorem 13.9.

Finally, recall that $(R, g)^{\lambda'}$ is a maximal $\big(Q^{\lambda'}\cdot C_{G'}(Q^{\lambda'}), f^{\lambda'}\big)$-Brauer pair (cf. 19.10) and therefore $(T^{\lambda'}, g^{\lambda'})$ is still a maximal $\big(C_{G'}(Q^{\lambda'}), f^{\lambda'}\big)$-Brauer pair, so that we have

19.11.23 $$N_{C_{G'}(Q^{\lambda'})}\big((T^{\lambda'})_{\varepsilon'_Q}\big) = N_{C_{G'}(Q^{\lambda'})}(T^{\lambda'}, g^{\lambda'}) \qquad ;$$

denote by N'_Q the set of elements x' in this group such that $(1, x')$ normalizes $C_{R^{\cdot\cdot}}(Q^\lambda)$ and choose $X' \subset N'_Q$ such that $1 \times X'$ is a set of representatives for $(1 \times N'_Q)/\mathrm{Ker}(\tau)$. Then, it follows from Theorem 17.8 applied to the $\mathrm{Ker}(\rho_{\lambda_Q})$-basic $\mathcal{D}C_{R^{\cdot\cdot}}(Q^\lambda)$-interior algebra $D^{\cdot\cdot}$, to the point ε'_Q of $C_{P'}(Q^{\lambda'}) = T^{\lambda'}$ on D' or, equivalently, on $kC_{G'}(Q^{\lambda'})$(recall that we have a canonical exoembedding from D' to $kC_{G'}(Q^{\lambda'})$), and to the p-group T, that the sums (cf. Theorem 13.9)

19.11.24 $$l_\nu = \sum_{x' \in X'} e^{D^{\cdot\cdot}}_{\nu^{x'}}\big(1 \otimes (1 \otimes 1)\big) \quad \text{and} \quad l_{\nu_Q} = \sum_{x' \in X'} e^{D^{\cdot\cdot}}_{\nu^{x'}_Q}\big(1 \otimes (1 \otimes 1)\big)$$

are central idempotents in the $\mathcal{D}Z(R)$-interior algebra

19.11.25 $$\Big(\mathrm{Ind}_{\rho_{\lambda_Q}}\big(D^{\cdot\cdot} \underset{k}{\otimes} \mathrm{Res}_{\rho'_{\lambda_Q}}(D_{\varepsilon'_Q})\big)\Big)(T) \qquad ;$$

moreover, they are not orthogonal since, by the factorization of $\tilde{h}^{\lambda_Q,\varepsilon'_Q}_{\hat{\varepsilon}}$ (cf. 19.11.8 and 19.11.10) and the commutativity of diagram 19.11.20, we have

19.11.26 $$\tilde{e}^{D^{\cdot\cdot}}_\nu \circ \tilde{h}^{\lambda_R,\varepsilon'_Q}_{\hat{\varepsilon}} = \tilde{h}^{\lambda_Q,\varepsilon'_Q}_{\hat{\varepsilon}}(T) = \big(\mathrm{Ind}_{\rho_{\lambda_Q}}(\tilde{g}_{\beta^{\cdot\cdot}_Q})\big)(T) \circ \tilde{h}^{\beta^{\cdot\cdot}_Q,\varepsilon'_Q}_{\hat{\varepsilon}}(T)$$

up to suitable identifications; but, by expliciting exoembedding 19.11.12, we get

19.11.27 $$\tilde{h}_{\hat{\varepsilon}}^{\beta^{\cdot\cdot}_Q,\varepsilon'_Q} = \mathrm{Ind}_{\rho_{\lambda_Q}}\big(\tilde{h}^{\delta^{\cdot\cdot}_Q}_{\beta^{\cdot\cdot}_Q} \underset{k}{\otimes} \mathrm{Res}_{\rho'_{\lambda_Q}}(\widetilde{\mathrm{id}}_{D'_{\varepsilon'_Q}})\big) \circ \tilde{h}_{\hat{\varepsilon}}^{\delta^{\cdot\cdot}_Q,\varepsilon'_Q} \quad ;$$

consequently, from the commutativity of diagrams 19.11.14, 19.11.21 and 19.11.22, and from the equalities 19.11.26 and 19.11.27, it follows that

19.11.28 $$\begin{aligned}\tilde{e}^{D^{\cdot\cdot}}_{\nu} \circ \tilde{h}_{\hat{\varepsilon}}^{\lambda_R,\varepsilon'_Q} &= \tilde{g}_r \circ \tilde{h}_r \circ \tilde{h}_{\hat{\varepsilon}}^{\delta^{\cdot\cdot}_Q,\varepsilon'_Q}(T) = \tilde{g}_r \circ \tilde{h}_r \circ \tilde{e}^{\delta^{\cdot\cdot}_Q}_{\nu_Q} \circ \tilde{h}_{\hat{\varepsilon}}^{\nu_Q} \\ &= \tilde{e}^{D^{\cdot\cdot}}_{\nu_Q} \circ \tilde{g}_l \circ \tilde{h}_l \circ \tilde{h}_{\hat{\varepsilon}}^{\nu_Q}\end{aligned}$$

and therefore, for any representative h of $\tilde{h}_{\hat{\varepsilon}}^{\lambda_Q,\varepsilon'_Q}$, we have

19.11.29 $$l_\nu h(1) = h(1) = l_{\nu_Q} h(1) \quad .$$

In conclusion, these central idempotents coincide, so that there is $x' \in N'_Q$ such that $\nu'_Q(u)^{x'} = \lambda'(u)$ for any $u \in T$.

Now, setting $U^{\cdot\cdot} = (Q^\lambda{\cdot}T^{\cdot\cdot})^{(1,x')}$, it is quite clear that we have (recall that $\lambda(Z(Q)) = \nu_Q(Z(Q)) \subset T^{\cdot\cdot}$)

19.11.30 $$R^\lambda \subset U^{\cdot\cdot} \subset Q^\lambda{\cdot}C_{R^{\cdot\cdot}}(Q^\lambda) \quad \text{and} \quad C_{U^{\cdot\cdot}}(Q^\lambda) = (T^{\cdot\cdot})^{(1,x')}$$

and that, for a suitable point $\omega^{\cdot\cdot}$ of $U^{\cdot\cdot}$ on $A^{\cdot\cdot}_{(\varepsilon^{\cdot\cdot})^{(1,x')}}$, the local point $(\delta^{\cdot\cdot}_Q)^{(1,x')}$ of $C_{U^{\cdot\cdot}}(Q^\lambda)$ on $A^{\cdot\cdot}_{(\varepsilon^{\cdot\cdot})^{(1,x')}}(Q^\lambda)$ already comes from $A^{\cdot\cdot}_{\omega^{\cdot\cdot}}(Q^\lambda)$; in particular, we have $A^{\cdot\cdot}_{\omega^{\cdot\cdot}}(Q^\lambda) \neq \{0\}$ and therefore Q^λ is contained in any defect group of $\omega^{\cdot\cdot}$ in $U^{\cdot\cdot}$ (cf. 2.11.3 and Corollary 14.11); then, it is easily checked that the localness of $(\delta^{\cdot\cdot}_Q)^{(1,x')}$ forces $\omega^{\cdot\cdot}$ to be local too. Moreover, by 19.10, for a suitable $k \otimes_{\mathcal{O}} \mathcal{D}C_{U^{\cdot\cdot}}(Q^\lambda)$-module $M^{\cdot\cdot}_{\omega^{\cdot\cdot},Q^\lambda}$, we have a $\mathcal{D}C_{U^{\cdot\cdot}}(Q^\lambda)$-interior algebra isomorphism

19.11.31 $$A^{\cdot\cdot}_{\omega^{\cdot\cdot}}(Q^\lambda) \cong \mathrm{End}_k(M^{\cdot\cdot}_{\omega^{\cdot\cdot},Q^\lambda})$$

and therefore $(\delta^{\cdot\cdot}_Q)^{(1,x')}$ determines, up to isomorphisms, an indecomposable direct summand $N^{\cdot\cdot}_Q$ of $M^{\cdot\cdot}_{\omega^{\cdot\cdot},Q^\lambda}$ with vertex $C_{U^{\cdot\cdot}}(Q^\lambda)$ such that, denoting by $\varphi^{\cdot\cdot}_{x'} : C_{U^{\cdot\cdot}}(Q^\lambda) \cong T^{\cdot\cdot}$ the group isomorphism mapping $(u, u') \in C_{U^{\cdot\cdot}}(Q^\lambda)$ on $(u, x'u'x'^{-1})$, we have a $\mathcal{D}C_{U^{\cdot\cdot}}(Q^\lambda)$-interior algebra isomorphism (cf. 2.10)

19.11.32 $$\mathrm{End}_k(N^{\cdot\cdot}_Q) \cong \mathrm{Res}_{\varphi^{\cdot\cdot}_{x'}}(D^{\cdot\cdot}_{\delta^{\cdot\cdot}_Q}) \quad ;$$

on the other hand, since x' normalizes $C_{P'}(Q^{\lambda'})_{\varepsilon'_Q} = T'_{\varepsilon'_Q}$ (cf. 19.11.2 and 19.11.23) and belongs to $C_{G'}(Q^{\lambda'})$, denoting by $\varphi'_{x'}$ the group automorphism of $C_{P'}(Q^{\lambda'})$ determined by x', $\tilde{\varphi}'_{x'}$ belongs to (cf. [20, Proposition 2.14 and Theorem 3.1])

19.11.33 $$E_{C_{G'}(Q^{\lambda'})}\big(C_{P'}(Q^{\lambda'})_{\varepsilon'_Q}\big) = F_{D'}\big(C_{P'}(Q^{\lambda'})_{\varepsilon'_Q}\big)$$

and therefore we have a $\mathcal{D}C_{P'}(Q^{\lambda'})$-interior algebra isomorphism (cf. 2.13.1)

19.11.34 $$D'_{\varepsilon'_Q} \cong \mathrm{Res}_{\varphi'_{x'}}(D'_{\varepsilon'_Q}) \quad .$$

Consequently, respectively denoting by

$$\eta : C_{U^{\cdot\cdot}}(Q^{\lambda}) \to C_P(Q) \quad \text{and} \quad \eta' : C_{U^{\cdot\cdot}}(Q^{\lambda}) \to C_{P'}(Q^{\lambda'})$$

the group homomorphisms determined by π and π', we get a $\mathcal{D}C_{U^{\cdot\cdot}}(Q^{\lambda})$-interior algebra isomorphism

19.11.35 $$\mathrm{Res}_{\varphi^{\cdot\cdot}_{x'}}\big(D^{\cdot\cdot}_{\delta^{\cdot\cdot}_Q} \otimes_k \mathrm{Res}_{\tau'}(D'_{\varepsilon'_Q})\big) \cong \mathrm{End}_k(N^{\cdot\cdot}_Q) \otimes_k \mathrm{Res}_{\eta'}(D'_{\varepsilon'_Q})$$

and therefore, from exoembedding 19.11.12 and equalities 19.11.13, we obtain the announced $\mathcal{D}C_P(Q)$-interior algebra exoembedding (cf. Proposition 12.12)

19.11.36 $$\hat{A}_{\hat{\varepsilon}}(Q) \longrightarrow \mathrm{Ind}_{\eta}\big(\mathrm{End}_k(N^{\cdot\cdot}_Q) \otimes_k \mathrm{Res}_{\eta'}(D'_{\varepsilon'_Q})\big) \quad .$$

Appendix 1 A proof of Weiss' criterion for permutation $\mathcal{O}P$-modules

A1.1 Let $\mathcal{O}$ be a complete discrete valuation ring with a perfect residue field k of characteristic p and a quotient field $\mathcal{K}$ of characteristic zero. As announced in the introduction, we prove here the following Weiss' criterion guaranteeing that, for a finite p-group P, an $\mathcal{O}$-free $\mathcal{O}P$-module M of finite $\mathcal{O}$-rank is a *permutation* $\mathcal{O}P$-module (cf. 2.2).

Theorem A1.2 *Let P be a finite p-group, M an $\mathcal{O}$-free $\mathcal{O}P$-module of finite $\mathcal{O}$-rank and Q a normal subgroup of P. If $\mathrm{Res}^P_Q(M)$ is a projective $\mathcal{O}Q$-module and M^Q is a permutation $\mathcal{O}(P/Q)$-module then M is a permutation $\mathcal{O}P$-module.*

A1.3 We need the following well-known lemma; we owe to Jean-Pierre Serre the elegant ending argument of our proof. Here, we need also Lemma 13.6 which plays such an important role in proving our results on basic Rickard equivalences between blocks (see Section 13, 17 and 19); as a matter of fact, we have found this lemma when trying to find a proof of Weiss' criterion without any restriction on the ramification of $\mathcal{O}$ for our main result in Section 8. If G is a finite group and M an $\mathcal{O}$-free $\mathcal{O}G$-module of finite $\mathcal{O}$-rank, recall that $\mathcal{K} \otimes_{\mathcal{O}} M$ is a semisimple $\mathcal{K}G$-module; we call the converse image in M of each $\mathcal{K}G$-isotypic component of $\mathcal{K} \otimes_{\mathcal{O}} M$ *$\mathcal{O}G$-isotypic component* of M.

Lemma A1.4 *Assume that $\mathcal{O}$ contains a primitive p-th root of unity ζ and let M be an $\mathcal{O}$-free $\mathcal{O}$-module of finite $\mathcal{O}$-rank. Then, any finite subgroup G of $\mathrm{id}_M + (1-\zeta)\,\mathrm{End}_{\mathcal{O}}(M)$ is p-elementary abelian, and M is the direct sum of its $\mathcal{O}G$-isotypic components.*

Proof: First of all, we claim that $\mathrm{id}_M + (1-\zeta)^2\,\mathrm{End}_{\mathcal{O}}(M)$ has no nontrivial finite subgroups; indeed, arguing by contradiction, we may assume that there is $a \in (1-\zeta)^2\,\mathrm{End}_{\mathcal{O}}(M)$ such that $\mathrm{id}_M + a$ has order p; that is to say, we have $a \neq 0$ and

A1.4.1 $$\mathrm{id}_M = (\mathrm{id}_M + a)^p = \mathrm{id}_M + pa + pa^2b + a^p \quad ,$$

where $b \in \mathrm{End}_{\mathcal{O}}(M)$; but p belongs to $(1-\zeta)^{p-1}\mathcal{O}$ and therefore, denoting by h the biggest natural number such that $a \in (1-\zeta)^h\mathrm{End}_{\mathcal{O}}(M)$ (so that $h \geq 2$), pa does not belong to $(1-\zeta)^{h+p}\mathrm{End}_{\mathcal{O}}(M)$, whereas we have clearly

A1.4.2 $$pa^2 \in (1-\zeta)^{p-1+2h}\mathrm{End}_{\mathcal{O}}(M) \quad \text{and} \quad a^p \in (1-\zeta)^{ph}\mathrm{End}_{\mathcal{O}}(M) \quad ;$$

since $p-1+2h$ and ph are not smaller than $h+p$, the equality $-pa = pa^2b + a^p$ is impossible.

Consequently, G is injectively mapped into

A1.4.3 $$\big(\mathrm{id}_M + (1-\zeta)\mathrm{End}_{\mathcal{O}}(M)\big)/\big(\mathrm{id}_M + (1-\zeta)^2\mathrm{End}_{\mathcal{O}}(M)\big)$$

which is a p-elementary abelian group. Now, arguing by induction on $d = \mathrm{rank}_{\mathcal{O}}(M)$, we may assume that $d > 1$; set $V = K \otimes_{\mathcal{O}} M$, identify $\mathrm{End}_{\mathcal{O}}(M)$ to its canonical image in $\mathrm{End}_{\mathcal{K}}(V)$ and consider the action of G on V; we may also assume that there is $x \in G$ having at least two different eigenvalues λ and λ' on V, and then it suffices to prove that the idempotent

A1.4.4
$$i = \frac{1}{p}\sum_{h=0}^{p-1} \lambda^h x^{-h}$$

belongs to $\mathrm{End}_{\mathcal{O}}(M)$; indeed, in that case, we have

A1.4.5
$$M = i(M) + (\mathrm{id}_M - i)(M)$$

where both terms in the right member are nonzero G-stable $\mathcal{O}$-modules, so that they are the direct sum of their $\mathcal{O}G$-isotypic components by the induction hypothesis.

Since x belongs to $\mathrm{id}_M + (1-\zeta)\mathrm{End}_{\mathcal{O}}(M)$ and λ to $1 + (1-\zeta)\mathcal{O}$, we have $\lambda x^{-1} = \mathrm{id}_M + (1-\zeta)c$ for a suitable $c \in \mathrm{End}_{\mathcal{O}}(M)$ and therefore we get

A1.4.6
$$\begin{aligned} i &= \frac{1}{p}\sum_{h=0}^{p-1}(\mathrm{id}_M + (1-\zeta)c)^h = \frac{1}{p}\sum_{h=0}^{p-1}\left(\sum_{l=0}^{h}\binom{h}{l}(1-\zeta)^l c^l\right) \\ &= \sum_{l=0}^{p-1}\left(\frac{1}{p}\sum_{h=1}^{p-1}\binom{h}{l}\right)(1-\zeta)^l c^l \quad ; \end{aligned}$$

but, it is easily checked by induction that, for any natural numbers, n, m we have

A1.4.7
$$\sum_{h=m}^{n}\binom{h}{m} = \binom{n+1}{m+1} \quad ;$$

consequently, we obtain

A1.4.8
$$\begin{aligned} i &= \sum_{l=0}^{p-1}\frac{1}{p}\binom{p}{l+1}(1-\zeta)^l c^l \\ &= \sum_{l=1}^{p-1}\frac{1}{p}\binom{p}{l}(1-\zeta)^{l-1}c^{l-1} + \frac{1}{p}(1-\zeta)^{p-1}c^{p-1} \end{aligned}$$

and any term of the sum in the right member belongs to $\mathrm{End}_{\mathcal{O}}(M)$.

A1.5 From now on, we prove Theorem A1.2; we argue by induction on $|P|$ first and on $|Q|$ after, and we may assume that M is a faithful indecomposable $\mathcal{O}P$-module and that Q is not trivial. Let Z be a subgroup of $Z(P) \cap Q$ of order p and set $\tilde{P} = P/Z$ and $\tilde{Q} = Q/Z$; since $\mathrm{Res}^P_Q(M)$ is projective, $\mathrm{Res}^P_Z(M)$ and $\mathrm{Res}^{\tilde{P}}_{\tilde{Q}}(M^Z)$

are projective too; hence, since $(M^Z)^{\tilde{Q}} = M^Q$, it follows from the induction hypothesis that M^Z is a permutation $\mathcal{O}\tilde{P}$-module and then, if $Z \neq Q$, applying again the induction hypothesis, M is a permutation $\mathcal{O}P$-module.

A1.6 From now on, we assume that $Z = Q$. Assume moreover that $\tilde{P}$ acts non-trivially on M^Z; in that case, since M^Z is a permutation $\mathcal{O}\tilde{P}$-module, an indecomposable direct summand of M^Z has a vertex $\tilde{R}$ different from $\tilde{P}$; then, by the induction hypothesis applied to the converse image R of $\tilde{R}$ in $\tilde{P}$, $\mathrm{Res}^P_R(M)$ is a permutation $\mathcal{O}R$-module and therefore, according to Lemma 13.6, the inclusion $M^Z \subset M$ induces a $k\overline{N}_P(R)$-module isomorphism (notice that $\overline{N}_P(R) = N_P(R)/R \cong N_{\tilde{P}}(\tilde{R})/\tilde{R}$)

A1.6.1 $$(M^Z)(\tilde{R}) \cong \bigoplus_T M(T)^Z \quad ,$$

where T runs on the set of complements of Z in R and $\overline{N}_P(R)$ acts on the right member of that isomorphism by permuting this set and, for any such a complement T, by the natural action of $N_P(T)$ on $M(T)^Z$. But, according to our choice of $\tilde{R}$, since $(M^Z)(\tilde{R})$ is isomorphic to the multiplicity module on the $\mathcal{O}\tilde{P}$-interior algebra $\mathrm{End}_{\mathcal{O}}(M^Z)$ corresponding to the trivial $\mathcal{O}\tilde{R}$-module (cf. 2.2.5 and 2.8), the $k\overline{N}_P(R)$-module $(M^Z)(\tilde{R})$ has a direct summand isomorphic to $k\overline{N}_P(R)$ (cf. 2.9.3). Consequently, there is a complement T of Z in R such that the $k\big(N_P(T)/R\big)$-module $M(T)^Z$ has a direct summand isomorphic to $k\big(N_P(T)/R\big)$ and, in particular, we get (cf. 2.2.6)

A1.6.2 $$\big(M(T)^Z\big)_1^{N_P(T)/R} \neq \{0\} \quad .$$

A1.7 Moreover, since $\mathrm{Res}^P_Z(M)$ is projective and $\mathrm{Res}^P_R(M)$ is a permutation $\mathcal{O}R$-module, Z acts freely on an R-stable $\mathcal{O}$-basis X of M; hence, since $\{\mathrm{Br}_T(x)\}_{x\in X^T}$ is a k-basis of $M(T)$ (cf. 2.2.5), we have actually

A1.7.1 $$M(T)^Z = M(T)_1^Z = M(T)_1^{R/T} \quad ;$$

so, it follows from A1.6.2 and A1.7.1 that

A1.7.2 $$M(T)_1^{\overline{N}_P(T)} \neq \{0\} \quad .$$

But, it is easily checked that the natural action of $\mathrm{End}_k\big(M(T)\big)^{\overline{N}_P(T)}$ on $M(T)_1^{\overline{N}_P(T)}$ induces a surjective k-algebra homomorphism

A1.7.3 $$\mathrm{End}_k\big(M(T)\big)_1^{\overline{N}_P(T)} \longrightarrow \mathrm{End}_k\big(M(T)_1^{\overline{N}_P(T)}\big) \quad .$$

Consequently, denoting by α the unique point of P on $\mathrm{End}_{\mathcal{O}}(M)$ (so that $\alpha = \{\mathrm{id}_M\}$) and by τ the local point of T on $\mathrm{End}_{\mathcal{O}}(M)$ determined by the trivial $\mathcal{O}T$-module $\mathcal{O}$, it follows from 2.2.5, A1.7.2 and A1.7.3 that

A1.7.4 $$\big(\mathrm{End}_{\mathcal{O}}(M)\big)(T_\tau)_1^{\overline{N}_P(T)} \neq J\Big(\big(\mathrm{End}_{\mathcal{O}}(M)\big)(T_\tau)_1^{\overline{N}_P(T)}\Big)$$

and therefore T_τ is a defect pointed group of P_α (cf. 2.9.2 and 2.9.4), so that we have $M \cong \mathrm{Ind}_T^P(\mathcal{O})$ (cf. 2.12.2) and $T = 1$.

A1.8 Finally, assume that $M^Z = M^P$ and, setting $\overline{\mathcal{O}Z} = \mathcal{O}Z/\mathcal{O}(\Sigma_{z\in Z} z)$ and $\overline{\mathcal{O}} = \mathcal{O}/p\mathcal{O}$, consider the evident commutative diagram

A1.8.1
$$\begin{array}{ccc} \overline{\mathcal{O}Z} & \xrightarrow{\bar{\varepsilon}} & \overline{\mathcal{O}} \\ \uparrow & & \uparrow \\ \mathcal{O}Z & \xrightarrow{\varepsilon} & \mathcal{O} \end{array} ,$$

where ε is the augmentation map; we claim that this diagram is a pull-back; indeed, any pair $(\Sigma_{z\in Z}\lambda_z\bar{z}, \lambda)$, where $\lambda \in \mathcal{O}$ and, for any $z \in Z$, $\lambda_z \in \mathcal{O}$ and $\bar{z}$ is the image of z in $\overline{\mathcal{O}Z}$, such that $\Sigma_{z\in Z}\bar{\lambda}_z = \bar{\lambda}$ in $\overline{\mathcal{O}}$ comes from the following element of $\mathcal{O}Z$

A1.8.2
$$\sum_{z\in Z}\lambda_z z + \frac{1}{p}\left(\lambda - \sum_{z\in Z}\lambda_z\right)\sum_{z\in Z} z \quad .$$

Now, since $\mathrm{Res}_Z^P(M)$ is projective, for a suitable n we have an $\mathcal{O}Z$-module isomorphism

A1.8.3
$$\mathrm{End}_{\mathcal{O}Z}(M) \cong (\mathcal{O}Z)^n \quad ;$$

hence, by considering diagram A1.8.1 as a pull-back of $\mathcal{O}Z$-modules, the tensor product of this diagram by the $\mathcal{O}Z$-module $\mathrm{End}_{\mathcal{O}Z}(M)$ remains a pull-back of $\mathcal{O}Z$-modules; moreover, always because of $\mathrm{Res}_Z^P(M)$ is projective, it is easily checked that the canonical maps

A1.8.4
$$\overline{\mathcal{O}Z} \underset{\mathcal{O}Z}{\otimes} \mathrm{End}_{\mathcal{O}Z}(M) \longrightarrow \mathrm{End}_{\overline{\mathcal{O}Z}}(M/M^Z) \text{ and } \mathcal{O} \underset{\mathcal{O}Z}{\otimes} \mathrm{End}_{\mathcal{O}Z}(M) \longrightarrow \mathrm{End}_{\mathcal{O}}(M^Z)$$

are bijective. Consequently, we get the following pull-back of surjective canonical $\mathcal{O}$-algebra homomorphisms

A1.8.5
$$\begin{array}{ccc} \mathrm{End}_{\overline{\mathcal{O}Z}}(M/M^Z) & \xrightarrow{\bar{\varepsilon}\otimes \mathrm{id}} & \mathrm{End}_{\overline{\mathcal{O}}}(M^Z/pM^Z) \\ \uparrow & & \uparrow \\ \mathrm{End}_{\mathcal{O}Z}(M) & \xrightarrow{\varepsilon\otimes \mathrm{id}} & \mathrm{End}_{\mathcal{O}}(M^Z) \end{array} ,$$

where $\bar{\varepsilon} \otimes \mathrm{id}$ is determined by the canonical action of $\mathrm{End}_{\mathcal{O}Z}(M)$ on M/M^Z and on M^Z/pM^Z, and, choosing $z \in Z - \{1\}$, it is easily checked that

A1.8.6
$$\mathrm{Ker}(\varepsilon \otimes \mathrm{id}) = (1-z)\cdot\mathrm{End}_{\mathcal{O}Z}(M) \quad ;$$

in particular, since $M^Z = M^P$, the image of P in $\mathrm{End}_{\mathcal{O}Z}(M)$ is contained in $\mathrm{id}_M + \mathrm{Ker}(\varepsilon \otimes \mathrm{id})$ and therefore, setting $\overline{M} = M/M^Z$ and denoting by $\overline{P}$ the image of P in $\mathrm{End}_{\overline{\mathcal{O}Z}}(\overline{M})$, we have

A1.8.7
$$\overline{P} \subset \mathrm{id}_{\overline{M}} + (1-\bar{z})\mathrm{End}_{\overline{\mathcal{O}Z}}(\overline{M}) \quad .$$

A1.9 Now, let $\mathcal{O}_u$ be the absolutely unramified subring of $\mathcal{O}$ having the same residue field k (cf. [37, Chap. II, §5, Theorem 4]) and set $X = 1 + J(\mathcal{O})$; since $\mathcal{O} = \mathcal{O}_u + J(\mathcal{O})$, the $\mathcal{O}_u$-algebra $\mathcal{O}$ is generated by X and, in particular, we have

A1.9.1 $$\mathrm{End}_{\overline{\mathcal{O}Z}}(\overline{M}) = \mathrm{End}_{\overline{\mathcal{O}_u Z}}(\overline{M})^X \quad ,$$

where $\overline{\mathcal{O}_u Z}$ denotes the image of $\overline{\mathcal{O}_u Z}$ in $\overline{\mathcal{O}Z}$ which is actually a complete discrete valuation ring, namely it is, up to isomorphism, the extension of $\mathcal{O}_u$ by the group of all the p-th roots of unity (cf. [37, Chap. I, §6, Proposition 18]); moreover, since $\mathrm{Res}^P_Z(M)$ is a free $\mathcal{O}Z$-module, $\overline{M}$ is clearly a free $\overline{\mathcal{O}Z}$-module and therefore it is also a free $\overline{\mathcal{O}_u Z}$-module of finite rank since $\mathcal{O}$ is a finitely generated $\mathcal{O}_u$-module (cf. [37, Chap. II, §5, Theorem 4]). Hence, according to Lemma A1.4 applied to the free $\overline{\mathcal{O}_u Z}$-module $\overline{M}$, the subgroup $\overline{P}$ of $\mathrm{id}_{\overline{M}} + (1 - \bar{z})\mathrm{End}_{\overline{\mathcal{O}_u Z}}(\overline{M})$ is p-elementary abelian and $\overline{M}$ is the direct sum of its $(\overline{\mathcal{O}_u Z})\overline{P}$-isotypic components; moreover, since X centralizes $\overline{P}$, X stabilizes each $(\overline{\mathcal{O}_u Z})\overline{P}$-isotypic component of $\overline{M}$ and, in particular, it centralizes the set $\overline{I}$ of idempotents in $\mathrm{End}_{\overline{\mathcal{O}_u Z}}(\overline{M})^{\overline{P}}$ determined by that direct sum decomposition of $\overline{M}$, so that by A1.9.1 we get

A1.9.2 $$\overline{I} \subset \mathrm{End}_{\overline{\mathcal{O}Z}}(\overline{M})^{\overline{P}} \quad .$$

A1.10 Since all the maps in diagram A1.8.5 are surjective, the family $\{(\bar{\varepsilon} \otimes \mathrm{id})(\bar{\imath})\}_{\bar{\imath} \in \overline{I}}$ of pairwise orthogonal idempotents in $\mathrm{End}_{\overline{\mathcal{O}}}(M^Z/pM^Z)$ can be lifted to a family $\{l_{\bar{\imath}}\}_{\bar{\imath} \in \overline{I}}$ of pairwise orhogonal idempotents of $\mathrm{End}_{\mathcal{O}}(M^Z)$ and then, since diagram A1.8.5 is a pull-back, $\{(\bar{\imath}, l_{\bar{\imath}})\}_{\bar{\imath} \in \overline{I}}$ determines a family of pairwise orthogonal idempotents in $\mathrm{End}_{\mathcal{O}Z}(M)$; moreover, since P centralizes $\overline{I}$ and acts trivially on M^Z, P centralizes this family of pairwise orthogonal idempotents in $\mathrm{End}_{\mathcal{O}Z}(M)$ and therefore, since M is an indecomposable $\mathcal{O}P$-module, we have necessarily $\overline{I} = \{\mathrm{id}_{\overline{M}}\}$, so that $\overline{P} = \overline{Z}$; but, since diagram A1.8.5 is a pull-back, we have $P \cong \overline{P}$; consequently, we get $P = Z$ and $M \cong \mathcal{O}Z$.

Appendix 2 Tensor induction of $\mathcal{D}$-interior G-algebras

A2.1 Let G be a finite group, H a subgroup of G and N a $\mathcal{D}H$-module (as usual, we assume and omit to say that N is $\mathcal{O}$-finite); recall that, considering the restriction of N to $\mathcal{O}H$, the *tensor induction* of this restriction from H to G is the $\mathcal{O}G$-module $\mathrm{Ten}_H^G(N)$ defined, following Serre, as follows: for any class $C \in G/H$, we denote by $\mathcal{O}C$ the $\mathcal{O}$-free $\mathcal{O}$-module over C endowed with its natural right $\mathcal{O}H$-module structure, and then it is clear that G acts by left multiplication on the family of $\mathcal{O}$-modules $\{\mathcal{O}C \otimes_{\mathcal{O}H} N\}_{C\in G/H}$, so that G acts also on the tensor product of this family (cf. 2.5), namely

A2.1.1
$$\mathrm{Ten}_H^G(N) = \bigotimes_{C\in G/H} (\mathcal{O}C \underset{\mathcal{O}H}{\otimes} N)$$

is indeed an $\mathcal{O}G$-module. In order to endow $\mathrm{Ten}_H^G(N)$ with a reasonable $\mathcal{D}$-module structure from the $\mathcal{D}$-module structure of N, the problem is that, although for any $C \in G/H$ the $\mathcal{O}$-module $\mathcal{O}C \otimes_{\mathcal{O}H} N$ has an evident $\mathcal{D}$-module structure as a tensor product of $\mathcal{D}H$-modules (cf. 10.2), just by endowing $\mathcal{O}C$ with the trivial $\mathcal{D}$-module structure, the definition of a $\mathcal{D}$-module structure on the tensor product of a family of $\mathcal{D}$-modules needs (unless $\mathrm{char}(\mathcal{O}) = 2$!) the choice of a total order relation in the set of indices: obviously, if $H \neq G$, G does not stabilize such a choice in G/H. However, we show in this appendix how Proposition 9.18 provides a reasonable definition of tensor induction of $\mathcal{D}H$-modules.

A2.2 Actually, with no extra-effort, we discuss here on the more general case of the $\mathcal{D}$-interior H-algebras (cf. 11.4) which we think it provides a better framework. If B is a $\mathcal{D}$-interior H-algebra, in particular it has an $\mathcal{O}H$-module structure and, for any class $C \in G/H$, the $\mathcal{O}$-module $\mathcal{O}C \otimes_{\mathcal{O}H} B$ has the $\mathcal{O}$-algebra structure given by

A2.2.1
$$(y \otimes b)(y \otimes b') = y \otimes bb'$$

for any $y \in C$ and any $b, b' \in B$; hence, the $\mathcal{O}G$-module $\mathrm{Ten}_H^G(B)$ has the $\mathcal{O}$-algebra structure corresponding to the tensor product of $\mathcal{O}$-algebras, which is G-stable since, for any $C \in G/H$ and any $x \in G$, the left multiplication by x determines an $\mathcal{O}$-algebra isomorphism

A2.2.2
$$\mathcal{O}C \underset{\mathcal{O}H}{\otimes} B \cong \mathcal{O}(xC) \underset{\mathcal{O}H}{\otimes} B \quad ,$$

and consequently, $\mathrm{Ten}_H^G(B)$ becomes a G-algebra; notice that the tensor factor corresponding to H determines an H-subalgebra canonically isomorphic to B and we denote by

A2.2.3
$$n_H^G(B) : B \longrightarrow \mathrm{Res}_H^G(\mathrm{Ten}_H^G(B))$$

the corresponding H-algebra homomorphism. If B' is a second $\mathcal{D}$-interior H-algebra, it is clear that any $\mathcal{D}$-interior H-algebra homomorphism $g : B \to B'$ induces an

$\mathcal{O}$-algebra homomorphism

A2.2.4 $$\mathrm{id}_{\mathcal{O}C} \underset{\mathcal{O}H}{\otimes} g : \mathcal{O}C \underset{\mathcal{O}H}{\otimes} B \longrightarrow \mathcal{O}C \underset{\mathcal{O}H}{\otimes} B'$$

for any $C \in G/H$, and therefore it determines a G-algebra homomorphism

A2.2.5 $$\mathrm{Ten}_H^G(g) : \mathrm{Ten}_H^G(B) \longrightarrow \mathrm{Ten}_H^G(B')$$

fulfilling $\mathrm{Ten}_H^G(g) \circ n_H^G(B) = n_H^G(B')$.

Remark A2.3 Even when B is a $\mathcal{D}H$-interior algebra, $\mathrm{Ten}_H^G(B)$ need not come from an $\mathcal{O}G$-interior algebra; for instance, if H is a nontrivial abelian group then $\mathrm{Ten}_H^G(\mathcal{O}H)$ is a commutative $\mathcal{O}$-algebra whereas G acts nontrivially on it unless $H = G$.

A2.4 Consider the structural $\mathcal{O}$-algebra homomorphism

A2.4.1 $$s : \mathcal{D} \longrightarrow B^H \qquad ;$$

for any $C \in G/H$, s determines the $\mathcal{O}$-algebra homomorphism

A2.4.2 $$s_C : \mathcal{D} \longrightarrow \mathcal{O}C \underset{\mathcal{O}H}{\otimes} B$$

mapping $c \in \mathcal{D}$ on $y \otimes s(c)$ for any $y \in C$, so that $\mathcal{O}C \otimes_{\mathcal{O}H} B$ becomes a $\mathcal{D}$-interior algebra, and it is clear that isomorphisms A2.2.2 become $\mathcal{D}$-interior algebra isomorphisms; now, according to 9.17, for any total order relation $R \subset G/H \times G/H$ on G/H, we have a canonical $\mathcal{O}$-algebra homomorphism

A2.4.3 $$r_R : \mathcal{D} \longrightarrow \mathrm{Ten}_H^G(B)$$

given by (cf. 9.17.1)

A2.4.4 $$r_R = \Delta^*_{\mathrm{Ten}_H^G(B),R}\big(\bigotimes_{C\in G/H} s_C\big) \qquad ,$$

so that $\mathrm{Ten}_H^G(B)$ becomes a $\mathcal{D}$-interior algebra, and moreover, if H is the first class with respect to R, it is not difficult to check that $n_H^G(B)$ is a $\mathcal{D}$-algebra homomorphism (cf. 11.6); notice that the restriction of r_R to $\mathcal{F}$ does not depend on the choice of R, which allows us to consider $\mathrm{Ten}_H^G(B)$ as an $\mathcal{F}$-interior G-algebra; notice also that G acts on the set $\mathrm{Ord}(G/H)$ of all the total order relations on G/H and that, for any $x \in G$ and any $c \in \mathcal{D}$, we have

A2.4.5 $$r_R(c)^{x^{-1}} = r_{x\cdot R}(c) \qquad .$$

A2.5 On the other hand, denoting by $s_1 : \mathcal{D}_1 \to B$ the restriction of s (cf. 9.10.3) and considering $\mathcal{D}_1$ endowed with the trivial action of H as an H-algebra, we have the G-algebra homomorphism

A2.5.1 $$\mathrm{Ten}_H^G(s_1) : \mathrm{Ten}_H^G(\mathcal{D}_1) \longrightarrow \mathrm{Ten}_H^G(B)$$

and consequently, denoting by $\mathcal{T}_1$ the subgroup of all the involutions in $\mathrm{Ten}_H^G(\mathcal{F}_1)^*$, which is clearly G-stable, setting $\overline{\mathcal{T}}_1 = \mathcal{T}_1/\{1,-1\}$ and considering the semidirect product $\overline{\mathcal{T}}_1 \rtimes G$, $\mathrm{Ten}_H^G(B)$ becomes actually a $\overline{\mathcal{T}}_1 \rtimes G$-algebra; moreover, this semidirect product acts already on $\mathrm{Ten}_H^G(\mathcal{D}_1)$, so that it acts on the set of $\mathcal{O}$-algebra homomorphisms (cf. 9.16 and 9.17)

A2.5.2 $$(\Delta_{1,R})^{t'} : \mathcal{D}_1 \longrightarrow \mathrm{Ten}_H^G(\mathcal{D}_1)$$

when t' runs on $\mathcal{T}_1$, R runs on $\mathrm{Ord}(G/H)$ and, for any $R \in \mathrm{Ord}(G/H)$, $\Delta_{1,R}$ is the iterated coproduct defined in 9.16 with $X = G/H$; then, it follows from Proposition 9.18 that, for any $R \in \mathrm{Ord}(G/H)$, the stabilizer G_R of $\Delta_{1,R}$ in that semidirect product is a complement of $\overline{\mathcal{T}}_1$ and therefore the canonical map $\overline{\mathcal{T}}_1 \rtimes G \to G$ determines a group isomorphism

A2.5.3 $$\varphi_R : G_R \cong G \quad .$$

A2.6 Now, for any $R \in \mathrm{Ord}(G/H)$, the $\mathcal{O}$-algebra $\mathrm{Ten}_H^G(B)$, endowed with the homomorphism r_R and with the *new* action of G defined throughout the action of G_R and the isomorphism $(\varphi_R)^{-1}$, becomes a $\mathcal{D}$-interior G-algebra that we denote by $\mathrm{Den}_H^G(B, R)$ to avoid confusion (notice that the G-algebra structure on $\mathrm{Den}_H^G(B, R)$ need not be isomorphic to $\mathrm{Ten}_H^G(B)$); moreover, for any $R' \in \mathrm{Ord}(G/H)$, it follows from Proposition 9.18 that there is a unique $\bar{t}_{R,R'} \in \overline{\mathcal{T}}_1$ such that

A2.6.1 $$(\Delta_{1,R})^{\bar{t}_{R,R'}} = \Delta_{1,R'}$$

and then, since $\mathcal{D} = \mathcal{F} \cdot \mathcal{D}_1$ and the $\mathcal{O}$-algebra homomorphisms r_R and $r_{R'}$ coincide over $\mathcal{F}$, the action of $\bar{t}_{R,R'}$ on $\mathrm{Ten}_H^G(B)$ defines a $\mathcal{D}$-interior G-algebra isomorphism

A2.6.2 $$t_{R,R'} : \mathrm{Den}_H^G(B, R) \cong \mathrm{Den}_H^G(B, R') \quad .$$

Notice that, if H is the first class with respect to both R and R', it follows also from Proposition 9.18, applied this time to the set $G/H - \{H\}$, that $t_{R,R'}$ centralizes the image of $n_H^G(B)$; in particular, setting

A2.6.3 $$\mathrm{Ord}_\circ(G/H) = \big\{R \in \mathrm{Ord}(G/H) | (H, C) \in R \text{ for any } C \in G/H\big\} \quad ,$$

it is not difficult to check that, for any $R \in \mathrm{Ord}_\circ(G/H)$, the map

A2.6.4 $$n_H^G(B) : B \to \mathrm{Res}_H^G\big(\mathrm{Den}_H^G(B, R)\big)$$

is actually a $\mathcal{D}H$-algebra homomorphism (cf. 11.6). Then, following Pierre Deligne (cf. [11, 1.1.4.2.1]), in order to avoid any choice we define the *differential tensor induction* $\mathrm{Den}_H^G(B)$ of B from H to G as the $\mathcal{D}$-interior G-subalgebra of the direct product (cf. 2.5 and 11.11)

A2.6.5
$$\prod_{R\in \mathrm{Ord}_\circ(G/H)} \mathrm{Den}_H^G(B, R)$$

formed by the elements $(a_R)_{R\in \mathrm{Ord}_\circ(G/H)}$ of it fulfilling

A2.6.6
$$t_{R,R'}(a_R) = a_{R'}$$

for any $R, R' \in \mathrm{Ord}_\circ(G/H)$; evidently, for any $R \in \mathrm{Ord}_\circ(G/H)$, the corresponding projection map determines a $\mathcal{D}$-interior G-algebra isomorphism

A2.6.7
$$f_R : \mathrm{Den}_H^G(B) \cong \mathrm{Den}_H^G(B, R)$$

and we have a $\mathcal{D}H$-algebra homomorphism

A2.6.8
$$m_H^G(B) : B \longrightarrow \mathrm{Res}_H^G\big(\mathrm{Den}_H^G(B)\big)$$

such that $f_R \circ m_H^G(B) = n_H^G(B)$ for any $R \in \mathrm{Ord}_\circ(G/H)$. Next proposition gives a handy description of $\mathrm{Den}_H^G(B)$.

Proposition A2.7 *Let A be a $\mathcal{D}$-interior G-algebra, B a $\mathcal{D}$-interior H-algebra and $m : B \to \mathrm{Res}_H^G(A)$ a $\mathcal{D}H$-algebra homomorphism. There is a $\mathcal{D}$-interior G-algebra isomorphism $f : A \cong \mathrm{Den}_H^G(B)$ such that $f \circ m = m_H^G(B)$ if and only if there is an $\mathcal{F}$-interior algebra isomorphism $g : A \cong \mathrm{Ten}_H^G(B)$ such that $g \circ m = n_H^G(B)$, fulfilling the following two conditions*

A2.7.1 Denoting by $r_1 : \mathcal{D}_1 \to A$ and $s_1 : \mathcal{D}_1 \to B$ the restriction of the structural maps, we have $g \circ r_1 = \mathrm{Ten}_H^G(s_1) \circ \Delta_{1,R}$ for a suitable R in $\mathrm{Ord}_\circ(G/H)$.

A2.7.2 For a suitable map $\varphi : G \to \mathcal{T}_1$ we have $\varphi(x)\cdot g(a^x) = g(a)^x \cdot \varphi(x)$ for any $a \in A$ and any $x \in G$.

Moreover, in that case f is unique.

Proof: If $A \cong \mathrm{Den}_H^G(B)$ is a $\mathcal{D}$-interior G-algebra isomorphism such that $f \circ m = m_H^G(B)$, it suffices to choose $R \in \mathrm{Ord}_\circ(G/H)$ and to consider $g = f_R \circ f$ from A to $\mathrm{Den}_H^G(B, R)$; since $\mathrm{Den}_H^G(B)$ is just $\mathrm{Ten}_H^G(B)$ endowed with its canonical $\mathcal{F}$-interior algebra structure (cf. A2.4), with the $\mathcal{O}$-algebra homomorphism $\mathrm{Ten}_H^G(s_1) \circ \Delta_{1,R}$ (cf. 9.17.2) and with the action of G induced by the inclusion $G_R \subset \overline{\mathcal{T}_1} \rtimes G$, g fulfills clearly all the requirements. Moreover, a $\mathcal{D}$-interior G-algebra automorphism of $\mathrm{Den}_H^G(B)$ which induces the identity on $\mathrm{Im}\big(m_H^G(B)\big)$, induces also the identity on $t'\cdot\mathrm{Im}\big(m_H^G(B)\big)^x\cdot t'$ for any $t' \in \mathcal{T}_1$ and any $x \in G$, and therefore it is the identity map since the union of these $\mathcal{O}$-subalgebras generates $\mathrm{Den}_H^G(B)$ as $\mathcal{O}$-algebra.

Conversely, if $g : A \cong \mathrm{Ten}_H^G(B)$ is an $\mathcal{F}$-interior algebra isomorphism fulfilling $g \circ m = n_H^G(B)$ and both conditions above, it follows from the equality $\mathcal{D} = \mathcal{F} \cdot \mathcal{D}_1$ and condition A2.7.1 that g is a $\mathcal{D}$-interior algebra isomorphism from A to $\mathrm{Den}_H^G(B, R)$ for a suitable $R \in \mathrm{Ord}_\circ(G/H)$; moreover, according to condition A2.7.2, the action g_x of $x \in G$ on $\mathrm{Ten}_H^G(B)$ induced by g and by the action of x on A is contained in the actions on $\mathrm{Ten}_H^G(B)$ of the elements in the converse image of x in $\overline{\mathcal{T}}_1 \rtimes G$ and certainly fixes the $\mathcal{O}$-algebra homomorphism $\mathrm{Ten}_H^G(s_1) \circ \Delta_{1,R}$ (cf. 9.17.2). But, if the element $\mathfrak{s}$ ($\in \mathcal{F}_1$) acts trivially on B then $\overline{\mathcal{T}}_1$ acts trivially on A, and otherwise it is not difficult to check that the restriction of $\mathrm{Ten}_H^G(s_1)$ to the set

$$\left\{(\Delta_{1,R})^{t'}(d)\right\}_{t' \in \mathcal{T}_1, R \in \mathrm{Ord}(G/H)}$$

is injective. Consequently, in all the cases, g_x coincides with the action of $(\varphi_R)^{-1}(x)$ on $\mathrm{Ten}_H^G(B)$ for any $x \in G$. In conclusion, g is actually a $\mathcal{D}$-interior G-algebra isomorphism from A to $\mathrm{Den}_H^G(B, R)$ and therefore

A2.7.3 $$(f_R)^{-1} \circ g : A \cong \mathrm{Den}_H^G(B)$$

is a $\mathcal{D}$-interior G-algebra isomorphism such that

A2.7.4 $$(f_R)^{-1} \circ g \circ m = (f_R)^{-1} \circ n_H^G(B) = m_H^G(B) \quad .$$

Proposition A2.8 *Let B and B' be $\mathcal{D}$-interior H-algebras and $g : B \to B'$ a $\mathcal{D}$-interior H-algebra homomorphism. For any $R \in \mathrm{Ord}(G/H)$, the G-algebra homomorphism $\mathrm{Ten}_H^G(g)$ is also a $\mathcal{D}$-interior G-algebra homomorphism*

A2.8.1 $$\mathrm{Ten}_H^G(g) : \mathrm{Den}_H^G(B, R) \longrightarrow \mathrm{Den}_H^G(B', R) \quad .$$

Proof: For any $C \in G/H$, it is clear that the $\mathcal{O}$-algebra homomorphism

A2.8.2 $$\mathrm{id}_{\mathcal{O}C} \underset{\mathcal{O}H}{\otimes} g : \mathcal{O}C \underset{\mathcal{O}H}{\otimes} B \longrightarrow \mathcal{O}C \underset{\mathcal{O}H}{\otimes} B'$$

is actually a $\mathcal{D}$-interior algebra homomorphism (cf. A2.2.2); hence, it is easily checked from 9.16 and 9.17 that the map

A2.8.3 $$\mathrm{Ten}_H^G(g) : \mathrm{Den}_H^G(B, R) \longrightarrow \mathrm{Den}_H^G(B', R)$$

is a $\mathcal{D}$-interior algebra homomorphism for any $R \in \mathrm{Ord}(G/H)$. On the other hand, it is easily checked from A2.4 that $\mathrm{Ten}_H^G(g)$ is a $\overline{\mathcal{T}}_1 \rtimes G$-algebra homomorphism from $\mathrm{Ten}_H^G(B)$ to $\mathrm{Ten}_H^G(B')$ and, in particular, for any $R \in \mathrm{Ord}(G/H)$, it is a G_R-algebra homomorphism.

A2.9 Assume now that $B = \mathrm{End}_{\mathcal{O}}(N)$, where N is a $\mathcal{D}H$-module, so that B is actually a $\mathcal{D}H$-interior algebra and, in order to solve our initial problem, we have to exhibit a suitable $\mathcal{D}G$-interior algebra structure in $\mathrm{End}_{\mathcal{O}}\big(\mathrm{Ten}_H^G(N)\big)$. First of all, for any class $C \in G/H$, we have the $\mathcal{D}$-interior algebra isomorphism

A2.9.1 $$\mathcal{O}C \underset{\mathcal{O}H}{\otimes} B \cong \mathrm{End}_{\mathcal{O}}(\mathcal{O}C \underset{\mathcal{O}H}{\otimes} N)$$

which, for any $y \in C$ and any $b \in B$, maps $y \otimes b$ on the $\mathcal{D}$-module endomorphism of $\mathcal{O}C \otimes_{\mathcal{O}H} N$ mapping $y \otimes n$ on $y \otimes b(n)$ for any $n \in N$; consequently, since this family of $\mathcal{D}$-interior algebra isomorphisms is clearly compatible with the left multiplication by G, we get a G-algebra isomorphism

A2.9.2 $$\mathrm{Ten}_H^G(B) \cong \bigotimes_{C \in G/H} \mathrm{End}_{\mathcal{O}}(\mathcal{O}C \underset{\mathcal{O}H}{\otimes} N) \quad .$$

Moreover, we have a canonical $\mathcal{O}$-algebra homomorphism (cf. A2.1.1)

A2.9.3 $$\bigotimes_{C \in G/H} \mathrm{End}_{\mathcal{O}}(\mathcal{O}C \underset{\mathcal{O}H}{\otimes} N) \longrightarrow \mathrm{End}_{\mathcal{O}}\big(\mathrm{Ten}_H^G(N)\big)$$

which, as it is easily checked, is actually a G-algebra homomorphism (and it is bijective whenever N is $\mathcal{O}$-free). Hence, by composition, we get a G-algebra homomorphism

A2.9.4 $$\mathrm{Ten}_H^G(B) \longrightarrow \mathrm{End}_{\mathcal{O}}\big(\mathrm{Ten}_H^G(N)\big)$$

which endows $\mathrm{Ten}_H^G(N)$ both with a $\mathcal{D}$-module structure via the homomorphism r_R for any $R \in \mathrm{Ord}(G/H)$, and with an $\mathcal{O}\mathcal{T}_1$-module structure via the homomorphism $\mathrm{Ten}_H^G(s_1)$, so that $\mathrm{Ten}_H^G(N)$ becomes an $\mathcal{O}(\mathcal{T}_1 \rtimes G)$-module.

A2.10 Since we can identify $\mathrm{Ten}_H^G(\mathcal{F}_1)$ to the commutative $\mathcal{O}$-free G-algebra of all the $\mathcal{O}$-valued functions on $(\mathbf{Z}/2\mathbf{Z})^{G/H}$ (cf. 9.3), the evaluation on the zero element of $(\mathbf{Z}/2\mathbf{Z})^{G/H}$ provides a G-algebra homomorphism

A2.10.1 $$\varepsilon : \mathrm{Ten}_H^G(\mathcal{F}_1) \longrightarrow \mathcal{O}$$

and therefore $\mathcal{T}_1 \cap \big(1+\mathrm{Ker}(\varepsilon)\big)$ is a G-stable complement in $\mathcal{T}_1$ of the subgroup $\{1, -1\}$ (but notice that there are still another G-algebra homomorphism and another G-stable complement). Consequently, $\mathrm{Ten}_H^G(N)$ has actually a canonical $\mathcal{O}(\overline{\mathcal{T}}_1 \rtimes G)$-module structure and therefore, for any $R \in \mathrm{Ord}(G/H)$, restricting it to G_R and considering r_R and $(\varphi_R)^{-1}$ as above, $\mathrm{Ten}_H^G(N)$ becomes a $\mathcal{D}G$-module (with a *new* action of G) that we denote by $\mathrm{Den}_H^G(N, R)$; moreover, for any $R' \in \mathrm{Ord}(G/H)$, $\bar{\imath}_{R,R'} \in \overline{\mathcal{T}}_1$ defines clearly a $\mathcal{D}G$-module isomorphism

A2.10.2 $$\mathrm{Den}_H^G(N, R) \cong \mathrm{Den}_H^G(N, R') \quad .$$

As above, we obtain finally the *differential tensor induction* $\mathrm{Den}_H^G(N)$ of N from H to G.

A2.11 If N and N' are $\mathcal{D}H$-modules and $g : N \to N'$ is a $\mathcal{D}H$-module homomorphism, we claim that, as in Proposition A2.8, the ordinary $\mathcal{O}G$-module homomorphism $\mathrm{Ten}_H^G(g)$ is also a $\mathcal{D}G$-module homomorphism

A2.11.1 $$\mathrm{Ten}_H^G(g) : \mathrm{Den}_H^G(N, R) \longrightarrow \mathrm{Den}_H^G(N', R)$$

for any $R \in \mathrm{Ord}(G/H)$. Indeed, up to replace N and N' by their direct sum $N \oplus N'$ and g by the corresponding endomorphism of $N \oplus N'$, we may assume that $N = N'$; in that case, setting $B = \mathrm{End}_{\mathcal{O}}(N)$ as above, it is quite clear that, for any $C \in G/H$, $\mathrm{id}_{\mathcal{O}C} \otimes_{\mathcal{O}H} g$ determines an element of $\mathrm{C}_\circ(\mathcal{O}C \otimes_{\mathcal{O}H} B)$ via isomorphism A2.9.1 and therefore $\otimes_{C\in G/H}(\mathrm{id}_{\mathcal{O}C} \otimes_{\mathcal{O}H} g)$ determines an element of $\mathrm{Ten}_H^G(B)^G$ which centralizes both the image of $\mathrm{Ten}_H^G(s_1)$ and the image of r_R for any $R \in \mathrm{Ord}(G/H)$, and is mapped on $\mathrm{Ten}_H^G(g)$ by homomorphism A2.9.4, so that $\mathrm{Ten}_H^G(g)$ is a $\mathcal{D}G$-module endomorphism of $\mathrm{Den}_H^G(N, R)$ for any $R \in \mathrm{Ord}(G/H)$. Actually, this is a particular case of the fact that, for *any* $\mathcal{D}$-interior H-algebra B, the *norm map*

A2.11.2 $$\mathrm{Nr}_H^G : B^H \longrightarrow \mathrm{Ten}_H^G(B)^G$$

defined by $\mathrm{Nr}_H^G(b) = \otimes_{x\in X}(x \otimes b)$ for any $b \in B^H$ and any set of representatives X for G/H in G, induces a *differential norm map*

A2.11.3 $$\mathrm{Dr}_H^G : \mathrm{C}_\circ(B)^H \longrightarrow \mathrm{C}_\circ\big(\mathrm{Den}_H^G(B)\big)^G$$

such that $f_R \circ \mathrm{Dr}_H^G = \mathrm{Nr}_H^G$ for any $R \in \mathrm{Ord}_\circ(G/H)$. Finally, we prove the transitivity property for the differential tensor induction and, as above, the module case follows easily from the algebra case in our last result.

Proposition A2.12 *Let L be a subgroup of H and C a $\mathcal{D}$-interior L-algebra. There is a unique $\mathcal{D}$-interior G-algebra isomorphism*

A2.12.1 $$\mathrm{Den}_H^G\big(\mathrm{Den}_L^H(C)\big) \cong \mathrm{Den}_L^G(C)$$

mapping $m_H^G\big(\mathrm{Den}_L^H(C)\big)\big(m_L^H(C)(c)\big)$ on $m_L^G(C)(c)$ for any $c \in C$.

Proof: According to Proposition A2.7, we can reformulate this statement on the following way. Assume that A and B are respectively $\mathcal{D}$-interior G- and H-algebras that $m : B \to \mathrm{Res}_H^G(A)$ and $n : C \to \mathrm{Res}_L^H(B)$ are respectively $\mathcal{D}H$-and $\mathcal{D}L$-algebra homomorphisms and that we have $\mathcal{F}$-algebra isomorphisms

A2.12.2 $$h : B \cong \mathrm{Ten}_L^H(C) \quad \text{and} \quad f : A \cong \mathrm{Ten}_H^G(B)$$

fulfilling $h \circ n = n_L^H(C)$, $f \circ m = n_H^G(B)$ and conditions A2.7.1 and A2.7.2; then, the claim is that there is a $\mathcal{F}$-algebra isomorphism

A2.12.3 $$g : A \cong \mathrm{Ten}_L^G(C)$$

fulfilling $g \circ m \circ n = n_L^G(C)$ and conditions A2.7.1 and A2.7.2.

Let Z be a set of representatives for G/H in G, and consider the $\mathcal{O}$-algebra isomorphism

A2.12.4 $$\ell : \mathrm{Ten}_H^G(B) \cong \mathrm{Ten}_H^G\big(\mathrm{Ten}_L^H(C)\big)$$

which, for any family $\{b_z\}_{z\in Z}$ in B, maps $\otimes_{z\in Z}(z \otimes b_z)$ on $\otimes_{z\in Z}\big(z \otimes h(b_z)\big)$; since h maps the image of $f \in \mathcal{F}$ in B on its image in $\mathrm{Ten}_L^H(C)$ (cf. A2.4), ℓ maps the image of any element of $\mathrm{Ten}_H^G(\mathcal{F}_1)$ in $\mathrm{Ten}_H^G(B)$ on its image in $\mathrm{Ten}_H^G\big(\mathrm{Ten}_L^H(C)\big)$ and is a $\mathcal{F}$-interior algebra homomorphism; moreover, notice that the iterate coproduct $\mathcal{F}_1 \to \mathrm{Ten}_L^H(\mathcal{F}_1)$ induces an $\mathcal{O}$-algebra homomorphism

A2.12.5 $$\mathrm{Ten}_H^G(\mathcal{F}_1) \longrightarrow \mathrm{Ten}_H^G\big(\mathrm{Ten}_L^H(\mathcal{F}_1)\big) \cong \mathrm{Ten}_L^G(\mathcal{F}_1)$$

which, in particular, determines a group homomorphism

A2.12.6 $$\theta : \mathcal{T}_1(G/H) \longrightarrow \mathcal{T}_1(G/L) \quad ,$$

where $\mathcal{T}_1(G/H)$ and $\mathcal{T}_1(G/L)$ are respectively the groups of all the involutions in $\mathrm{Ten}_H^G(\mathcal{F}_1)^*$ and in $\mathrm{Ten}_L^G(\mathcal{F}_1)^*$, and consequently, for any $t', t'' \in \mathcal{T}_1(G/H)$ and any $a \in \mathrm{Ten}_H^G(B)$, we get

A2.12.7 $$\ell(t'\cdot a\cdot t'') = \theta(t')\cdot\ell(a)\cdot\theta(t'') \quad .$$

On the other hand, denoting by $\mathcal{T}_1(H/L)$ the group of all the involutions in $\mathrm{Ten}_L^H(\mathcal{F}_1)^*$, it is quite clear that we have a canonical group isomorphism (cf. 2.)

A2.12.8 $$\mathcal{T}_1(H/L)^Z \cong \mathcal{T}_1(G/L) \quad .$$

We are ready to prove that the composed $\mathcal{F}$-interior algebra isomorphism

A2.12.9 $$A \overset{f}{\cong} \mathrm{Ten}_H^G(B) \overset{\ell}{\cong} \mathrm{Ten}_H^G\big(\mathrm{Ten}_L^H(C)\big) \cong \mathrm{Ten}_L^G(C)$$

fulfills our claim; first of all, for any $c \in C$, setting $b_z = n(c)$ or 1 according to $z \in H$ or $z \notin H$ for any $z \in Z$, it is clear that, up to suitable identifications, we have

A2.12.10 $$\begin{aligned}\ell\Big(f\big(m(n(c))\big)\Big) &= \ell\Big(n_H^G(B)\big(n(c)\big)\Big) = \ell\big(\bigotimes_{z\in Z}(z\otimes b_z)\big)\\ &= \bigotimes_{z\in Z}\big(z\otimes h(b_z)\big) = n_L^G(C)(c) \quad .\end{aligned}$$

Now, denote respectively by Q and S the total order relations on G/H and on H/L involved in condition A2.7.1, and by $\varphi : G \to \mathcal{T}_1(G/H)$ and $\eta : H \to \mathcal{T}_1(H/L)$ the maps involved in condition A2.7.2, always with respect to isomorphisms A2.12.2; for any $z \in Z$, it is clear that $z\cdot S$ is a total order relation in the subset zH/L of G/L and we denote by R the lexicographic order relation in G/L obtained from Q and the family $\{z\cdot S\}_{z\in Z}$; then, denoting by $r_1 : \mathcal{D}_1 \to A$ and $t_1 : \mathcal{D}_1 \to C$ the restriction of the structural maps, it is not difficult to check from condition A2.7.1 applied to isomorphisms A2.12.2 that, up to suitable identifications, we have

A2.12.11 $$\ell \circ f \circ r_1 = \mathrm{Ten}_L^G(t_1) \circ \Delta_{1,R} \quad ,$$

so that isomorphism A2.12.9 fulfills condition A2.7.1.

Finally, consider the map $\omega : G \to \mathcal{T}_1(G/L)$ sending $x \in G$ to the image in $\mathcal{T}_1(G/L)$ by isomorphism A2.12.8 of $\big(\eta(\sigma_x(z)^{-1}xz)\big)_{z\in Z}$, where σ_x is the permutation of Z determined by the action of x on G/H; then, for any $x \in G$ and any family $\{b_z\}_{z\in Z}$ in B, we have

A2.12.12

$$\begin{aligned}\ell\Big(\big(\bigotimes_{z\in Z}(z\otimes b_z)\big)^x\Big) &= \ell\big(\bigotimes_{z\in Z}(x^{-1}z\otimes b_z)\big)\\ &= \ell\Big(\bigotimes_{z\in Z}\big(z\otimes (b_{\sigma_x(z)})^{\sigma_x(z)^{-1}xz}\big)\Big) = \bigotimes_{z\in Z}\Big(z\otimes h\big((b_{\sigma_x(z)})^{\sigma_x(z)^{-1}xz}\big)\Big)\\ &= \bigotimes_{z\in Z}\Big(z\otimes\big(\eta(\sigma_x(z)^{-1}xz)\,.\,h(b_{\sigma_x(z)})^{\sigma_x(z)^{-1}xz}\,.\,\eta(\sigma_x(z)^{-1}xz)\big)\Big)\\ &= \omega(x)\,.\Big(\bigotimes_{z\in Z}\big(z\otimes h(b_z)\big)\Big)^x.\,\omega(x) = \omega(x)\,.\,\ell\Big(\bigotimes_{z\in Z}(z\otimes b_z)\Big)^x.\,\omega(x) \quad ;\end{aligned}$$

consequently, for any $x \in G$ and any $a \in A$, up to suitable identifications, we have

A2.12.13
$$\begin{aligned}\ell(f(a^x)) &= \ell\big(\varphi(x)\cdot f(a)^x\cdot\varphi(x)\big) = \theta\big(\varphi(x)\big)\cdot\ell\big(f(a)^x\big)\cdot\theta\big(\varphi(x)\big)\\ &= \theta\big(\varphi(x)\big)\omega(x)\cdot\ell\big(f(a)\big)^x\cdot\omega(x)\theta\big(\varphi(x)\big)\\ &= \psi(x)\cdot\ell\big(f(a)\big)^x\cdot\psi(x)\end{aligned}$$

where $\psi : G \to \mathcal{T}_1(G/L)$ maps $x \in G$ on $\theta\big(\varphi(x)\big)\omega(x)$, so that isomorphism A2.12.9 fulfills condition A2.7.2 too.

References

[1] Barker Laurence, Induction, restriction and G-algebras, *Comm. in Algebra*, 22(1994), 6349–83

[2] Barker Laurence, G-algebras, Clifford theory and Green correspondence, *J. of Algebra*, 172(1995), 335–53

[3] Broué Michel, Les ℓ-blocs des groupes GL(n,q) et U(n,q^2) et leurs structures locales (d'après M. Broué et L. Puig), Astérisque 133–34(1986), 159–88

[4] Broué Michel, Isométries parfaites, types de blocs, catégories dérivées, Astérisque 181–82(1990), 61–92

[5] Broué Michel, Isométries de caractères et équivalences de Morita ou dérivées, *Publ. Math. IHES*, 71(1990), 45–63

[6] Broué Michel, Equivalences of blocks of group algebras, *Proc. Int. Conf. Representations of Algebras*, Ottawa, 1992, Holland 93

[7] Broué Michel, Rickard equivalences and block theory, Groups' 93 Galway-Saint Andrews, *London Math. Soc. Lecture Notes 211*, Cambridge Univ. Press, 1995

[8] Broué Michel and Puig Lluis, Characters and local structure in G-algebras, *J. of Algebra*, 63(1980), 306–17

[9] Broué Michel and Puig Lluis, A Frobenius theorem for blocks, *Inventiones Math.*, 56 (1980), 117–28

[10] Cartan Henri and Eilenberg Samuel, "Homological algebra", Princeton Univ. Press, 1956

[11] Deligne Pierre, Cohomologie à supports propres, SGA 4, Vol. 3, Exp. XVII, 250–480, *Lecture Notes in Math.* 305, Springer-Verlag

[12] Demazure Michel and Gabriel Pierre, "Groupes algébriques", Vol. 1, Masson and Cie., Paris

[13] Keller Bernhard, Deriving DG categories, *Ann. Scient. Ec. Norm. Sup.*, 27(1994), 63–102

[14] Külshammer Burkhard and Puig Lluis, Extensions of nilpotent blocks, *Inventiones Math.*, 102(1990), 17–71

[15] Linckelmann Markus, Stable equivalences of Morita type for self-injective algebras and p-groups, *Math. Zeit.*, 223(1996), 87–100

[16] Linckelmann Markus, On derived equivalences and local structure of blocks of finite groups, Preprint, 1996

[17] Morita Kiiti, Duality for modules and its applications to the theory of rings with minimum conditions, *Sci. Rep. Tokyo Kyoiku Daigaku*, A6, 150(1958), 85–142

[18] Pareigis Bodo, A non-commutative non-cocommutative Hopf algebra in "nature", *J. of Algebra*, 70(1981), 356–74

[19] Puig Lluis, Pointed groups and construction of characters, *Math. Zeit.*, 176(1981), 265–92

[20] Puig Lluis, Local fusions in block source algebras, *J. of Algebra*, 104(1982), 358–69

[21] Puig Lluis, Nilpotent blocks and their source algebras, *Inventiones Math.*, 93(1988), 77–116

[22] Puig Lluis, Pointed groups and construction of modules, *J. of Algebra*, 116(1988), 7–129

[23] Puig Lluis, Affirmative answer to a question of Feit, *J. of Algebra*, 131(1990), 513–26

[24] Puig Lluis, Une correspondance de modules pour les blocs à groupes de défaut abéliens, *Geometriae Dedicata*, 37(1991), 9–43

[25] Puig Lluis, Block source algebras in p-solvable groups, *Letter to Morton Harris*, 1993

[26] Puig Lluis, On Thévenaz' parametrization of interior G-algebras, *Math. Zeit.*, 215 (1994), 321–35

[27] Puig Lluis, On Joanna Scopes' criterion of equivalence for blocks of symmetric groups, *Algebra Colloq.*, 1(1994), 25–55

[28] Puig Lluis, A survey on the local structure of Morita and Rickard equivalences between Brauer blocks in "Representation Theory of Finite Groups", *Proc. Sp. Res. Quarter at OSU*, Spring 1995, Walter de Gruyter, Berlin-New York, 1997

[29] Rickard Jeremy, Derived categories and stable equivalences, *J. Pure and Appl. Alg.*, 61(1989), 307–17

[30] Rickard Jeremy, Morita theory for derived categories, *J. London Math. Soc.*, 39 (1989), 436–56

[31] Rickard Jeremy, Derived equivalences as derived functors, *J. London Math. Soc.*, 43(1991), 37–48

[32] Rickard Jeremy, Finite group action and etale cohomology, *Publ. Math. IHES*, 80 (1994), 81–84

[33] Rickard Jeremy, Splendid equivalences: Derived categories and permutation modules, *Proc. London Math. Soc.*, 72(1996), 331–58

[34] Roggenkamp Klaus, Subgroup rigidity of p-adic group rings (Weiss arguments revisited), *J. London Math. Soc.*, 46(1992), 432–48

[35] Rouquier Raphaël, From stable equivalences to Rickard equivalences for blocks with cyclic defect, Groups' 93 Galway-Saint Andrews, *London Math. Soc. Lecture Notes 211*, Cambridge Univ. Press, 1995

[36] Scott Leonard, Defect groups and the isomorphism problem, in "Représentations linéaires des groupes finis", Astérisque, 181–82(1990), 257–62

[37] Serre Jean-Pierre, "Corps locaux", *Publ. Inst. Math. Univ. Nancago*, VIII(1968), Hermann, Paris

[38] Thévenaz Jacques, "G-algebras and modular representation theory", *Oxford Math. Monographs*, Clarendon Press, Oxford, 1995

[39] Weiss Alfred, Rigidity of p-adic p-torsion, *Annals of Math.*, 127(1988), 317–32

Index

Mathematics with Birkhäuser

Progress in Mathematics

Edited by
Hyman Bass, Columbia University New York, NY, USA
Joseph Oesterlé, Institut Henri Poincaré, Université Paris VI, France
Alan Weinstein, University of California, Berkeley, CA, USA

Progress in Mathematics is a series of books intended for professional mathematicians and scientists, encompassing all areas of pure mathematics. This distinguished series, which began in 1979, includes research level monographs, polished notes arising from seminars or lecture series, graduate level textbooks, and proceedings of focused and refereed conferences. It is designed as a vehicle for reporting ongoing research as well as expositions of particular subject areas.

PM 145 Albert, C. / Brouzet, R. / Dufour, J.P., **Integrable Systems and Foliations.** Feuilletages et Systèmes Intégrables
1997. 224 pages. Hardcover. ISBN 3-7643-3894-6

PM 146 Jardine, J.F., **Generalized Etale Cohomology Theories.**
1997. 328 pages. Hardcover. ISBN 3-7643-5494-1

PM 147 Di Biase, F., **Fatou Type Theorems.** Maximal Functions and Approach Regions
1998. 166 pages. Hardcover. ISBN 3-7643-3976-4

PM 148 Huang, Y.-Z., **Two-Dimensional Conformal Geometry and Vertex Operator Algebras.**
1997. 296 pages. Hardcover. ISBN 3-7643-3829-6

PM 149 Souriau, J.-M., **A Symplectic View of Physics.** Structure of Dynamical Systems
1997. 440 pages. Hardcover. ISBN 3-7643-3695-1

PM 150 Shiota, M., **Geometry of Subanalytic and Semialgebraic Sets.**
1997. 444 pages. Hardcover. ISBN 3-7643-4000-2

PM 151 Hummel, Ch., **Gromov's Compactness Theorem for Pseudo-holomorphic Curves.**
1997. 148 pages. Hardcover. ISBN 3-7643-5735-5

PM 152 Gromov, M. / Bates, S.M., **Metric Structures for Riemannian and Non-Riemannian Spaces.**
1999. 608 pages. Hardcover. ISBN 3-7643-3898-9

PM 153 Buescu, J., **Exotic Attractors.** From Liapunov Stability to Riddled Basins
1997. 144 pages. Hardcover. ISBN 3-7643-5793-2

PM 154 Böttcher, A. / Karlovich, Y.I., **Carleson Curves, Muckenhoupt Weights, and Toeplitz Operators.**
1997. 416 pages. Hardcover. ISBN 3-7643-5796-7

PM 155 Dragomir, S. / Ornea, L., **Locally Conformal Kähler Geometry.**
1998. 344 pages. Hardcover. ISBN 3-7643-4020-7

PM 156 Guivarc'h, Y. / Ji, L. / Taylor, J.C. , **Compactifications of Symmetric Spaces.**
1998. 300 pages. Hardcover. ISBN 3-7643-3899-7

PM 157 Murty, M.R. / Murty, V.K., **Non-vanishing of L-Functions and Applications.**
1997. 208 pages. Hardcover. ISBN 3-7643-5801-7

PM 158 Tirao, J. / Vogan, D.A. / Wolf, J.A., **Geometry and Representation.** Theory of Real and *p*-Adic Groups
1997. 336 pages. Hardcover. ISBN 3-7643-3931-4

PM 159 Thangavelu, S., **Harmonic Analysis on the Heisenberg Group.**
1998. 208 pages. Hardcover. ISBN 3-7643-4050-9

PM 160 Kashiwara, M. et al. (Ed.), **Topological Field Theory, Primitive Forms and Related Topics.**
1998. 494 pages. Hardcover. ISBN 3-7643-3975-6

PM 161 Sagan, B.E. / Stanley, R.P.(Ed.), **Mathematical Essays in Honor of Gian-Carlo Rota.**
1998. 480 pages. Hardcover. ISBN 3-7643-3872-5

PM 162 Arnold, V.I. / Greuel, G.M. / Steenbrink, J.H.M. (Ed.), **Singularities.** The Brieskorn Anniversary Volume
1998. 495 pages. Hardcover. ISBN 3-7643-5913-7

PM 163 Berndt, R. / Schmidt, R., **Elements of the Representation Theory of the Jacobi Group.**
1998. 235 pages. Hardcover. ISBN 3-7643-5922-6

PM 164 Roussarie, R., **Bifurcations of Planar Vector Fields and Hilbert's Sixteenth Problem.**
1998. 224 pages. Hardcover. ISBN 3-7643-5900-5

PM 165 Migliore, J.C., Univ. of Notre Dame, India, **Introduction to Liaison Theory and Deficiency Modules.**
1998. 232 pages. Hardcover. ISBN 3-7643-4027-4

PM 166 Elias, J. et al. (Ed.), **Six Lectures on Commutative Algebra.**
1998. 416 pages. Hardcover. ISBN 3-7643-5951-X

PM 167 Facchini, A., **Module Theory.** Endomorphism Rings and Direct Sum Decompositions in some Classes of Modules
1998. 290 pages. Hardcover. ISBN 3-7643-5908-0

PM 168 / 169 Balog, A. et al. (Ed.), **European Congress of Mathematics, Budapest, July 22-26, 1996.** Set: Vols. I + II
1998. 736 pages. Hardcover. ISBN 3-7643-5496-8

PM 170 Paterson, A., **Groupoids, Inverse Semigroups and their Operator Algebras.**
1998. 288 pages. Hardcover. ISBN 3-7643-4051-7

PM 171 Reznikov, A. / Schappacher, N. (Ed.), **Regulators in Analysis, Geometry, and Number Theory.**
1999. Approx. 400 pages. Hardcover. ISBN 3-7643-4115-7

PM 172 Brylinski, J-L. et al., **Advances in Geometry.**
1998. 404 pages. Hardcover. ISBN 3-7643-4044-4

PM 173 Dräxler, P. / Michler, G. / Ringel, C.M. (Ed.), **Computational Methods for Representations of Groups and Algebras.** Euroconference in Essen (Germany), 1997. 1999. Approx. 372 pages. Hardcover. ISBN 3-7643-6063-1

PM 174 Goerss, P.G. / Jardine, J.F., **Simplicial Homotopy Theory.**
1999. Approx. 500 pages. Hardcover. ISBN 3-7643-6064-X

PM 175 Bañuelos, R. / Moore, Ch. N., **Probabilistic Behavior of Harmonic Functions.**
1999. Approx. 220 pages. Hardcover. ISBN 3-7643-6062-3

PM 176 Bass, H. / Lubotsky, A., **Tree Lattices.**
1999. Approx. 232 pages. Hardcover. ISBN 3-7643-4120-3

PM 177 Birkenhage, C. / Lange, H., **Complex Tori.**
1999. 272 pages. Hardcover. ISBN 3-7643-4103-3

PM 9 Springer, T.A., **Linear Algebraic Groups.**
1998. 350 pages. Hardcover. ISBN 3-7643-4021-5

PM 5 Helgason, S., **The Radon Transform.** Second Edition
1999. Approx. 208 pages Hardcover. ISBN 3-7643-4109-2